DEHYDROGENASES

Edited by **Rosa Angela Canuto**

Dehydrogenases
http://dx.doi.org/10.5772/2903
Edited by Rosa Angela Canuto

Contributors

Adil M. Allahverdiyev, Malahat Bagirova, Olga Nehir Oztel, Serkan Yaman, Emrah Sefik Abamor, Rabia Cakir Koc, Sezen Canim Ates, Serhat Elcicek, Serap Yesilkir Baydar, Ellen Friday, Robert Oliver III, Francesco Turturro, Tomas Welbourne, Paul M. Bingham, Zuzana Zachar, P. Naik, S. Prasad, L. Cucullo, Nina Atanassova, Yvetta Koeva, Nwaoguikpe Reginald Nwazue, Agnieszka Wolińska, Zofia Stępniewska, Dorota Kregiel, Marie Kondrashova, Marina Zakharchenko, Andrej Zakharchenko, Natalya Khunderyakova, Eugen Maevsky, Małgorzata Kotula-Balak, Anna Hejmej, Barbara Bilińska, Ines Canosa, Fernando Govantes, Marco Cardosi, Zuifang Liu

Published by InTech

Janeza Trdine 9, 51000 Rijeka, Croatia

Publishing Process Manager Oliver Kurelic
Technical Editor InTech Prepress, Novi Sad
Cover InTech Design team

Additional hard copies can be obtained from orders@intechopen.com

Dehydrogenases, Edited by Rosa Angela Canuto
p. cm.
ISBN 978-953-307-019-3

For EU product safety concerns:
IN TECH d.o.o., Prolaz Marije Krucifikse Kozulić 3, 51000 Rijeka, Croatia,
info@intechopen.com or visit our website at intechopen.com.

Meet the editor

A Graduate in Biological Science and in Medicine and Surgery, prof. Canuto is a full professor of General Pathology at the Faculty of Medicine and Surgery, Turin University. She was in charge of national research projects, and the scientific organizer of conferences. Her research activity is in two fields: experimental oncology and tissue regeneration. The aim of the first field is to increase the knowledge on the importance of PUFA as preventive factors or as coadjuvant drugs in cancer. The aim of the second field is to study the biocompatibility and the colonization of new bone substitutes or hernia prostheses, able to induce the bone or peritoneum production. Her activity has been the subject of several full-length publications, and several conference papers.

Contents

Preface

The dehydrogenases are NAD(P)H/NAD(P)+-dependent oxidoreductases with different functions in the cells and with different substrates. They are present in human, plant and micro-organism tissues, and are of enormous biochemical interest. They are associated with both catabolic and anabolic pathways linked to normal functioning and homeostasis. The book does not aim to examine all these enzymes, but to take into consideration only a small number of them.

- Hydroxysteroid dehydrogenases (HSDs), which interconvert ketones and the corresponding secondary alcohols. As their names imply, they regio- and stereo-selectively catalyze the oxidoreduction in different positions of steroidal substrates (3α-, 3β-, 11β-, 17β-, 20α- and 20β-position). The steroid-converting HSDs play central roles in the biosynthesis and inactivation of steroid hormones, but some of them are also involved in the metabolism of diverse non-steroidal compounds. The HSDs are integral parts of systemic (endocrine) and local (intracrine) mechanisms. In target tissues, they convert inactive steroid hormones to their corresponding active forms, and vice-versa, thus modulating the transactivation of steroid hormone receptors or other elements of the non-genomic signal transduction pathways. HSDs thus act as molecular switches. allowing for pre-receptor modulation of steroid hormone action.

- Aldehyde dehydrogenases (ALDH), which catalyze the conversion of aldehydes to their corresponding acids. Many types of ALDH have been distinguished, according to their physico-chemical characteristics, enzymological properties, subcellular localization, and tissue distribution. They are involved in several cell functions, including proliferation, differentiation and survival, as well as the cellular response to oxidative stress. The aldehyde dehydrogenase gene superfamily plays an important role in the enzymatic detoxification of endogenous and exogenous aldehydes. They are also responsible for the formation of molecules, including retinoic acid, betaine, and gamma aminobutyric acid, which are important in cellular processes. The most significant role played by aldehyde dehydrogenases is to detoxifiy aldehydes, since these are potentially cytotoxic, mutagenic and genotoxic to healthy cells. Mutations in ALDH genes cause severe diseases (e.g., Sjögren-Larsson syndrome, pyridoxine-dependent seizures, type II hyperprolinemia, and cancer). Since the enzyme ALDH has been proven to possess a vital role in somatic cells, researchers have recently focused on its presence and functions in stem cells. It has been demonstrated that

ALDH is an important marker for identifying stem cells, especially cancer stem cells. ALDH Bright ($ALDH^{br}$) cells are derived from different tissues expressing high ALDH activity and that have progenitor cell activity. Studies have shown that $ALDH^{br}$ populations may be useful in several cell therapy applications: thus the $ALDH^{br}$ population may play an important role in regenerative medicine, and could become a new therapeutic target in cancer studies.

- Glucose-6-phosphate dehydrogenase (G6PD), whose deficiency is the most common enzymopathological disease in humans. This disease is described as a widespread, heritable, X-chromosome-linked abnormality. It is estimated that it affects approximately 400 million people worldwide. This disease is reported most frequently in Africa, Asia, and the countries bordering on the Mediterranean Sea. The enzyme G6PD has been demonstrated to play an active role in erythrocyte survival. It is known that in the pentose phosphate pathway of erythrocytes, G6PD is responsible for the production of NADPH and GSH. GSH, produced by the pentose phosphate pathway, can react with H_2O_2 and reduce it to H_2O. This prevents the generation of oxidative stress within red blood cells; oxidative stress can be induced in erythrocytes whose G6PD enzymes are deficient. In this situation, GSH is not produced and H_2O_2 is not reduced to H_2O, leading to oxidative stress and hemolysis. This is the only mechanism available to erythrocytes for generating reducing equivalence, therefore making it essential for erythrocyte survival. In individuals whose G6PD enzyme is deficient, different kinds of hemolysis, ranging from mild to severe, occur that are linked with differences in variants of the disease

- The pyruvate dehydrogenase complex (PDH), which occupies a unique position in cancer biology. First, PDH catalyzes the irreversible step introducing glucose carbon into the TCA cycle. Thus, PDH governs the flux of carbon through the TCA cycle under many circumstances. This flux sustains both catabolic and anabolic processes. Second, regulation of the PDH complex is exceptionally well understood, including diverse sources of allosteric modification both of PDH enzymatic activities and of the kinases (PDKs) that exert post-translational control of PDH. Third, the regulatory properties of PDH are systematically altered in cancer cells. The functional logic of these regulatory changes is apparently straightforward. Specifically, the TCA cycle is substantially repurposed in tumor cells, in which it takes on a more extensively anabolic role, with concomitant reduction in its catabolic function, shifting ATP production more heavily to glycolysis. Fourth, the cancer-specific regulatory reconfiguration of the PDH complex includes substantial changes in the spectrum of associated PDKs. Moreover, these changes in PDK spectrum are directly implicated in sustaining the transformed state in cancer cells. Fifth, several classes of agents have been developed to target PDKs and, thus, to potentially selectively attack tumor cell mitochondrial metabolism. Pre-clinical studies indicate that these novel agent classes work in ways that are distinct from one another, and different from the mechanisms of action of other chemotherapeutic agents. Further, early human clinical trial results suggest that at least some of these agents may have clinical potential. These early indications of possible success in targeting the PDH complex further suggest that targeting other cancer-central dehydrogenases might add to the chemotherapeutic armamentarium. Collectively and most generally, these observations indicate that targeting the *mitochondrial* features of tumor-specific metabolism may be especially effective in therapeutic targeting of cancer cells.

- Glutamate dehydrogenase (GDH), which is present in all domains of life. GDH has a physiological role in different organisms, which may include nitrogen assimilation, glutamate catabolism, osmotic balance or survival at high temperatures. It catalyses the reductive amination of 2-oxoglutarate to yield glutamate or the opposite reaction, the oxidative deamination of glutamate, using NAD(P) as cofactor. Therefore, the main role of GDH is glutamate catabolism and ammonium assimilation. The synthesis of both glutamate and glutamine are key steps in the cell metabolism in all organisms, because they represent the only means of incorporating inorganic nitrogen into carbon backbones. Inorganic nitrogen is assimilated in the form of ammonium, which is incorporated as an amino group to glutamate, or as an amido group to glutamine. These amino acids in turn act as amino group donors for the synthesis of most nitrogen-containing compounds in the cell. In particular, the amino group of glutamate is used in the synthesis of purines, pyrimidines, amino sugars, histidine, tryptophan, and asparagines.

- In the liver, glutamate is the source of excess ammonium release, and the concentration of glutamate modulates the rate of ammonia detoxification into urea. In pancreatic β-cells, GDH is involved in insulin homeostasis, and oxidation of glutamate mediates amino acid-stimulated insulin secretion. In the central nervous system, glutamate serves as a neurotransmitter and also as the precursor of the inhibitory neurotransmitter □γaminobutyric acid (GABA), as well as glutamine, a potential mediator of hyperammonemic neurotoxicity.

- The imbalance of GDH activity may lead to disturbances of clinical relevance for humans. Free ammonia is highly toxic to organisms that excrete urea, leading to inhibition of brain respiration and an excess ketone body formation from acetyl-CoA in the liver.

- GDH is also involved in tumours. Epigallocatechin gallate (EGCG) is an inhibitor of GDH whose ability to induce cancer-cell death is dependent upon the extent to which GDH flux supports cell survival. In glucose and glutamine replete media, EGCG induces minimal cancer cell death, which is greatly accelerated with glucose depletion. However, in combination with troglitazone, EGCG induces maximal cell death, associated with depletion of TCA cycle intermediates and cellular acidosis.

- Succinate dehydrogenase is a key enzyme in intermediary metabolism and aerobic energy production in living cells. This enzyme catalyses the oxidation of succinate into fumarate in the Krebs cycle, derived electrons being fed to the respiratory chain complex III to reduce oxygen and form water. This builds up an electrochemical gradient across the mitochondrial inner membrane, allowing for the synthesis of ATP. Alternatively, electrons can be diverted to reduce the ubiquinone pool, and provide reducing equivalents necessary to reduce superoxide anions, originating either from an exogenous source or from the respiratory chain itself. Deficiencies in TCA cycle enzymes have been shown to cause a wide spectrum of human diseases: Encephalopathy (Leigh syndrome), pheochromocytoma and paraganglioma.

Considering the importance of the dehydrogenases taken into consideration, the book is divided in three chapters:

1. Dehydrogenases and cancer
2. Dehydrogenases and some diseases
3. Dehydrogenases and physiological role.

I thank Dr. Marina Maggiora and Dr. Elena Paiuzzi for their help in reviewing the articles.

Rosa Angela Canuto
Dept. of Experimental Medicine and Oncology, University of Turin, Italy

Section 1

Dehydrogenases and Cancer

Chapter 1

Aldehyde Dehydrogenase: Cancer and Stem Cells

Adil M. Allahverdiyev, Malahat Bagirova, Olga Nehir Oztel, Serkan Yaman, Emrah Sefik Abamor, Rabia Cakir Koc, Sezen Canim Ates, Serhat Elcicek and Serap Yesilkir Baydar

Additional information is available at the end of the chapter

http://dx.doi.org/10.5772/48591

1. Introduction

Aldehyde dehydrogenases (ALDH) belong to the oxidoreductase family, which catalyze the conversion of aldehydes to their corresponding acids. As a group of NAD(P)+-dependent enzymes, aldehyde dehydrogenases (ALDHs) are involved in oxidation of a large number of aldehydes into their weak carboxylic acids (Moreb, *et al.*, 2012). ALDH is found in every subcellular region such as cytosol, endoplasmic reticulum, mitochondria, and the nucleus, with some even found in more than one location (Marchitti, *et al.*, 2008).

ALDH is also found in stem cells. During early life and growth, stem cells (SCs) have a spectacular potential to develop into several cell types in the body. In many tissues, SCs behave as a kind of internal repair system, dividing essentially without limit to replenish other cells (Fuchs & Segre, 2000). Stem cells are distinguished from other cell types by two important characteristics: (1) Their unspecialized properties and renewal potencies; and (2) differentiation into other cell types under certain physiologic or experimental conditions (Discher, *et al.*, 2009, Solis, *et al.*, 2012). These cells are identified by their expression of a particular panel of surface molecules, with the presence of CD73, CD90, CD105, and the absence of CD14, CD34, CD45, and HLA-DR. They show no proliferative response from alloreactive lymphocytes because of the negligible levels of extracellular MHC class I and II determinants and they also have important immunomodulatory functions in all the cells involved in both the innate and adaptive immune responses (Nauta & Fibbe, 2007). On the other hand, cancer stem cell theory is supported by biological reason for aging. The theory postulates that cancer SCs, a small subset of tumor cells also have stem cell-like properties (epithelial-to-mesenchymal progression, differentiation and self-renewing capacity). ALDH expression has demonstrated itself to be a possibly relevant prognostic marker. For this reason, the subpopulation of cancer SCs (CSCs) can present a therapeutic target for poor-prognostic, treatment-resistant and recurrent breast cancer. Through its role in oxidizing retinol to RA, which is a modulator of cell proliferation, ALDH1 might have a role in early differentiation of SCs and stem cell proliferation (Mieog, *et al.*, 2012).

There are several isoforms of ALDH (ALDH1A1, ALDH1A2, ALDH1A3 and ALDH8A1) that play a role in RA formation by oxidation of all-trans-retinal and 9-cis-retinal in RA cell signaling, which has been related to the "stemness" characteristics of SCs (Marcato, *et al.*, 2011). ALDH1 is better as a marker of breast cancer SCs than CD44+/CD24- (Tanei, *et al.*, 2009). While cellular markers including CD133 have been used to identify tumor SCs, especially for glioblastomas (GBMs) ALDH1 was desrcribed as a marker for the identification of non-neoplastic SCs and tumor stem cells (TSCs) (Corti, *et al.*, 2006, Ginestier, *et al.*, 2007, Huang, *et al.*, 2009).

After CD133- GBMs are characterized to behave as brain TSCs(Beier, *et al.*, 2007). ALDH1 has also been described as a stem cell marker in various solid neoplasms including lung cancer (Jiang, *et al.*, 2009), breast carcinoma (Ginestier, *et al.*, 2007), colorectal cancer (Huang, *et al.*, 2009), and GBM. ALDH1B1 and ALDH1A1 are differentially expressed in normal human tissues. ALDH1B1 is expressed at higher levels than ALDH1A1 in human epithelial cancers. ALDH1B1 was abundantly expressed in adenocarcinomas originating from the tissue and particularly in colonic adenocarcinoma (Chen, *et al.*, 2011).

$ALDH^{br}$ cells can be detected with ALDEFLUOR reagent by using flow cytometry or fluorescent microscopy. Aldefluor assay is based on the conversion of fluorescent non-toxic substrate for ALDH substrate to the fluorescent reaction product. Non-toxic substrate for ALDH can freely diffuse into intact, viable cells. The BODIPY aminoacetaldehyde is converted to the fluorescent product BODIPY aminoacetate by ALDH activity. These cell populations, which are known as ALDH bright ($ALDH^{br}$) cells are isolated from adult tissues by flow sorting. $ALDH^{br}$ cells were also found in various cancer tissues including breast, liver, colon, pancreas, prostate, lung, ovarian and acute myelogenous leukemia and are related to cancer chemo resistance (Siclari & Qin, 2010).

$ALDH^{br}$ population may play an important role in regenerative medicine. The regenerative potential of $ALDH^{br}$ cells obtained from different tissues was investigated in various disease models such as ischemic tissue damage, hind limb model, brain damage (spinal motor atrophy, etc.) and pancreatitis.

Thus, as mentioned above, ALDH is an important enzyme for cancer and stem cells. This chapter aims to represent the important role of aldehyde dehydrogenases in stem cells, cancer stem cells, therapy and regenerative medicine.

1.1. Aldehydes

Aldehydes are formed in various physiological processes such as catabolism of transmitters like GABA, serotonin, adrenaline, noradrenaline and dopamine, as well as catabolism of amino acids. In addition, there are more than 200 different aldehydes that are produced through lipid, and aldehydic intermediates through carbohydrate metabolism. Along with these endogenous aldehydes, there are also exogenously present aldehydes in a variety of industrial processes, including the production of polyester plastics (formaldehyde, acetaldehyde, acrolein, etc.), polyurethane, smog, cigarette smoke or motor vehicle exhaust. With their malodorous properties, some dietary and aromatic aldehydes are accepted as additives in food and cosmetics (e.g., citral, cinnamaldehyde, benzaldehyde, and retinal), though many others are cytotoxic (Chen, *et al.*, 2010). Aldehydes could interact with thiol compounds of some proteins, leading to structural and functional alterations of these molecules (Weiner, *et al.*, 2008). In order to protect the human body from the deleterious effects of aldehydes in general, and myocardium and the brain in particular, a fast aldehyde detoxification mechanism is essential. Aldehydes are significantly reactive and possess high diffusion capacities in cells, thus they can easily form complexes with DNA, proteins and lipids, of which they can alter the function and cause their inactivation. As a result of DNA damage induced by these complexes, many aldehydes are classified as mutagenic or carcinogenic, including acetaldehyde, which is derived from ethanol consumption. Over–consumption of ethanol has been related to liver disease and several gastrointestinal and upper aerodigestive cancers. Numerous other cytotoxic and reactive aldehydes have been shown to be linked with other types of diseases (Hofseth & Wargovich, 2007, Perluigi, *et al.*, 2009, Chen, *et al.*, 2010).

1.2. Aldehyde dehydrogenases

Aldehyde dehydrogenases [EC 1.2.1.3; *systematic name*: aldehyde: $NAD(P)^{+}$oxidoreductase] catalyze aldehyde conversion into their matching acids by $NAD(P)^{+}$-dependent nearly irreversible reaction. In 1949, mammalian ALDH was first discovered in ox liver. After that, many types of ALDH were distinguished according to their physico-chemical characteristics, enzymological properties, subcellular localization, and tissue distribution (Oraldi, *et al.*, 2011). They are involved in several cell functions such as proliferation, differentiation, survival as well as cellular response to oxidative stress (Jackson, *et al.*, 2011). ALDHs are commonly delivered from bacteria and humans (Moreb, 2008). Based on their physico-chemical characteristics, subcellular localization, tissue distribution and enzymological properties, a number of types of ALDH have been distinguished since the 1960s, around the time when mammalian ALDH activity was observed in ox for the first time. In 1985, 2 ALDH genes were cloned and characterized. Genes or cDNAs for more than 50 animals, fungi and bacterial ALDHs in addition to protein sequences have been discovered (Yoshida, *et al.*, 1998).

The human genome contains 19 ALDH functional genes and 3 pseudogenes (Black, *et al.*, 2009). At least 5 ALDH isozymes function in the mitochondria, and all the ALDH genes are encoded in the nucleus (Chen, *et al.*, 2010).

All of the ALDH gene superfamily plays an important role in the enzymic detoxification of endogenous and exogenous aldehydes. They are also involved in the formation of molecules that are important in cellular processes like RA, betaine and gamma aminobutyric acid formation. Furthermore, ALDHs also have several non-enzymic functions such as binding to some hormones and other small molecules and decreasing the effects of ultraviolet irradiation in the cornea (Pappa, *et al.*, 2003, Wymore, *et al.*, 2004). The most important role of ALDHs is detoxification of aldehydes, which caused cytotoxicity, mutagenicity, genotoxicity, and carcinogenesis in healthy cells. Mutations in ALDH genes cause severe diseases including Sjögren-Larsson syndrome, pyridoxine-dependent seizures, and type II hyperprolinemia, and also plays a role in cancer and Alzheimer's disease (Black, *et al.*, 2009).

Functions of some of these ALDHs in endobiotic and xenobiotic metabolisms have been highly reviewed before and the distinctive metabolic pathways' influences have been depicted. Because of their chemical reactivity, many distinct aldehydes are pervasive in nature and are toxic at low levels. Hence, levels of metabolic-intermediate aldehydes should be cautiously regulated. The presence of several distinct ALDH families in most studied organisms seem to have wide fundamental tissue distribution. A wide range of allelic variants within the ALDH gene family have been identified, leading to heterogeneity in pharmacogenetic characteristics between individuals, resulting distinctive phenotypes including intolerance to alcohol and increased risk of ethanol-induced cancers in most cases (ALDH2 and ALDH1A1), Sjogren-Larson Syndrome (ALDH3A1), type II hyperprolinemia (ALDH4A1), 4-hydroxybutyric aciduria, mental retardation and seizures (ALDH5A1), developmental delay (ALDH6A1), hyperammonemia (ALDH18A1), Pyridoxine-dependent epilepsy (ALDH7A1), and late-onset Alzheimer's disease (ALDH2).

ALDH dysfunction could also be caused by drugs and environmental substances, substrate inhibition, as well as oxidative and metabolic stress. ALDH activity in drug resistance to oxazaphosphorines is one of the most vigorously studied pathways. The role of ALDH1A1 in drug resistance has been studied first in hematopoietic progenitors and more recently in lung cancer (Marchitti, *et al.*, 2008).

2. Stem cells

During early life and growth, SCs have a spectacular potential to develop into several cell types in the body. In many tissues, SCs behave as a kind of internal repair system, dividing essentially without limit to replenish other cells (Weissman, 2000). Stem cells are distinguished from other cell types by two important characteristics: First, they are unspecialized cells and, sometimes after long periods of inactivity, they can renew themselves through cell division; second, under certain physiologic or experimental conditions, they are naturally sensitive to their environment, responding to chemical, physical, and mechanical features of their matrices or substrates (Discher, *et al.*, 2009, Solis, *et al.*, 2012).

Until recently, scientists primarily worked with two kinds of SCs from animals and humans: embryonic SCs and non-embryonic "somatic" or "adult" SCs (Feng, *et al.*, 2009).

In 1981, scientists discovered ways to derive embryonic SCs from mouse embryos. In 1998, a detailed study of the biology of mouse SCs led to the discovery of a method to derive SCs from human embryos and grow the cells in the laboratory, and these cells are called human embryonic SCs.

In 2006, genetically "reprogrammed" stem-cell-like cells were identified by using specialized adult cells. This new type of stem cell is called **induced pluripotent SCs (iPSCs)** (Krishna, *et al.*, 2011).

2.1. Cancer stem cells

Cancer is a class of diseases characterized by unregulated cell growth(Deisboeck, *et al.*, 2011). Cancer initiation depends on genetic mutations in series that affects cellular programming. Many cancer researches have focused on the identification and characterization of these genetic and molecular properties of cancer cells (Balmain, *et al.*, 2003). Tumors are also heterogeneous cellular entities whose growth is dependent upon dynamic interactions among the cancer cells themselves, and between cells and the constantly changing microenvironment (Bissell & Radisky, 2001). That kind of interaction is depent on signaling through cell adhesion molecules and different cell responses to growth factors and other external signals. All of these interactive processes act together to control cell phenotypic behaviors such as proliferation, apoptosis, and migration. There are over 100 different types of cancer, and each is classified by the type of cell that is initially affected (Lakshmi Prasanna & Sathish Kumar, 2011).

According to recent statistics, cancer accounts for about 23% of the total deaths in the USA and is the second most common cause of death after heart disease (Jemal, *et al.*, 2007).

Cancer is caused by many internal and external factors. Inherited mutations, hormones, and immune conditions are internal factors while tobacco, diet, radiation, and infectious organisms are environmental/acquired factors (Kalluri & Weinberg, 2009, Nagy, *et al.*, 2010, Langley & Fidler, 2011, Mantel & Schmidt-Weber, 2011, Noman, *et al.*, 2011). In recent years, a particular sub-population of tumor cells are said to have a critical role in cancer; these cells are commonly called **CSCs** or **tumor initiating cells (TICs).** In most cancer types, CScs have been identified. CSCs are characterised by their two important properties: (1) Enhanced tumorigenicity; and (2) the capacity for self-renewal/differentiation (Bonnet & Dick, 1997, Al-Hajj, *et al.*, 2003). Thus, isolating CSCs is important in analyzing their characteristics *in vitro*. The isolated CSC population will not only give rise to de novo tumors with high efficiency, but will also recapitulate the tumor with both CSC and non-CSC populations.

One potential human CSC marker is the membrane antigen CD133 (Prominin) identified in subpopulations of cells in brain, colon and lung tumors (Singh, *et al.*, 2004, Ricci-Vitiani, *et al.*, 2007, Eramo, *et al.*, 2008). CD133+ tumor cells are also a marker identifying lung CSCs (Wang, *et al.*, 2008, Salnikov, *et al.*, 2010).

The expression and activity of ALDHs is determined as another potential CSC marker (Ginestier, *et al.*, 2007). ALDH1 is a marker of normal and malignant human mammary SCs and a predictor of poor clinical outcome (Huang, *et al.*, 2009). Aldehyde dehydrogenase enzymes participate in cellular detoxification, differentiation and drug resistance through the oxidation of cellular aldehydes (Moreb, *et al.*, 1996).

The functional activity of ALDH has been widely used to identify and isolate CSCs found in the bone marrow (Ran, *et al.*, 2009), breast (Ginestier, *et al.*, 2007), lung (Ucar, *et al.*, 2009), ovary (Deng, *et al.*, 2010), colon (Huang, *et al.*, 2009), prostate (van den Hoogen, *et al.*, 2010), and pancreas (Dembinski & Krauss, 2009).

2.2. Stem cell markers

The derivation of SCs from adult tissues, their relative ease of isolation and enormous expansion potential in culture make them attractive therapeutic candidates (Prockop, *et al.*, 2010). These cells are identified by their expression of a particular panel of surface molecules, with the presence of CD73, CD90, CD105, and the absence of CD14, CD34, CD45, and HLA-DR. They show no proliferative response from alloreactive lymphocytes because of the negligible levels of extracellular MHC class I and II determinants. SCs also have important immunomodulatory functions in all the cells involved in both the innate and adaptive immune responses (Nauta & Fibbe, 2007).

3. ALDH as a stem cell marker

In theory, ALDH isozymes including ALDH1A, ALDH1A2, ALDH1A3, and ALDH3A1, which are involved in drug resistance and RA formation, are vital in protecting SCs against toxic endogenous and exogenous aldehydes and for SCs' ability to differentiate, respectively. It is unknown what ALDH isozymes are responsible for the ALDH activity that are used to identify stem cell progenitors. In the overlap gene profile of different stem cell populations, ALDH7A1, known as antiquitin, and ALDH2 were identified, consequently, and are worthy of further investigation. There is more about ALDH to be explored as a cause of its full physiological function has remained elusive. ALDH7A1 is a green pea 26g protein, which has function in regulation of turgor pressure, and has ≥50% amino acid identity with the 3 pseudogenes in the ALDH family. It also has 69% equity with ALDH2, but nevertheless has considerably lower affinity for acetaldehyde than ALDH2. However, ALDH2, which is a mitochondrial enzyme, has been widely studied mostly for its affilitation with ethanol metabolism. Yet, there might be an extent of confusion as to how ALDH2 is associated in gene profiling studies. According to the nomenclature, this enzyme indeed is ALDH1A, related to a series of events linked to the development of dopaminergic neurons through its ability to produce RA. It was reported that ALDH2 or AHD2 expression changes during differentiation of NIH-3T3 cells into adipocytes. These studies continue to focus on ALDH1A1's role in SCs and stem cell differentiation. For hematopoietic stem cell progenitors, ALDH1A1 has been a thoroughly established marker for many years. Reseach on the role of RA in granulocyte differentiation of hematopoietic SCs discovered that ALDH1A1 and ALDH1B1 catalyze cellular RA synthesis and are expressed in CD34+ hematopoietic progenitors (Russo, *et al.*, 2002, Luo, *et al.*, 2007).They also showed that ALDH1A2 or 1A3 do not show those characteristics. For the differentiation to mature granulocytes, these 2 enzymes' expressions are necessary, however their expressions are lost once the differentiation is complete. The *in vitro* disulfiram treatment in which disulfiram acts as an ALDH inhibitor may inhibit granulocytic differentiation. ALDH1A1 is found in erythrocytes and has been pointed to contribute to the aldophosphamide detoxification. A study that inhibited ALDH and retinoid signaling with diethylamino-benzaldehyde (DEAB) that reported the expansion of human HSCs probably by blocking differentiation and assisting self-renewal and HSC expansion (Marchitti, *et al.*, 2008).

The cancer stem cell theory is supported by current evidence in tumor biology, which may also provide a biological reason for the age-related survival difference. The theory demonstrates that CSCs, a small subset of tumor cells with stem cell-like properties such as epithelial-to-mesenchymal progression, are capable of differentiation and self-renewal, after which leads to formation of a heterogeneous tumor cell population. Including aldehyde dehydrogenase-1 (ALDH1) activity, CD44+/CD24-, CD133, and ITGA6, a wide range of putative breast cancer stem cell markers have been proposed. ALDH1 expression has especially demonstrated an assurance of a clinically relevant prognostic marker. In addition, the subset of CSCs is shown to be relatively insusceptible to chemo and radiotherapy by various studies. For this reason, the subpopulation of CSCs can present a statement and a therapeutic target for poor-prognostic, treatment-resistant and recurrent breast cancer. Through its role in oxidizing retinol to RA, which is a modulator of cell proliferation, ALDH1 might have a role in early differentiation of SCs and stem cell proliferation (Mieog, *et al.*, 2012).

It is possible to isolate leukemia SCs depending on the elevated ALDH activity by using the aldefluor assay. In patient samples, the researchers encountered a population of ALDH+ acute myeloid leukemia (AML) cells (Rollins-Raval, *et al.*, 2012). In most cases, the ALDH+ AML cells coexpressed CD34+ (formerly determined leukemia stem cell marker), and were introduced considerably better than the ALDH- AML cells in immunocompromised mice. In the same year, ALDH+ cells from breast cancers, which had the tumorigenic and self-renewal features of CSCs, were shown to be possibly isolated. This innovative study displayed the potential applicability of quantifying ALDH activity in solid tumors. ALDH activity would be used successfully as a CSC marker for abundant cancers including liver, colon, lung, bone, prostate, pancreatic, head and neck, thyroid, bladder, brain, cervical and melanoma in the proceeding years. With one exception of a current study for melanoma, 35 demonstrate growing evidence recommending ALDH's activity to be a universal CSC marker. Nonetheless, as amounted by the aldefluor assay in various tissues and cancers, the cause of ALDH activity may differ. Essentially, determination of specific ALDH isoforms carried out commonly in certain cancers might have prognostic suitability. Besides their valuable function in detoxification of aldehydes, ALDHs carry out other functions such as serving as binding proteins for various molecules (e.g., androgens and cholestorol), potentially act as antioxidants by NAD(P)H production, ultraviolet light absorption and/or hydroxyl radical scavenging and ester hydrolysis.

Lastly, several isoforms (ALDH1A1, ALDH1A2, ALDH1A3 and ALDH8A1), take place via RA formation by oxidation of all-trans-retinal and 9-cis-retinal in RA cell signaling, which has been related to the "stemness" characteristics of CSCs. Consequently, its supported by widening evidence that ALDH may be more than just a CSC marker and have an accomplishable role in CSC biology (Marcato, *et al.*, 2011).

3.1. ALDH family members as stem cell markers

ALDH proteins can be found in every subcellular region such as cytosol, endoplasmic reticulum, mitochondria, and the nucleus, with some even found in more than one location. ALDH isozymes found in organelles besides cytosol carry signal or leader sequences that make their translocation to specific subcellular regions possible. After translocation or import, while nuclear and microsomal signals remain intact, mitochondrial sequences might be removed (causing mature proteins to be shorter). Most of the ALDHs have a large tissue distribution and show distinct substrate specificity (Marchitti, *et al.*, 2008).

3.1.1. ALDH1A1

ALDH1A1 encodes a homotetramer that is ubiquitously distributed in the adult epithelia of several organs such as brain, testis, kidney, eye lens, retina, liver and lungs. ALDH1A1 takes its position among the three highly-conserved cytosolic isozymes (see ALDH1A2 and ALDH1A3), which catalyze the oxidation of the retinol metabolite, retinal (retinaldehyde), to RA. ALDH1A1 has great affinity for the oxidation of both all-*trans*-($K_m < 0.1$ μM) and 9-*cis*-retinal. By serving as a ligand for nuclear RA receptors (RAR) and retinoid X receptors (RXR), RA regulates gene expression; therefore its synthesis is crucial for normal growth, differentiation, development and the maintenance of adult epithelia in vertebrate animals. In retinoid-dependent tissues (including the retina), retinal-oxidizing ALDHs have been shown to display differential expression patterns during organogenesis in rodents, reflecting that RA signaling is indeed important for embryogenesis. The *in vivo* function of ALDH1A1 in RA synthesis is proven by the fact that after retinol treatment, while *Aldh1a1*$^{-/-}$mice are viable and possess normal morphology of the retina, the livers of *Aldh1a1*$^{-/-}$ mice have reduced RA synthesis and increased serum retinal levels. Surprisingly, it appeared that *Aldh1a1*$^{-/-}$ mice are protected against both diet-induced obesity and insulin resistance and this demonstrates that retinal might regulate the metabolic response to high-fat diets transcriptionally, and that the *ALDH1A1* could be a candidate gene for therapeutic targeting. Supression of *ALDH1A1* in cultured hepatocytes reduces both the omega-oxidation of free fatty acids and the production of reactive oxygen species (ROS). Liver ALDH1A1 levels were shown to be decreased in *RXRα*$^{-/-}$ mice, which suggests that RA binding is an activating factor in *ALDH1A1* gene expression. The androgen receptor might also be included in modulation of ALDH1A1, which is recognized to be an androgen binding protein. RA is required for testicular development and *ALDH1A1* is absent in genital tissues of humans with androgen receptor-negative testicular feminization. ALDH1A1 is significantly expressed in dopaminergic neurons that are known to require RA for their differentiation and development in the human brain. In these neurons, *ALDH1A1* is under the control of Pitx3, a homeodomain transcription factor that, possibly through *ALDH1A1* upregulation, regulates the particularization and maintenance of disassociated populations of dopaminergic neurons. Decreased levels of ALDH1A1 takes place in dopaminergic neurons of the substantia nigra of patients with Parkinson's disease (PD), as well as the ventral tegmental area in schizophrenic patients. In the central nervous system (CNS), monoamine oxidase (MAO) metabolizes dopamine to aldehyde, as it's metabolite 3,4-dihydroxyphenylacetaldehyde (DOPAL), which growing evidence suggests might be neurotoxic, and it may lead to cell death in relation to neurological pathologies when accumulated. In maintaining low intraneuronal levels of DOPAL, ALDH1A1 may undertake a critical role by catalyzing its metabolism to 3,4-dihydroxyphenylacetic acid (DOPAC). Being one of 139 genes that are differentially expressed in primary human HSCs, and through the production of RA, ALDH1A1 has been shown to promote their differentiation.

These data suggest that for the therapeutic amplification of HSCs, ALDH1A1 inhibition could potentially be used (Marchitti, *et al.*, 2008, Moore, *et al.*, 2009).

3.1.2. ALDH1A2

ALDH1A2 is a cytosolic homotetramer expressed in several embryonic and adult tissues such as brain, kidney, intestine, testis, liver, retina, lung. As ALDH1A1, ALDH1A2 also catalyzes the reaction in which both all-*trans*-retinal and 9-*cis*-retinal oxidize to RA. However, when compared with other ALDH isozymes, ALDH1A2 appears to acquire the highest specificity (V_{max}/K_m = 49 $nmol{\cdot}min^{-1}{\cdot}mg^{-1}{\cdot}\mu M^{-1}$) for all-*trans*-retinal. This characteristic may be because of an uncommon discreate loop in its active site that binds all-*trans*-retinal in a unique manner.

Taking action in several developmental processes, ALDH1A2 might be a key regulator of RA synthesis in developing tissues. Due to defects in early heart morphogenesis, $Aldh1\alpha2^{-/-}$ mice die in early embryonic stages in which they seem to lack axial rotation, incomplete neural tube closure, reduction of the trunk region and many of the properties of human DiGeorge/velocardiofacial syndrome, a disorder characterized by cleft palate, heart abnormalities and learning disabilities. During early vascular development, aberrations in endothelial cell cycle progression have also been determined in $Aldh1\alpha2^{-/-}$ embryos. *Aldh1a2* has been determined as a key regulator in the development of many tissues including kidney, retina, lung, forebrain, pancreas, and spinal cord by miscellanous animal models (Marchitti, *et al.*, 2008, El Kares, *et al.*, 2010).

3.1.3. ALDH1A3

ALDH1A3 is a cytosolic homodimer that participates in RA synthesis, oxidizes both all-*trans*-retinal and 9-*cis*-retinal (K_m 0.2 μM for all-*trans*-retinal) to RA, and has an important role in embryonic development; including brain, retina, skeletal muscle, tooth buds, intestine, kidney, prostate, lung, liver and pancreas, it is expressed in various late-stage embryonic and adult rodent tissues. In humans, ALDH1A3 expression has been noted in stomach, salivary gland, breast, kidney and fetal nasal mucosa. $Aldh1a3^{-/-}$ mouse embryos die as a result of defects in nasal development.

It's been shown that ALDH1A3 takes part in the development of the eye, nucleus accumbens and olfactory bulbs, the forebrain, hair follicles and the cerebral cortex.

ALDH1A3 deficiency has been shown to play a critical role in cancer by a number of studies. For instance, in human breast cancer MCF-7 cells, ALDH1A3 expression is downregulated, whereas in cultured human colon cancer cells, *ALDH1A3* is one of two genes that are upregulated by induction of wild type *p*53. In mammary tumor-susceptible BALB/cJ mice that are heterozygous for *p*53, *Aldh1a3* is one of five candidate genes located within a region determined for its linkage to mammary tumorigenesis. In mice resistant to induced mammary tumors, (C57BL/6J), *Aldh1a3* is one of the two upregulated genes. *ALDH1A3* is silenced by methylation in gastric cancer cells, whereas in glioblastoma cells, it is triggered by the antitumor agent IL-13 cytotoxin (Marchitti, *et al.*, 2008).

3.1.4. ALDH2

ALDH2 is a tetrameric enzyme expressed profusely in lungs and liver; it is also present in organs that obligate high mitochondrial capacity for oxidative ATP generation including heart and brain. Apart from that, ALDH2 is also important in the aldehydic substrate oxidation such as 4-HNE, acrolein, and short-chain, aromatic or polycyclic carbons. To add to its dehydrogenase activity, depending on the substrates, ALDH2 can function as an esterase and reductase. More recent attention has also been focused on ALDH2 in regards to its function in the biotransformation of nitroglycerin, reducing it to 1,2-glyceryl dinitrate for the production of nitric oxide, which is a critical vasodilator (Chen, *et al.*, 2010).

3.1.5. ALDH7A1

ALDH7A1 is a homotetramer that's expressed in a large number of tissues; in rat heart, liver and kidney, increased levels of ALDH7A1 are noted, whereas in black seabream fish (sbALDH7A1), ALDH7A1 is significantly formed in the liver and the kidney, excluding the heart. In human fetal tissues, ALDH7A1 has been encountered at elevated levels in the cochlea, eye, ovary, heart and kidney. In contrast, balanced levels are detected in the liver, spleen, muscle, lung and brain.

Human ALDH7A1's primary role happens in the pipecolic acid pathway of lysine catabolism, in which it catalyzes the oxidation reaction of alpha-aminoadipic semialdehyde (AASA) (K_m180 μM) to alpha-aminoadipate. *ALDH7A1* mutations form the molecular basis for pyridoxine-dependent epilepsy (PDE), an autosomal recessive disorder characterized by the aggression of tenacious seizures during infancy and early childhood and are avoidable by daily use of high-dose pyridoxine (Vitamin B_6) supplementation.

Remarkably, ADH7A1 expression in the cochlea of the ear, the region dependent on the healthy upkeep of internal hydrostatic pressure, clarifies that mammalian ALDH7A1 might have an accomplishable function in osmotic regulation and in hearing disorders. However, no connection has been revealed yet, including patients with the inner-ear disorder Ménière's disease, which effects hearing and balance.

ALDH7A1 is notably and differentially expressed within the first and second meiotic stages of porcine oocyte development. Screening of the promoter region *sbALDH7A1* has discovered *cis*-elements linked with cell cycle regulation (Marchitti, *et al.*, 2008).

4. ALDH in cancer and cancer stem cells

4.1. Adenocarcinoma

Adenocarcinoma is an epithelium cancer that is generated from glandular tissue. Epithelial tissue includes, but is not limited to, the surface layer of skin, glands and a variety of other tissues that line the cavities and organs of the body. Epithelium can be derived from the three germ layers ectoderm, mesoderm and endoderm during embryologic period. Adenocarcinoma classification depends on not only being a part of the gland, but also depends on having the same secretory characteristics. But, this form of carcinoma can occur in some higher mammals, including humans (Fauquier, *et al.*, 2003).

Adenocarcinomas can arise in many tissues of the body due to the ubiquitous nature of glands within the body. While each gland may not be secreting the same substance, as long as there is an exocrine function to the cell, it is considered glandular and its malignant form is therefore named adenocarcinoma. Endocrine gland tumors, such as a VIPoma, an insulinoma, a pheochromocytoma, etc. are typically not referred to as adenocarcinomas, but rather, are often called neuroendocrine tumors. If the glandular tissue is abnormal, but benign, it is called an adenoma. Benign adenomas typically do not invade other tissues and rarely metastasize, whereas malignant adenocarcinomas do both. Colon, urogenital (cervical (Tewari, *et al.*, 2002), prostate, urachus and vagina), breast (Buchholz, 2009), esophagus, pancreas, stomach and throat are several examples of adenocarcinoma (Subramanian & Govindan, 2007).

It is reported that ALDH expression marks pancreatic cancer stem cells. Also, they have mentioned that the enhanced clonogenic growth and migratory properties of ALDH-positive pancreatic cancer cells suggest a key role in the development of metastatic disease that negatively affects the overall survival of patients with pancreatic adenocarcinoma (Rasheed, *et al.*, 2010).

4.2. Breast cancer

From high-grade, absence of hormone receptor expression to positive HER2 status and the basal-like molecular subtype, the expression of ALDH1 is in direct relation with undesired tumor characteristics in breast cancer (Mieog, *et al.*, 2012).

Breast cancer cells with stem-cell-like properties are suggested to be responsible for metastatic spread. Aldehyde dehydrogenase 1 (ALDH1) and cluster of differentiation 44 (CD44) in addition to RhoC GTPase are among the stem cell markers that are expressed by these cells (Chaterjee & van Golen, 2011).

Breast CSCs were initially isolated, established on cell surface marker with CD24/lowCD44 expression. More currently, ''functional'' markers depending on stem cell properties are investigated for their plausible applications in the breast CSCs isolation. By this method, applying the aldefluor assay (Stemcell Technologies), originally designed to isolate viable HSCs and is an enzyme-based assay that recognizes ALDH activity, Ginestier et al. isolated breast CSCs. The assay is thought to precisely recognize ALDH isoform ALDH1A1 activity degree. Besides its application as a prognostic and CSC marker, ALDH activity that is primarily carried out by ALDH1A3 might be functional in breast cancer progression.

Expression of genes and tumor sphere formation in self-renewal and differentiation could be changed by adding chemical RA signaling inducers or inhibitors in breast cancer cell lines (Marcato, *et al.*, 2011).

ALDH1 could work as a marker of breast CSCs better than CD44+/CD24-. Though we could not maintain a conclusion that ALDH1 expression was significantly related with any conventional clinicopathologic attributes, nevertheless, there is a compelling relation between ALDH1-positive breast tumors and resistance to neoadjuvant chemotherapy, because of the pCR rates being obtained, which are lower in ALDH1-positive tumors (9.5%) than ALDH1-negative tumors (32.2%). Moreover, after neoadjuvant chemotherapy, a considerable increase in the proportion of ALDH1-positive tumor cells was observed. These results are an indication of ALDH1-positive tumor cells playing an important role in resistance to chemotherapy. Because of tumor cells being more tumorigenic than CD44+/CD24-, tumor cells of breast CSCs are thought to be richer in ALDH1-positive tumor cells than in CD44+/CD24- tumor cells. As a matter of course, we have shown that ALDH1-positive, in contrast with CD44+/CD24-, is closely associated with colony formation in the collagen gel as well. The subset of ALDH1-positive and CD44+/CD24- tumor cells has been reported to contain the largest proportion of breast cancer stem cells (BCSCs); consequently, it is speculated to have the strongest resistantance to chemotherapy. However, in our current study, pCR rates in the ALDH1-positive and CD44+/CD24- high subset (20%, 2 of 10), are not the lowest among all the subsets consisting of the ALDH1-positive and CD44+/CD24- low subset (0%, 0 of 11), the ALDH1-negative and CD44+/CD24- high subset (34.1%, 15 of 44), and the ALDH1-negative and CD44+/CD24- low subset (30.2%, 13 of 43). Adding CD44/CD24 status to ALDH1 status does not seem to positively improve the prediction of response to chemotherapy. Together, these results direct us to assume that, at least for the prediction of resistance to chemotherapy, ALDH1-positive tumor cells serve as a better marker for BCSCs than CD44+/CD24- tumor cells. Because such tumors contain a higher proportion of CSCs, we suppose that ALDH1- positive tumors are resistant to chemotherapy. However, because ALDH1 has been shown to play an important role in the resistance to chemotherapy in hematopoietic cells, ALDH1-positive tumor cells might be involved in resistance to chemotherapy, regardless of whether they are CSCs or not. In addition to deeper illumination of ALDH1's function in chemotherapy resistance in breast cancers, obtaining a significantly specific marker for BCSCs is necessary to enlighten an authentic role of BCSCs' chemotherapy resistance.

ALDH1-positive, in contrast to CD44+/CD24-, was tremendously related to sequential paclitaxel- and epirubicin-based chemotherapy resistance, and the expression of ALDH1 increased after neoadjuvant chemotherapy, which stands for an indication of BCSCs, determined by ALDH1, indeed having played a significant role in chemotherapy resistance. This means that ALDH1-positive appears to be a better marker than CD44+/CD24- in identifying BCSCs, at least for the prediction of resistance to chemotherapy (Tanei, *et al.*, 2009).

4.3. Lung cancer

Each year, approximately 171,000 new cases of lung cancer are diagnosed, and 160,000 individuals do not survive from the disease in the United States. This high incidence and mortality makes lung cancer one of the most common cancers and the leading cause of cancer death in men. Lung cancer is still the leading cause of death from malignant diseases worldwide in spite of the advances in surgical treatment and multimodality treatments (Hibi, *et al.*, 1998).

Cancer stem cells have attributed resistance of a smaller fraction of cells in the tumor bulk against chemotherapeutics. The isolation of CSCs is important for these reasons and have been isolated using a variety of stem cell markers and phenotypes. CD133 has recently been reported to identify tumor-initiating cells in non-small cell lung cancer (NSCLC). ABCG2 is also a stem cell marker of a variety of tissues and transporter responsible for the multidrug-resistance phenotype. However, it was demonstrated that many cells in NSCLC and SCLC cell lines show tumorigenic potential, regardless of ABCG2 and CD133 expression. Recently, ALDH activity has been used for isolation of these kinds of cells. Normal SCs were shown to contain higher levels of ALDH activity than their more differentiated progeny. ALDH-positive cells of tumors have higher proliferation rates, migration and adhesion ability, and metastatic potential than ALDH-negative cells. This may occur because that RA product of ALDHs is thought to participate in cellular differentiation and stem cell self-protection (Serrano, *et al.*, 2011).

4.4. Ovarian cancer

Epithelial ovarian cancer is the sixth most common cancer in women worldwide and it is still the most lethal gynecologic malignancy (Iorio, *et al.*, 2007). Application of new technologies for detection of ovarian cancer could have an important effect on public health, but to achieve this goal, specific and sensitive molecular markers are essential (Petricoin, *et al.*, 2002). Aldehyde dehydrogenase-1A1 (ALDH1A1) has been a valid marker among several malignant and non-malignant tissues in spite of several stem cell markers to identify CSCs. ALDH plays a role in the biology of TICs as well as being a stem cell marker. Because ALDH1A1 is implicated in chemo resistance pathways, it is questioned that targeting ALDH1A1 can effect cells resistant to chemotherapy and represent a potential target for cancer stem-cell-directed therapy. In a study, ALDH1A1 was investigated in ovarian cancer cell lines and patient samples and examined whether targeting ALDH1A1 sensitizes cells to chemotherapy in both *in vitro* and *in vivo* ovarian cancer models. They showed that ALDH1A1 expression and activity have increased chemo resistant ovarian cancer cell lines. Most importantly, down-regulation of ALDH1A1 expression has sensitized normally chemo resistant tumors to both docetaxel and cisplatin *in vitro* and in mouse models. Besides being a stem cell marker, ALDH1A1 is also a viable target for therapy and a mediator of the aggressive phenotype (Landen, *et al.*, 2010).

4.5. Pancreatic cancer

Pancreatic adenocarcinoma is a highly lethal disease, which is usually diagnosed in an advanced state, and for which there is little or no effective therapies (Li, *et al.*, 2007). Therefore, finding markers to detect a malignant cell transformation at an early stage is very important. Researches demonstrated that the pancreas possesses ALDH activity, and ALDH is also present in the pancreatic cancer cells. Different from other cancer tissues (such as ovarian and lung cancer), the activity of ALDH does not differ in pancreatic carcinoma tissue compared to normal pancreatic tissue. Additionally, serum levels of ALDH were not significantly elevated in patients with pancreatic cancer in comparison to healthy controls (Jelski, *et al.*, 2011).

4.6. Prostate cancer

The latest estimates of global cancer incidence show that prostate cancer has become the third most common cancer in men, with half a million new cases each year, constituting almost 10% of all cancers in men (Quinn & Babb, 2002). Identifying the origin of cells in prostate cancer and its distant metastases may be important for the improvement of more effective treatment strategies and preventive therapies. Measurement of ALDH activity provides great contribution to functional identification and characterization of normal SCs and their malignant counterparts. ALDH activity is important for drug resistance, cell proliferation, differentiation, and response to oxidative stress of prostate cancer like other important cancers.

ALDH enzyme activity is used for the isolation of "stem-like" cells based on a developmentally conserved stem/progenitor cell function. In a study, high ALDH activity was used to isolate human prostate cancer cells with significantly enhanced clonogenic and migratory properties both *in vitro* and *in vivo*. Similar to other cancer tissues, the percentage of ALDHhi cells in prostate cancer cell lines are also related to tumorigenicity and metastatic behavior.

Although high expression of ALDH7A1 is shown in prostate cancer cell lines, primary cultures, and in primary prostate cancer tissue and matched bone metastases, ALDH3A2 and ALDH18A1 are not observed high ALDH activity in human prostate cancer (van den Hoogen, *et al.*, 2010).

4.7. Brain cancer

Glioblastoma (GBM) is the most common primary brain tumor in adults with an approximately 15-month survival (Stupp, *et al.*, 2005). Although there are several studies to improve the postoperative therapeutic applications within the last few years, there is not enough succes for this highly aggressive tumor. After resection, radiation, and chemotherapy regimens, relapses occur regularly. Thus, it is thought that this can be a clue to the presence of tumor stem cells (TSCs). This cellular subfraction within GBM causes continuous tumor growth and resistance to drugs and radiation (Rasper, *et al.*, 2010). TSCs are believed to nestle in the tumor, keeping it alive and growing, providing pluripotency, self-renewal, and resistance to chemo and radiation therapy(Reya, *et al.*, 2001). The first malignancies from which cells could be isolated and showed the potential to self-renew and to drive tumor formation and growth were leukemias (Bonnet & Dick, 1997). After that, a stem cell subfraction was described in brain tumors (Singh, *et al.*, 2003). This was the first study that identified and showed a population with stem cell properties in pediatric solid brain tumors. Those cells were identified by their ability to proliferate under serum-free cell culture conditions and by the expression of CD133 and nestin. CD133 has long remained the most important TSC marker in malignant glioma. On the other hand, ALDH1 is a cytoplasmatic stem cell marker in a variety of malignant tumors and catalyzes the oxidation of intracellular aldehydes including the transformation of retinol to RA. As mentioned above, RA is a modulator of cell proliferation and differentiation that possibly contributes to the maintenance of an undifferentiated stem cell phenotype. Jones et al. presented a method to isolate human cells via flow cytometry depending on the amount of cytosolic ALDH (Jones, *et al.*, 1995). Recently, Ginestier et al. found ALDH1 to be a stem cell marker in breast carcinoma associated with poor clinical outcomes (Ginestier, *et al.*, 2007). Since then, ALDH1 has been described as a marker of stemness in other solid malignancies including lung cancer (Jiang, *et al.*, 2009) and colorectal cancer (Huang, *et al.*, 2009).

Therefore, identification and isolation of these cells seem crucial for a better understanding of tumor behavior, origin, and therapy. Recently, ALDH1 has been described as a marker for the identification of non-neoplastic SCs and TSCs (Ginestier, *et al.*, 2007).

So far, cellular markers including CD133 have been used to identify TSCs in GBMs, but recently, CD133-negative GBMs are characterized to behave as brain TSCs (Beier, *et al.*, 2007).

Therefore, ALDH1 has also been described as a stem cell marker in various solid neoplasms including lung cancer (Jiang, *et al.*, 2009), breast carcinoma (Ginestier, *et al.*, 2007), and colorectal cancer (Huang, *et al.*, 2009) and GBM (Rasper, *et al.*, 2010).

4.8. Colon cancer

Most colon cancers are adenocarcinomas that release mucus and other cellular secretions. In the United States in 2012, estimated new cases and deaths from colon and rectal cancer are reported as: 103,170 colon cancers and 51,690 deaths (Levin, *et al.*, 2008). Studies showed that ALDH1B1 and ALDH1A1 are differentially expressed in normal human tissues, but ALDH1B1 is expressed at higher levels than ALDH1A1 in human epithelial cancers. ALDH1B1 was abundantly expressed in adenocarcinomas originating from the tissue and particularly in colonic adenocarcinoma (Chen, *et al.*, 2011). Thus it can be deduced that ALDH1B1 may be a marker for colon cancer diagnosis.

5. Aldefluor activity in stem cells and cancer stem cells

$ALDH^{br}$ (ALDH-bright) cells can be detected with ALDEFLUOR reagent by using flow cytometry or fluorescent microscopy. These $ALDH^{br}$ cell populations are isolated from adult tissues by flow sorting.

ALDH activity was shown in human and mouse bone marrow hematopoietic progenitor cells (HPCs) by Jones et al. for the first time (Jones, *et al.*, 1995). ALDH was assayed by using a new substrate with low light scatter properties with flow cytometry. Now this method was improved and known as Aldefluor assay. Aldefluor assay can be used for to measure ALDH activity of adult tissue cells, primary cancer cells and cultured cells. Aldefluor assay is based on the conversion of fluorescent non-toxic substrate for ALDH substrate to the fluorescent reaction product. Non-toxic substrate for ALDH freely diffuses into intact and viable cells. The BODIPY aminoacetaldehyde is converted to the fluorescent product BODIPY aminoacetate by ALDH activity (Figure 1). In this assay, a specific inhibitor of this reaction (diethylaminobenzaldehyde-DEAB) is used to control for background fluorescence. Aldehyde dehydrogenase plays a role as a cancer stem cell marker comes down to the specific isoform.

Stem and progenitor cells are identified as cells with low side scatter and high expression of ALDH. DEAB allows to distinguish between ALDH-bright cells and cells with low ALDH activity. Generally, 10^5-10^6 cells are suspended in Aldefluor assay buffer containing BODIPY aminoacetaldehyde with/without DEAB. Aldefluor was excited at 488 nm and fluorescence emission was detected at 530/30 (van den Hoogen, *et al.*, 2010). This assay provides a successful isolation of viable HSCs and more recently ALDH positive CSCs. However, aldefluor assay detects the ALDH activity of several ALDH isoforms expressed in the cells. ALDH1A1 is not the only isoform responsible from aldeflour activity. In some studies, it was demonstrated that ALDH1A1-deficient hematopoietic cells showed aldefluor activity owing to ALDH2, ALDH3A1 and ALDH9A1 isoforms (Marcato, *et al.*, 2011).

6. ALDH bright ($ALDH^{br}$) cell

Intracellular ALDH enzymes are responsible for oxidizing aldehydes to carboxylic acids in the cell. $ALDH^{br}$ cells from different tissues express high ALDH activity and have progenitor cell activity (Gentry, *et al.*, 2007). Firstly, HSC were defined as $SSC^{lo}ALDH^{br}$ – reflecting their low orthogonal light scattering and bright fluorescence intensity by using flow cytometry (Lioznov, *et al.*, 2005). After that, high levels of the enzyme ALDH ($ALDH^{br}$) have proven to be a novel marker for the identification and isolation of SCs (Mitchell, *et al.*, 2006). In the same time angiogenic activity of $ALDH^{br}$ cells were discovered and these cells were used for regenerative medicine with preclinical models and have been used safely to treat patients in early clinical trials (White, *et al.*, 2011).

$ALDH^{br}$ cells were found in various cancer tissues including breast, liver, colon, and acute myelogenous leukemia and related with cancer chemo resistance. Human and murine HSCs and neural stem and progenitor cells have increased ALDH activity compared to non-stem-cells (Siclari & Qin, 2010).

Therefore, recently the importance of ALDH activity in normal and malignant stem cell functions, and the potential diagnostic and therapeutic implications gain importance (Moreb, 2008).

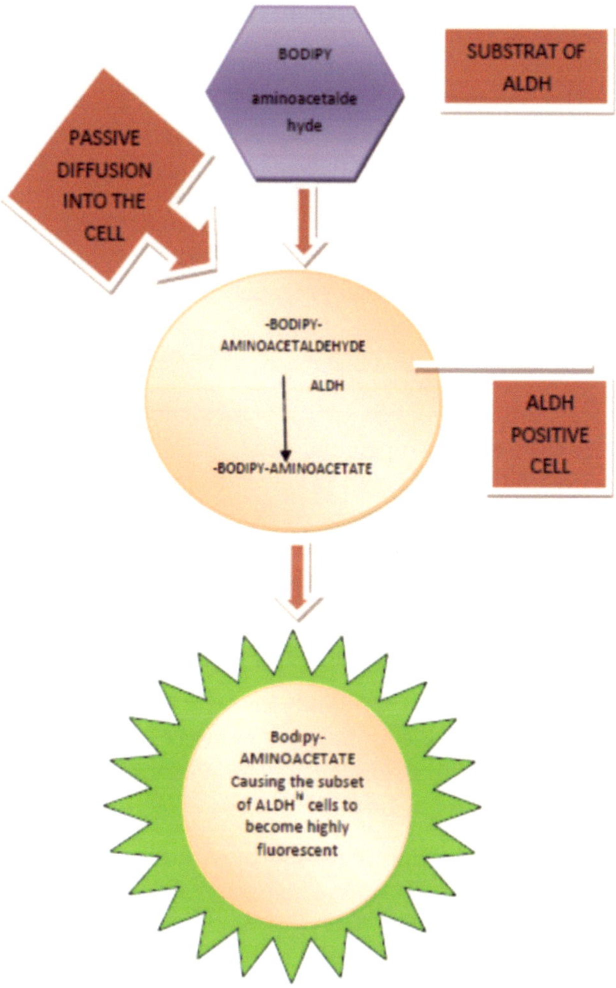

Figure 1. The Aldefluor® Assay. Firstly, ALDH positive cell will uptake BODIPY-aminoacetaldehyde by passive diffusion and then convert BODIPY-aminoacetaldehyde into BODIPY-aminoacetate. Then BAA is retained inside cells, causing the subset of ALDHhi cells to become highly fluorescent (Marcato, *et al.*, 2011).

7. Aldehyde dehydrogenase in regenerative medicine

Until today, studies showed that BM ALDH^{br} populations may be useful in several cell therapy applications (Gentry, *et al.*, 2007). According to this information, ALDH^{br} population may play an important role in regenerative medicine owing to RAs ALDH product (Balber, 2011). Retinoic acids could influence tissue repair by binding to transcription factors and regulating developmental programs, especially ALDH1A1 and ALDH3A1 of enzyme isoforms that produce RAs from oxidize retinaldehyde (Moreb, 2008). Therefore, ALDH1a1 and ALDH1A3 may influence cell activity and proliferation by controlling intracellular retinoid concentrations and play important roles in stem cell biology (Balber, 2011).

The studies about value of $ALDH^{br}$ cells in regenerative medicine were conducted by different researchers. The regenerative potential of $ALDH^{br}$ cells obtained from different tissues were investigated in various disease models such as ischemic tissue damage hind limb model, brain damage and pancreatitis (Balber, 2011).

In the beginning of studies, $ALDH^{br}$ cells were obtained from bone marrow and umbilical cord blood and normal peripheral blood (Sondergaard, *et al.*, 2010). Multipotent mesenchymal progenitors and endothelial progenitor cells are concentrated in human $ALDH^{br}$ populations. Because of potential progenitor and paracrine activities of $ALDH^{br}$ cells, these cells especially obtained from bone marrow are important for tissue repair.

Manipulation of the graft to selectively concentrate or expand hematopoietic and/or neural stem cells prior to transplant may be a potential strategy in the future. UCBT using ALDH bright cells from the CB units have shown faster and higher engraftment in preliminary study and is being explored further (Prasad & Kurtzberg, 2010). One of these studies showed that human cord blood progenitors with high ALDH activity improve vascular density in a model of acute myocardial infarction. In this study, $ALDH^{br}$ cells were homed to the infracted anterior surface of the heart, while ALDH-low cells were in the spleen after intravenously administration.

Another study with animal model of hindlimb ischemia demonstrated that the isolated $ALDH^{br}$ cells effectively restored blood flow to ischemic areas by mediation of local formation of new blood vessels with largeer diameter and increasing capillary density even if there was no improvement in cardiac functions (Keller, 2009).

The reason for the restoration of tissue perfusion by $ALDH^{br}$ cells were attempted to be explained with angiogenic properties of these cell groups. Angiogenic factors secreted by transplanted $ALDH^{br}$ cells stimulate formation of new blood vessels at sites of ischemic injury (human cord blood progenitors with high ALDH activity improve vascular density in a model of acute myocardial infarction). Paracrine mechanisms of $ALDH^{br}$ cells can protect endothelial cells from ischemic damage and respond to ischemic tissue damage (Balber, 2011, White, *et al.*, 2011).

Another exciting finding is that $ALDH^{br}$ cells improve formation of new vessels and increase capillary density, while $ALDH^{br}$ cells together with ALDH-low cells did not restore tissue perfusion at all. It is suggested that ALDH-low cells can inhibit the homing and/or angiogenic activity of $ALDH^{br}$ cells. This situation showed the importance of isolating $ALDH^{br}$ cells from bone marrow tissue for therapeutic uses (Balber, 2011). As a result, $ALDH^{br}$ cells may be promising for patients with ischemic heart failure and critical limb ischemia (Keller, 2009).

TISSUE	OBTAINED FROM	BENEFITS
Neural Tissue	Rat embryonic neural tube Fetal mouse brain Subventricular and subcortical zones of adult mouse brain	Ability to form neurospheres and retained multipotency Transplantation significantly ameliorated disease progression and extended life, but did not rescue the animals.
Skeletal Muscle	Biopsies or primary explants of human skeletal muscle	Strong myogenic potential on IM transplantation
Mammary Epithelium	Mammary epithelium	Myoepithelial, luminal epithelial and mixed colonies, and ducts, when transplanted into mammary fat pads.
Pancreatic Cells	Central acinar/terminal duct cells from peripheral acinar duct units of adult mice	Contributed to both exocrine and endocrine lineages in the developing pancreas
Prostate Epithelium	-	Express basal epithelial and characteristic prostate progenitor cell markers
Corneal Limbic Cells	Cadaveric human limbic tissue	Protects the cornea from oxidative damage

Table 1. Different tissue repair models including human $ALDH^{br}$ cells (Balber, 2011).

8. Conclusion

Since ALDH enzyme has been proven to possess a vital role in somatic cells and their deficiency cause various diseases, research has focused on the presence and functions of the enzyme in SCs. It was demonstrated that ALDH is an important marker for identification of SCs and has several functions in these cells just as they possess in somatic cells.

Exploring some of the isoforms of ALDH for use as a marker of CSCs improved the importance of ALDH. Thus, there are several methods to detect ALDHs and their levels (Marcato, *et al.*, 2011). After the discovery of ALDH activity in human and mouse bone marrow hematopoietic progenitor cells (HPCs) by Jones et al. (Jones, *et al.*, 1995), the properties and locations of ALDH-positive cells have started to be investigated.

Recently, $ALDH^{br}$ cells were found in cancer tissues including breast, liver, colon, and acute myelogenous leukemia. It was demonstrated that proliferation rates, migration and adhesion ability, and metastatic potential of ALDHbr CSCs were more than ALDH low cells and $ALDH^{br}$ cells related with cancer chemo resistance. ALDHbr cells became one of new therapeutic target against cancer and anti-cancer studies based on targeting ALDHbr cells have started recently (Serrano, *et al.*, 2011). It is expected that the anti-cancer studies with this perspective may intensively continue.

On the other hand, studies showed that BM ALDH^{br} populations may be useful in several cell therapy applications (Gentry, *et al.*, 2007). It is suggested that ALDH^{br} population may play an important role in regenerative medicine owing to Ras, which are one of the ALDH products. Paracrine effects of products of ALDH activity may influence tissue repair by binding to transcription factors and regulating developmental programs (Balber, 2011).

Therefore, regenerative potential of ALDHbr SCs were investigated in various disease models such as ischemic tissue damage hind limb model, brain damage and pancreatitis (Balber, 2011).

Studies on ALDH^{br} cells provide restoration of tissue perfusion and stimulation of formation of new blood vessels in ischemic tissue damage (Keller, 2009). These promising findings showed that ALDH^{br} cells may gain importance in different areas; however, there are still many things to investigate about potential properties of ALDH^{br} cells for use in regenerative medicine. Thus, ALDH have many roles such as a marker of many disease and cell lines for detection of them also can using for therapy and have potential for use in regenerative medicine.

However, there are few studies about ALDH as a marker of SCs and potential usage in regenerative medicine. Therefore, we suggested that studies should focus on this and this review aims to consider the roles of ALDH in SCs and their potential use in regenerative medicine. We believe that constructing a review including current studies related to this subject will guide future studies.

Author details

Adil M. Allahverdiyev, Malahat Bagirova, Olga Nehir Oztel,
Serkan Yaman, Emrah Sefik Abamor, Rabia Cakir Koc,
Sezen Canim Ates and Serap Yesilkir Baydar
Department of Bioengineering, Yildiz Technical University, Istanbul, Turkey

Serhat Elcicek
Department of Bioengineering, Yildiz Technical University, Istanbul, Turkey
Department of Bioengineering, Firat University, Elazig, Turkey

9. References

Al-Hajj M, Wicha MS, Benito-Hernandez A, Morrison SJ & Clarke MF (2003) Prospective identification of tumorigenic breast cancer cells. *Proc Natl Acad Sci U S A* Vol.100(No. 7):3983-3988.

Balber AE (2011) Concise review: aldehyde dehydrogenase bright stem and progenitor cell populations from normal tissues: characteristics, activities, and emerging uses in regenerative medicine. *Stem Cells* Vol.29(No. 4):570-575.

Balmain A, Gray J & Ponder B (2003) The genetics and genomics of cancer. *Nat Genet* Vol.33 Suppl(No.:238-244.

Beier D, Hau P, Proescholdt M, *et al.* (2007) CD133(+) and CD133(-) glioblastoma-derived cancer stem cells show differential growth characteristics and molecular profiles. *Cancer Res* Vol.67(No. 9):4010-4015.

Bissell MJ & Radisky D (2001) Putting tumours in context. *Nat Rev Cancer* Vol.1(No. 1):46-54.

Black WJ, Stagos D, Marchitti SA, Nebert DW, Tipton KF, Bairoch A & Vasiliou V (2009) Human aldehyde dehydrogenase genes: alternatively spliced transcriptional variants and their suggested nomenclature. *Pharmacogenet Genomics* Vol.19(No. 11):893-902.

Bonnet D & Dick JE (1997) Human acute myeloid leukemia is organized as a hierarchy that originates from a primitive hematopoietic cell. *Nat Med* Vol.3(No. 7):730-737.

Buchholz TA (2009) Radiation therapy for early-stage breast cancer after breast-conserving surgery. *N Engl J Med* Vol.360(No. 1):63-70.

Chaterjee M & van Golen KL (2011) Breast cancer stem cells survive periods of farnesyl-transferase inhibitor-induced dormancy by undergoing autophagy. *Bone Marrow Res* Vol.2011(No.:362938.

Chen CH, Sun L & Mochly-Rosen D (2010) Mitochondrial aldehyde dehydrogenase and cardiac diseases. *Cardiovasc Res* Vol.88(No. 1):51-57.

Chen Y, Orlicky DJ, Matsumoto A, Singh S, Thompson DC & Vasiliou V (2011) Aldehyde dehydrogenase 1B1 (ALDH1B1) is a potential biomarker for human colon cancer. *Biochem Biophys Res Commun* Vol.405(No. 2):173-179.

Corti S, Locatelli F, Papadimitriou D, *et al.* (2006) Identification of a primitive brain-derived neural stem cell population based on aldehyde dehydrogenase activity. *Stem Cells* Vol.24(No. 4):975-985.

Deisboeck TS, Wang Z, Macklin P & Cristini V (2011) Multiscale cancer modeling. *Annu Rev Biomed Eng* Vol.13(No.:127-155.

Dembinski JL & Krauss S (2009) Characterization and functional analysis of a slow cycling stem cell-like subpopulation in pancreas adenocarcinoma. *Clin Exp Metastasis* Vol.26(No. 7):611-623.

Deng S, Yang X, Lassus H, *et al.* (2010) Distinct expression levels and patterns of stem cell marker, aldehyde dehydrogenase isoform 1 (ALDH1), in human epithelial cancers. *PLoS One* Vol.5(No. 4):e10277.

Discher DE, Mooney DJ & Zandstra PW (2009) Growth factors, matrices, and forces combine and control stem cells. *Science* Vol.324(No. 5935):1673-1677.

El Kares R, Manolescu DC, Lakhal-Chaieb L, Montpetit A, Zhang Z, Bhat PV & Goodyer P (2010) A human ALDH1A2 gene variant is associated with increased newborn kidney size and serum retinoic acid. *Kidney Int* Vol.78(No. 1):96-102.

Eramo A, Lotti F, Sette G, *et al.* (2008) Identification and expansion of the tumorigenic lung cancer stem cell population. *Cell Death Differ* Vol.15(No. 3):504-514.

Fauquier D, Gulland F, Haulena M & Spraker T (2003) Biliary adenocarcinoma in a stranded northern elephant seal (Mirounga angustirostris). *J Wildl Dis* Vol.39(No. 3):723-726.

Feng B, Ng JH, Heng JC & Ng HH (2009) Molecules that promote or enhance reprogramming of somatic cells to induced pluripotent stem cells. *Cell Stem Cell* Vol.4(No. 4):301-312.

Fuchs E & Segre JA (2000) Stem cells: a new lease on life. *Cell* Vol.100(No. 1):143-155.

Gentry T, Foster S, Winstead L, Deibert E, Fiordalisi M & Balber A (2007) Simultaneous isolation of human BM hematopoietic, endothelial and mesenchymal progenitor cells by flow sorting based on aldehyde dehydrogenase activity: implications for cell therapy. *Cytotherapy* Vol.9(No. 3):259-274.

Ginestier C, Hur MH, Charafe-Jauffret E, *et al.* (2007) ALDH1 is a marker of normal and malignant human mammary stem cells and a predictor of poor clinical outcome. *Cell Stem Cell* Vol.1(No. 5):555-567.

Ginestier C, Hur MH, Charafe-Jauffret E, *et al.* (2007) ALDH1 is a marker of normal and malignant human mammary stem cells and a predictor of poor clinical outcome. *Cell Stem Cell* Vol.1(No. 5):555-567.

Hibi K, Liu Q, Beaudry GA, *et al.* (1998) Serial analysis of gene expression in non-small cell lung cancer. *Cancer Res* Vol.58(No. 24):5690-5694.

Hofseth LJ & Wargovich MJ (2007) Inflammation, cancer, and targets of ginseng. *J Nutr* Vol.137(No. 1 Suppl):183S-185S.

Huang EH, Hynes MJ, Zhang T, *et al.* (2009) Aldehyde dehydrogenase 1 is a marker for normal and malignant human colonic stem cells (SC) and tracks SC overpopulation during colon tumorigenesis. *Cancer Res* Vol.69(No. 8):3382-3389.

Iorio MV, Visone R, Di Leva G, *et al.* (2007) MicroRNA signatures in human ovarian cancer. *Cancer Res* Vol.67(No. 18):8699-8707.

Jackson B, Brocker C, Thompson DC, Black W, Vasiliou K, Nebert DW & Vasiliou V (2011) Update on the aldehyde dehydrogenase gene (ALDH) superfamily. *Hum Genomics* Vol.5(No. 4):283-303.

Jelski W, Kutylowska E, Laniewska-Dunaj M & Szmitkowski M (2011) Alcohol dehydrogenase (ADH) and aldehyde dehydrogenase (ALDH) as candidates for tumor markers in patients with pancreatic cancer. *J Gastrointestin Liver Dis* Vol.20(No. 3):255-259.

Jemal A, Siegel R, Ward E, Murray T, Xu J & Thun MJ (2007) Cancer statistics, 2007. *CA: a cancer journal for clinicians* Vol.57(No. 1):43-66.

Jiang F, Qiu Q, Khanna A, *et al.* (2009) Aldehyde dehydrogenase 1 is a tumor stem cell-associated marker in lung cancer. *Mol Cancer Res* Vol.7(No. 3):330-338.

Jones RJ, Barber JP, Vala MS, *et al.* (1995) Assessment of aldehyde dehydrogenase in viable cells. *Blood* Vol.85(No. 10):2742-2746.

Kalluri R & Weinberg RA (2009) The basics of epithelial mesenchymal transition. *J Clin Invest* Vol.119(No. 6):1420-1428.

Keller LH (2009) Bone marrow-derived aldehyde dehydrogenase-bright stem and progenitor cells for ischemic repair. *Congest Heart Fail* Vol.15(No. 4):202-206.

Krishna KA, Krishna KS, Berrocal R, Tummala A, Rao KS & Rao KR (2011) A review on the therapeutic potential of embryonic and induced pluripotent stem cells in hepatic repair. *J Nat Sci Biol Med* Vol.2(No. 2):141-144.

Lakshmi Prasanna N & Sathish Kumar D (2011) A Compendium of Cancer Therapeutic Strategies and their Modality. *J Cancer Sci Ther S* Vol.17(No.:2.

Landen CN, Jr., Goodman B, Katre AA, *et al.* (2010) Targeting aldehyde dehydrogenase cancer stem cells in ovarian cancer. *Mol Cancer Ther* Vol.9(No. 12):3186-3199.

Langley RR & Fidler IJ (2011) The seed and soil hypothesis revisited--the role of tumor-stroma interactions in metastasis to different organs. *Int J Cancer* Vol.128(No. 11):2527-2535.

Levin B, Lieberman DA, McFarland B, *et al.* (2008) Screening and surveillance for the early detection of colorectal cancer and adenomatous polyps, 2008: a joint guideline from the American Cancer Society, the US Multi-Society Task Force on Colorectal Cancer, and the American College of Radiology. *CA Cancer J Clin* Vol.58(No. 3):130-160.

Li C, Heidt DG, Dalerba P, *et al.* (2007) Identification of pancreatic cancer stem cells. *Cancer Res* Vol.67(No. 3):1030-1037.

Lioznov MV, Freiberger P, Kroger N, Zander AR & Fehse B (2005) Aldehyde dehydrogenase activity as a marker for the quality of hematopoietic stem cell transplants. *Bone Marrow Transplant* Vol.35(No. 9):909-914.

Luo P, Wang A, Payne KJ, *et al.* (2007) Intrinsic retinoic acid receptor alpha-cyclin-dependent kinase-activating kinase signaling involves coordination of the restricted proliferation and granulocytic differentiation of human hematopoietic stem cells. *Stem Cells* Vol.25(No. 10):2628-2637.

Mantel PY & Schmidt-Weber CB (2011) Transforming growth factor-beta: recent advances on its role in immune tolerance. *Methods Mol Biol* Vol.677(No.:303-338.

Marcato P, Dean CA, Giacomantonio CA & Lee PW (2011) Aldehyde dehydrogenase: its role as a cancer stem cell marker comes down to the specific isoform. *Cell Cycle* Vol.10(No. 9):1378-1384.

Marcato P, Dean CA, Pan D, *et al.* (2011) Aldehyde dehydrogenase activity of breast cancer stem cells is primarily due to isoform ALDH1A3 and its expression is predictive of metastasis. *Stem Cells* Vol.29(No. 1):32-45.

Marchitti SA, Brocker C, Stagos D & Vasiliou V (2008) Non-P450 aldehyde oxidizing enzymes: the aldehyde dehydrogenase superfamily. *Expert Opin Drug Metab Toxicol* Vol.4(No. 6):697-720.

Mieog JS, de Kruijf EM, Bastiaannet E, *et al.* (2012) Age determines the prognostic role of the cancer stem cell marker aldehyde dehydrogenase-1 in breast cancer. *BMC Cancer* Vol.12(No.:42.

Mitchell JB, McIntosh K, Zvonic S, *et al.* (2006) Immunophenotype of human adipose-derived cells: temporal changes in stromal-associated and stem cell-associated markers. *Stem Cells* Vol.24(No. 2):376-385.

Moore SM, Liang T, Graves TJ, McCall KM, Carr LG & Ehlers CL (2009) Identification of a novel cytosolic aldehyde dehydrogenase allele, ALDH1A1*4. *Hum Genomics* Vol.3(No. 4):304-307.

Moreb J, Schweder M, Suresh A & Zucali JR (1996) Overexpression of the human aldehyde dehydrogenase class I results in increased resistance to 4-hydroperoxycyclophosphamide. *Cancer Gene Ther* Vol.3(No. 1):24-30.

Moreb JS (2008) Aldehyde dehydrogenase as a marker for stem cells. *Curr Stem Cell Res Ther* Vol.3(No. 4):237-246.

Moreb JS, Ucar D, Han S, Amory JK, Goldstein AS, Ostmark B & Chang LJ (2012) The enzymatic activity of human aldehyde dehydrogenases 1A2 and 2 (ALDH1A2 and

ALDH2) is detected by Aldefluor, inhibited by diethylaminobenzaldehyde and has significant effects on cell proliferation and drug resistance. *Chem Biol Interact* Vol.195(No. 1):52-60.

Nagy JA, Chang SH, Shih SC, Dvorak AM & Dvorak HF (2010) Heterogeneity of the tumor vasculature. *Semin Thromb Hemost* Vol.36(No. 3):321-331.

Nauta AJ & Fibbe WE (2007) Immunomodulatory properties of mesenchymal stromal cells. *Blood* Vol.110(No. 10):3499-3506.

Noman MZ, Benlalam H, Hasmim M & Chouaib S (2011) Cytotoxic T cells - stroma interactions. *Bull Cancer* Vol.98(No. 2):E19-24.

Oraldi M, Saracino S, Maggiora M, *et al.* (2011) Importance of Inverse Correlation Between ALDH3A1 and PPAR [gamma] in Tumor Cells and Tissue Regeneration. *Chemico-Biological Interactions* (No.

Pappa A, Estey T, Manzer R, Brown D & Vasiliou V (2003) Human aldehyde dehydrogenase 3A1 (ALDH3A1): biochemical characterization and immunohistochemical localization in the cornea. *Biochem J* Vol.376(No. Pt 3):615-623.

Perluigi M, Sultana R, Cenini G, *et al.* (2009) Redox proteomics identification of 4-hydroxynonenal-modified brain proteins in Alzheimer's disease: Role of lipid peroxidation in Alzheimer's disease pathogenesis. *Proteomics Clin Appl* Vol.3(No. 6):682-693.

Petricoin EF, Ardekani AM, Hitt BA, *et al.* (2002) Use of proteomic patterns in serum to identify ovarian cancer. *Lancet* Vol.359(No. 9306):572-577.

Prasad VK & Kurtzberg J (2010) Cord blood and bone marrow transplantation in inherited metabolic diseases: scientific basis, current status and future directions. *Br J Haematol* Vol.148(No. 3):356-372.

Prockop DJ, Kota DJ, Bazhanov N & Reger RL (2010) Evolving paradigms for repair of tissues by adult stem/progenitor cells (MSCs). *J Cell Mol Med* Vol.14(No. 9):2190-2199.

Quinn M & Babb P (2002) Patterns and trends in prostate cancer incidence, survival, prevalence and mortality. Part I: international comparisons. *BJU Int* Vol.90(No. 2):162-173.

Ran D, Schubert M, Pietsch L, *et al.* (2009) Aldehyde dehydrogenase activity among primary leukemia cells is associated with stem cell features and correlates with adverse clinical outcomes. *Exp Hematol* Vol.37(No. 12):1423-1434.

Rasheed ZA, Yang J, Wang Q, *et al.* (2010) Prognostic significance of tumorigenic cells with mesenchymal features in pancreatic adenocarcinoma. *J Natl Cancer Inst* Vol.102(No. 5):340-351.

Rasper M, Schafer A, Piontek G, *et al.* (2010) Aldehyde dehydrogenase 1 positive glioblastoma cells show brain tumor stem cell capacity. *Neuro Oncol* Vol.12(No. 10):1024-1033.

Reya T, Morrison SJ, Clarke MF & Weissman IL (2001) Stem cells, cancer, and cancer stem cells. *Nature* Vol.414(No. 6859):105-111.

Ricci-Vitiani L, Lombardi DG, Pilozzi E, Biffoni M, Todaro M, Peschle C & De Maria R (2007) Identification and expansion of human colon-cancer-initiating cells. *Nature* Vol.445(No. 7123):111-115.

Rollins-Raval MA, Fuhrer K, Marafioti T & Roth CG (2012) ALDH, CA I, and CD2AP: novel, diagnostically useful immunohistochemical markers to identify erythroid precursors in bone marrow biopsy specimens. *Am J Clin Pathol* Vol.137(No. 1):30-38.

Russo J, Barnes A, Berger K, Desgrosellier J, Henderson J, Kanters A & Merkov L (2002) 4-(N,N-dipropylamino)benzaldehyde inhibits the oxidation of all-trans retinal to all-trans retinoic acid by ALDH1A1, but not the differentiation of HL-60 promyelocytic leukemia cells exposed to all-trans retinal. *BMC Pharmacol* Vol.2(No.:4.

Salnikov AV, Gladkich J, Moldenhauer G, Volm M, Mattern J & Herr I (2010) CD133 is indicative for a resistance phenotype but does not represent a prognostic marker for survival of non-small cell lung cancer patients. *Int J Cancer* Vol.126(No. 4):950-958.

Serrano D, Bleau AM, Fernandez-Garcia I, Fernandez-Marcelo T, Iniesta P, Ortiz-de-Solorzano C & Calvo A (2011) Inhibition of telomerase activity preferentially targets aldehyde dehydrogenase-positive cancer stem-like cells in lung cancer. *Mol Cancer* Vol.10(No.:96.

Siclari VA & Qin L (2010) Targeting the osteosarcoma cancer stem cell. *J Orthop Surg Res* Vol.5(No.:78.

Singh SK, Clarke ID, Terasaki M, Bonn VE, Hawkins C, Squire J & Dirks PB (2003) Identification of a cancer stem cell in human brain tumors. *Cancer Res* Vol.63(No. 18):5821-5828.

Singh SK, Hawkins C, Clarke ID, *et al.* (2004) Identification of human brain tumour initiating cells. *Nature* Vol.432(No. 7015):396-401.

Solis MA, Chen YH, Wong TY, Bittencourt VZ, Lin YC & Huang LL (2012) Hyaluronan regulates cell behavior: a potential niche matrix for stem cells. *Biochem Res Int* Vol.2012(No.:346972.

Sondergaard CS, Hess DA, Maxwell DJ, *et al.* (2010) Human cord blood progenitors with high aldehyde dehydrogenase activity improve vascular density in a model of acute myocardial infarction. *J Transl Med* Vol.8(No.:24.

Stupp R, Mason WP, van den Bent MJ, *et al.* (2005) Radiotherapy plus concomitant and adjuvant temozolomide for glioblastoma. *N Engl J Med* Vol.352(No. 10):987-996.

Subramanian J & Govindan R (2007) Lung cancer in never smokers: a review. *J Clin Oncol* Vol.25(No. 5):561-570.

Tanei T, Morimoto K, Shimazu K, *et al.* (2009) Association of breast cancer stem cells identified by aldehyde dehydrogenase 1 expression with resistance to sequential Paclitaxel and epirubicin-based chemotherapy for breast cancers. *Clin Cancer Res* Vol.15(No. 12):4234-4241.

Tewari DS, McHale MT, Kuo JV, Monk BJ & Burger RA (2002) Primary invasive vaginal cancer in the setting of the Mayer-Rokitansky-Kuster-Hauser syndrome. *Gynecol Oncol* Vol.85(No. 2):384-387.

Ucar D, Cogle CR, Zucali JR, *et al.* (2009) Aldehyde dehydrogenase activity as a functional marker for lung cancer. *Chem Biol Interact* Vol.178(No. 1-3):48-55.

van den Hoogen C, van der Horst G, Cheung H, *et al.* (2010) High aldehyde dehydrogenase activity identifies tumor-initiating and metastasis-initiating cells in human prostate cancer. *Cancer Res* Vol.70(No. 12):5163-5173.

Wang J, Sakariassen PO, Tsinkalovsky O, *et al.* (2008) CD133 negative glioma cells form tumors in nude rats and give rise to CD133 positive cells. *Int J Cancer* Vol.122(No. 4):761-768.

Weiner D, Levy Y, Khankin EV & Reznick AZ (2008) Inhibition of salivary amylase activity by cigarette smoke aldehydes. *J Physiol Pharmacol* Vol.59 Suppl 6(No.:727-737.

Weissman IL (2000) Stem cells: units of development, units of regeneration, and units in evolution. *Cell* Vol.100(No. 1):157-168.

White H, Smith L, Gentry T & Balber A (2011) Mechanisms of Action of Human Aldehyde Dehydrogenase Bright Cells in Therapy of Cardiovascular Diseases: Expression Analysis of Angiogenic Factors and Aldehyde Dehydrogenase Isozymes. *J Stem Cell Res Ther S* Vol.1(No.:2.

Wymore T, Hempel J, Cho SS, Mackerell AD, Jr., Nicholas HB, Jr. & Deerfield DW, 2nd (2004) Molecular recognition of aldehydes by aldehyde dehydrogenase and mechanism of nucleophile activation. *Proteins* Vol.57(No. 4):758-771.

Yoshida A, Rzhetsky A, Hsu LC & Chang C (1998) Human aldehyde dehydrogenase gene family. *Eur J Biochem* Vol.251(No. 3):549-557.

Chapter 2

Role of Glutamate Dehydrogenase in Cancer Growth and Homeostasis

Ellen Friday, Robert Oliver III, Francesco Turturro and Tomas Welbourne

Additional information is available at the end of the chapter

http://dx.doi.org/10.5772/48606

1. Introduction

Glutamate Dehydrogenase (GDH) catalyzes the oxidative conversion of glutamate to alpha ketoglutarate and ammonium supplying the TCA cycle with intermediates in support of anaplerosis (**Figure 1 Rxn1**). Conversely GDH catalyzes the reductive amination of alpha ketoglutarate and ammonium producing glutamate when the TCA cycle pool is filled. The net GDH flux resulting from these bidirectional fluxes can be obtained from the conversion of either 15N labeled glutamate and analyzing 15N ammonium or from 15N labeled $NH_{4}+$ and monitoring 15N labeled glutamate. Besides deamination and 15N NH_4^+ production, 15N labeled glutamate can be converted to 15N labeled amino acids, most prominently alanine, via transamination reactions (**Figure 1, Rxn2**). In contrast to glutamate deamination which yields net keto acid production for anaplerosis [1], transamination does not yield net keto acid production (consuming a keto acid e.g. pyruvate in the process of generating alpha ketoglutarate). Under physiological conditions plasma glutamate concentration, 10-20uM, is limiting for GDH flux supplying TCA intermediates while plasma glutamine concentration, 600uM, is not [2]. The conversion of 15N amide labeled glutamine to 15N ammonium (**Figure1, Rxn1**) approximates the net glutaminase flux generating glutamate and ammonium, both potential substrates for GDH. Indeed GDH can also incorporate the amide derived 15N ammonium and alpha ketoglutarate into glutamate (**Figure 1, Rxn3**, reductive amination) which can subsequently transaminate with pyruvate generating 15N alanine [3]. Noteworthy this glutaminolytic anabolic pathway providing glutamate has been proposed as the primary metabolic transformation in tumor cells [4]. **Figure 1, Rxn 3** also illustrates how ammonium production from the 15N amide of glutamine may underestimate the true glutaminase flux; to the extent that this occurs, it contributes to differences in estimated net glutaminase fluxes between the chemically measured glutamine disappearance and 15N amide ammonium appearance. Glutamine labeled with 15N in the amino position provides an assessment of net GDH flux (**Figure 1,Rxn1)** as 15N NH_4^+ and, or, ALT flux as 15N alanine produced(**Figure1,Rxn2**).

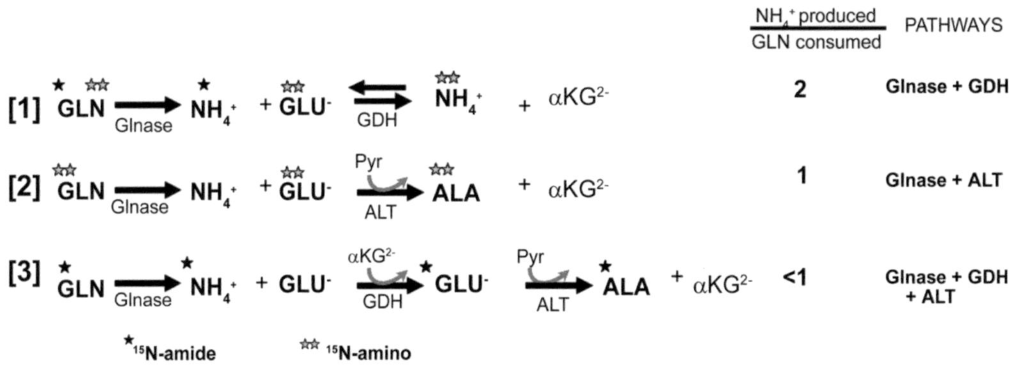

Figure 1. GDH determines the fate of 15N glutamine. Pathways of glutaminolysis, net keto acid production and NH_4^+ produced per glutamine consumed ratio. [1] Deamidation coupled to GDH deamination yielding 2NH4+/Gln and net keto acid (αKG). [2] Deamination coupled to ALT-mediated transamination yielding 1 NH4+/Gln and no net acid production. [3] Deamidation coupled to GDH reductive amination and transamination yielding < 1 NH4+/Gln and net keto acid

2. Glutamine is the major source of ammonium produced in cultured cells

Because the bulk (≈90%),[1,3-5] of the ammonium produced by cells in culture derives from glutamine's 2 nitrogen moieties (preformed DMEM media glutamate is <50uM) chemical measures of ammonium produced (after subtracting any ammonium produced in the absence of glutamine) and glutamine consumed offers an index of the GDH pathway activity. A ratio of 1, for example, would be consistent with glutamine metabolized by glutaminase with NH_4^+ released to the media and glutamate either released to the media or transaminated to amino acids e.g. alanine (**Figure 1, Rxn2**). In either case there is no net GDH flux and therefore no net generation of keto acids. In contrast a ratio of 2 NH_4^+/glutamine (**Figure 1, Rxn1**) is consistent with glutamine metabolized by glutaminase and the glutamate produced metabolized by GDH yielding NH_4^+ and net keto acid (anaplerosis) for the TCA cycle(running in either the normal forward or reverse direction,[6]). On the other hand, an NH_4^+ produced per glutamine consumed less than 1 (**Figure 1, Rxn3**) is consistent with glutaminase generating NH_4^+ from the amide nitrogen with reductive amination [3,4] catalyzed by GDH reducing the NH_4^+/Gln ratio to less than 1 producing glutamate and consuming net keto acid (alpha ketoglutarate); subsequent transamination yields amino acids(e.g. alanine, aspartate) consuming keto acids (cataplerosis) containing the amide nitrogen of glutamine[3,4]. Normal breast cell line exhibits an NH_4^+/glutamine ratio less than 1[7] whereas cancerous breast cell lines exhibit a ratio greater than 1[7] consistent with a quantitative difference in bidirectional GDH fluxes between normal (reductive amination) and tumorigenic cells (oxidative deamination). In addition, in vivo administered [2-(14)C] labeled glutamate was taken up by tumors with 14C distributed more into protein and lactate than in normal tissues [8]. In line with this early finding, a recent study[9] showed[U-(13)C] glucose contributed less than 50% of the acetyl COA pool in human brain tumors consistent with glutaminolysis and GDH's role in maintaining TCA pool homeostasis (anaplerosis).

3. Intracellular glutamate and alpha ketoglutarate are in near equilibrium

In most cells intracellular glutamate and alpha ketoglutarate are in near equilibrium[4,10], and changes in TCA cycle intermediates(αKG) as well as the redox state(NADPH/NADP), energy charge(ADP,GTP) and cell pH shift the GDH catalyzed flux to net production or consumption of αKG (**Figure 2**). Normally pyruvate (glucose) provides the TCA cycle with pool intermediates while generated glutamate is transaminated (NH_4^+/GLN ratio<1, **Figure 1 Rxn3**). In cancer cells, glucose is shunted into aerobic glycolysis (Warburg effect, [11]) and the TCA cycle intermediates are reduced as the result of cataplerosis as evidenced by lower intracellular glutamate [7]. This reduction in TCA cycle intermediates "pulls" glutamate through GDH generating αKG as evidenced by the higher steady state NH_4^+/GLN ratio>1, **Figure 1, Rxn1** and consistent with glutamate (glutaminolysis) supporting anaplerosis (**Figure 2**). As a corollary, the ammonium to alanine produced ratio increases [7] reflecting the increased GDH and decreased ALT flux as the result of reduced intramitochondrial pyruvate(metabolized in cytosol to lactate, **Figure 2**). Thus the increased glutamate flux through GDH generates αKG while sparing keto acid consumption (reduced transamination).

4. Glutamate is generated by extra- and intracellular glutaminases

Glutaminolysis as illustrated in **Figure 2** is associated with the increased expression of both the extrinsic cell membrane phosphate independent glutaminase/gamma glutamyltransferase/gamma glutamyltranspeptidase (PIG, GGT, GGTP) which generates extracellular glutamate [2,12] and intracellular phosphate dependent glutaminases, Phosphate dependent glutaminases (PDG,GLS1 and GAC, [13,14]) which generates glutamate cytosolically [2,13]; extracellular glutamate can be transported(GLAST, **Figure2**) into the cytosol functioning as an inhibitor of the intracellular glutaminases[2]. Noteworthy, c-myc signaling up-regulates both the cell membrane glutamine transporter (ASC, **Figure 2**) and the intracellular glutaminases in cancer cells [15]. On the other hand, increased expression of the extracellular PIG is also a hallmark of cancer cells [16] and PIG hydrolysis of ϒ-glutamyl-tagged fluorescent markers can be used to delineate tumor boundaries [16]. However, in contrast to glutamine uptake, cell membrane glutamate transport (GLAST1) is shifted from the cell membrane to an intracellular location in breast cancer cells as shown in **Figure 3**, effectively uncoupling extracellular glutamate from inhibiting the intracellular glutaminases; this allows full blown expression of intracellular glutamate generation(**Figure 1RXI**) and, if the relocated glutamate transporter, GLAST1 transports glutamate from the outer surface of inner mitochondrial membrane into the into the mitochondria matrix [17],

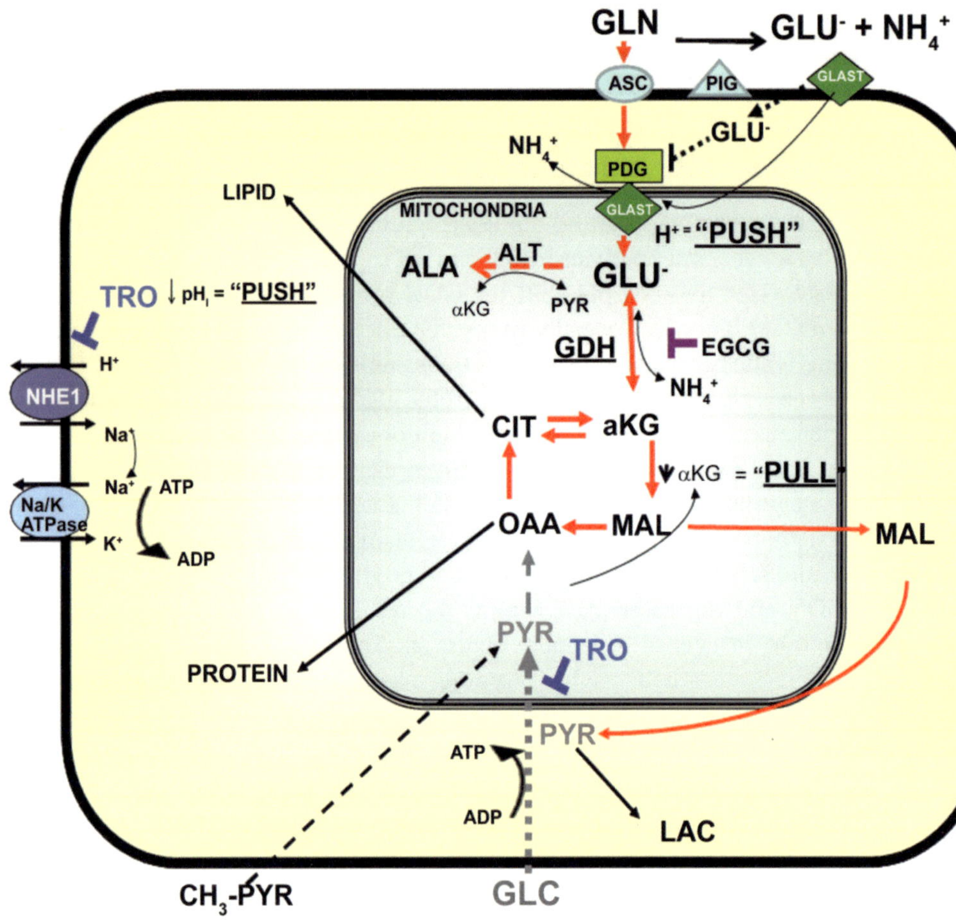

Figure 2. Central role of GDH flux in cancer cells. Glucose (**GLC**) derived pyruvate(**PYR**) is metabolized(Warburg effect) to lactic acid(LAC) at the expense of the **TCA** (tricarboxylic acid) cycle intermediate pool(αKG) "pulling" glutamate through GDH to supply anaplerosis. **NHE(sodium hydrogen ion exchanger 1)** mediated acid extrusion is up-regulated coupled to anaerobic glycolysis acidifying extracellular milieu while cell membrane glutamate transporter(**GLAST1**) relocates to mitochondria with **PIG** produced GLU^- accumulating extracellularly and contributing with reduced pHe to host defense barrier. Glutamine(**GLN**) transported into cell by **ASC** and hydrolyzed to glutamate on outer surface of inner mitochondrial membrane by **GAC [13]** coupled with **GLAST1.** GLC removal (dashed line) accentuates GDH flux by "pull" mechanism while NHE mediated acid extrusion supported by GDH and accelerated αKG input with cytosolic **malate**(MAL) conversion to **PYR** supplying anaplerosis. TRO blocks PYR entry into mitochondria and accelerates GDH flux by exaggerated pull mechanism in conjunction with reduced pHi as result of NHE1 inhibition ("push" mechanism). **EGCG** inhibits GDH and induces cell death that can be partly rescued with methyl pyruvate(**CH3-PYR**) and restored TCA cycle pool while correction of cellular acidosis requires GDH flux pointing to the dual role for GDH in anaplerotic and acid base homeostasis.

then it would supply GDH glutamate in support of anaplerosis . Noteworthy overexpression of PIG promotes tumorigenesis [16] presumably by building up extracellular glutamate and suppressing local immune responses [18] . In addition NHE mediated acid extrusion is up-regulated in cancer cells [19,20] importing a Na^+ load requiring Na+/K+ ATPase - ATP expenditure and ATP regeneration associated with acidogenic aerobic glycolysis(Warburg effect) and by substrate level phosphorylation. Because PIG (GGT/GGTP), NHE, glutamine transporter and glutaminase activities are all up-regulated in rapidly growing tumors, tagging molecular target inhibitors [21-24] with a ϒ-glutamyl moiety offers a tumor specific vehicle specific for limiting anaplerosis and preventing elevated cell pH, prerequisites for rapid tumor growth.

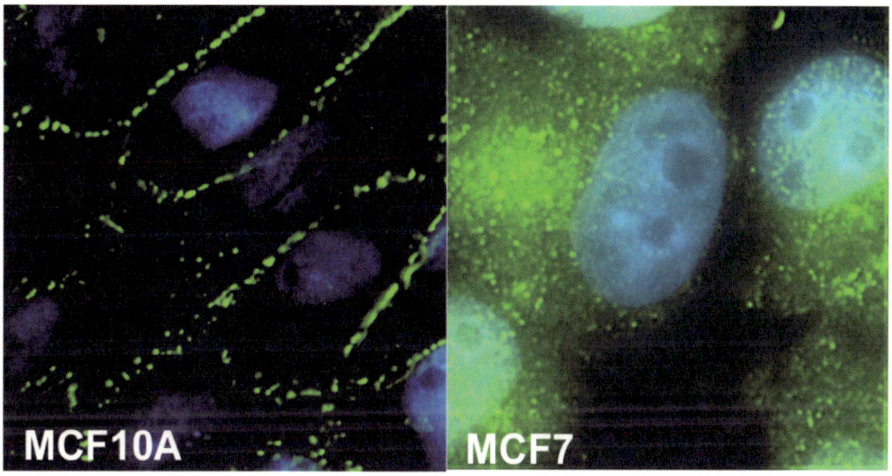

Figure 3. GLAST localization in normal versus cancer cells. MCF10A and MCF7 cells cultured on coverglass were stained with monoclonal antibodies to GLAST-1 . MCF10A demonstrate almost complete membrane localization of the transporter, while MCF7 have a cytoplasmic distribution pattern.

5. Glucose removal lowers TCA cycle intermediates and "pulls" glutamate through GDH

Removal of glucose from the media (Figure 2, dotted gray line from GLC) deprives cells of pyruvate input into the TCA cycle and a fall in the intermediate (αKG) pool level[5] as reflected by a drop in glutamate [7]. As a consequence, GDH flux (**Figure 1, Rxn1**) increases [5] supplying anaplerosis as malate exits the cycle forming pyruvate which in turn supports citrate formation (**Figure 2**). Noteworthy this increased glutamate flux through GDH("pulling" effect) is maintained by 2 responses:1] by a small increase in glutaminase flux [5,7] and, 2] a large fall in glutamate transamination [5,7,25].Under glucose deprivation cell survival is dependent on GDH flux at least in part to supply anaplerosis [5,26]. Surprisingly cell number actually increase in the glutamine (1.3mM) alone compared to the glucose(5mM) plus glutamine media (**Figure 4A)** because of reduced cell death; this increased survival is attributed to the increased GDH flux [26] which besides supporting anaplerosis also enhances NHE mediated acid extrusion (**Figure 4D**) although proliferation rate decreases (**Figure 4B**). Noteworthy is the increase in cell biomass(protein, nucleic acids and lipid) dependent upon glutaminolysis supported anaplerosis as shown by the increased incorporation of 14C-U-glutamine into cell biomass (**Figure 4C**).The critical role of GDH flux in cell survival is evident from the massive cell death induced by GDH inhibition under glucose depleted conditions with 100uM EGCG [5,26], an inhibitor of GDH [5]. Although supplying TCA cycle intermediates e.g. methyl pyruvate (10mM,**Figure 2**) rescued cells with GDH inhibited [5,26], a significant fraction of the population succumbed associated with a reduced cell pH [26]. Parenthetically, methylpyruvate is a strong acid constituting a large acid load which requires supplementing the media with equal moles of bicarbonate (10mM). Nevertheless, even after the above base compensation, supplying anaplerotic substrates does not restore NHE activity [26] pointing to an important dual role for GDH in maintaining both anaplerosis and pH homeostasis [22] for cell survival.

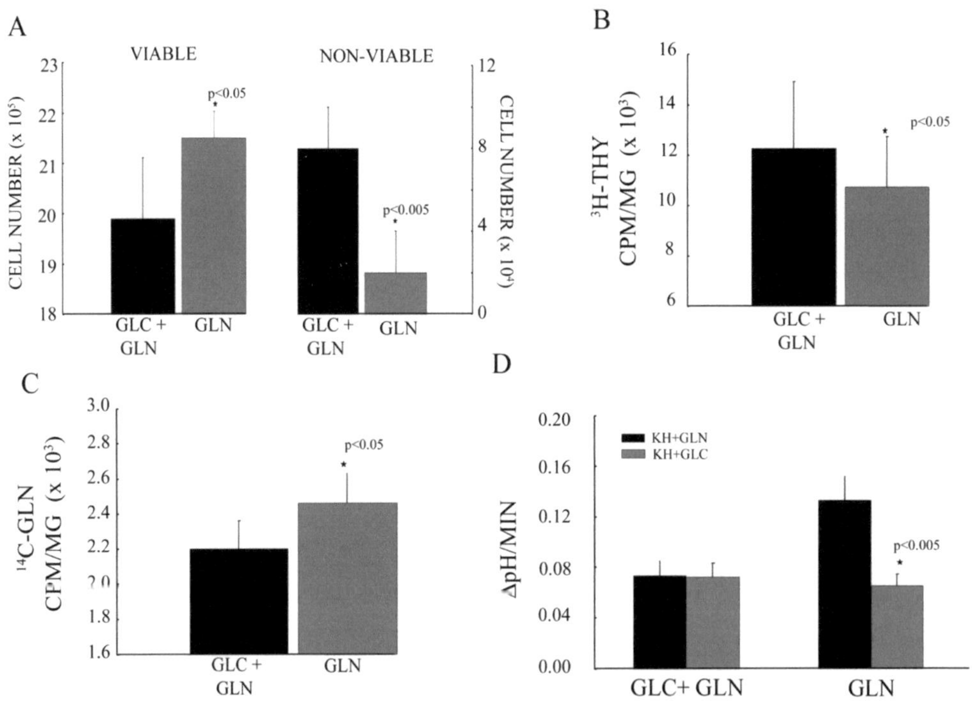

Figure 4. Physiological(1.3mM) glutamine concentration alone supports breast cancer cell survival associated with increased anaplerotic and acid extrusion function. 4A GLN increased cell number and decreased cell death compared to GLN plus 5mM glucose; 4B Gln slows cell proliferation compared with GLC+GLN; 4C GLN supports anaplerotic function[^{14}C-U-L-glutamine incorporation into TCA precipitated cellular protein and lipid 4D GLN supports accelerated NHE activity when assayed with K 10mM GLN as opposed to 10mM GLC

6. Cellular acidosis "pushes" glutamate through GDH

Glutamate flux through GDH can be also be "pushed" by a fall in intracellular pH [27]. Whether this reflects a shift from GHD1 to GDH2 isoform [28] is not known but, if so, this "pushing effect" of reduced pH effect could be additive with the above "pulling effect" of a reduced TCA pool (**Figure 2**). Indeed in metabolic acidosis, the ambient condition surrounding cancer cells in vivo, kidney cells' glutaminolysis is both "pushed"(reduced cell pHi, [27]) and "pulled"(inhibition of TCA, [29]) as a result of reduced TCA cycle pool size associated with true renal growth [30]. Interestingly enough, the in vivo kidney switches fuels from lactate to glutamine oxidation in metabolic acidosis[31] so that the anaplerotic glutaminolysis-GDH reactions matches [32] the cataplerotic reactions(CO_2, biomass formation, [30,31] as does acid excretion ($2NH_4^+$/glutamine) and base($2HCO_3^-$/glutamine) generation. Furthermore the pH-dependent enlistment of GDH2 isoform alone (push mechanism) or accompanying GDH 1 flux (anaplerosis driven pull mechanism) would provide regulatory options in responding to anaplerotic/cataplerotic and, or, acid /base demands in tumors.

7. Glitazones accelerate GDH flux via the push/pull mechanism: A strategy for therapeutic intervention

Fortuitously there are agents that can be employed to impose this push/pull mechanism on the GDH flux in cancer cells and thereby present a window of vulnerability (targeted inhibitors). The antihyperglycemic agents, troglitazone (Rezulin) and rosiglitazone (Avandia) block pyruvate entrance into the TCA cycle(25,33] lowering αKG(glutamate,7,25] and accelerating GDH flux via this "pull" mechanism (**Figure 2**). Simultaneously, both troglitazone and rosiglitazone directly inhibit NHE [25,34] lowering pHi and driving GDH via the "push" mechanism (**Figure 2**). Noteworthy the glutaminase flux (glutamine disappearance) remains unchanged while the NH_4^+ production increases as the result of the increase in deamination flux (**GDH Figure 1, Rxn1**). Although resembling glucose deprivation ("pull" mechanism), troglitazone further increases the NH_4^+ production (additive "push+pull" mechanism) exceeding the fall in alanine production ("pull" mechanism alone). More specifically the accelerated GDH flux ("push+pull") induced by troglitazone can be demonstrated using 15N amino labeled glutamine as shown in **Figure 5**; in contrast, another glitazone pioglitazone(Actos) activates GDH flux [34] solely by reducing cell pH("push" mechanism) and consequently does not reduce alanine production [34]. Noteworthy troglitazone acutely inhibits GDH flux (0-3hrs) as the result of a fall in mitochondrial membrane potential(Ψm) requiring accelerated GDH flux(3-24hrs) to fully restore the Ψm [7], a response that is PPARγ independent [7,25] and possibly mitoNEET [35] dependent. Little recognized is the direct inhibition of NHE[20,34] by both troglitazone and rosiglitazone as well as indirect inhibition mediated through PPARγ suppression of NHE gene expression[20,36]; in contrast, pioglitazone does not inhibit NHE directly[34] rather acts indirectly through PPARγ[36]lowering cell pH[34] and accelerating GDH flux("push" mechanism). In combination, TRO +PIO together exert an additive effect on GDH flux presumably reflecting both TRO's "pull" action and PIO's PPARγ- mediated down-regulation of NHE gene expression. Significantly, the dual effect of glitazones to increase GDH flux while reducing NHE activity decreases proliferation(NHE) but increases cell survival(GDH) resulting in only a slight decrease in cell number[26]. Nor does adding troglitazone to glucose deleted cells induce massive cell death since the effect on GDH flux is additive (further reducing TCA intermediates and cell pH, **Figure 2**) and although proliferative rates are decreased, survival is enhanced [26]. Under these conditions, e.g. cell survival mechanisms, inhibition of GDH is most effective in causing massive cell death as occurs with the GDH inhibitor EGCG [5] combined with troglitazone [26]. Although rescue of cells is partly possible by restoring anaplerosis with methyl pyruvate, failure to restore NHE activity and the cellular acidosis preclude full recovery underlining the importance of both GDH and NHE in cell survival[26].

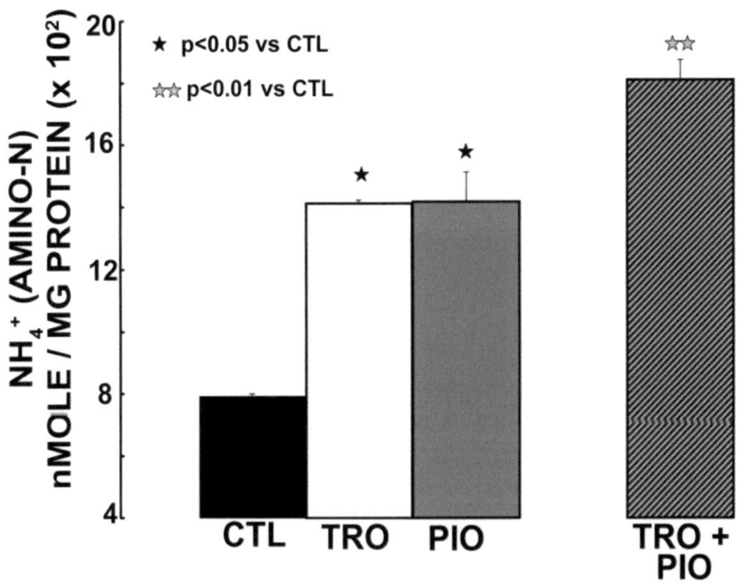

Figure 5. Ammonium production from amino nitrogen of glutamine. Cells were incubated for 18 hours in [2-15N] glutamine. TRO was used at 20 uM, PIO 10 uM. Results are from 3 experiments.

Author details

Ellen Friday
Department of Medicine- Feist- Weiller Cancer Center
Department of Cellular and Molecular Physiology, Louisiana State University Health Sciences Center, Shreveport, LA, USA

Robert Oliver III and Tomas Welbourne
Department of Cellular and Molecular Physiology, Louisiana State University Health Sciences Center, Shreveport, LA, USA

Fancesco Turturro
Department of Lymphoma; Myeloma, MD Anderson Cancer Center, Houston, TX, USA

Acknowledgement

The authors would like to acknowledge the support from the Feist-Weiller Cancer Center (EF and FT) and The Southern Arizona Research Foundation (TW).

8. References

[1] DeBerardinis RJ, Mancuso A, Daikhin E, Nissim I, Yudkoff M, Wehrli S, Thompson CB.(2007) Beyond aerobic glycolysis: Transformed cells can engage in glutamine metabolism that exceeds the requirement for protein and nucleotide synthesis. Proc Natl Acad Sci. 104:19345-19350.

[2] Welbourne T, Routh R, Yudkoff M, Nissim I.(2001) The glutamine/glutamate couplet and cellular function. News Physiol Sci.16:157-160.

[3] Welbourne, T., Friday E, Fowler R, Turturro F, Nissim I.(2004)Troglitazone acts by PPARgamma dependent and PPARgamma-independent pathways on LLC-PK1-F+ acid base metabolism. Am J Physiol Renal Physiol. 286:F100-110.

[4] Meng M, Chen S, Lao T, Liang D, Sang N. (2010) Nitrogen anabolism underlies the importance of glutaminolysis in proliferating cells. Cell Cycle 9:3921-3932.

[5] Yang C, Sudderth J, Dang T, Bachoo RM, McDonald JG, DeBerardinis RJ.(2009) Glioblastoma cells require glutamate dehydrogenase to survive impairments of Akt signaling. Cancer Res 69:7986-7893.

[6] Mullen AR, Wheaton WW, Jin ES, Chen PH, Sullivan LB, Cheng T, Yang Y, Linehan WM, Chandel NS, DeBerardinis RJ. (2011) Reductive carboxylation supports growth in tumor cells with defective mitochondria. Nature 481:385-38

[7] Friday E, Oliver R 3rd, Welbourne T, Turturro F. (2011) Glutaminolysis and glycolysis regulation by troglitazone in breast cancer cells: relationship to mitochondrial membrane potential. J. Cell Physiol 226:511-519.

[8] Nyhan WL, Busch H. (1958) Metabolic patterns for glutamate-2-C14 in tissues of tumor bearing rats. Cancer Res. 18:385-393.

[9] Maher EA, Marin-Valencia I, Bachoo RM, Mashimo T, Raisanen J, Hatanpaa KJ, Jindal A, Jeffrey FM, Choi C, Madden C, Mathews D, Pascual JM, Mickey BE, Malloy CR, Deberardinis RJ. (2012) Metabolism of [U-(13)C] glucose in human brain tumors in vivo. NMR Biomed 15 doi 10.1002/nbm.2794

[10] Silverstein E, Sulebele G.(1974) Equilibrium kinetic study of the catalytic mechanism of oxidative deamination of alanine by bovine liver glutamate dehydrogenase. Biochemistry 13:1815-1818.

[11] Warburg, O. (1956) On the origin of cancer cells. Science. 123: 309-314

[12] Wong KT, Lee YY, Brusic V, Tan J, Yap MG, Nissom PM (2005) Elevation of gamma glutamyltransferase activity in 293 HEK cells constitutively expressing antisense glutaminase mRNA. Metab Eng 7:375-383.

[13] Kvamme E, Nissen-Meyer LS, Roberg BA, Torgner IA. (2008) Novel form of phosphate activated glutaminase in cultured astrocytes and human neuroblastoma cells, PAG in brain pathology and localization in the mitochondria. Neurochem Res 33:1341-1345.

[14] Cassago A, Ferreira AP, Ferreira IM, Fornezari C, Gomes ER, Greene KS, Pereira HM, Garratt RC, Dias SM, Ambrosio AL. (2012) Mitochondrial localization and structure-based phosphate activation mechanism of glutaminase C with implications for cancer metabolism. Proc Natl Acad Sci USA 109:1092-1097.

[15] Wise DR, DeBerardinis RJ, Mancuso A, Sayed N, Zhang XY, Pfeiffer HK, Nissim I, Daikhin E, Yudkoff M, McMahon SB, Thompson CB.(2008) Myc regulates a transcriptional program that stimulates mitochondrial glutaminolysis and leads to glutamine addiction. Proc Natl Acad Sci U S A. 105: 18782-18787.

[16] Urano Y, Sakabe M, Kosaka N, Ogawa M, Mitsunaga M, Asanuma D, Kamiya M, Young MR, Nagano T, Choyke PL, Kobayashi H.(2011) Rapid cancer detection by topically spraying a γ-glutamyltranspeptidase-activated fluorescent probe. Sci Trend Med 31:110 -119

[17] Bauer DE, Jackson JG, Genda EN, Montoya MM, Yudkoff M, Robinson MB.(2012) The glutamate transporter ,GLAST, participates in a macromolecular complex that supports glutamate metabolism. Neurochem Int. doi 10.1016/jneuint.2012.01.013

[18] Dröge W, Eck HP, Betzler M, Schlag P, Drings P, Ebert W.(1988) Plasma glutamate concentration and lymphocyte activity. J Cancer Res Clin Oncol 114:124-128.

[19] Reshkin SJ, Bellizzi A, Caldeira S, Albarani V, Malanchi I, Poignee M, Alunni-Fabbroni M, Casavola V, Tommasino M.(2000) Na+/H+ exchanger-dependent intracellular alkalinization is an early event in malignant transformation and plays an essential role in the development of subsequent transformation-associated phenotypes. FASEB J 14:2185-97.

[20] Turturro F, Friday E, Fowler R, Surie D, Welbourne T.(2004) Troglitazone acts on cellular pH and DNA synthesis through a peroxisome proliferator-activated receptor gamma-independent mechanism in breast cancer-derived cell lines. Clin Cancer Res. 10:7022-7030.

[21] García-Cañero R, Trilla C, Pérez de Diego J, Díaz-Gil JJ, Cobo JM.(1999) Na+/H+ exchange inhibition induces intracellular acidosis and differentially impairs cell growth and viability of human and rat hepatocarcinoma cells. Toxicol Letts 106:215-228.

[22] Schelling, JR, Abu Jawdeh, BG. (2008) Regulation of cell survival by Na+/H+ exchanger-1. Am J Physiol Renal Physiol. 295:F265-632.

[23] Fuchs BC, Finger RE, Onan MC, Bode BP.(2007) ASCT2 silencing regulates mammalian target-of-rapamycin growth and survival signaling in human hepatoma cells. Am J Physiol Cell Physiol.. 293:C55-63

[24] Hartwich,EW., Curthoys,NP. (2011) BPTES inhibition of hGA(124-551), a truncated form of human kidney –type glutaminase. J Enzyme Inhib Med Chem doi:10.3109/ 14756366.2011.622272)

[25] Oliver R 3rd, Friday E, Turturro F, Welbourne T.(2010) Troglitazone regulates anaplerosis via a pull/push affect on glutamate dehydrogenase mediated glutamate deamination in kidney derived epithelial cells; implications for the Warburg effect. Cell Physiol Biochem 26:619-628.

[26] Friday E, Oliver R, III, Turturro F, Welbourne T. (2011) Enhancing glutaminolysis in glutamine dependent breast cancer cells with TRO provides a therapeutic window for GDH inhibition by EGCG inducing massive cell death. FASEB J. 25: 915.8.

[27] Nissim I, Sahai A, Sandler RS, Tannen RL (1994) The intensity of acidosis differentially alters the pathways of ammoniagenesis in LLC-PK1 cells. Kidney Int. 45:1014-1019.

[28] Spanaki,C., Plaitakis,A.(2012) The role of glutamate dehydrogenase in mammalian ammonia metabolism . Neurotox Res 21:117-127.

[29] Nissim, I, Nissim I, Yudkoff M. (1990) Carbon flux through TCA cycle in rat renal tubules.Biochim Biophys Acta 1033:194-200.

[30] Lotspeich, W. (1967) Metabolic aspects of acid base change. Science 155:1066-1075.

[31] Pitts,RF. (1975) Production of CO2 by the intact functioning kidney of the dog Med Clin N Am 59:507-517.

[32] Owen OE, Kalhan SC, Hanson RW. (2002) The key role of anaplerosis and cataplerosis for citric acid cycle function. J Biol Chem.. 277:30409-30412

[33] Fediuc S, Pimenta AS, Gaidhu MP, Ceddia RB (2008) Activation of AMP-activated protein kinase, inhibition of pyruvate dehydrogenase activity, and redistribution of substrate partitioning mediate the acute insulin-sensitizing effects of troglitazone in skeletal muscle cells. J Cell Physiol. 215:392-400.

[34] Turturro F, Oliver R 3rd, Friday E, Nissim I, Welbourne T. (2007) Troglitazone and pioglitazone interactions via PPARγ independent and –dependent pathways in regulating physiological responses in renal tubule-derived cell lines. Am J Physiol Cell Physiol 292:C1137-1146.

[35] Feinstein DL, Spagnolo A, Akar C, Weinberg G, Murphy P, Gavrilyuk V, Dello Russo C.(2005) Receptor-independent actions of PPAR thiazolidinedione agonists: is mitochondrial function the key. Biochem Pharm. 70:177-188.

[36] Kumar AP, Quake AL, Chang MK, Zhou T, Lim KS, Singh R, Hewitt RE, Salto-Tellez M, Pervaiz S, Clément MV (2009) Repression of NHE1 expression by PPARgamma activation is a potential new approach for specific inhibition of the growth of tumor cells in vitro and in vivo. Cancer Res. 69:8636-8644.

Chapter 3

The Pyruvate Dehydrogenase Complex in Cancer: Implications for the Transformed State and Cancer Chemotherapy

Paul M. Bingham and Zuzana Zachar

Additional information is available at the end of the chapter

http://dx.doi.org/10.5772/48582

1. Introduction

The problem of finding effective chemotherapies for advanced cancer remains largely unsolved. We review here the role of a specific class of central metabolic dehydrogenases whose regulatory properties are remodeled significantly in cancer cells. These remodeled properties may provide an attractive set of targets for the development of new chemotherapeutic drugs. The design features of the first agents to be developed attacking these new targets [1,2] may illustrate ways to exploit this potentially valuable therapeutic opportunity.

2. The central role of the pyruvate dehydrogenase (PDH) and alpha-ketoglutarate (KDGH) complexes in cell metabolism

The evolution of the high-energy, oxygen-dependent metabolism of eukaryotes [4,5] has produced mitochondrial metabolic pathways whose control is centrally focused on a series of dehydrogenases. Two of these pivotal dehydrogenases, the pyruvate dehydrogenase (PDH) and alpha-ketoglutarate or 2-oxoglutarate (KDGH) complex will be our central focus here. These dehydrogenases control the entry of carbon into the TCA cycle from two major sources, carbohydrate and gluconeogenic amino acids (pyruvate; PDH) and glutamine (KGDH), the most abundant serum amino acid and a central carbon source for normal and pathological tissues (Figure 1). The TCA cycle, in turn, is almost exclusively responsible for the capture of reducing potential from nutrients for the purpose of driving the oxygen-dependent mitochondrial electron transport system for ATP production [6].

In the solid, three dimensional structure of animal bodies the availability of nutrients and molecular oxygen is locally contingent and dynamically variable. Thus, continuous real-time control of PDH and KGDH activities is essential to proper function. Moreover, in various pathological contexts these regulatory processes are substantially altered in ways that are essential to that pathology and this may reflect targets for therapeutic intervention (see sections 5 and 8 below for detailed references).

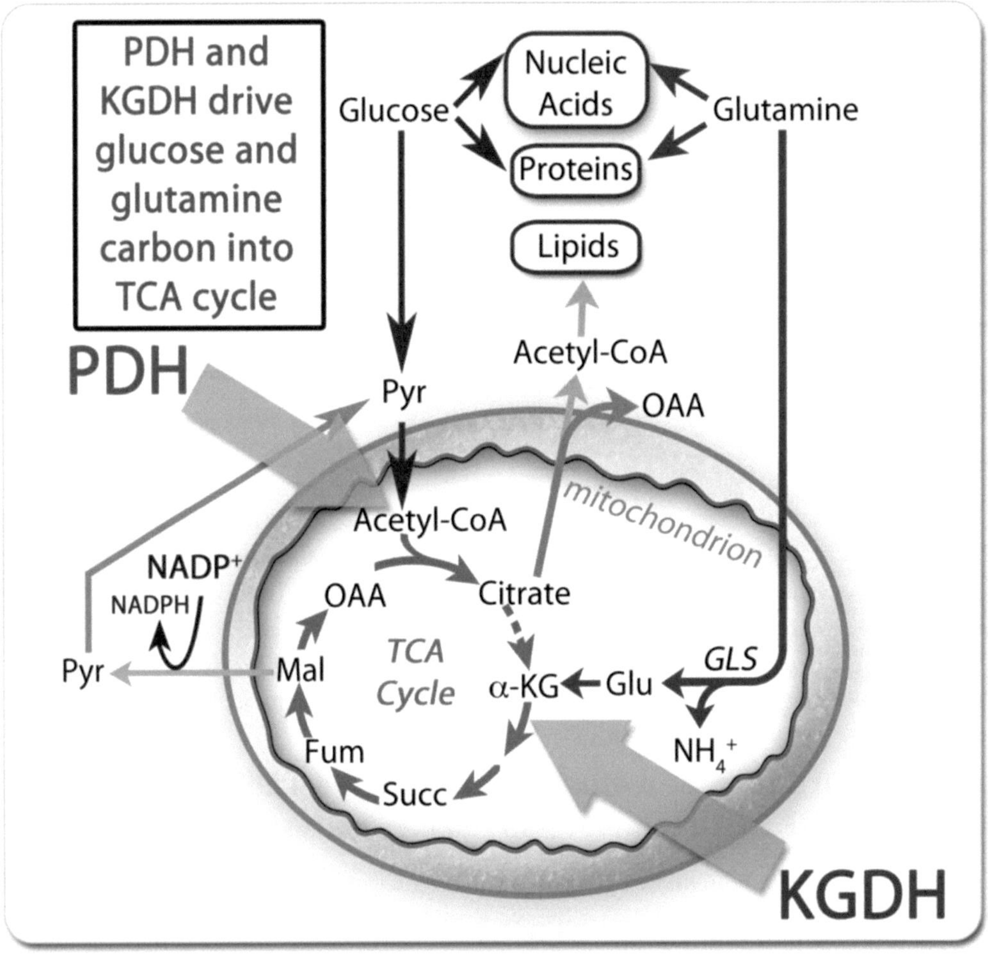

Figure 1. The pyruvate dehydrogenase (PDH) and alpha-ketoglutarate dehydrogenase (KGDH) complexes govern the entry of carbohydrate- and glutamine-derived carbon, respectively, into the TCA cycle. Glucose and glutamine are two major carbon sources feeding many mammalian cells, including tumor cells. These molecules are metabolized to support ATP generation, through cytosolic glycolysis and oxygen-dependent mitochondrial metabolism in which most reducing potential is derived from the TCA cycle. As well, both carbon and reducing potential from glucose and glutamine are delivered to anabolic, biosynthetic functions. Among these are provision of citrate and NADPH for lipid biosynthesis as illustrated and delivery of carbon skeletons from glycolysis and the TCA cycle for amino acid biosynthesis (not illustrated). The entry of glucose derived carbon (either from dietary glucose or from amino acids used for gluconeogenesis) is controlled by the PDH complex and the entry of glutamine derived carbon is controlled by the KGCH complex. Figure redrawn from several sources, including [3].

PDH decarboxylates pyruvate, releasing carbon dioxide, capturing reducing potential and transferring the two-carbon acyl unit (acetate) to Co-enzyme A (CoA) (Figure 2). Structurally homologous KGDH catalyzes the analogous reaction, decarboxylating alpha-ketoglutarate and generating succinyl-CoA. Each of these reactions is highly exergonic and also generates rapidly dispersed or consumed products (carbon dioxide and acyl-CoA). Thus, the initial decarboxylation reaction in each case is effectively irreversible, reflecting a forward commitment step.

The catalytic details of PDH and KGDH are as follows (Figures 2 and 3). Each complex contains enzymatic activities conventionally referred to as E1, E2, and E3. E1 catalyzes oxidative decarboxylation (of pyruvate or alpha-ketoglutarate, respectively) using a thiamine pyrophosphate co-enzyme. The activated acyl unit created by this process is transferred to sulfhydryl groups of a lipoic acid (lipoate) residue of an E2 subunit. E2 lipoate is in the form of lipoamide, in which the carboxylic acid moiety of lipoate is joined in amide linkage to epsilon amino group of an E2 active site lysine. The two sulfurs on the lipoate moiety (Figures 3) exist in a disulfide configuration which is reduced and acylated (thioesters) to produce activated acetate or succinate residues. The E2 subunit further catalyzes the transfer of the lipoate-activated acyl residue to CoA leaving dihydro-lipoamide. Finally, the E3 subunit catalyzes the transfer of the reducing potential in the dihydro-lipoamide residues of E2 to NAD+ (through enzyme disulfide to FAD to NAD steps) [7,8] to generate NADH for transfer of the reducing potential to the electron transport complex and regenerating oxidized lipoate (Figure 2).

Both PDH and KGDH are very large complexes containing multiple copies of each of these three key enzymatic activities (Figure 3). For example, mammalian PDH consists of 60 copies of the E2 subunit (48 of E2 itself and 12 of the E2 homolog called the E3 binding protein, E3BP; Figure 3) [9].

This organization allows efficient channeling of the products of the intermediate reactions [10]. Flux through these complexes is subject to extensive regulation as expected from their central role in governing carbon flow in the face of variable supply and demand. The end products of PDH and KGDH activity, NADH and acyl-CoA, inhibit complex activity directly, apparently by binding to the E2 and E3 active sites, respectively [11,12]. However, PDH is much more extensively regulated through its associated kinases and phosphatases [13].

The four PDKs that regulate the mammalian PDH complex are a novel class of kinases apparently unrelated to the large families of serine-threonine and tyrosine kinases so prominent in other mammalian regulatory processes as reviewed in [7,8]. These kinases are named on the basis of their order of discovery, PDKs1-4. Though the PDKs represent a related family of kinases, the sequence divergence between the four different isoforms (61-69%) is consistent with their significantly divergent regulatory behavior [14,15]. Moreover, PDK isoform function is apparently ancient and essential as the corresponding isoforms in rodents and humans are at least 94% conserved as reviewed in [15].

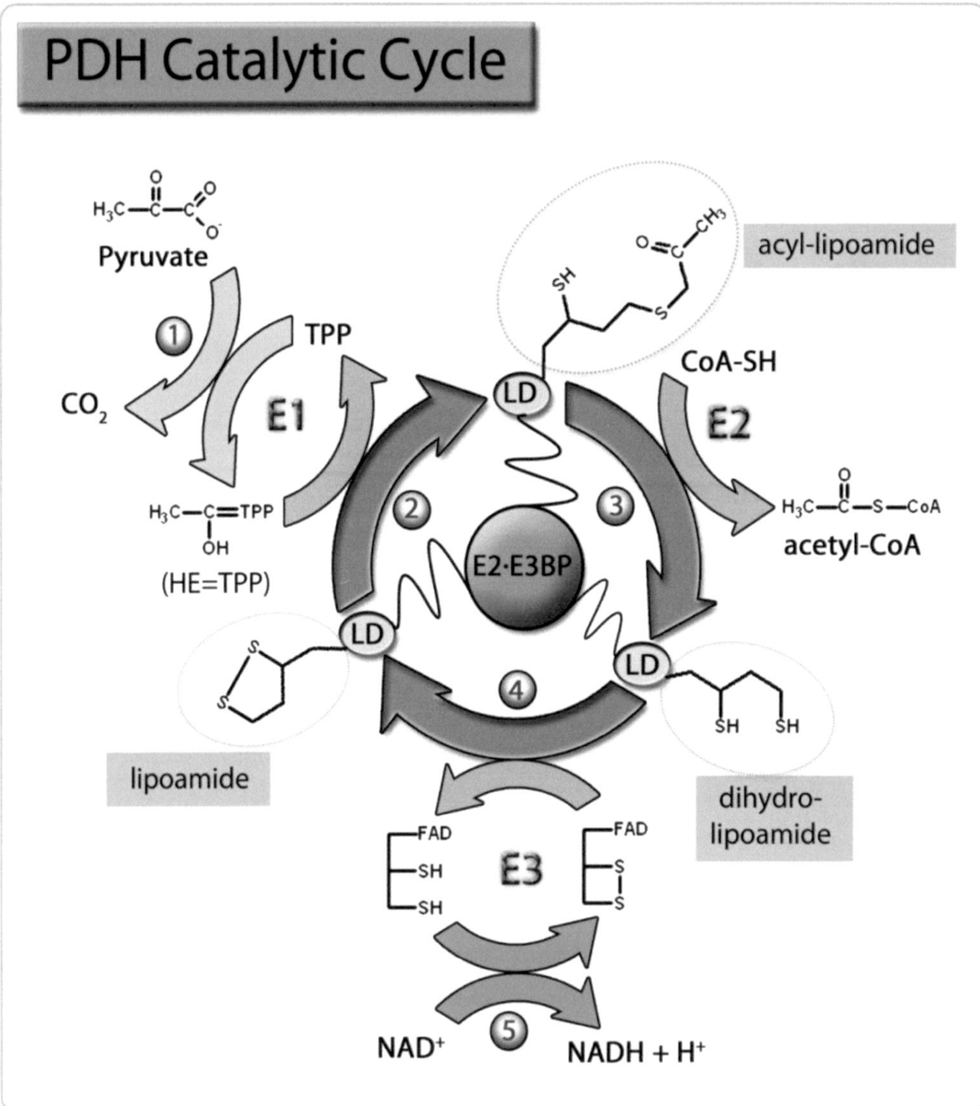

Figure 2. The catalytic cycle of the pyruvate dehydrogenase (PDH) complex. This cycle begins with the oxidative decarboxylation of pyruvate catalyzed by the E1 subunit, generating carbon dioxide and an activated two carbon unit (hydroethyl thiamine pyrophosphate, HE-TPP) as illustrated at the 10 o'clock position. E1 then catalyzes the transfer of this two carbon unit to one of the lipoate residues of the E2 subunits, creating acyl-lipoamide as illustrated at the 12 o'clock position. The E2 catalytic activity transfers this two carbon unit to CoA, creating acetyl-CoA and dihydro-lipoamide as illustrated at the 4 o'clock position. Finally, E3 catalyzes the transfer of the dihydro-lipoamide to NADH (through protein sulfhydryl and FAD intermediates) to recreate the oxidize form, lipoamide, as illustrated at the 8 o'clock position. Both the E2 catalyzed transfer of the lipoamide acyl group to CoA and the E3 catalyzed transfer of dihydro-lipoamide reducing potential to NADH are highly reversible reactions. Figure redrawn from several sources, including [8].

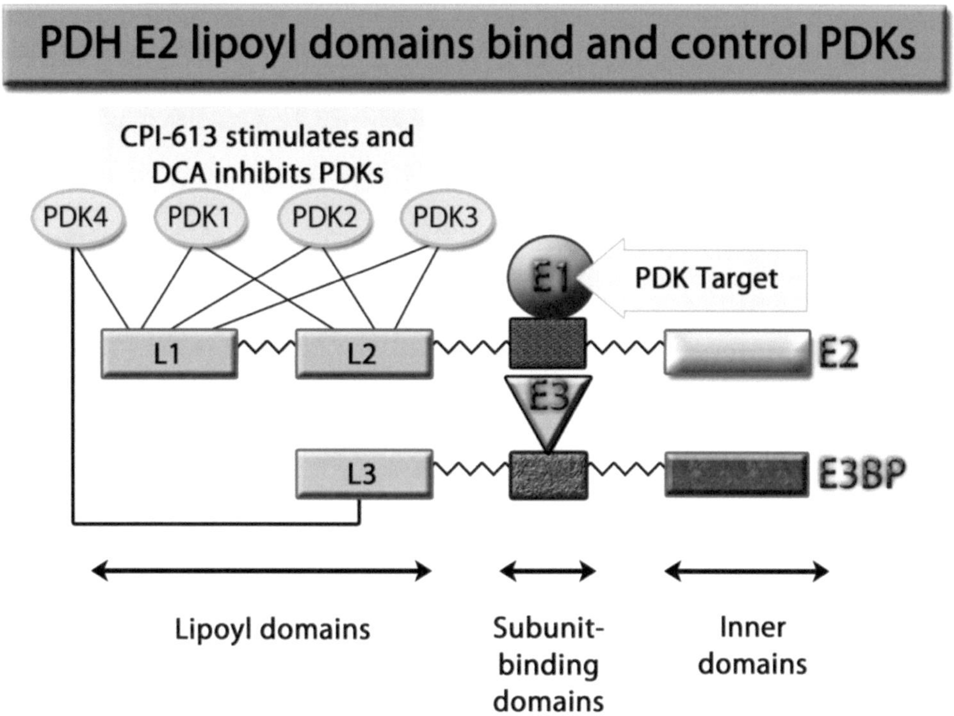

Figure 3. Functional organization of the pyruvate dehydrogenase (PDH) complex in mammals. This complex catalyzes the oxidative decarboxylation of pyruvate to yield acetyl-CoA to feed carbohydrate carbon into the TCA cycle and to capture reducing potential in NADH for transfer to the electron transport complex (ETC) to support ATP synthesis during step down transfer to molecular oxygen (Figure 2). The mammalian PDH complex is built around a core consisting of the inner domains of ca. 48 copies of the catalytically active E2 protein and 12 copies of the catalytically inactive E2 homolog, E3BP, The E1 subunit binds to E2 and the E3 subunit binds to E3BP. Both E2 and E3BP contain lipoate residues joined in amide linkage to the epsilon amino groups of specific lysine residues in the lipoyl domains (L1-L3). The lipoamide residues on the two lipoyl domains of E2 and the single domain of E3BP are acylated, reduced and re-oxidized during the PDH catalytic cycle (Figure 2). The zig-zag lines indicate highly flexible proline-rich domains allowing the E2 subunits high mobility throughout the complex and, perhaps, beyond. The regulatory kinases (PDKs) control PDH activity by phosphorylating (inactivating) the E1 subunit of the complex. Their activity is antagonized by two phosphatases (PDPs). The status of E2 lipoate residues strongly modulates the activity of the PDKs with acylation/reduction stimulating PDK activity. Figure redrawn from several sources, including [7].The PDKs normally function as dimers [see, for example 16,17]. Of particular potential importance is the observation that at least two of the PDK isoforms (PDK1 & 2) readily form heterodimers [18]. This raises the possibility that there might be as many as 10 different PDK isoforms (four homodimers and six heterodimers). In view of the substantial changes in PDK levels associated with malignancy (below), this potential complexity may be very important and is worthy of more investigation.

The PDKs regulate PDH complex activity by responding to diverse allosteric modulators. High ratios of acetyl-CoA to CoASH and NADH to NAD+ represent signals of saturation of mitochondrial demand. Elevations of these ratios are strong allosteric activators of PDKs, shutting down PDH activity in response to demand saturation [19, 20, 21]. Though saturation of mitochondrial demand is most often produced in response to high fat or carbohydrate intake in healthy animals, as reviewed in [8], the altered metabolism of tumor cells also creates a very new metabolic environment wherein these regulatory processes may be entrained for different purposes (below).

PDKs are also subject to allosteric regulation by pyruvate [22] . Elevated pyruvate levels interact synergistically with ADP to inhibit PDK activity (activating PDH). This PDK inhibition apparently results from the binding of pyruvate to PDK-ADP inhibiting exchange for ATP and, thus, phosphorylation. Dichloroacetate (DCA) is a pyruvate analog and acts very similarly to pyruvate in this allosteric modulation of PDK activity [22] DCA was subsequently investigated as an anticancer drug (below) on the basis of this observation.

The final relevant feature of PDK regulation is the response of these regulatory kinases to the lipoyl domains of the E2 subunits of PDH [7, 8, 23]. PDK1-3 bind strongly to the L2 lipoate domain of E2 and more weakly to the E2 L1 lipoyl domain (Figure 3). PDK4 binds to the L1 E2 domain and to the sole lipoyl domain of E3BP (Figure 3). This lipoyl domain binding requires the lipoamide component of the domain. Moreover, the well characterized binding of PDK3 to the E2 L2 domain defines the binding site in detail including the amino acids interacting with the lipoamide residue [16]. These residues are highly conserved in PDK structure [17] indicating that this binding mode is likely universal to the PDKs. This binding of PDKs to E2 lipoyl domains juxtaposes them to their E1 target (associated with the E2 subunit binding domain; Figure 3), producing a large increase in the kinase reaction rate. PDK dimers apparently interact simultaneously with two lipoyl domains, producing very high binding affinity to the PDH complex and supporting hand-over-hand migration of the PDKs through the complex [24].

Most importantly for our purposes here, the reduction and acylation state of the E2 lipoates strongly modulates associated PDK activity [25, 26]. Specifically, acylation and reduction of lipoate enhances PDK2 activity [7,8, 27]. Moreover, acetyl-CoA and NADH can be used as substrates to run elements of the PDH reaction (above) in reverse, thereby creating reduced and/or acylated lipoate residues in response to elevation in the levels of acetyl-CoA and NADH in the mitochondrial matrix [25,26]. In other words, the lipoate residues of PDH represent a real-time census of these crucial features of the energy status of the mitochondrial matrix.

Several additional details of lipoate regulation of PDH and KGDH are also relevant for this discussion. The lipoates in PDH and KGDH comprise a dense cloud of highly mobile residues [28], potentially interacting with crucial regulatory functions. This condition results from several features of the structure of these complexes. PDH illustrates the crucial issues. Each of the ca. 48 E2 subunits contains 2 lipoate (lipoamide) residues and each of the 12 E3BP proteins contains 1 such residue, for a total of 108 lipoate residues in each complex. The regulatory kinases (PDKs-1-4) interact with these lipoate residues embedded in the lipoyl domains (Figure 3). Moreover, these lipoyl domains not only have the boom-like structure resulting from their connection to lysine epsilon amino groups in the E3 lipoyl binding domains, but also each of these domains is separated by highly flexible (alanine/proline-rich) linker domain.

These details will be important to us below in considering the possible mechanisms of action to pursue in developing new drugs based on lipoate structure. Specifically, each PDK can potentially interact simultaneously with two lipoate residues, as well as rapidly migrate between residues. This creates an environment in which PDKs may effectively sample the lipoate cloud, regulating their activity in response to what is effectively a measurement of the aggregate cloud status, especially the acylation/redox status of lipoate as determined by both forward flux through the complex and by backward charging of lipoate from NADH and acetyl-CoA products (above).

Such a sampling procedure would yield especially robust assessment of enzyme saturation and, thus, allow PDK control of PDH to be more reliably connected to the global mitochondrial energy status. *In vitro* studies demonstrate that free lipoate can interact with the PDH (and KGDH) complexes [29]. Thus, we anticipate that exogenous lipoate analogs might be effective in modulating the activity of PDK bound to the PDH complex (below).

3. Clinical implications of the reconfiguration of metabolism in cancer cells

The hypothesis that metabolism may be altered in important ways in cancer is longstanding [30]. However, our understanding of cancer metabolism and its relationship to malignancy and clinical outcome has only recently begun to receive extensive attention. Though our knowledge about cancer metabolism remains quite incomplete, some important global insights have emerged.

First, fluxes of both glucose and glutamine, two major carbon sources, are substantially up-regulated in many tumors as reviewed in [31-36]. More specifically, mutational changes altering signaling through the Akt pathway enhance glucose uptake and glycolytic metabolism [37]. Likewise, alterations of Myc expression substantially up-regulate glutamine utilization [3, 38, 39].

Second, evidence from multiple perspectives supports the view that one of the major functions of these tumor-specific changes is the redirection of metabolism toward provision of increased levels of anabolic substrates. For example, large fractions of glucose carbon are diverted into the pentose phosphate pathway in support of nucleotide synthesis and into amino acid biosynthesis. Likewise, a large portion of glutamine carbon is siphoned off of the TCA cycle as citrate to support lipid biosynthesis (including fatty acids and sterols) (Figure 1). We can call this the *anabolic shift* in tumor cells.

The therapeutic implications of the alterations in cancer metabolism depend heavily on how unique these changes are to tumor cells. There is good evidence that some elements of the anabolic shift are not unique to tumors cells, but may also be shared with rapidly dividing normal cells. For example, activated lymphocytes show some of these same metabolic properties [40]. However, there is also reason to suspect that other features of the cancer cell version of the anabolic shift will not be shared with most normal, rapidly dividing cells. In particular, the poor vascularization of solid tumors reflects an environment rather different than rapidly dividing normal cells typically experience [41, 42]. Tumor cells are likely to have additional metabolic adaptations to this extreme environment.

Of course, the existence of such metabolic adaptations to the extreme tumor microenvironment may still not be helpful therapeutically if these adaptations are idiosyncratic to each individual case of cancer. Fortunately, there is both empirical and theoretical reason to believe that these special adaptations will, in fact, be generally consistent and predictable, as we now discuss. It has long been recognized that solid tumors bear a striking resemblance to healing wounds [43-45]. Specifically, solid tumors resemble wounds that continue the early steps in wound healing, but never resolve as normal wound healing does. Thus, it is plausible to speculate that metabolic modifications in tumor cells usually or always involve the pathological, uncontrolled activation of regulatory pathways normally accommodating cells to the rigors of the wound environment. On this working hypothesis, tumor metabolism is expected to have features not shared with normal cells (except potentially in healing wounds) and to be potentially universal to most or all cancers.

The possibility that tumor metabolism might have unique features is particularly important therapeutically. The dramatic increase in our knowledge about the genetic changes in cancer and our experience with genetically targeted therapies has produced some sobering potential insights as illustrated by the following examples. Specifically, in many or even most cancers, the "driver" mutations may represent loss-of-function in tumor suppressors rather than gain-of-function mutations [46-48]. Moreover, the fraction of tumor-producing changes that are epigenetic may be high. All of these features mean that targeting specific oncogene-targeted products, such as small molecules like gleevec, may be viable in only a minority of tumors [49].

Finally, the levels of redundancy in the regulatory circuits mutated in cancers mean that targeting an individual gain-of-function driver, even when it is possible, may be incompletely successful and subject to evolved resistance.

The upshot of these genetic considerations is that chemotherapy might be better focused on *properties* of the transformed state rather than on the genetic drivers producing that state. Moreover, non-redundant, indispensable properties of the transformed state are expected to represent the most propitious of such targets. Cancer metabolism may represent an environment in which such desirable targets exist. Below we will develop the argument that the dehydrogenases at key control points in metabolism are especially attractive candidates for such targets.

4. Modification of PDH regulation is vital to malignant transformation

If cancer metabolism is to be a therapeutic target it is vital to identify likely targets for intervention. In this section we will review some of the evidence that the PDKs regulating

PDH activity are excellent potential candidates for such targets. Later we will return to the question of whether the other major dehydrogenase entry point for mitochondrial metabolism, KGDH, might also be an attractive therapeutic target.

PDKs have long been recognized as key regulators of PDH function (above). Moreover, in view of the central role of modification of carbohydrate metabolism in cancer (above), alteration of PDK regulation in malignancy is likely. Initially, analysis of clinical samples demonstrated that up-regulation of PDK levels was a frequent correlate of advanced malignancy [50] a pattern that has persisted in subsequent studies [51].

These and other observations stimulated systematic detailed study to great effect. Kim, et al., [52] and Papandreou, et al. [53] demonstrated that HIF-1, a global regulator of the hypoxic response, up-regulated PDK1, with the expected effect of down-regulating carbon flow through PDH. Moreover, PDK up-regulation correlated with increased resistance to programmed cell death, a hallmark of cancer as reviewed in [54]. In view of the consistent hypoxia in solid tumors as reviewed in [41, 42] and the frequent activation of HIF1 in tumors as reviewed in [55] these observations suggested that PDK up-regulation might be a central element of the cancer metabolic reconfiguration.

The next challenge was to ask if more direct evidence for the importance of PDK up-regulation in cancer metabolism could be found. Was PDK up-regulation required for malignancy or marginal, epiphenomenal? Several observations argue strongly that PDK up-regulation is essential to the malignant phenotype.

First, McFate, et al. [56] demonstrated that PDK1 up-regulation was essential for the malignant phenotype in head and neck squamous cell carcinoma cells (HNSC). Specifically, PDK1 expression contributed to induction of limited flux of carbon through mitochondrial metabolism and PKD1 over-expression further reduced this flux (as measured by glucose consumption and diversion of carbohydrate carbon to lactate formation). Confirming this relationship, PKD1 knockdown increased TCA processing of carbohydrate carbon. PDK1 knockdown reduced several indicators of malignant potential, including ability to resist cell death under hypoxia or to form colonies in soft agar in culture. Most importantly, PDK1 knockdown dramatically, reduced the rate of tumor growth in xenograft models.

Earlier work from this group indicated that accumulation of glycolytic intermediates might, itself, be sufficient to up-regulate HIF1 [57]. Thus, McFate [56] proposed a positive feedback loop central to the malignant state wherein PDK1 up-regulation by HIF1 enhanced production of glycolytic intermediates (by blocking pyruvate consumption) further up-regulating HIF1 and so on.

Second, Lu, et al. [58] found a similar role for HIF1-dependent up-regulation of PDK3 in malignant metabolic state. In this case also, reduction in mitochondrial carbohydrate metabolism with the corresponding up-regulation of oxygen-independent cytosolic glycolysis was driven by elevated PDK3 expression. Of particular importance here, the resistance of tumor cells to killing by chemotherapeutic agents paclitaxel and cisplatin was strongly improved by PDK3 up-regulation, while PDK3 knockdown sensitized to these agents. Finally, Lu, et al. [58] observed that both PDK1 and PDK3 collaborated to produce this drug resistance.

Collectively, these results clearly indicate that PDK regulation of PDH activity is central to the malignant phenotype. Moreover, there is apparently no metabolic bypass of the requirement for properly managed mitochondrial metabolism. Mitochondria are not turned off in hypoxic tumor cells. Rather, they are redeployed to provide the limited carbon traffic they can sustain to the provision of anabolic intermediates (above). There appears to be no other way to provide some of these biosynthetic resources (citrate for lipid biosynthesis, for example). Further, mitochondria are central controllers of Ca++ signaling [see 59 and references therein] and of cell death control as reviewed in [60,61]. Thus, their continued properly controlled function is likely to be essential to cancer cell homeostasis and survival. These results and considerations are consistent with the hypothesis that this re-regulation of PDH cannot be dispensed with. Without this metabolic pattern cancer cells are likely to be unable to survive and prosper in the hypoxic environment crucial to malignant progression. Thus, PDK re-regulation is a candidate for the kind of essential, non-redundant therapeutic target needed if we are to have fundamentally new clinical options. We will argue below that this target can be therapeutically exploited in several different ways.

Before turning to assessment of how we might attack this potential target it is important to emphasize a particular implication of PDK re-regulation in cancer cells. Because of the possibility of PDK heterodimer formation (above), tumor-specific quantitative changes in individual PDK expression may be substantially amplified in their regulatory effects. For example, if PDK1 and PDK3 were both up-regulated 10 fold in the *in vivo* tumor environment, this could represent a 100-fold increase in the levels of the hypothetical PDK1/PDK3 heterodimer. Such very large effects might represent targets that are essentially qualitatively unique to tumor cells.

5. DCA, attempting to reverse tumor-specific PDK repression of PDH activity

Dichloroacetate (DCA) is a small molecule with a long history of clinical use for treatment of elevated serum lactate levels in some inherited metabolic conditions [62]. Moreover, DCA is well characterized as being a non-metabolizable pyruvate analog that can strongly inhibit PDK function [17,22, 63] .Thus, DCA is a candidate to reverse PDK repression of PDH important to cancer cell tumorigenicity (above).

These considerations influenced Bonnet [1] to investigate DCA's potential anti-cancer effects. These authors observed that DCA treatment could reverse the mitochondrial membrane hyper-polarization characteristic of tumor cells without noticeable effect on the lower level of polarization in normal cells. Further, DCA substantially increased glucose oxidation while decreasing fatty acid oxidation in A549 tumors cells [1]. These and other effects correlate with DCA inhibition of cancer cell growth and induction of a modest increase in the rate of cancer cell apoptosis. Finally and most importantly, DCA very significantly inhibits tumor growth in mouse xenograft models. Specifically, DCA administered ad libitum in drinking water at 75mg/l produced both inhibition of xenograft tumor growth and modest, but significant reduction in tumor volume in larger tumors. Subsequent reports from other groups corroborate these preclinical findings [64, 65].

The clinical follow up on these preclinical results has been generally weak. This largely results from economic factors. DCA is a cheap, generic drug and no corporate actors have an incentive to develop the agent. To date clinical data consist of several anecdotal case reports [66,67] suggesting possible efficacy, including in a poorly differentiated metastatic carcinoma. In addition, in view of its low cost and wide availability, DCA is being prescribed off-label or self- administered. Anecdotal reports of efficacy from this use group can be found in patient/physician webshops (see, for example, www.thedcasite.com). It will be of considerable interest to see if sufficient evidence can be accumulated to assess the clinical potential of this agent.

An apparently common side effect of DCA use is peripheral neuropathy. Though this is generally limited and reversible, this may not always be the case. A patient report describes relatively severe encephalopathy probably induced by DCA doses within the range currently being experimented with by patients and their physicians [68]. These problems emphasize the importance for individual patients of using DCA only under professional medical supervision.

Finally, a series of small molecule PDK inhibitors have been developed, initially for the treatment of diabetes as reviewed in [8]. These compounds are far more potent than DCA on a mole basis and it will be of interest to see if they have significant anti-cancer activity. To date, no clinical trial data are available on any possible anti-cancer activities of these agents. Moreover, the design strategy for these compounds renders them relatively selective for individual PDK homodimers. We do not currently know which homo- or heterodimer(s) might be the most propitious anti-cancer target(s) for this class of agents.

6. Thioctoid lipoate analogs exaggerate cancer-specific, PDK-mediated repression of PDH to shut down tumor cell mitochondria and induce cell death

The PDH complex includes a dense cloud of lipoate residues making up the functional catalytic co-factor domain of the E2 activity (above). Acyl groups and reducing potential can be back-transferred to this lipoate cloud from the surrounding mitochondrial matrix pool as well as being generated by the conventional PDH forward reaction (above). Moreover, these acyl groups and reducing potential can apparently be rapidly shuttled between these residues before ultimately being transferred to CoA and NAD+[28]. Finally, the oxidation/acylation state of the elements of the lipoate cloud strongly effect the activity of the PDK regulatory kinases, with acylation/reduction stimulating PDK activity (above). Thus, each PDK homo- or heterodimer docked at any point on the PDH complex is apparently continuously exposed to the result of an ongoing, real-time poll of the mitochondrial matrix energy status. Further, these PDKs are then able to act rapidly on all their E1 targets in the PDH complex in response to these polling results as a result of the combination of the highly flexible linkers surrounding the lipoyl domains to which they are bound (Figure 3) and their ability to move hand-over-hand through the complex.

In vitro studies demonstrate that free lipoate and lipoate-containing lipoyl domain fragments can interact productively with the intact PDH complex (above). This suggests that lipoate analogs might be able to perturb PDH regulation *in vivo*. Moreover, this perturbation might show strong selectivity for tumor cells in view of the substantial reconfiguration of PDK regulation in cancer (above). Based on the original PDH regulatory biochemistry and the presumption, subsequently confirmed (above), that PDH regulation was likely to be substantially altered in tumor cells, we began a systematic study of lipoate analogs as anti-cancer agents in the late 1990s (Patent # 6331559, filed 1999). The concept was that appropriately designed lipoate analogs might mimic the effects of specific intermediate states of lipoate (Figures 3 and 4) on PDK function and, thereby, misinform the PDH regulatory machinery in ways that were selectively toxic to tumor cells.

These studies initially demonstrated that lipoate analogs, designated thioctoids, had strong anti-cancer activity in cell culture [2; our unpublished results]. Many of our initial analyses were carried out using our current lead clinical investigational compound, CPI-613 (Figure 4). Indeed, these agents induce cell death in cancer cells extremely efficiently, with all cells dying within 24-48 hours by a mixture of apoptosis and necrosis-like mechanisms. This is in contrast to DCA which typically induces cell death in only a small fraction of tumor cells in culture [1; our unpublished results]. Each of the many human tumor cell lines we have examined is comparably sensitive to thioctoids, with tumor genotype at oncogenes and multidrug resistance loci having little or no effect on response [2].

Though the precise biochemical mechanism of action of thioctoids remains to be established, the *in vivo* behavior of these compounds demonstrates that they modulate PDK regulation of PDH [2; our unpublished results]. Moreover, these tumor effects are diametrically opposite to the effects of DCA and of other small molecule PDK inhibitors. Specifically, DCA induces *de-phosphorylation* of PDH E1 *in vivo* [55] while thioctoids induce *hyper-phosphorylation in vivo* [2; our unpublished results]. This thioctoid-induced PDH hyper-phosphorylation is selective for tumor cells and correlates with inactivation of PDH activity as assessed by analysis of the oxidative flux of pyruvate carbon through the complex. Coincident with this inhibition of PDH activity, thioctoids induce rapid shut down of mitochondrial ATP synthesis, typically reducing mitochondrial ATP production from pyruvate and glutamine carbon sources to less that 10% of controls within 15-30 minutes [2]. Finally, the capacity of CPI-613 to kill tumor cells is significantly attenuated in cell culture systems by RNAi knockdown of regulatory PDKs, supporting a role for PDKs in the response to these agents.

Collectively, these results indicate that thioctoid lipoate analogs kill tumor cells by addressing tumor-specific PDK regulation of PDH. Moreover, the consequence of these effects is apparently to exploit tumor-specific PDH regulation, in part or entirely, to shut down tumor mitochondrial function. As continued mitochondrial function is required for the survival of tumor cells, even in the hypoxic solid tumor environment (above), these effects are expected to engender the observed response, rapid commitment to and execution of mitochondrially controlled cell death pathways.

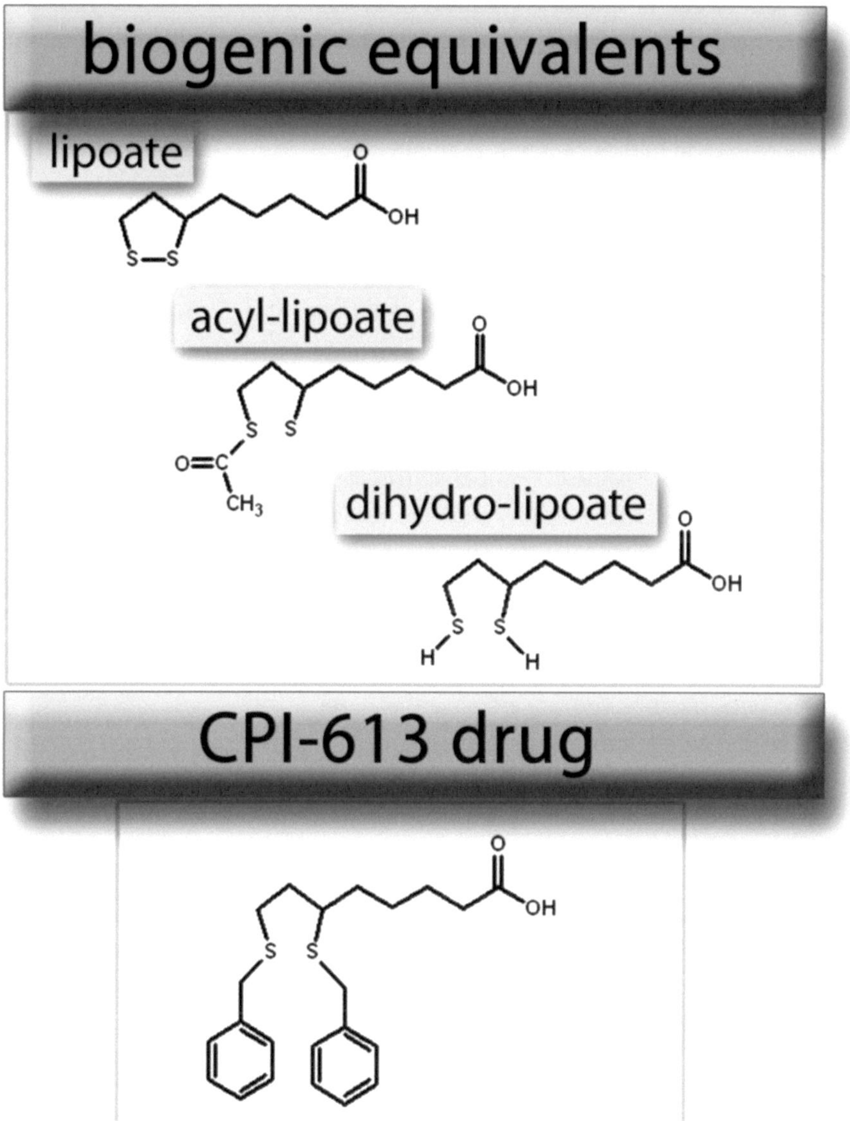

Figure 4. Structure of thioctoid CPI-613 compared to biogenic lipoates. The structure of the thioctoid CPI-613 is shown at bottom. This molecule is a non-redox active analog of lipoate designed to have some resemblance to the acylated form of lipoate as it occurs in the normal catalytic cycle of PDH (Figure 2). The corresponding biogenic lipoate intermediates in the PDH catalytic cycle are illustrated at top.

These properties suggested that thioctoids might be promising anti-cancer agents. This possibility is further supported by their behavior in human tumor xenograft mouse models. Specifically, CPI-613 produces strong tumor growth inhibition in several tumor types. For example, the growth of BxPC-3 human pancreatic tumor xenografts implanted subcutaneously on posterior flanks is largely or entirely suppressed by intraperitoneal dosing with CPI-613 at 25-75mg/kg. Moreover, ca. 40% of these animals survive for more than 9 months (termination of study) without regrowth of tumors after an initial three week treatment regime. [A large fraction of the 60% mortality in this study is attributable to normal attrition in these genetically immunocompromised *nude* mice. Thus, cancer-free post-treatment survival in these animals is apparently substantially higher than 40%.]

Titration studies demonstrate that CPI-613 produces very substantial tumor growth inhibition in xenograft model systems at doses (0.1-1mg/kg) very substantially lower than the maximum tolerated dose (ca. 100mg/kg)[2; our unpublished results]. This indicates a very high possible therapeutic index. Large-animal toxicology studies further corroborate the very low toxicity of CPI-613 [69]. Moreover, drug metabolism studies indicate that CPI-613 breakdown products are unlikely to pose a barrier to clinical development [70].

Based on this favorable mechanistic and pre-clinical animal data, Phase I/II clinical trials are currently underway at several locations (see comments at http://clinicaltrials.gov/ct2/show/NCT01520805). These studies indicate that adverse events of CPI-613 in humans are low and mostly mild (well tolerated to at least 3,000mg/m^2, when infused over one hour) and suggest efficacy against several advanced cancers, including refractory/relapsed AML, in several patients [71-73]. It will be of interest to continue to pursue the clinical investigation of these investigational drugs.

7. The central role of lipoate as a metabolic regulatory signal, other potential thioctoid tumor targets

In view of the capacity of thioctoids to attack tumor cell PDH it may be of value to consider the potential of these agents to address tumor metabolism more generally. Indeed, the properties of lipoate suggest that it may act as a global mitochondrial regulatory signaling molecule, possibly addressing the entire flow of carbon through this compartment. On this view, lipoate analogs may address multiple regulatory pathways, some or all altered in tumor cells analogously to the well-understood PDH case.

Specifically, two other enzyme complexes are recognizably homologous to PDH and catalyze analogous reactions. One of these is the branched chain oxoacid dehydrogenase complex (BCDH) as reviewed in [74]. The other is the KGDH complex (above). BCDH governs the entry of carbon from a series of amino acids into the TCA cycle. KGDH governs the entry of glutamine-derived carbon (Figure 1). Together with PDH control of carbohydrate derived carbon, these three enzymes directly control initial access of all carbon to the TCA cycle, except for that derived from fatty acids. Moreover, even fatty acid-derived carbon flow is implicitly governed by lipoate-dependent dehydrogenases because no fatty acid carbon can be oxidatively released without first passing through the KGDH complex [6]. Thus, all carbon flow through mitochondria is ultimately controlled by lipoate-containing enzymes.

The role of lipoate residues in controlling PDH function and the unique capacity of these residues to poll and reflect the mitochondrial energy status strongly suggests that lipoate residues will play a role in regulating BCDH and KGDH function. While this possibility remains to be investigated, it is noteworthy that the BCDH complex is regulated by kinases homologous to PDKs. It is plausible that these regulatory kinases are also responsive to lipoate redox/acylation state. Moreover, in view of the rather dramatic reprogramming of tumor mitochondrial metabolism, it is also plausible that BCDH regulatory kinases may be reprogramed in malignancy.

KGDH is apparently not regulated by PDK-like kinases and the details of its metabolic regulation remain to be investigated. However, the complex shows strong allosteric regulation by end products, consistent with important regulatory objectives [75,_76]. Thus, it is likely that KGDH lipoates also participate in this regulation in some way that largely remains to be determined.

As reviewed in section 3 above, the lipoate residues of PDH, BCDH and KGDH exist as a cloud polling crucial features of the mitochondrial energy state, including the aggregate levels of free reducing potential (NADH) and of their respective acyl-CoA products. Moreover, these details of the mitochondrial energy state are likely to be among the state variables most informative about the moment-to-moment metabolic needs of the mitochondrion. Further, these regulatory goals of mitochondria are so substantially altered in tumor cells that we anticipate that all these potential lipoate-sensitive regulatory targets might be substantially altered in tumor cells, as they clearly are in the PDH case. Finally, the long boom-like structure of the enzyme linked lipoamide moieties (Figures 2) and their attachment to highly mobile E2 enzyme subdomains (Figure 3) raises an additional important regulatory possibility. It is conceivable that these actively polling lipoate residues are not merely reporting their results to their individual complexes but also to other mitochondrial consumers of regulatory information.

Collectively, these considerations suggest that thioctoid lipoate analogs may achieve their dramatic inhibition of tumor cell mitochondrial metabolism by addressing multiple, essential metabolic regulatory processes in a tumor-specific fashion. The promising pre-clinical and early clinical properties of thioctoid CPI-613 indicate that it will be of interest to explore this possibility in detail going forward.

8. Conclusions

The central role of lipoate-dependent dehydrogenases in governing carbon flow through mammalian mitochondria is reflected in their extensive regulation. Moreover, both empirical evidence and theoretical considerations indicate that the regulation of these complexes is strongly controlled by the dynamic status of their lipoate residues, reflecting a moment-to-moment polling of the mitochondrial matrix. Finally, evidence and theory also indicate that some or most of these lipoate-dependent regulatory processes are very significantly altered in support of the substantial repurposing of tumor cell mitochondrial metabolism relative to its normal cell condition.

Collectively, these results indicate that the regulatory reprogramming of these dehydrogenases may represent a target-rich environment for developing new anti-tumor drugs that have the crucial properties that may be required to impart new efficacy to cancer chemotherapy – targets that are both essential to the malignant condition and non-redundant in their function. The preclinical and early clinical properties of agents directed at these targets support the possibility of useful promise in this therapeutic domain.

Author details

Paul M. Bingham* and Zuzana Zachar
Biochemistry and Cell Biology, Stony Brook University, Stony Brook, NY
Cornerstone Pharmaceuticals, 1 Duncan Drive, Cranbury, NJ

Acknowledgement

This work was largely supported by Cornerstone Pharmaceuticals, Inc. with additional support from the Carol M. Baldwin Breast Cancer Research Fund; Stony Brook University Center for Biotechnology, and from the Department of Biochemistry and Cell Biology and the School of Medicine at Stony Brook University. We are grateful to many colleagues for helpful discussions, including Robert Shorr, Robert Rodriguez, King Lee, Shawn D. Stuart, Sunita Gupta, and Alexandra Schauble. Both authors have a financial stake in Cornerstone Pharmaceuticals, Inc. the clinical developer of thioctoids, including CPI-613.

9. References

[1] Bonnet S, Archer SL, Allalunis-Turner J, Haromy A, Beaulieu C, Thompson R, Lee CT, Lopaschuk GD, Puttagunta L, Bonnet S, Harry G, Hashimoto K, Porter CJ, Andrade MA, Thebaud B, Michelakis ED. (2007) A Mitochondria-K+ Channel Axis Is Suppressed in Cancer and its Normalization Promotes Apoptosis and Inhibits Cancer Growth. Cancer cell 11: 37-51.

[2] Zachar, Z, Marecek, J, Maturo, C, Gupta, S, Stuart, S D, Howell, K, Bingham, P M (2011) Non-Redox-Active Lipoate Derivatives Disrupt Cancer Cell Mitochondrial Metabolism and Are Potent Anticancer Agents *In Vivo*. J. mol. med. 89: 1137-1148.

[3] DeBerardinis, R J, Cheng, T (2010) Q's Next: the Diverse Functions of Glutamine in Metabolism, Cell Biology and Cancer. Oncogene 29: 313-324.

* Corresponding Authors

[4] Nealson, K H, Conrad, P G (1999) Life: Past, Present and Future Philos trans r. soc Lond B biol sci, 354: 1923-1939.

[5] Raymond, J, Segre, D (2006) The Effect of Oxygen on Biochemical Networks and the Evolution of Complex Life. Science, 311:, 1764-1767.

[6] Garrett, R, Grisham, C M (2010) Biochemistry (4th ed) Belmont, CA: Brooks/Cole, Cengage Learning

[7] Patel, M S, Korotchkina, L G (2006) Regulation of the Pyruvate Dehydrogenase Complex. Biochem. soc. trans., 34: 217-222.

[8] Roche, T E, Hiromasa, Y (2007) Pyruvate Dehydrogenase Kinase Regulatory Mechanisms and inhibition in Treating Diabetes, Heart Ischemia, and Cancer. Cell. mol. life sci. 64: 830-849.

[9] Hiromasa, Y, Hu, L Y, Roche, T E (2006) Ligand-Induced Effects on Pyruvate Dehydrogenase Kinase Isoform 2. J. biol. chem. 281: 12568-12579.

[10] Perham, R N (2000) Swinging Arms and Swinging Domains in Multifunctional Enzymes: Catalytic Machines for Multistep Reactions Annu. rev.biochem., 69: 961-1004.

[11] Garland, P B, Randle, P J (1964) Control of Pyruvate Dehydrogenase in Perfused Rat Heart by Intracellular Concentration of Acetyl-Coenzyme. A. biochem. j. 91: C6-C12.

[12] Kanzaki, T, Hayakawa, T, Hamada, M, FukuyoshY, Koike, M (1969) Mammalian Alpha-Keto Acid Dehydrogenase Complexes. IV. Substrate Specificities and Kinetic Properties of Pig Heart Pyruvate and 2-Oxoglutarate Dehydrogenase Complexes. J.biol. chem. 244: 1183-1187.

[13] Linn, T C, Pettit, F H, Reed, L J (1969) Alpha-Keto Acid Dehydrogenase Complexes X Regulation of Activity of Pyruvate Dehydrogenase Complex From Beef Kidney Mitochondria by Phosphorylation and Dephosphorylation. Proc. nat. acad. sci. USA 62: 234-241.

[14] Baker, J C, Yan, X H, Peng, T, Kasten, S, Roche, T E (2000) Marked Differences Between Two Isoforms of Human Pyruvate Dehydrogenase Kinase. J. biol. chem. 275: 15773-15781.

[15] Roche, T E, Baker, J C, Yan, Y H, Hiromasa, Y, Gong, X M, Peng, T, Kasten, S A (2001) Distinct Regulatory Properties of Pyruvate Dehydrogenase Kinase and Phosphatase Isoforms. Prog. nucleic acid res. mol. biol., 70: 33-75.

[16] Kato, M, Chuang, J L, Tso, S C, Wynn, R M, Chuang, D T (2005) Crystal Structure of Pyruvate Dehydrogenase Kinase 3 Bound To Lipoyl Domain 2 of Human Pyruvate Dehydrogenase Complex. Embo j. 24: 1763-1774.

[17] Knoechel, T R, Tucker, A D, Robinson, C M, Phillips, C, Taylor, W, Bungay, P J, Brown, D G (2006) Regulatory Roles of the N-Terminal Domain Based on Crystal Structures of Human Pyruvate Dehydrogenase Kinase 2 Containing Physiological and Synthetic Ligands. Biochemistry, 45: 402-415.

[18] Boulatnikov, I, Popov, K A (2003) Formation of Functional Heterodimers by Isozymes 1 and 2 of Pyruvate Dehydrogenase Kinase. Biochim. biophys. acta 1645: 183-192.

[19] Pettit, F H, Pelley, J W, Reed, L J (1975) Regulation of Pyruvate-Dehydrogenase Kinase and Phosphatase by Acetyl-CoA-CoA and NADH-NAD Ratios. Biochem. biophys. res. commun. 65: 575-582.

[20] Batenburg JJ, Olson MS. (1976) Regulation of Pyruvate Dehydrogenase by Fatty Acid in Isolated Rat Liver Mitochondria. J. biol. chem. 251: 1364-1370.

[21] Hansford, R G (1976) Studies on Effects of Coenzyme A-Sh - Acetyl Coenzyme A, Nicotinamide Adenine-Dinucleotide - Reduced Nicotinamide Adenine-Dinucleotide, and Adenosine-Diphosphate - Adenosine-Triphosphate Ratios on Interconversion of Active and Inactive Pyruvate-Dehydrogenase in Isolated Rat-Heart Mitochondria. J. biol. chem. 251: 5483-5489.

[22] Pratt, M L, Roche, T E (1979) Mechanism of Pyruvate Inhibition of Kidney Pyruvate-Dehydrogenase Kinase and Synergistic Inhibition by Pyruvate and ADP. J.biol. chem. 254: 7191-7196.

[23] Bao, H Y, Kasten, S A, Yan, X H, Hiromasa, Y, Roche, T E (2004) Pyruvate Dehydrogenase Kinase Isoform 2 Activity Stimulated by Speeding Up the Rate of Dissociation of ADP. Biochemistry 43: 13442-13451.

[24] Hiromasa, Y, Roche, T E (2003) Facilitated Interaction Between the Pyruvate Dehydrogenase Kinase Isoform 2 and the Dihydrolipoyl Acetyltransferase. J.biol. chem. 278: 33681-33693.

[25] Cate, R L, Roche, T E (1978) Unifying Mechanism for Stimulation of Mammalian Pyruvate Dehydrogenase Kinase by Reduced Nicotinamide Adenine-Dinucleotide, Dihydrolipoamide, Acetyl Coenzyme-A, or Pyruvate. J. biol. chem. 253: 496-503.

[26] Cate, R L, Roche, T E (1979) Function and Regulation of Mammalian Pyruvate-Dehydrogenase Complex - Acetylation, Interlipoyl Acetyl Transfer, and Migration of the Pyruvate-Dehydrogenase Component. J. biol. chem. 254: 1659-1665.

[27] Korotchkina, L G, Patel, M S (2001) Site Specificity of Four Pyruvate Dehydrogenase Kinase Isoenzymes Toward the Three Phosphorylation Sites of Human Pyruvate Dehydrogenase. J.biol. chem. 276: 37223-37229.

[28] Collins, J H, Reed, L J (1977) Acyl Group and Electron Pair Relay System - Network of Interacting Lipoyl Moieties in Pyruvate and Alpha-Ketoglutarate Dehydrogenase Complexes from Escherichia coli. Proc. nat. acad. sci. USA 74: 4223-4227.

[29] Liu, S J, Baker, J C, Andrews, P C, Roche, T E (1995) Recombinant Expression and Evaluation of the Lipoyl Domains of the Dihydrolipoyl Acetyl Transferase Component of the Human Pyruvate –Dehydrogenase Complex. Arch. biochem. biophys. 316: 926-940.

[30] Warburg, O (1930) The Metabolism of Tumors London, UK: Constable Press

[31] Tennant, D A, Duran, R V, Gottlieb, E (2010) Targeting Metabolic Transformation for Cancer Therapy. Nat. rev. canc. 10: 267-277.

[32] Cairns, R A, Harris, I S, Mak, T W (2011) Regulation of Cancer Cell Metabolism. Nat. rev. canc. 11: 85-95.

[33] Vander Heiden, M G (2011) Targeting Cancer Metabolism: A Therapeutic Window Opens. Nat. rev. drug disc. 10: 671-684.

[34] Ramsay, E E, Hogg, P J, Dilda, P J (2011) Mitochondrial Metabolism inhibitors for Cancer Therapy Pharm. res., 28: 2731-2744.

[35] Metallo, C M, Gameiro, P A, Bell, E L, Mattaini, K R, Yang, J J, Hiller, K, Stephanopoulos, G (2012) Reductive Glutamine Metabolism by IDH1 Mediates Lipogenesis Under Hypoxia. Nature, 481, 380-384.

[36] Mullen, A R, Wheaton, W W, Jin, E S, Chen, P H, Sullivan, L B, Cheng, T, DeBerardinis, R J (2012) Reductive Carboxylation Supports Growth in Tumour Cells with Defective Mitochondria. Nature, 481: 385-388.

[37] Elstrom, R L, Bauer, D E, Buzzai, M, Karnauskas, R, Harris, M H, Plas, D R, Thompson, C B (2004) Akt Stimulates Aerobic Glycolysis in Cancer Cells. Canc. res. 64: 3892-3899.

[38] Wise, D R, DeBerardinis, R J, Mancuso, A, Sayed, N, Zhang, X Y, Pfeiffer, H K, Thompson, C B (2008) Myc Regulates a Transcriptional Program That Stimulates Mitochondrial Glutaminolysis and Leads To Glutamine Addiction. Proc. nat. acad. sci. USA 105: 18782-18787.

[39] Wise, D R, Thompson, C B (2010) Glutamine Addiction: A New Therapeutic Target in Cancer. Trends biochem. sci. 35: 427-433.

[40] DeBerardinis R J, Lum, J J, Thompson C B (2008) The Biology of Cancer: Metabolic Reprogramming Fuels Cell Growth and Proliferation. Cell metabolism. 7:11-20.

[41] Tatum, J L, Kelloff, G J, Gillies, R J, Arbeit, J M, Brown, J M, Chao, K S C, Sullivan, D (2006) Hypoxia: Importance in Tumor Biology, Noninvasive Measurement by Imaging, and Value of its Measurement in the Management of Cancer Therapy. int. j. rad. biol. 82: 699-757.

[42] Wilson, W R, Hay, M P (2011) Targeting Hypoxia in Cancer Therapy. Nat. rev. canc. 11(6), 393-410.

[43] Dvorak HF (1986) Tumors: Wounds That Do Not Heal. Similarities Between Tumor Stroma Generation and Wound Healing. N. engl. j. med. 315: 1650-1659.

[44] Schafer M, Werner S.(2008) Cancer as an Overhealing Wound: An Old Hypothesis Revisited. Nature rev. mol. cell biol. 9: 628-638.

[45] Nagy JA, Chang SH, Dvorak AM, Dvorak HF. (2009) Why Are Tumour Blood Vessels Abnormal and Why Is It Important to Know? Br. j. cancer. 100: 865-869.

[46] Insider, S (2010) UPDATED: A Skeptic Questions Cancer Genome Projects. Science online, http://newssciencemagorg/scienceinsider/2010/04/a-skeptic-questions-cancer-genomhtml

[47] Stratton, M R (2011) Exploring the Genomes of Cancer Cells: Progress and Promise. Science, 331: 1553-1558.

[48] MacDougall, R (2012) Bert Vogelstein Considers the Cancer Genome at 10th Annual NHGRI Trent Lectureship National Human Genome Research institute, http://wwwgenomegov/27547785

[49] Vaidya, S., K. Ghosh, et al. (2011). Recent Developments in Drug Resistance Mechanism in Chronic Myeloid Leukemia: A Review. European Journal of Haematology 87(5): 381-393

[50] Koukourakis, M I, Giatromanolaki, A, Sivridis, E, Gatter, K C, Harris, A L, Tumor Angiogenesis Res, G (2005) Pyruvate Dehydrogenase and Pyruvate Dehydrogenase Kinase Expression in Non-Small Cell Lung Cancer and Tumor-Associated Stroma. Neoplasia, 7: 1-6.

[51] Lu, C W, Lin, S C, Chien, C W, Lee, C T, Lin, B W, Lee, J C, Tsai, S J (2011) Overexpression of Pyruvate Dehydrogenase Kinase 3 Increases Drug Resistance and Early Recurrence in Colon Cancer American j. pathol., 179: 1405-1414.

[52] Kim JW, Tchernyshyov I, Semenza GL, Dang CV.(2006) HIF-1-Mediated Expression of Pyruvate Dehydrogenase Kinase: A Metabolic Switch Required for Cellular Adaptation to Hypoxia. Cell metabolism. 3: 177-185.

[53] Papandreou I, Cairns RA, Fontana L, Lim AL, Denko NC. (2006) HIF-1 Mediates Adaptation to Hypoxia by Actively Downregulating Mitochondrial Oxygen Consumption. Cell metabolism 3: 187-197.

[54] Hanahan, D, Weinberg, R A (2011) Hallmarks of Cancer: The Next Generation. Cell 144: 646-674.

[55] Harris, A L (2002) Hypoxia - A Key Regulatory Factor in Tumour Growth. Nat. rev. canc. 2: 38-47.

[56] McFate, T, Mohyeldin, A, Lu, H, Thakar, J, Henriques, J, Halim, N D, Verma, A (2008) Pyruvate Dehydrogenase Complex Activity Controls Metabolic and Malignant Phenotype in Cancer Cells. J.biol. chem. 283: 22700-22708.

[57] Lu, H S, Forbes, R A, Verma, A (2002) Hypoxia-Inducible Factor 1 Activation by Aerobic Glycolysis Implicates the Warburg Effect in Carcinogenesis. J.biol. chem. 277: 23111-23115.

[58] Lu, C W, Lin, S C, Chen, K F, Lai, Y Y, Tsai, S J (2008) Induction of Pyruvate Dehydrogenase Kinase-3 by Hypoxia-Inducible Factor-1 Promotes Metabolic Switch and Drug Resistance. J.biol. chem. 283: 28106-28114.

[59] Visch, H J, Koopman, W J H, Zeegers, D, Vries, S, van Kuppeveld, F J M, van den Heuvel, L, Willems, P (2006) Ca2+-Mobilizing Agonists increase Mitochondrial ATP Production to Accelerate Cytosolic Ca2+ Removal: Aberrations in Human Complex I Deficiency. Am. j. physiol. cell physiol., 291: C308-C316.

[60] Krampe, B, Al-Rubeai, M (2010) Cell Death in Mammalian Cell Culture: Molecular Mechanisms and Cell Line Engineering Strategies. Cytotechnology, 62: 175-188.

[61] Nagley, P, Higgins, G C, Atkin, J D, Beart, P M (2010) Multifaceted Deaths Orchestrated by Mitochondria in Neurons. Biochim. biophys. acta, 1802: 167-185.

[62] Stacpoole, P W, Kerr, D S, Barnes, C, Bunch, S T, Carney, P R, Fennell, E M, Valenstein, E (2006) Controlled Clinical Trial of Dichloroacetate for Treatment of Congenital Lactic Acidosis in Children. Pediatrics, 117: 1519-1531.

[63] Bowker-Kinley, M M, Davis, W I, Wu, P F, Harris, R A, Popov, K M (1998) Evidence for Existence of Tissue-Specific Regulation of the Mammalian Pyruvate Dehydrogenase Complex. Biochem. j.329: 191-196.

[64] Cao, W G, Yacoub, S, Shiverick, K T, Namiki, K, Sakai, Y, Porvasnik, S, Rosser, C J (2008) Dichloroacetate (DCA) Sensitizes Both Wild-Type and over Expressing Bcl-2 Prostate Cancer Cells *in Vitro* To Radiation. Prostate 68: 1223-1231.

[65] Wong, J Y Y, Huggins, G S, Debidda, M, Munshi, N C, De Vivo, I (2008) Dichloroacetate induces Apoptosis in Endometrial Cancer Cells. Gynecol. oncol. 109: 394-402.

[66] Michelakis, E D, Sutendra, G, Dromparis, P, Webster, L, Haromy, A, Niven, E, Petruk, K C (2010) Metabolic Modulation of Glioblastoma With Dichloroacetate. Sci. transl. med., 2: 31-34.

[67] Khan, A (2011) Use of Oral Dichloroacetate for Palliation of Leg Pain Arising From Metastatic Poorly Differentiated Carcinoma: A Case Report. J. palliat. med, 14: 973-977.

[68] Brandsma, D, Dorlo, T P C, Haanen, J H, Beijnen, J H, Boogerd, W (2010) Severe Encephalopathy and Polyneuropathy Induced by Dichloroacetate. J. neurol. 257: 2099-2100.

[69] Lee K, Maturo C, Shorr R, Rodriguez R. (2010) Uncommon Toxicologic Profile at Toxic Doses of CPI-613 (an Agent Selectively Alters Tumor Energy Metabolism) in Rats and Minipigs Reflects Novel Mechanism. American Journal of Pharmacology and Toxicology 5: 183-208.

[70] Lee, K, Shorr, R, Rodriguez, R, Maturo, C, Boteju, LW and Sheldon A (2011) Formation and Anti-Tumor Activity of Uncommon *in Vitro* and *in Vivo* Metabolites of CPI-613, A Novel Anti-Tumor Compound that Selectively Alters Tumor Energy Metabolism. Drug metab. lett. 5: 163-182.

[71] Pardee TS, Levitan DA, Hurd DD: (2011) Altered Mitochondrial Metabolism as a Target in Acute Myeloid Leukemia [abstract]. J Clin Oncol 29: suppl; abstr 6590

[72] Pardee TS, DeFord-Watts LM, Peronto E, Levitan D, Hurd D, Kridel SK, Powell B. (2012) The 1st in Class Tumor Specific Anti-Mitochondrial Metabolism Agent CPI-613 is Well Tolerated and Has Activity in Several Hematologic Malignancies [abstract]. J clin oncol 30: suppl

[73] Lee K, Khaira D, Rodriguez R, Maturo C, O'Donnell K, Shorr R. (2011) Long-Term Stable Disease of Stage IV Pancreatic Neuroendocrine Tumors and Without Significant Adverse Effect by CPI-613, an Investigational Novel Anti-Cancer Agent. Case Study and Case Report, 1:137-145.1

[74] Harris RA, Joshi M, Jeoung NH, Obayashi M. (2005) Overview of the Molecular and Biochemical Basis of Branched-Chain Amino Acid Catabolism. J. nutr. 135:1527S-1530S

[75] Strumilo, S (2005) Short-Term Regulation of the Alpha-Ketoglutarate Dehydrogenase Complex by Energy-Linked and Some Other Effectors. Biochemistry-Moscow, 70: 726-729.

[76] Bunik, V I, Fernie, A R (2009) Metabolic Control Exerted by the 2-Oxoglutarate Dehydrogenase Reaction: A Cross-Kingdom Comparison of the Crossroad Between Energy Production and Nitrogen Assimilation. Biochem. j. 422: 405-421.

Section 2

Dehydrogenases and Some Diseases

Chapter 4

Glucose-6-Phosphate Dehydrogenase Deficiency and Malaria: A Method to Detect Primaquine-Induced Hemolysis *in vitro*

Adil M. Allahverdiyev, Malahat Bagirova, Serhat Elcicek, Rabia Cakir Koc, Sezen Canim Ates, Serap Yesilkir Baydar, Serkan Yaman, Emrah Sefik Abamor and Olga Nehir Oztel

Additional information is available at the end of the chapter

http://dx.doi.org/10.5772/48403

1. Introduction

Glucose-6-phosphate dehydrogenase (G6PD) deficiency is the most common enzymopathological disease in humans. This disease is described as a widespread, heritable, X-chromosome linked abnormality (Reclos, *et al.*, 2000). It is estimated that it affects approximately 400 million people worldwide (Noori-Daloii, *et al.*, 2004). This disease is seen most frequently in approximately all of Africa, Asia, and the countries near the Mediterranean Sea (Frank, 2005). G6PD enzyme was demonstrated to play an active role in survival of erythrocytes. It is known that in the pentose phosphate pathway of erythrocytes, glucose-6 phosphate dehydrogenase (G6PD) enzyme provides the production of NADPH and GSH. GSH, produced by pentose phosphate pathway can react with H_2O_2 and reduce it to H_2O. This prevents the generation of oxidative stress within red blood cells; oxidative stress can be induced in erythrocytes whose G6PD enzymes are deficient. In this situation, GSH is not produced and H_2O_2 is not reduced to H_2O, leading to oxidative stress and hemolysis. This is the only mechanism available for the erythrocyte in order to generate reducing equivalence, therefore making it essential for the survival of erythrocytes. In individuals whose G6PD enzyme is deficient, different kinds of hemolysis from mild to severe are seen bound to differences in variants of the disease (Beutler, 1983, Luzzatto, 1989).

In epidemiological studies, it was shown that the prevalence of G6PD deficiency significantly related to malaria. Malaria is known as a parasitic disease that affects 300-500 million people all over the world. It is widespread in tropical and subtropical regions of Asia, Africa and American continents. Five different types of *Plasmodium* species—*P.*

falciparum, P. vivax, P. ovalae, P. malariae and P.knowlesi—lead to this disease by infecting erythrocytes. Malaria can become a life-threatening condition when it is not treated. Each year, malaria leads to deaths of millions of people all around the world and a large percentage of deaths are seen in Sub-Saharan regions of Africa. As it can be easily seen, malaria and G6PD deficiency share the same geographic distribution. It was shown that G6PD enzyme has various genetic variants and polymorphic frequencies. Highly polymorphic frequencies, which are indicators of G6PD deficiency, are seen in endemic regions for malaria such as Asia, Africa, Central and South America, while in non-endemic regions, these rates decrease, suggesting the relationship between G6PD deficiency and malaria (Haworth, *et al.*, 1988, Organization, 2009, Sutherland, *et al.*, 2010). This relationship reveals two important results. One of them is that G6PD deficiency provides great protection from malaria infection, especially for falciparum infections (Motulsky, 1961, Siniscalco & Bernini, 1961, Ganczakowski, *et al.*, 1995). On the other hand, G6PD deficiency has been recently demonstrated to cause serious problems in fighting against malaria. Primaquine, which is the only drug currently, used in the treatment of Plasmodium infections leads to severe hemolysis in G6PD-deficient patients. This drug may even cause death in G6PD-deficient patients. When primaquine is administered to individuals with G6PD deficiency, its metabolites lead to more severe hemolysis by inducing oxyhemoglobin generation, GSH depletion and stimulation of the hexose monophosphate pathway (Beutler, *et al.*, 1955, Bolchoz, *et al.*, 2002, Beutler & Duparc, 2007).

Therefore, investigations on detection of G6PD deficiency have a vital importance for malaria patients before their treatment with primaquine. On the other hand, the methods that are used for diagnosing G6PD deficiency are unreliable. Even worse is that it is very difficult to distinguish heterozygously-deficient patients from healthy individuals (Peters & Noorden, 2009). All these data indicate that there is an urgent need to develop new methods for reliable detection of G6PD deficiency in order to prevent hemolysis in patients treated with primaquine. Current methods cannot determine primaquine sensitivity in patients with G6PD deficiency every time. However, in our previously researches, we developed a new method for the determination primaquine induced hemolysis *in vitro*. This method provides the determination of G6PD deficiency patients that are susceptible to primaquine independently from the variants of G6PD deficiency. In our studies, it was determined that this method demonstrated high sensitivity for detection of primaquine-induced hemolysis before treatment of malaria patients with primaquine. This chapter aims to represent the relationship between G6PD deficiency and malaria, and to demonstrate the method that has high sensitivity for detection of primaquine-induced hemolysis in patients with malaria whose G6PD enzyme is deficient before their treatment with primaquine.

This chapter aims to represent the problems in treatment of malaria patients with G6PD deficiency by using primaquine, different methods for determination of G6PD deficiency and a new method to determine primaquine induced hemolysis before treatment of patients with G6PD deficiency.

2. Genetic basis of G6PD

G6PD deficiency was identified in 1956 by Carson *et al.* (Alving, *et al.*, 1956), and its X-chromosomal inheritance was discerned in the 1950s by Childs *et al.* (Childs, *et al.*, 1958). G6PD was cloned and sequenced by Persico *et al.*(Persico, *et al.*, 1986, Persico, *et al.*, 1986) in 1986 and independently by Takizawa and Yoshida (Takizawa, *et al.*, 1986) G6PD (Misumi, *et al.*, 1982) is in the hexose monophosphate pathway, the only NADPH-generation process in mature erythrocytes, which lack the citric acid cycle. Deficiency of this enzyme in erythrocytes causes various forms of illnesses such as favism, anemia, chronic nonspherocytic hemolytic anemia, drug-sensitive hemolytic anemia, primaquine sensitivity and jaundice in newborns (Beutler, *et al.*, 1968).

By virtue of fact that G6PD is found in all cells, functional and structural studies have revealed properties of this housekeeping gene (Luzzatto, 2006). G6PD expression level is regulated by hormonal and nutritional factors in only a few tissues. G6PD expression is regulated in liver and adipose tissue, and its activity depends on the rate of fatty acid biosynthesis (Greene, 1993). The G6PD gene region is one of the first regions of the human genome to be completely sequenced (Chen, *et al.*, 1996). The gene encoding G6PD is located near the telomeric region of the distal arm of the X chromosome (Pai, *et al.*, 1980, Szabo, *et al.*, 1984, Patterson, *et al.*, 1987) (band Xq28) and a valuable X-linked genetic marker for determination of X chromosome inactivations (Migeon, 1983). G6PD has various polymorphism sites at the G6PD locus like the colorblindness, Xg blood group and the hemophilia A locus and has close linkage at the X chromosome (Boyer & Graham, 1965, Adam, *et al.*, 1967). G6PD is one of a group of genes including fragile X, (Oberle, *et al.*, 1987) color vision (Motulsky, 1988, Filosa, *et al.*, 1993) hemophilia A (Boyer & Graham, 1965) clasped-thumb mental retardation syndrome (MASA), (Macias, *et al.*, 1992) and dyskeratosis congenita (Arngrimsson, *et al.*, 1993) existing on the distal long arm of the X chromosome.

The X-linkage of the G6PD gene has important implications. This linkage is very stable and linkage with other group locuses is similar in all mammals (Luzzatto & Battistuzzi, 1985, Group, 1989, Luzzatto, 1989, Beutler, 1990). In mice, X-linkage of G6PD was shown by Epstein (Epstein, 1969). Epstein concluded that the G6PD gene is X-linked in the mouse; its synthesis occurs in the oocyte and is dosage-dependent. G6PD is a sex-linked and very polymorphic gene in populations in which males have only one allele (hemizygous) and females have two G6PD alleles. Thus, females can be either normal or deficient (homozygous), or intermediate (heterozygous) phenotypes, whereas males can be either normal or G6PD-deficient phenotype (Luzzatto, 2006). The frequency of the deficient phenotype is higher in males than females owing to males being hemizygous, in which one allele of the gene expresses the deficient phenotype; to arise in females, G6PD-deficiency needs two deficient alleles. However, hemizygous deficient males and homozygous express the same degree of enzyme deficiency level. Since deactivation of one X-chromosome in embryological development in heterozygous females have two populations of red cells (G6PD-normal and G6PD-deficient), with a wide range of total G6PD enzyme activity depending on the relative proportions. If one of the alleles contains deficiency, as a result of random deactivation of X-chromosomes, about half of the cells will be normal and the other half will be deficient, although there is a wide range of variation around that average (Nance, 1964, Rinaldi, *et al.*, 1976). For this reason, total G6PD activity in heterozygous females can show variability between near-normal to near-deficient (Luzzatto & Battistuzzi, 1985, Segel, 2000). Deactivation of X-chromosome actualizes at random. Correspondingly, binomial distribution would be expected in deficiency level; the extent of this distribution depends on X-inactivation time in embryonic tissue and the number of cells in the embryo. Furthermore, random deactivation of one X-chromosome engenders genetic mosaics in heterozygous females (Luzzatto, 2006). As a result, G6PD mutations show the typical mendelian X-linked inheritance (Adam, 1961), severe G6PD deficiency is much more common in males than in females, and X-chromosome inactivation in heterozygous females for two different G6PD alleles indicate somatic cell mosaicism (Beutler, *et al.*, 1962, Gall, *et al.*, 1965).

The total length of the gene is about 18.5 kb on the X chromosome (xq28) and contains 13 exons. Exon 13 is about 800 nucleotides long and contains the translation stop codon (Nagel & Roth, 1989, Greene, 1993). The protein-coding region is divided into 12 segments, ranging in size from 12 to 236 bp (Martini, *et al.*, 1986). Exon and intron numbers and the exon sizes and sequences are conserved in higher eukaryotes (Nagel & Roth, 1989, Greene, 1993). The first exon contains no coding sequence and intron 2 between exons 2 and 3 is extraordinarily long, extending for 9,857 bp. The function of this long intron is unknown; it may be important for transcription or processing because compressed versions of the G6PD gene still have this largest intron in some species (Chen, *et al.*, 1991, Mason, *et al.*, 1995).

The sequence of the whole G6PD gene is known (Chen, *et al.*, 1991). G6PD sequence analogy between humans and mice or rats is 87%. The analogy between the mouse and rat cDNA sequences is greater than humans with 93% similarity. Most of the sequence dissimilarity is in the 3′- UTR region, which has 600 nucleotides on average and contains a single polyA site (Nagel & Roth, 1989, Greene, 1993). G6PD gene promoter is embedded in a CpG island that starts about 680 nucleotides upstream of the transcription initiation site, extending about 1,050 nucleotides downstream of the initiation site, and ends at the start of the first intron (Chen, *et al.*, 1991). CpG island is conserved between some species (Martini, *et al.*, 1986, Toniolo, *et al.*, 1991), and has highly enriched guanine and cytosine residues, like characteristically in other housekeeping genes and this island appears to be preserved between humans and mice (Toniolo, *et al.*, 1991).

The promoter of the G6PD gene contains a TATA-like, TTAAAT sequence, and a great number of stimulatory protein 1 (Sp1) elements (Philippe, *et al.*, 1994, Rank, *et al.*, 1994, Franze, *et al.*, 1998, Hodge, *et al.*, 1998). These Sp1-binding sites are essential for promoter activity (Philippe, *et al.*, 1994). Deletion analysis has uncovered that the "essential" segment of the promoter is only about 150 bp (Ursini, *et al.*, 1990).

The transcribed region from the initiation site to the poly(A) addition site covers 15,860 bp. (Chen, *et al.*, 1991). The major 5′-end of mature G6PD mRNA in several cell lines is located

177 bp upstream of the translation-initiating codon (Martini, *et al.*, 1986). The G6PD activity and mRNA quantity differ between tissues (Nagel & Roth, 1989, Greene, 1993). S1 nuclease and primer extension analyses of mouse G6PD mRNA indicate that when the transcriptional start site regulated with lipogenesis in liver and adipose tissue, in kidney G6PD is expressed constitutively (Ho, *et al.*, 1988); this quantity potentially depends on oxidative stress, tissue specific differences and reductive biosynthesis reactions (Nagel & Roth, 1989, Greene, 1993). Some different mRNA forms of G6PD mRNA have been found, but their functions are completely unknown. The alternatively spliced form has been documented (Hirono & Beutler, 1988, Hirono & Beutler, 1989, Cappellini, *et al.*, 1993), but this mRNA frame contains 138 nucleotides (Mason, *et al.*, 1988, Persico, *et al.*, 1989, Bautista, *et al.*, 1992, Tang, *et al.*, 1994). Some researchers (Kanno, *et al.*, 1989) suggested that in reality, G6PD translation product made from two separate mRNAs as a result of study to be based on an artifact (Henikoff & Smith, 1989, Beutler, *et al.*, 1990, Mason, *et al.*, 1990, Yoshida & Kan, 1990).

Up to 450 G6PD variants have been identified depending on the enzyme kinetics, physicochemical characteristics, and other parameters (Luzzatto & Battistuzzi, 1985, Chen, *et al.*, 1991). Nearly 300 variants of these have been confirmed by the World Health Organization (1967). Point mutations and small deletions trigger defects in the enzyme structure. These structural defects cause altered activity, instability of the enzyme or decreased affinity of G6PD for its substrates (Luzzatto, 2006).

3. Structure of G6PD and enzymatic properties

G6PD is a typical cytoplasmic, housekeeping enzyme and has been found in all cells from liver to kidney and organisms, from prokaryotes to yeasts, to protozoa, to plants and animals (Luzzatto & Battistuzzi, 1985, Antonenkov, 1989, Glader, 1999, Notaro, *et al.*, 2000). Inactive form of G6PD is a monomer with 515 amino acids and has a molecular weight of over 59 kDa (Rattazzi, 1968). The primary structure of the G6PD enzyme in humans has been determined from the sequence of full-length cDNA clones (Persico, *et al.*, 1986). Furthermore, the tertiary structure of the enzyme has been determined (Au, *et al.*, 1999). Dimer structure of the two subunits in the enzyme are symmetrically located across a complex interface of β-sheets (Au, *et al.*, 1999).

Activation of the enzyme requires $NADP^+$ tightly binding to dimer or tetramer formation of enzyme. G6PD catalyses the first step of the oxidative pentose phosphate pathway and controls reaction velocity (Wrigley, *et al.*, 1972). In this first step, while G6PD catalyses the conversion of glucose 6-phosphate (G6P) to 6-phosphogluconolactone, at the same time it reduces NADP to NADPH (Au, *et al.*, 1999, Turner, 2000). Human G6PD has no activity with nicotinamide adenine dinucleotide (NAD) as coenzyme. Also, G6P is very specific for its substrate compared to other hexose phosphates (e.g., galactose 6-phosphate or mannose 6-phosphate) (Luzzatto & Battistuzzi, 1985, Glader, 1999). The G6P binding site is nearby lysine 205 in tertiary structure of the enzyme, and this amino acid has a critical role in electron transfer (Bautista, *et al.*, 1995). The NADP binding site is located nearby 38 to 43 amino acids; this region constitutes the N-terminus in tertiary structure encoded in exon 3. This site is important for stability of G6PD (Au, *et al.*, 1999). As an inhibitory effect, one of the products of G6PD reaction NADPH is an effective inhibitor (Luzzatto, 1967). Increase in NADP and decrease in NADPH as a result of whichever oxidative event in cells effect prepotently to increase G6PD activity (Luzzatto & Testa, 1978). Consequently, G6PD is the most important enzyme in biosynthesis reactions owing to enzyme property as NADPH reducer in its critical role in the cytoplasm (Koehler & Van Noorden, 2003).

4. The effect of G6PD on erythrocyte metabolism

4.1. Erythrocytes

Erythrocytes, which contain hemoglobin, are blood cells that perform the transfer of oxygen and carbon dioxide between tissues. G6PD is an important enzyme that performs vital functions within all cells of the body (Greene, 1993). The quantity of active G6PD decreases during the life of an erythrocyte and also the older erythrocytes become vulnerable to oxidative stress. G6PD, an enzyme in the oxidative pentose phosphate pathway, converts the nicotinamide adenine dinucleotide phosphate ($NADP^+$) into its reduced form NADPH. It is necessary for the protection against oxidative stress in erythrocytes. The cells cannot eliminate this stress, which causes hemolysis of erythrocytes. Because H_2O_2 and other reactive oxygen species cannot be reduced, oxidation of hemoglobin to methemoglobin and membrane damage occur (Ruwende & Hill, 1998, Peters & Van Noorden, 2009).

4.2. The importance of pentose phosphate pathway for erythrocytes

G6PD is the key enzyme in the oxidative pentose phosphate pathway. The first step of the pentose phosphate pathway is catalyzed by G6PD. In this step, $NADP^+$ is reduced to NADPH, and ribulose-5-phosphate, a precursor of DNA, RNA, and ATP, emerge from G6P (Turner, 2000). The most important reducing agent in the cytoplasm of cells is NADPH (Koehler & Van Noorden, 2003). The second enzymatic step in this pathway is NADPH production as a consequence of reactions that reduce oxidized glutathione (GSSG) to reduced glutathione (GSH). The only defense against oxidant stress in the red blood cell (RBC) is GSH production (Friedman, 1979, Group, 1989, Peters & Van Noorden, 2009). In unstressed, normal erythrocytes, the G6PD activity is only about 2% of total capacity (Group, 1989, Peters & Van Noorden, 2009). The pentose phosphate pathway's main function is the generation of reducing capacity through the production of NADPH and ultimately, GSH. This is essential for cell survival and is available in the erythrocyte for generating reducing capacity (Greene, 1993).

4.3. Classification of Glucose-6-Phosphate Dehydrogenase variants

More than 400 variants of G6PD have been distinguished based on their biochemical characteristics, enzyme kinetics, physicochemical characteristics, and other parameters (Luzzatto & Battistuzzi, 1985, Chen, *et al.*, 1991, Greene, 1993). G6PD B^+ is the most commonly found enzyme type and it is used as a standard for normal enzyme activity and electrophoretic mobility. For identification of other variants, G6PD B^+ is used. The rate at which $NADP^+$ is reduced by glucose-6-phosphate with G6PD B^+ as the catalyst is the standard for activity. Based on this, enzyme activity relative to G6PD B^+ variants are classified as fast, normal, and slow in terms of electrophoretic mobility and as Classes I—V (Luzzatto, 1989, Beutler, 1990, Greene, 1993, Segel, 2000, Betke K, Brewer GJ, Kirkman HN, Luzzatto L,Motulsky AC, Ramot B, and Siniscalco M 1967). There are 5 classes for these variants. Class I includes chronic nonspherocytic hemolytic anemia with a severe enzyme deficiency (e.g., G6PD Minnesota, G6PD Tokyo, G6PD Campinas). Class II variants have severe enzyme deficiency without chronic nonspherocytic hemolytic anemia (e.g., G6PD mediterrian, G6PD Canton, G6PD Union, G6PD Kaiping). Class III variants includes medium or mild enzyme deficiency, with the activity at 10-60% of G6PD B^+ (e.g., G6PD A^-). Class IV variants have a weak or no enzyme deficiency. The activity is 60-100 % of G6PD B^+ (e.g., G6PD A^+). Class V variants have increased enzyme activity (e.g., G6PD Hektoen) (Beutler, 1994, Segel, 2000).

4.3.1. Some Important G6PD Variants

4.3.1.1 G6PD A^+ is the most widely seen variant worldwide and also the first variant in which the nucleotide mutation and amino acid substitution were determined (Beutler, 1990). This Class IV variant has 90% of the enzyme activity of G6PD B^+ (Luzzatto, 1989). This variant also called for the African variant cause widely seen in Africa; 20-40% of African men and 20% of African American men have this variant. It is faster than G6PD B^+ electrophoretically and it does not cause hemolysis (Beutler, 1989). G6PD A^+ derives from a single amino acid substitution of aspartic acid for asparagine at amino acid number 126, and this was the result of an adenine to guanine mutation at nucleotide number 376.

4.3.1.2 G6PD A^- is a Class II variant that has 10 to 20% of the activity of G6PD B^+ and the same electrophoretically mobility as G6PD A^+ (Luzzatto, 1989); 11% of African American men have this variant. Its half-life is 13 days. Three types of mutations have arisen with molecular studies. The most common mutation being at nucleotide number 202 is a result of a guanine to adenine mutation at amino acid number 68 substitution of valine to methionine (Beutler, 1989, Luzzatto, 1989, Beutler, 1990). The second one occurs at nucleotide number 680 as a result of a guanine to thymine mutation at amino acid number 227 substitution of arginine to leucine. And the third mutation occurs at the nucleotide number 968, as a result of a thymine to cytosine mutation at amino acid number 323 substitution of leucine to proline (Beutler, 1989). G6PD A^+ and G6PD A^- variants are defined as unique to Africa, but they can also be seen in Caucasian populations from Italy, Spain, Southeast Asia, Middle East and South America (Beutler, 1990).

4.3.1.3 G6PD Mediterranean is a widely seen variant in the Mediterranean region and Middle East. In addition, it is seen in the Indian subcontinent and other regions of the Americas (Beutler, 1991). This Class II variant has less than 10% of the enzyme activity of G6PD B^+ and the electrophoretical mobility is similar with G6PD B^+ (Luzzatto, 1989). Its half-life is only 8 days and DNA analysis identified two different point mutations. The first mutation is a result of a cytosine to thymine mutation at nucleotide number 563, at amino acid number 188 substitution of serine to phenylalanine (Vulliamy, *et al.*, 1988). Second is a silent mutation result of a cytosine to thymine mutation at nucleotide number 1311 (Beutler, 1990). There are many similar Class II variants in the Mediterranean region (Cagliari, Sassari, El Fayoum), South Asia (Hong Kong, Canton, Mahidol), and elsewhere. Most of these emerge as a consequence of point mutations resulting in single amino acid mutations that have variable effects on activity and electrophoretic mobility (Luzzatto & Battistuzzi, 1985, Luzzatto, 1989, Beutler, 1990, Beutler, 1991, Beutler, 1992).

5. Clinical tables on G6PD deficiency

Depending on G6PD enzyme deficiency are: Hemolytic Anemia (Drug-induced hemolysis), Diabetes mellitus-induced hemolysis and Infection-induced hemolysis; chronic nonspherocytic anemia, Favism and Neonatal jaundice.

5.1. Hemolytic anemia

5.1.1 Mechanism of hemolysis. In some people, for example, the Mediterranean-type, G6PD deficiency from drug intake occurs, although not a permanent hemolytic condition. In erythrocytes, NADPH cannot form with G6PD deficiency and unformed NADPH creates a deficiency in conversion of the oxidized form of glutathione (GSSG), to its reduced form (GSH) (Lachant, *et al.*, 1984, Beutler, 1994). There is normally plenty of GSH in erythrocytes and it protects the cell from oxidizing agents. If G6PD is deficient, hemoglobin is oxidized by oxidative substances to be eliminated and it returns methemoglobin that cannot function normally. Also, hemoglobin precipitates with denaturation in the cytoplasm forms Heinz bodies. These structures attach to the membrane with disulfide bonds and disrupt its normal structure. The erythrocytes that contain Heinz bodies in their cytoplasm are sequestered by macrophages in the spleen and removed from the circulation. G6PD deficiency hemolysis occurs like that in the extravascular compartment and also occurs again as a result of membrane defects (Alving, *et al.*, 1956). Thus, drug-induced hemolysis is the first and best-known morbid effect of G6PD deficiency. After a 1- or 2-day delay in such drug administration, a fall in the hemoglobin (Hb) concentration occurs.

The red blood cell (RBC) membrane was adhered to by Heinz bodies, which are particles of denatured protein. These appear in the early stages of drug administration and disappear as hemolysis progresses. Hemolysis usually occurs in blood vessels and hemoglobinuria follows. The increase of reticulocytes emerges in response to this situation and the hemoglobin level begins to increase again within 8-10 days (Beutler, 1994). In severe hemolysis, the patient may complain of back and stomach pain and the urine turns dark. The hemolytic anemia is self-limited when G6PD deficiency is relatively mild because only the older RBCs are destroyed and the young RBCs have normal or nearly-normal enzyme activity (Beutler, 1994).

Table 1 lists the drugs and chemicals that cause clinically significant hemolytic anemia.

5.2. Diabetes mellitus-induced hemolysis. Hemolysis in G6PD deficiency individuals might initiate diabetic ketoacidosis. This situation is not exactly accepted. However, hemolysis formation has been reported when blood glucose levels are normal in diabetic individuals (Beutler, 1994). It has also has been reported that hypoglycemia might precipitate hemolysis in patients with G6PD deficiency (Beutler, 1994).

5.3. Infection-induced hemolysis. Infections are probably the most common cause of hemolysis in people with G6PD deficiency. There are numerous reports about the importance of infection in causing hemolytic anemia. A large number of bacterial, viral and rickettsial infections have been reported as predisposing factors. Infectious hepatitis (hepatitis A), pneumonia and typhoid fever are known to trigger hemolysis. Involving the upper respiratory tract and gastrointestinal system, viral infections have been reported to cause a more severe hemolysis (Luzzatto, 2001). The mechanism of infection-induced hemolysis is not clear, but it is thought to be that during the infection, superoxide anion and H_2O_2 production by macrophages causes the hemolysis (Glader, 1999, Luzzatto, 2001).

5.4. Chronic nonspherocytic anemia

Class I G6PD variants, such as the absence of precipitating factors in the occurrence of excessive hemolytic anemia, lower still further the remaining enzyme activity. This is observed in people with chronic hemolytic anemia and oxidative stress, even if unstable conditions occur as a result of insufficient enzyme activity in erythrocytes. Granulocyte dysfunction is seen in some cases. In these cases, more severe hemolysis is due to increased susceptibility to infection (Beutler, 1994, Luzzatto, 2001).

5.5. Favism is an illness that occurs in G6PD deficiency individuals with acute hemolysis by eating raw beans (Vicia fabu). Wet, dry or frozen fava bean ingestion of grains, even if the mother eats fava beans can cause hemolysis in newborn infants through breast milk may occur (Luzzatto, 2001). Individuals with G6PD deficiency hemolytic effect caused by the beans contained many glycosides that are toxic due to the visin and konvisin (Beutler, 1994, Akhter, *et al.*, 2011). In addition, β-glucosides in bean seeds, maturity stage of fava beans attain very high amounts causing a severe course of hemolytic crisis (Katz & Schall, 1979, Greene, 1993, Beutler, 1994). Often, in the G6PD Mediterranean variant, acute and a very severe hemolytic crisis are seen due to fava bean ingestion, even capable of causing death (Fairbanks, 1999, Luzzatto, 2001). In favism, damage in erythrocytes is similar to oxidative damage of drugs. Fava beans include visin, konvisin, ascorbic acid and L-Dopa, which have oxidative properties. The most commonly cited konvisin and visin glycosides during digestion fava beans by β-glycosidase or acid hydrolysis demolished to the active agents, which are converted to "divisine" and "izouramil." Divisine and izouramilin reduce the level of the GSH and NADPH in vitro conditions and damage the cell membrane by the formation of cross-connection with Heinz bodies; it also has been shown to inhibit Ca^{2+}-ATPase and catalase (Arese & De Flora, 1990, Beutler, 1994, Gaetani, *et al.*, 1996, Luzzatto, 2001). 24-48 hours after ingesting foods like fava beans, characteristic symptoms occur in the form of pallor, jaundice and hemoglobinuria (Ninfali, *et al.*, 2000). In addition, jaundice, headache, backache, nausea, fever, and chills are all signs of acute hemolysis (Tyulina, *et al.*, 2000). Favism is most common seen in children between the ages of 2-5, and is also 2-3 times more common in boys than in girls (Luzzatto, 2001). Clinical signs of favism begin earlier and are more severe than drug-induced hemolytic crises. Rarely, as a result of pollen of fava inhalation, hemolysis may occur within hours (Beutler, 1994). While each favism patient must have G6PD deficiency, hemolytic reactions may not occur after ingestion of fava beans in each person with G6PD deficiency. Each individual with G6PD deficiency of the same family could not be affected in the same way when they eat fava bean. On the other hand, changes are observed in the same person at different times. Genetic variations between individuals, differences of fava bean active metabolites may be responsible for these variable characteristics (Meloni, *et al.*, 1983, Group, 1989, Luzzatto, 2001).

5.6. Neonatal jaundice

One of the most threatening consequences of G6PD deficiency is neonatal jaundice (Beutler, 1994). Jaundice in babies with G6PD enzyme deficiency could be mild or severe enough to cause kernicterus, a spastic type of cerebral palsy, and may even cause death (Luzzatto, 1993). In addition, infants with G6PD deficiency, hyperbilirubinemia is more remarkable than anemia. It facilitates this because of the inadequate physiological conjugation in liver in the neonatal period (Moskaug, *et al.*, 2004). G6PD A^-, G6PD mediterrian, G6PD Canton variants are known as types that cause kernicterus and hyperbilirubinemia (Luzzatto, 2001). Clinically, the jaundice, the level of G6PD in the normal physiological jaundice in newborns occur on the same days, or a little earlier, but it takes as long as 2-3 weeks (Tan, 1981, Luzzatto, 2001). There are two major differences between jaundice due to incompatibility of blood groups and jaundice due to G6PD deficiency. First, the presence of jaundice in G6PD deficiency is very rare during childbirth and usually it begins in the second or third day. Second, according to anemia, jaundice is more pronounced and it is encountered with severe anemia very rarely in the absence of the enzyme (Luzzatto, 1993, Luzzatto, 2001).

6. Malaria and glucose-6 phosphate dehydrogenase deficiency

As we mentioned above, there is a strong relationship between malaria and G6PD deficiency diseases. In several epidemiological studies, it was shown that distribution of malaria was nearly the same with distribution of G6PD deficiency (Motulsky, 1961, Siniscalco & Bernini, 1961, Ganczakowski, *et al.*, 1995). This situation reveals two important facts. One of them is that G6PD deficiency provides great protection from malaria, especially for falciparum infections. On the other hand, using antimalarial drugs can cause life-threatening hemolytic anemia in patients with G6PD deficiency. Hence, malaria patients should be screened for their tendency to G6PD deficiency before their treatment with antimalarial drugs. In this part, we will first summarize the importance of malaria for the world. Then, we will explore the relationship between these two diseases in detail.

As it is known, malaria is a parasitic disease that threatens 300-500 million people all over the world. Malaria can be defined as the most deadly vector-borne disease in the world

(Myrvang & Godal, 2000). It is widespread in tropical and subtropical regions of Asia, Africa and the American continents. Each year, malaria leads to deaths of millions of people all around the world and a large percentage of deaths are seen in Sub-Saharan regions of Africa. The causative agents of malaria are the Plasmodium parasites, which are transmitted to humans by the bites of infected mosquitoes. If patients are not treated with antimalarial drugs, malaria can easily lead to death. Five different types of Plasmodium species—*P. falciparum, P. vivax, P. ovalae, P. malariae* and *P.knowlesi*—lead to this disease (Wernsdorfer & McGregor, 1988, Sutherland, *et al.*, 2010).

Plasmodium falciparum (P. falciparum) is the most serious and life-threatening form of the disease. 80% of death cases are reported from patients that have been infected with *P. falciparum*. It was also demonstrated that resistance has been developed in this type of parasites against current antimalarial drugs. It is generally seen in Africa, specifically in sub-Saharan regions. Interestingly, falciparum-derived malaria cases have been recently reported in various parts of the world where this parasite species was believed to be completely eradicated.

Plasmodium vivax (P. vivax) constitutes a milder form of the disease. Vivax infections generally do not cause death. However, individuals that suffer from vivax infection also need to be treated. Among all *Plasmodium* species, *P. vivax* is the one that shows the broadest geographic distribution worldwide. Causative agents for 60% of malaria infections are reported as *P. vivax* infections in India. This parasite has a liver stage and can remain in the body for years without causing sickness. If the patient is not treated, the liver stage may re-activate and cause relapses—malaria attacks—after months, or even years without symptoms.

Plasmodium ovale (P. ovale) is known as one of the other milder form of the disease. Like *P. vivax*, it generally does not commonly lead to death. Nevertheless, infected individuals require medical therapy. This parasite, similar to *P.vivax*, can live in the liver for long periods without causing symptoms. Therefore, if it is not treated, reactivation of parasites can be observed in the liver and this leads to relapse of the disease

Plasmodium malariae (P. malariae) is also another milder form of the disease. It does not commonly lead to death. However, it still requires treatment. This type of *Plasmodium* parasites are reported to stay in the blood of some individuals for several decades.

Plasmodium knowlesi (P. knowlesi) causes malaria in macaques, but can also infect humans (Mendis, *et al.*, 2001, Singh, *et al.*, 2004, Mueller, *et al.*, 2007).

When life cycles of Plasmodium parasites are investigated, it is seen that the parasites multiply in the liver of the human body, and then infect erythrocytes. As we mentioned before, *Plasmodium* parasites enter the human body when bitten by an infective female mosquito, which is called Anopheles. These mosquitoes become infected with malaria when they take *Plasmodium*-containing blood from an infected person. Approximately one week later, these parasites mix with the mosquito's saliva when the mosquito takes its next blood meal from another person and this individual is injected with *Plasmodium* parasites when they are being bitten (Bozdech, *et al.*, 2003).

Multiplication of the parasites within erythrocytes enhances the severity of the disease and cause symptoms such as anemia, fever, chills, nausea, flu-like illness, and, in severe cases, coma, and death. Treatment of this disease can be achieved by using antimalarial drugs. Primaquine, which is the most common antimalarial drug, can be used as a primary prophylactic because it prevents primary parasitemia of *Plasmodium* species by destroying these parasites in the liver before they reach the bloodstream and cause disease (Yazdani, *et al.*, 2006).

As we pointed out before, according to epidemiological studies, the prevalence of malaria deeply relates to glucose-6 phosphate dehydrogenase (G6PD) enzyme deficiency. In these studies, it was demonstrated that 66 of 77 genetic variants that have reached polymorphic frequencies were seen in populations living in tropical and subtropical areas where malaria was endemic. On the other hand, this genetic diversity does not occur in populations living in non-endemic regions of the world for malaria, indicating that high polymorphism is the indicator of G6PD deficiency.

When investigated in terms of cellular biology, we can see that Plasmodium parasite that causes malaria use erythrocytes as host cells. Erythrocytes are also the most affected cells from G6PD deficiency. This situation also suggests the relationship between the two diseases. In several studies, it was demonstrated that G6PD deficiency provides a protection against malaria infections. In one of the early studies, it was indicated that *P. falciparum* and *P. vivax* parasites preferred to invade younger erythrocytes, which possessed high levels of G6PD enzyme. Since enzyme levels are diminished in older erythrocytes, parasites do not prefer to invade these erythrocytes. These studies suggested the protective effect of G6PD deficiency from parasitemia (Allison & Clyde, 1961, Kruatrachue, *et al.*, 1962). In the recent past, Ruwando et al. also carried out a case-control study on more than 2,000 African children and exhibited that risk of contracting malaria in patients that have the African form of G6PD deficiency decreased at a rate of 46 to 58%. In this study, it was suggested that the selective advantage of resistance to malaria was counterbalanced with selective disadvantageous results of G6PD deficiency, and this stopped the rise of malaria frequencies in endemic regions (Ruwende, *et al.*, 1995). In another study, Ninokata et al. (2006) investigated 345 healthy adults for G6PD deficiency on Phuket Island, which had been determined to be a malaria-endemic region and found out that 10% of these individuals had G6PD deficiency. Interestingly, it was observed that none of the individuals had molecular evidence of malaria infection. According to this study, researchers postulated that G6PD deficiency provided an advantageous genetic trait against malaria (Ninokata, *et al.*, 2006).

The exact mechanism of this protection is still unknown. However there are two postulated explanations. According to the first suggestion, it was found that parasites that cause malaria can only survive in conditions with low oxygen levels (Clark, *et al.*, 1989). This demonstrates that these parasites are very susceptible to oxidative stress. It is known that in the pentose phosphate pathway of erythrocytes, glucose-6 phosphate dehydrogenase (G6PD) enzyme has an important role in production of NADPH and GSH. This is the only mechanism for erythrocytes to survive. GSH that is produced by NADP+ reduction reacts with H_2O_2 and reduce it to H_2O. This prevents the generation of oxidative stress within red blood cells. Since oxidative stress is the most important factor for the disruption of red blood cells, these cells are protected from this effect. However, in G6PD deficient erythrocytes, G6PD activity is significantly reduced. In G6PD A (-) variant, enzyme activity level reduces to 10 or 20% of normal levels, while enzyme activity completely disappears in G6PD variant. Therefore, oxidative stress can be induced in erythrocytes whose G6PD enzymes are deficient. In this situation, GSH is not produced and H_2O_2 is not reduced to H_2O and leads to oxidative stress. Hence, it is thought that since malaria parasites are susceptible to oxidative stress, they do not live within the erythrocytes where their maturation occurs (Toncheva & Tzoneva, 1985, Greene, 1993). Additionally, during oxidative stress, the loss of potassium from the cell and from the parasite can cause the death of the parasite (Friedman & Trager, 1981).

According to the second suggestion, *Plasmodium* parasites oxidize NADPH and reduce the level of reduced glutathione (GSH) in erythrocytes. In the situation of G6PD deficiency, this effect becomes more severe and induces oxidative-induced damage within erythrocytes. Moreover, *Plasmodium* parasites break down hemoglobin and release toxic components like iron and these substances lead to hemolysis. Hence, the development rates of *Plasmodium* parasites are diminished. Additionally, red blood cells that are affected by oxidative stress and are damaged are eliminated by the immune system via phagocytosis. This elimination decreases the growth of parasites much more since it occurs during an early ring-stage of parasites' maturation. Therefore, all of these data indicate that G6PD deficiency can provide protection against malaria infections. Considering the relationship between G6PD deficiency and plasmodium infections, research has aimed to develop antimalarial drugs that decrease the level of GSH within erythrocytes and then produce hydrogen peroxide and the other free radical species in order to enhance the inhibition of Plasmodium species (Mehta, *et al.*, 2000, Fortin, *et al.*, 2002, Kwiatkowski, 2005, Prchal & Gregg, 2005).

Primaquine is the only effective antimalarial drug that provides inhibition of persistent liver stages of *P. falciparum, P. vivax,* and *P. ovalae* parasites that lead to relapses of malaria (Phompradit, *et al.*, 2011).

However, as we initially mentioned, using primaquine in order to prevent the relapse of malaria can be very dangerous for G6PD deficiency patients since its usage results in very severe hemolysis. In all G6PD variants, activity levels of the enzyme have been diminished and this partially prevents the defense of erythrocytes against oxidative attack. However, when primaquine is administered, its metabolites lead to more severe hemolysis than oxidative damage by inducing oxyhemoglobin generation, GSH depletion and stimulation of the hexose monophosphate pathway. Moreover, primaquine can also induce the generation of Heinz bodies, which are insoluble aggregates that attach to the surfaces of erythrocytes. The most probable mechanism of primaquine-induced hemolysis is the generation of oxyhemoglobin, which forms hydrogen peroxide. Since G6PD enzyme level is low in G6PD-deficient erythrocytes, these peroxides accumulate and lead to denaturation of hemoglobin. Peroxides also generate Heinz bodies that attach to cell membranes of red blood cells. Hemolysis occurs when damaged erythrocytes pass through the spleen. In each pass, red blood cells lose a portion of the cell membrane. After additional passes, membranes of cells completely lose their competency (Beutler, *et al.*, 1955, Bolchoz, *et al.*, 2002, Beutler & Duparc, 2007).

These conditions reach life-threatening scenarios for all G6PD deficiency patients with different genetic variants. Hence, individuals that are required to use antimalarial drugs should be screened very carefully for their tendency to have G6PD deficiency. For effective control and treatment, either a reliable test for detecting G6PD deficiency or an anti-malarial drug that can be safely given to G6PD deficiency patients is required.

7. Detection methods of G6PD deficiency

Currently, primaquine, which causes hemolysis in G6PD-deficient patients, is the only radical cure of *Plasmodium vivax* infections (Burgoine, *et al.*, 2010). Therefore, screening to detect G6PD deficiency is very important. Various tests can be used for the detection of G6PD deficiency, which are based on the assessment of the NADPH production capacity of G6PD. The most frequently used tests that measure NADPH production are the fluorescent spot test, cytochemical assay and spectrophotometric assay. However, fluorescent spot test and the spectrophotometric assay are not reliable for the detection of heterozygous females. In addition, DNA analysis can be done to detect G6PD deficiency for the homozygous, hemizygous, and heterozygous-deficient patients. However, we have to design primers for all mutations (Peters & Van Noorden, 2009).

7.1. Fluorescent spot test

Fluorescence is a form of luminescence that uses the physical change of emission of light upon excitation of molecules. There are various different types of luminescence, classified depending on the style of excitation: chemo-luminescence (ending in a chemical reaction) photo-luminescence (fluorescence, phosphorescence and delayed fluorescence), bio-luminescence (via a living organism) and others (Bernard, 2002).

Nicotinamide Adenine Dinucleotide Phosphate (NADPH) is the reduced form of NADP, with absorption maximum at 340 nm and a maximum emission at 460 nm. NADPH concentrations have been studied in great detail using optical methods. A parameter for direct measurements of the G6PD activity is the fluorescence of NADPH. When G6PD shows enough functional activity in erythrocytes, two molecules of $NADP^+$ are reduced to NADPH. After the addition of glucose 6-phosphate and $NADP^+$, blood spot fluoresces at *340 nm* if NADPH is produced (Beutler & Baluda, 1966).

7.2. Spectrophotometric assay

Spectrophotometric methods are greatly used in biological sciences for quantitative and qualitative measurements due to the fact that these methods do not break down the molecules analyzed and enable us to assay small quantities of matter fundamentally (Lehninger, 2000). Spectrophotometric techniques allow detection of the concentration of a solution by evaluating its absorbance of a specific wavelength by way of a spectrophotometer, which produces light at a chosen wavelength and passes it directly through the sample. Because every molecule have a specific absorption spectrum, we can recognize and characterize its properties or detect its current concentration in the presence of other compounds (Lehninger, 2000).

In the case of enzyme activity measurements, the assay solution contains some other compounds that are required for the reaction to occur. Other compounds in the reaction mix may absorb light at the same wavelength with the enzyme being analyzed. To eliminate the interference of other compounds, the absorbance of a sample solution is compared with blank solution, which is taken as the reference. The blank contains everything found in the sample solution except the substance to be assayed.

In the matter of protein (enzymatic activity or protein concentration) measurements, colorimetric methods are used. Colorimetric measurements are performed by way of quantitative assessment of a colored complex, which is mostly formed by the reaction of a colorless compound and a dye reagent. However, the compound that will be analyzed can be naturally colored and can be read directly spectrophotometrically.

Glucose-6-phosphate dehydrogenase catalyzes the first step in the pentose phosphate shunt, oxidizing glucose-6-phosphate (G-6-P) to 6-phosphogluconate (6-PG). The enzyme activity can be determined quantitatively by spectrophotometer assay method, which is based on the rate of NADPH production from NADP+ in G6PD-deficient patients (Kornberg, *et al.*, 1955, Lohr & Waller, 1974).

These reactions are illustrated below:

$$\text{Glucose-6-phosphate} + \text{NADP}^+ \xrightarrow{\text{G6PD}} \text{6- phosphogluconate} + \text{NADPH} + \text{H}^+$$

Nictotinamide adenine dinucleotide phosphate (NADP) is reduced by G6PD in the presence of G-6-P. The rate of formation of NADPH is proportional to the G6PD activity and is measured spectrophotometrically as in increase in absorbance at 340 nm. Production of a second molar equivalent of NADPH by erythrocyte 6-phosphogluconate dehydrogenase (6-PGDH) occurs according to the reaction:

$$\text{6- phosphogluconate} + \text{NADP}^+ \xrightarrow{\text{6PGD}} \text{Ribulose-5-Phosphate} + \text{NADPH} + \text{H}^+ + \text{CO}_2$$

This is prevented by use of maleimide, an inhibitor of 6-PGDH.

The Enzyme Commission of the International Union of Biochemistry recommends expressing this in international units (IU) and defines 1 IU as the amount of an enzyme that catalyzes the transformation of 1 micromole of substrate per minute under standard conditions of temperature, optimal pH, and optimal substrate concentration. Specific activity relates activity to total mass of protein to avoid bias through individual differences in weight (Bairoch, 1993). Therefore, G6PD activity was expressed as units (micromoles of NADP reduced per minute) per miligram of soluble protein at 37°C.

7.3. Cytochemical staining assay

The Cytochemical staining assay is based on the intracellular reduction of the tetranitro blue tetrazolium (TNBT) by the G6PD via exogenous electron carrier 1-methoxyphenazine methosulfate and TNBT is reduced to dark-colored water-insoluble formazan, which can be determined by light microscopy (Peters & Van Noorden, 2009).

7.4. *In vitro* primaquine-induced hemolysis methods

3 cc of venous blood anti-coagulated by 2% heparin solution (126 mM NaCl, 14 mM Na_2HPO_4, 1 mM KH_2PO_4, 13,2 mM glucose, pH 7.4) was collected from healthy and G6PD-deficient persons. The blood was washed three times with sterile heparin solution at 3000 rpm for 10 min. Erythrocytes were resuspended in PBS after that hematocrit was adjusted to 2%. This is the one of the most important steps for detection of *in vitro* primaquine-induced hemolysis. Primaquine solution was prepared in 0.1 M Tris buffer (pH 7.4). Primaquine concentration was used between 1 and 4 mM in experiments. Different concentrations of primaquine were added into tubes containing 2% erythrocytes that were prepared before. Tubes were then placed and rotated in a rotator tube for 2 hours at 37°C. The rotation speed was less than 2 rpm. This is another important step for detection of *in vitro* primaquine-induced hemolysis. After 2 hours, the supernatant was collected for the heme concentration, which was then determined spectrophotometrically. Hemoglobin released from erythrocyte induction of primaquine-induced hemolysis and compared with complete lysis (100% hemolysis, control group) obtained by adding 5 mM Tris-HCl (Fig. 1) (Allakhverdiev & Grinberg, 1981).

The principle of this method is based on conversion of Hemoglobin (Hb) to cyanmethemoglobin by the addition of KCN and ferricyanide, whose absorbance is measured spectrophotometrically as cyanmethemoglobin at 540 nm versus a standard solution. Supernatant of hemolyzed red blood was diluted four-fold (v/v) with distilled water. On the other hand, the control group was diluted twenty-fold (v/v) with distilled water. After that, 50 μL KCN (10% w/v) and 50 μL potassium ferricyanide (2% w/v) were directly mixed and the color was measured at 540 nm. The standard curve was constructed using the standard cyanmethemoglobin solutions in different concentrations (Bhaskaram, *et al.*, 2003).

This method demonstrated that the *in vitro* model of primaquine-induced hemolysis can be only maintained by using 2% hematocrit in physiological conditions. Primaquine leads to hemolysis at concentrations between 1 and 4 mM. Other factors that induce primaquine-derived hemolysis are exposure time, incubation temperature, drug

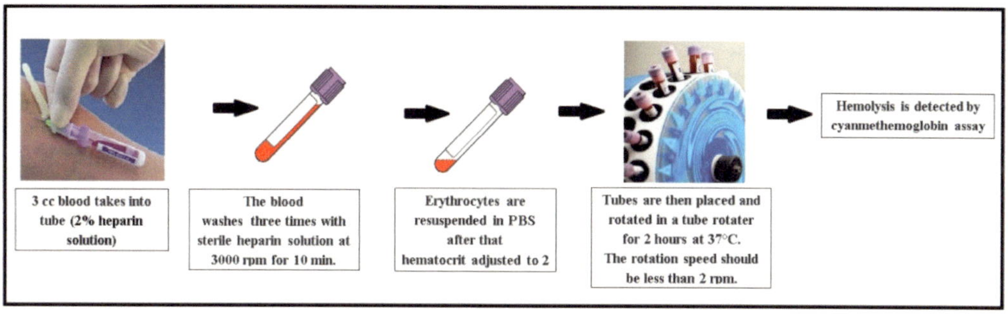

Figure 1. *In vitro* Primaquine-Induced hemolysis

concentration and amount of oxygen. Despite the fact that there are several methods in order to diagnose G6PD deficiency, these methods do not determine primaquine sensitivity in patients with G6PD deficiency every time. Therefore, lack of primaquine-based treatment by considering only G6PD deficiency can be very dangerous in terms of health of patients with malaria and the epidemiology of the disease. On the other hand, treatment of primaquine-sensitivive individuals with primaquine can cause death. Hence, in the Centers for Disease Control and Prevention (CDC) report (Hill, *et al.*, 2006), it was highlighted that there was an urgent need to develop new *in vitro* methods for determining hemolysis that indicate primaquine sensitivity before treatment of patients with this drug. By considering primaquine-induced hemolysis in patients with G6PD deficiency, it can be determined whether these patients may be treated with primaquine or not. The advantage of this method is that it can determine primaquine-induced hemolysis before treatment with primaquine and its capacity to determine G6PD deficiency.

8. Conclusion

This chapter has aimed to represent the relationship between G6PD deficiency and malaria and to suggest a sensitive method for detection of primaquine-induced hemolysis in patients with G6PD deficiency. As mentioned above, G6PD deficiency is the most common enzymopathologic disorder in humans and it affects 400 million people worldwide. In patients with G6PD deficiency, oxidative stress cannot be prevented since G6PD enzyme is the initial catalyst of the pentose phosphate pathway in erythrocytes that reduces the peroxides to H_2O. This situation leads to mild to severe hemolysis, changing depending on genetic variants of the disease. As we mentioned before, according to epidemiological studies, the prevalence of G6PD deficiency deeply relates to malaria. In these studies, it was demonstrated that 66 of 77 genetic variants, which have reached to polymorphic frequencies, were seen in populations living in tropical and subtropical places where malaria was endemic. On the other hand, this genetic diversity does not occur in populations living in non-endemic regions of the world for malaria, indicating that high polymorphism is the indicator of G6PD deficiency and distribution of malaria is nearly the same with distribution of G6PD deficiency. This situation reveals two important results.

One of them is that G6PD deficiency provides partial protection from malaria infections, especially for falciparum infections. In several studies, it was demonstrated that risk of contracting malaria in patients that have G6PD deficiency decreased at a rate of 46 to 58%. On the other hand, using antimalarial drugs can cause life-threatening hemolytic anemia in patients with G6PD deficiency. Since G6PD deficiency does not provide exact protection, these patients still have a risk of contracting malaria. However, using primaquine, which is the only radical cure of Plasmodium infections, can induce more severe hemolysis by generating oxyhemoglobin, GSH depletion and Heinz bodies and enhancing oxidative attack. This threatens the lives of patients with G6PD deficiency. Hence, patients with malaria should be screened for their tendency to G6PD deficiency before their treatment with antimalarial drugs. Common methods that are used for diagnosing G6PD deficiency are unreliable. Even worse is that it is very difficult to distinguish heterozygously-deficient patients from healthy individuals. Additionally, current methods cannot accurately indicate hemolysis, even though they give information about activity of the enzyme. Also, these methods do not determine primaquine sensitivity in patients with G6PD deficiency every time. However, the method that we developed provides the determination of primaquine sensitivity in patients with G6PD deficiency *in vitro* independently from the variants of G6PD deficiency. The principle of the method is based on the quantitative detection of hemolysis by incubation of erythrocytes obtained from G6PD-deficient patients with primaquine in low hematocrit while rotating the culture in a hybridization oven for 2 hours at 37°C. By considering primaquine-induced hemolysis in patients with G6PD deficiency, it can be determined whether these patients may be treated with primaquine or not. The advantages of this method are that it can determine the primaquine-sensitivity in patients with G6PD deficiency before treatment with primaquine. Using this method not only on G6PD deficiency patients but also on patients that suffer from other diseases that may cause primaquine-induced hemolysis constitutes another advantage of the method.

Author details

Adil M. Allahverdiyev, Malahat Bagirova,
Rabia Cakir Koc, Sezen Canim Ates, Serap Yesilkir Baydar,
Serkan Yaman, Emrah Sefik Abamor and Olga Nehir Oztel
Department of Bioengineering, Yildiz Technical University, Istanbul, Turkey

Serhat Elcicek
Department of Bioengineering, Yildiz Technical University, Istanbul, Turkey
Department of Bioengineering, Firat University, Elazig, Turkey

9. References

(1967) Standardization of procedures for the study of glucose-6-phosphate dehydrogenase. Report of a WHO Scientific Group. *World Health Organ Tech Rep Ser* Vol.366(No.:1-53.

(Betke K, Brewer GJ, Kirkman HN, Luzzatto L,Motulsky AC, Ramot B, and Siniscalco M 1967) Standardization of procedures for the study of glucose-6-phosphate dehydrogenase. Report of a WHO Scientific Group. *World Health Organ Tech Rep Ser* Vol.366(No.:1-53.

Adam A (1961) Linkage between deficiency of glucose-6-phosphate dehydrogenase and colour-blindness. *Nature* Vol.189(No.:686.

Adam A, Tippett P, Gavin J, Noades J, Sanger R & Race R (1967) The linkage relation of Xg to g-6-pd in Israelis: the evidence of a second series of families. *Annals of human genetics* Vol.30(No. 3):211-218.

Akhter N, Begum N & Ferdousi S (2011) Hematological Status in Neonatal Jaundice Patients and Its Relationship with G6PD Deficiency. *Journal of Bangladesh Society of Physiologist* Vol.6(No. 1):16-21.

Allakhverdiev AM & Grinberg LN (1981) Study of hemolysis in vitro for the purpose of detecting primaquine sensitivity. *Lab Delo* (No. 12):724-726.

Allison AC & Clyde DF (1961) Malaria in African children with deficient erythrocyte glucose-6-phosphate dehydrogenase. *Br Med J* Vol.1(No. 5236):1346-1349.

Alving AS, Carson PE, Flanagan CL & Ickes CE (1956) Enzymatic deficiency in primaquine-sensitive erythrocytes. *Science* Vol.124(No. 3220):484-485.

Antonenkov VD (1989) Dehydrogenases of the pentose phosphate pathway in rat liver peroxisomes. *Eur J Biochem* Vol.183(No. 1):75-82.

Arese P & De Flora A (1990) Pathophysiology of hemolysis in glucose-6-phosphate dehydrogenase deficiency. *Semin Hematol* Vol.27(No. 1):1-40.

Arngrimsson R, Dokal I, Luzzatto L & Connor J (1993) Dyskeratosis congenita: three additional families show linkage to a locus in Xq28. *Journal of medical genetics* Vol.30(No. 7):618-619.

Au SW, Naylor CE, Gover S, *et al.* (1999) Solution of the structure of tetrameric human glucose 6-phosphate dehydrogenase by molecular replacement. *Acta Crystallogr D Biol Crystallogr* Vol.55(No. Pt 4):826-834.

Bairoch A (1993) The ENZYME data bank. *Nucleic Acids Res* Vol.21(No. 13):3155-3156.

Bautista JM, Mason PJ & Luzzatto L (1992) Purification and properties of human glucose-6-phosphate dehydrogenase made in E. coli. *Biochim Biophys Acta* Vol.1119(No. 1):74-80.

Bautista JM, Mason PJ & Luzzatto L (1995) Human glucose-6-phosphate dehydrogenase. Lysine 205 is dispensable for substrate binding but essential for catalysis. *FEBS Lett* Vol.366(No. 1):61-64.

Bernard V (2002) Molecular fluorescence: principles and applications. ed.^eds.), p.^pp. Wiley-VCH, Weinheim.

Beutler E (1983) Glucose-6-phosphate dehydrogenase deficiency. *The Metabolic Basis of Inherited Disease,* Vol. Fifth edition (JB Stanbury JW, DS Fredrickson, and J Goldstein, ed.^eds.), p.^pp. 1629-1653. McGraw-Hill, New York.

Beutler E (1989) Glucose-6-phosphate dehydrogenase: new perspectives. *Blood* Vol.73(No. 6):1397-1401.

Beutler E (1990) The genetics of glucose-6-phosphate dehydrogenase deficiency. *Semin Hematol* Vol.27(No. 2):137-164.

Beutler E (1991) Glucose-6-phosphate dehydrogenase deficiency. *N Engl J Med* Vol.324(No. 3):169-174.

Beutler E (1992) The molecular biology of G6PD variants and other red cell enzyme defects. *Annu Rev Med* Vol.43(No.:47-59.

Beutler E (1994) G6PD deficiency. *Blood* Vol.84(No. 11):3613-3636.

Beutler E & Baluda MC (1966) A simple spot screening test for galactosemia. *J Lab Clin Med* Vol.68(No. 1):137-141.

Beutler E & Duparc S (2007) Glucose-6-phosphate dehydrogenase deficiency and antimalarial drug development. *The American journal of tropical medicine and hygiene* Vol.77(No. 4):779-789.

Beutler E & Duparc S (2007) Glucose-6-phosphate dehydrogenase deficiency and antimalarial drug development. *Am J Trop Med Hyg* Vol.77(No. 4):779-789.

Beutler E, Dern R & Alving A (1955) The hemolytic effect of primaquine. VI. An in vitro test for sensitivity of erythrocytes to primaquine. *The Journal of laboratory and clinical medicine* Vol.45(No. 1):40.

Beutler E, Dern RJ & Alving AS (1955) The hemolytic effect of primaquine. VI. An in vitro test for sensitivity of erythrocytes to primaquine. *J Lab Clin Med* Vol.45(No. 1):40-50.

Beutler E, Yeh M & Fairbanks VF (1962) The normal human female as a mosaic of X-chromosome activity: studies using the gene for G-6-PD-deficiency as a marker. *Proceedings of the national academy of sciences of the united states of america* Vol.48(No. 1):9.

Beutler E, Mathai CK & Smith JE (1968) Biochemical variants of glucose-6-phosphate dehydrogenase giving rise to congenital nonspherocytic hemolytic disease. *Blood* Vol.31(No. 2):131-150.

Beutler E, Gelbart T & Kuhl W (1990) Human red cell glucose-6-phosphate dehydrogenase: all active enzyme has sequence predicted by the X chromosome-encoded cDNA. *Cell* Vol.62(No. 1):7-9.

Bhaskaram P, Balakrishna N, Radhakrishna K & Krishnaswamy K (2003) Validation of hemoglobin estimation using Hemocue. *Indian journal of pediatrics* Vol.70(No. 1):25-28.

Bolchoz LJ, Morrow JD, Jollow DJ & McMillan DC (2002) Primaquine-induced hemolytic anemia: effect of 6-methoxy-8-hydroxylaminoquinoline on rat erythrocyte sulfhydryl status, membrane lipids, cytoskeletal proteins, and morphology. *J Pharmacol Exp Ther* Vol.303(No. 1):141-148.

Bolchoz LJC, Morrow JD, Jollow DJ & McMillan DC (2002) Primaquine-induced hemolytic anemia: effect of 6-methoxy-8-hydroxylaminoquinoline on rat erythrocyte sulfhydryl status, membrane lipids, cytoskeletal proteins, and morphology. *Journal of Pharmacology and Experimental Therapeutics* Vol.303(No. 1):141-148.

Boyer SH & Graham JB (1965) Linkage between the X chromosome loci for glucose-6-phosphate dehydrogenase electrophoretic variation and hemophilia A. *American journal of human genetics* Vol.17(No. 4):320.

Bozdech Z, Llinas M, Pulliam BL, Wong ED, Zhu J & DeRisi JL (2003) The transcriptome of the intraerythrocytic developmental cycle of Plasmodium falciparum. *PLoS Biol* Vol.1(No. 1):E5.

Burgoine KL, Bancone G & Nosten F (2010) The reality of using primaquine. *Malaria journal* Vol.9(No. 1):376.

Cappellini MD, Tavazzi D, Martinez di Montemuros F, Sampietro M, Gaviraghi A, Carandini D & Fiorelli G (1993) Alternative splicing of human G6PD messenger RNA in K562 cells but not in cultured erythroblasts. *Eur J Clin Invest* Vol.23(No. 3):188-191.

Chen EY, Cheng A, Lee A, *et al.* (1991) Sequence of human glucose-6-phosphate dehydrogenase cloned in plasmids and a yeast artificial chromosome. *Genomics* Vol.10(No. 3):792-800.

Chen EY, Zollo M, Mazzarella R, *et al.* (1996) Long-range sequence analysis in Xq28: thirteen known and six candidate genes in 219.4 kb of high GC DNA between the RCP/GCP and G6PD loci. *Hum Mol Genet* Vol.5(No. 5):659-668.

Childs B, Zinkham W, Browne EA, Kimbro EL & Torbert JV (1958) A genetic study of a defect in glutathione metabolism of the erythrocyte. *Bull Johns Hopkins Hosp* Vol.102(No. 1):21-37.

Clark I, Chaudhri G & Cowden W (1989) Some roles of free radicals in malaria. *Free Radical Biology and Medicine* Vol.6(No. 3):315-321.

Epstein CJ (1969) Mammalian oocytes: X chromosome activity. *Science* Vol.163(No. 3871):1078-1079.

Fairbanks VF, Klee, G.G. (1999) Biochemical aspects of hematology. *Burtıs CA, Ashwood ER,* Vol. 3rd edition (chemistry Ttoc, ed.^eds.), p.^pp. 1642-1655. WB Saunders Co, Philadelphia.

Filosa S, Calabro V, Lania G, *et al.* (1993) G6PD haplotypes spanning Xq28 from F8C to red/green color vision. *Genomics* Vol.17(No. 1):6-14.

Fortin A, Stevenson MM & Gros P (2002) Susceptibility to malaria as a complex trait: big pressure from a tiny creature. *Hum Mol Genet* Vol.11(No. 20):2469-2478.

Frank JE (2005) Diagnosis and management of G6PD deficiency. *American family physician* Vol.72(No. 7):1277.

Franze A, Ferrante MI, Fusco F, Santoro A, Sanzari E, Martini G & Ursini MV (1998) Molecular anatomy of the human glucose 6-phosphate dehydrogenase core promoter. *FEBS Lett* Vol.437(No. 3):313-318.

Friedman MJ (1979) Oxidant damage mediates variant red cell resistance to malaria. *Nature* Vol.280(No. 5719):245-247.

Friedman MJ & Trager W (1981) The biochemistry of resistance to malaria. *Sci Am* Vol.244(No. 3):154-155, 158-164.

Gaetani GF, Rolfo M, Arena S, Mangerini R, Meloni GF & Ferraris AM (1996) Active involvement of catalase during hemolytic crises of favism. *Blood* Vol.88(No. 3):1084-1088.

Gall JC, Brewer GJ & Dern RJ (1965) Studies of Glucose-6-Phosphate Dehydrogenase Activity of Individual Erythrocytes: The Methemoglobin-Elution Test for Identification of Females Heterozygous for G6PD Deficiency. *Am J Hum Genet* Vol.17(No. 4):359-368.

Ganczakowski M, Town M, Bowden D, *et al.* (1995) Multiple glucose 6-phosphate dehydrogenase-deficient variants correlate with malaria endemicity in the Vanuatu archipelago (southwestern Pacific). *American journal of human genetics* Vol.56(No. 1):294.

Ganczakowski M, Town M, Bowden DK, *et al.* (1995) Multiple glucose 6-phosphate dehydrogenase-deficient variants correlate with malaria endemicity in the Vanuatu archipelago (southwestern Pacific). *Am J Hum Genet* Vol.56(No. 1):294-301.

Glader BE, Lukens, J.N. (1999) Glucose-6-phosphate dehydrogenase deficiency and related disorders of hexose monophosphate shunt and glutathione metabolism. *Wintrobe's Clinical hematology,* Vol. 1 (Lee GR FJ, Lukens J, Paraskevas F, Greer JP, Rodgers GM, ed.^eds.), p.^pp. 1176-1190. WB Saunders Co, London.

Greene LS (1993) G6PD deficiency as protection against falciparum malaria: an epidemiologic critique of population and experimental studies. *American Journal of Physical Anthropology* Vol.36(No. S17):153-178.

Group WW (1989) Glucose-6-phosphate dehydrogenase deficiency. Vol. 67 ed.^eds.), p.^pp. 601-611.

Haworth J, Wernsdorfer W & McGregor I (1988) The global distribution of malaria and the present control effort. *Malaria: principles and practice of malariology. Volume 2.* (No.:1379-1420.

Henikoff S & Smith JM (1989) The human mRNA that provides the N-terminus of chimeric G6PD encodes GMP reductase. *Cell* Vol.58(No. 6):1021-1022.

Hill DR, Baird JK, Parise ME, Lewis LS, Ryan ET & Magill AJ (2006) Primaquine: report from CDC expert meeting on malaria chemoprophylaxis I. *The American journal of tropical medicine and hygiene* Vol.75(No. 3):402-415.

Hirono A & Beutler E (1988) Molecular cloning and nucleotide sequence of cDNA for human glucose-6-phosphate dehydrogenase variant A(-). *Proc Natl Acad Sci U S A* Vol.85(No. 11):3951-3954.

Hirono A & Beutler E (1989) Alternative splicing of human glucose-6-phosphate dehydrogenase messenger RNA in different tissues. *J Clin Invest* Vol.83(No. 1):343-346.

Ho YS, Howard AJ & Crapo JD (1988) Cloning and sequence of a cDNA encoding rat glucose-6-phosphate dehydrogenase. *Nucleic Acids Res* Vol.16(No. 15):7746.

Hodge DL, Charron T, Stabile LP, Klautky SA & Salati LM (1998) Structural characterization and tissue-specific expression of the mouse glucose-6-phosphate dehydrogenase gene. *DNA Cell Biol* Vol.17(No. 3):283-291.

Kanno H, Huang IY, Kan YW & Yoshida A (1989) Two structural genes on different chromosomes are required for encoding the major subunit of human red cell glucose-6-phosphate dehydrogenase. *Cell* Vol.58(No. 3):595-606.

Katz SH & Schall J (1979) Part three: Fava bean consumption and biocultural evolution. *Medical Anthropology* Vol.3(No. 4):459-476.

Koehler A & Van Noorden CJF (2003) Reduced nicotinamide adenine dinucleotide phosphate and the higher incidence of pollution-induced liver cancer in female flounder. *Environmental toxicology and chemistry* Vol.22(No. 11):2703-2710.

Kornberg A, Horecker B & Smyrniotis P (1955) Glucose-6-phosphate dehydrogenase 6-phosphogluconic dehydrogenase. *Methods in enzymology* Vol.1(No.:323-327.

Kruatrachue M, Charoenlarp P, Chongsuphajaisiddhi T & Harinasuta C (1962) Erythrocyte glucose-6-phosphate dehydrogenase and malaria in Thailand. *Lancet* Vol.2(No. 7267):1183-1186.

Kwiatkowski DP (2005) How malaria has affected the human genome and what human genetics can teach us about malaria. *Am J Hum Genet* Vol.77(No. 2):171-192.

Lachant NA, Tomoda A & Tanaka KR (1984) Inhibition of the pentose phosphate shunt by lead: a potential mechanism for hemolysis in lead poisoning. *Blood* Vol.63(No. 3):518-524.

Lehninger AL (2000) Principles of Biochemistry. *Advances in Enzymology,* Vol. 3rd Edition by Nelson DL and Cox MM. (A M, ed.^eds.), p.^pp. Worth Publishers, New York.

Lohr G & Waller H (1974) Glucose-6-phosphate dehydrogenase. *Methods of Enzymatic Analysis, Academic Press, New York* (No.:636-643.

Luzzatto L (1967) Regulation of the activity of glucose-6-phosphate dehydrogenase by NADP+ and NADPH. *Biochim Biophys Acta* Vol.146(No. 1):18-25.

Luzzatto L (1993) G6PD deficiency and hemolytic anemia. *Hematology of infancy and childhood,* Vol. ch 19, 4th ed. (Nathan DG OF, ed.^eds.), p.^pp. 674-695. WB Saunders Co, Philadelphia.

Luzzatto L (2006) Glucose 6-phosphate dehydrogenase deficiency: from genotype to phenotype. *Haematologica* Vol.91(No. 10):1303-1306.

Luzzatto L & Testa U (1978) Human erythrocyte glucose 6-phosphate dehydrogenase: structure and function in normal and mutant subjects. *Curr Top Hematol* Vol.1(No.:1-70.

Luzzatto L & Battistuzzi G (1985) Glucose-6-phosphate dehydrogenase. *Adv Hum Genet* Vol.14(No.:217-329, 386-218.

Luzzatto L, Mehta, A., Vulliamy, T. (2001) Glucose-6-phosphate dehydrogenase deficiency. *The Metabolic and molecular bases of inherited disease* Vol. 3 (Scriver CR BA, Sly WS, Valle D, ed.^eds.), p.^pp. 4517-4553. McGraw-Hill Co, New York.

Luzzatto LaM, A. (1989) Glucose-6-phosphate dehydrogenase deficiency. *The Metabolic of Inherited Disease,* Vol. Sixth edition (CR Scriver AB, WS Sly, and D Valle ed.^eds.), p.^pp. 2237-2265. McGraw-Hill, New York.

Macias VR, Day DW, King TE & Wilson GN (1992) Clasped-thumb mental retardation (MASA) syndrome: confirmation of linkage to Xq28. *Am J Med Genet* Vol.43(No. 1-2):408-414.

Martini G, Toniolo D, Vulliamy T, *et al.* (1986) Structural analysis of the X-linked gene encoding human glucose 6-phosphate dehydrogenase. *EMBO J* Vol.5(No. 8):1849-1855.

Mason PJ, Bautista JM, Vulliamy TJ, Turner N & Luzzatto L (1990) Human red cell glucose-6-phosphate dehydrogenase is encoded only on the X chromosome. *Cell* Vol.62(No. 1):9-10.

Mason PJ, Stevens DJ, Luzzatto L, Brenner S & Aparicio S (1995) Genomic structure and sequence of the Fugu rubripes glucose-6-phosphate dehydrogenase gene (G6PD). *Genomics* Vol.26(No. 3):587-591.

Mason PJ, Vulliamy TJ, Foulkes NS, Town M, Haidar B & Luzzatto L (1988) The production of normal and variant human glucose-6-phosphate dehydrogenase in cos cells. *Eur J Biochem* Vol.178(No. 1):109-113.

Mehta A, Mason PJ & Vulliamy TJ (2000) Glucose-6-phosphate dehydrogenase deficiency. *Baillieres Best Pract Res Clin Haematol* Vol.13(No. 1):21-38.

Meloni T, Forteleoni G, Dore A & Cutillo S (1983) Favism and hemolytic anemia in glucose-6-phosphate dehydrogenase-deficient subjects in North Sardinia. *Acta Haematol* Vol.70(No. 2):83-90.

Mendis K, Sina BJ, Marchesini P & Carter R (2001) The neglected burden of Plasmodium vivax malaria. *Am J Trop Med Hyg* Vol.64(No. 1-2 Suppl):97-106.

Migeon BR (1983) Glucose 6-phosphate dehydrogenase as a probe for the study of X-chromosome inactivation in human females. *Isozymes: Current Topics in Biological and Medical Research,* Vol. 9 (Rattazzi MC SJ, Whitt GS, ed.^eds.), p.^pp. 189-200. Alan R Liss, New York.

Misumi H, Wada H, Ichiba Y, Shohmori T & Kosaka M (1982) Separate detection of glucose-6-phosphate dehydrogenase from 6-phosphogluconate dehydrogenase by DEAE-paper chromatography. *Blut* Vol.45(No. 1):33-37.

Moskaug JØ, Carlsen H, Myhrstad M & Blomhoff R (2004) Molecular imaging of the biological effects of quercetin and quercetin-rich foods. *Mechanisms of ageing and development* Vol.125(No. 4):315-324.

Motulsky AG (1961) Glucose-6-phosphate-dehydrogenase deficiency, haemolytic disease of the newborn, and malaria. *The Lancet* Vol.277(No. 7187):1168-1169.

Motulsky AG (1988) Normal and abnormal color-vision genes. *Am J Hum Genet* Vol.42(No. 3):405-407.

Mueller I, Zimmerman PA & Reeder JC (2007) Plasmodium malariae and Plasmodium ovale--the "bashful" malaria parasites. *Trends Parasitol* Vol.23(No. 6):278-283.

Myrvang B & Godal T (2000) WHO's malaria program Roll Back Malaria. *Tidsskr Nor Laegeforen* Vol.120(No. 14):1661-1664.

Nagel RL & Roth EF, Jr. (1989) Malaria and red cell genetic defects. *Blood* Vol.74(No. 4):1213-1221.

Nance WE (1964) Genetic Tests with a Sex-Linked Marker: Glucose-6-Phosphate Dehydrogenase. *Cold Spring Harb Symp Quant Biol* Vol.29(No.:415-425.

Ninfali P, Perini MP, Bresolin N, Aluigi G, Cambiaggi C, Ferrali M & Pompella A (2000) Iron release and oxidant damage in human myoblasts by divicine. *Life Sci* Vol.66(No. 6):PL85-91.

Ninokata A, Kimura R, Samakkarn U, Settheetham-Ishida W & Ishida T (2006) Coexistence of five G6PD variants indicates ethnic complexity of Phuket islanders, Southern Thailand. *J Hum Genet* Vol.51(No. 5):424-428.

Noori-Daloii M, Najafi L, Ganji SM, Hajebrahimi Z & Sanati M (2004) Molecular identification of mutations in G6PD gene in patients with favism in Iran. *Journal of physiology and biochemistry* Vol.60(No. 4):273-277.

Notaro R, Afolayan A & Luzzatto L (2000) Human mutations in glucose 6-phosphate dehydrogenase reflect evolutionary history. *FASEB J* Vol.14(No. 3):485-494.

Oberle I, Camerino G, Wrogemann K, Arveiler B, Hanauer A, Raimondi E & Mandel JL (1987) Multipoint genetic mapping of the Xq26-q28 region in families with fragile X mental retardation and in normal families reveals tight linkage of markers in q26-q27. *Hum Genet* Vol.77(No. 1):60-65.

Organization WH (2009) The world health report 2003: shaping the future. Geneva: WHO, 2003. *Annex table* Vol.2(No.:156.

Pai GS, Sprenkle JA, Do TT, Mareni CE & Migeon BR (1980) Localization of loci for hypoxanthine phosphoribosyltransferase and glucose-6-phosphate dehydrogenase and biochemical evidence of nonrandom X chromosome expression from studies of a human X-autosome translocation. *Proc Natl Acad Sci U S A* Vol.77(No. 5):2810-2813.

Patterson M, Schwartz C, Bell M, *et al.* (1987) Physical mapping studies on the human X chromosome in the region Xq27-Xqter. *Genomics* Vol.1(No. 4):297-306.

Persico MG, Ciccodicola A, Martini G & Rosner JL (1989) Functional expression of human glucose-6-phosphate dehydrogenase in Escherichia coli. *Gene* Vol.78(No. 2):365-370.

Persico MG, Viglietto G, Martini G, *et al.* (1986) Isolation of human glucose-6-pbosphate debydrogenase (G6PD) cDNA clones: primary structure of the protein and unusual 5'non-coding region. *Nucleic acids research* Vol.14(No. 6):2511-2522.

Persico MG, Viglietto G, Martini G, *et al.* (1986) Isolation of human glucose-6-phosphate dehydrogenase (G6PD) cDNA clones: primary structure of the protein and unusual 5' non-coding region. *Nucleic Acids Res* Vol.14(No. 6):2511-2522.

Peters AL & Van Noorden CJ (2009) Glucose-6-phosphate dehydrogenase deficiency and malaria: cytochemical detection of heterozygous G6PD deficiency in women. *J Histochem Cytochem* Vol.57(No. 11):1003-1011.

Peters AL & Noorden CJFV (2009) Glucose-6-phosphate dehydrogenase deficiency and malaria: cytochemical detection of heterozygous G6PD deficiency in women. *Journal of Histochemistry & Cytochemistry* Vol.57(No. 11):1003.

Philippe M, Larondelle Y, Lemaigre F, *et al.* (1994) Promoter function of the human glucose-6-phosphate dehydrogenase gene depends on two GC boxes that are cell specifically controlled. *Eur J Biochem* Vol.226(No. 2):377-384.

Phompradit P, Kuesap J, Chaijaroenkul W, Rueangweerayut R, Hongkaew Y, Yamnuan R & Na-Bangchang K (2011) Prevalence and distribution of glucose-6-phosphate dehydrogenase (G6PD) variants in Thai and Burmese populations in malaria endemic areas of Thailand. *Malar J* Vol.10(No.:368.

Prchal JT & Gregg XT (2005) Red cell enzymes. *Hematology Am Soc Hematol Educ Program* (No.:19-23.

Rank KB, Harris PK, Ginsberg LC & Stapleton SR (1994) Isolation and sequence of a rat glucose-6-phosphate dehydrogenase promoter. *Biochim Biophys Acta* Vol.1217(No. 1):90-92.

Rattazzi MC (1968) Glucose 6-phosphate dehydrogenase from human erythrocytes: molecular weight determination by gel filtration. *Biochem Biophys Res Commun* Vol.31(No. 1):16-24.

Reclos G, Hatzidakis C & Schulpis K (2000) Glucose-6-phosphate dehydrogenase deficiency neonatal screening: preliminary evidence that a high percentage of partially deficient female neonates are missed during routine screening. *Journal of Medical Screening* Vol.7(No. 1):46-51.

Rinaldi A, Filippi G & Siniscalco M (1976) Variability of red cell phenotypes between and within individuals in an unbiased sample of 77 heterozygotes for G6PD deficiency in Sardinia. *American journal of human genetics* Vol.28(No. 5):496.

Ruwende C & Hill A (1998) Glucose-6-phosphate dehydrogenase deficiency and malaria. *J Mol Med (Berl)* Vol.76(No. 8):581-588.

Ruwende C, Khoo SC, Snow RW, *et al.* (1995) Natural selection of hemi- and heterozygotes for G6PD deficiency in Africa by resistance to severe malaria. *Nature* Vol.376(No. 6537):246-249.

Segel GB (2000) Enzymatic defects. *Nelson Textbook of pediatrics,* Vol. 16th (Behrman RE KR, Jenson HB, ed.^eds.), p.^pp. 1488-1491. WB Saunders Co, Philadelphia.

Singh B, Kim Sung L, Matusop A, *et al.* (2004) A large focus of naturally acquired Plasmodium knowlesi infections in human beings. *Lancet* Vol.363(No. 9414):1017-1024.

Siniscalco M & Bernini L (1961) Favism and thalassaemia in Sardinia and their relationship to malaria. *Nature* Vol.190(No.:1179-1180.

Sutherland CJ, Tanomsing N, Nolder D, *et al.* (2010) Two nonrecombining sympatric forms of the human malaria parasite Plasmodium ovale occur globally. *Journal of Infectious Diseases* Vol.201(No. 10):1544-1550.

Sutherland CJ, Tanomsing N, Nolder D, *et al.* (2010) Two nonrecombining sympatric forms of the human malaria parasite Plasmodium ovale occur globally. *J Infect Dis* Vol.201(No. 10):1544-1550.

Szabo P, Purrello M, Rocchi M, *et al.* (1984) Cytological mapping of the human glucose-6-phosphate dehydrogenase gene distal to the fragile-X site suggests a high rate of meiotic recombination across this site. *Proc Natl Acad Sci U S A* Vol.81(No. 24):7855-7859.

Takizawa T, Huang IY, Ikuta T & Yoshida A (1986) Human glucose-6-phosphate dehydrogenase: primary structure and cDNA cloning. *Proc Natl Acad Sci U S A* Vol.83(No. 12):4157-4161.

Tan KL (1981) Glucose-6-phosphate dehydrogenase status and neonatal jaundice. *Arch Dis Child* Vol.56(No. 11):874-877.

Tang TK, Yeh CH, Huang CS & Huang MJ (1994) Expression and biochemical characterization of human glucose-6-phosphate dehydrogenase in Escherichia coli: a system to analyze normal and mutant enzymes. *Blood* Vol.83(No. 5):1436-1441.

Toncheva D & Tzoneva M (1985) Prenatal selection and fetal development disturbances occurring in carriers of G6PD deficiency. *Hum Genet* Vol.69(No. 1):88.

Toniolo D, Filippi M, Dono R, Lettieri T & Martini G (1991) The CpG island in the 5' region of the G6PD gene of man and mouse. *Gene* Vol.102(No. 2):197-203.

Turner NJ (2000) Applications of transketolases in organic synthesis. *Curr Opin Biotechnol* Vol.11(No. 6):527-531.

Tyulina OV, Huentelman MJ, Prokopieva VD, Boldyrev AA & Johnson P (2000) Does ethanol metabolism affect erythrocyte hemolysis? *Biochimica et Biophysica Acta (BBA)-Molecular Basis of Disease* Vol.1535(No. 1):69-77.

Ursini MV, Scalera L & Martini G (1990) High levels of transcription driven by a 400 bp segment of the human G6PD promoter. *Biochemical and biophysical research communications* Vol.170(No. 3):1203-1209.

Vulliamy TJ, D'Urso M, Battistuzzi G, *et al.* (1988) Diverse point mutations in the human glucose-6-phosphate dehydrogenase gene cause enzyme deficiency and mild or severe hemolytic anemia. *Proc Natl Acad Sci U S A* Vol.85(No. 14):5171-5175.

Wernsdorfer WH & McGregor I (1988) *Malaria: principles and practice of malariology. Vols 1 and 2.* Churchill Livingstone.

Wrigley NG, Heather JV, Bonsignore A & De Flora A (1972) Human erythrocyte glucose 6-phosphate dehydrogenase: electron microscope studies on structure and interconversion of tetramers, dimers and monomers. *J Mol Biol* Vol.68(No. 3):483-499.

Yazdani SS, Mukherjee P, Chauhan VS & Chitnis CE (2006) Immune responses to asexual blood-stages of malaria parasites. *Curr Mol Med* Vol.6(No. 2):187-203.

Yoshida A & Kan YW (1990) Origin of "fused" glucose-6-phosphate dehydrogenase. *Cell* Vol.62(No. 1):11-12.

Chapter 5

Role and Function of Dehydrogenases in CNS and Blood-Brain Barrier Pathophysiology

P. Naik, S. Prasad and L. Cucullo

Additional information is available at the end of the chapter

http://dx.doi.org/10.5772/48338

1. Introduction

Dehydrogenase (DHO) is one of the most common types of enzyme that is crucial in oxidation reactions. This enzyme oxidizes its specific substrate by a redox reaction in which one or more hydrides (H^-) are transferred to an electron acceptor. Apart from energetics and ATP formation, DHOs are associated with both catabolic and anabolic pathways linked to normal functioning and homeostasis. In this chapter, we will cover different aspects of the major DHOs that play a role in the regulation of brain and blood-brain barrier (BBB) physiology starting from their role in bioenergetic metabolism. In born errors in metabolism (IEM) due to genetic deficiency in a single specific DHO have strong neurological implications. We will be covering some examples of such IEMs in this chapter. Furthermore, aging processes can impair the function or activity of DHOs. Recent studies show convincing evidence associating altered DHO activity with the pathogenesis and progression of several neurological disorders such as Alzheimer's and Parkinson's disease. Whether they are contributors to the etiology of the disease or symptomatic manifestation of these complex neurological disorders is still debatable; however, this link between DHOs and neurological disorders cannot be overlooked and will be further discussed in this chapter. We will also cover neuronal signaling, neurotransmitter release and degradation emphasizing localized region-specific expression of some brain DHOs in these processes. It is not possible to cover the detailed cerebral physiology and function in this chapter; however to summarize we will discuss the different types of DHOs in central nervous system (CNS) and BBB physiology, their key enzymatic action, their function in crucial metabolic pathways, and thus how their altered activity or expression can be linked to the underlying pathogenesis of various brain disorders.

2. Structural and functional complexity of blood brain barrier (BBB) and cerebral physiology – A need for high energy

The BBB is a dynamic interface between the peripheral blood and the brain which controls the influx and efflux of substrates and metabolites necessary for normal neuronal function (see **Figure 1**). The BBB is crucial in protecting the brain from harmful substances both endogenous and exogenous in nature. Any alteration in normal BBB functions can play a central role in the pathogenesis and progression of broad variety of CNS disorders such as multiple sclerosis, Alzheimer's disease, neoplasia, hypertension, dementia, epilepsy, infection and trauma (1-4).

At the cellular level, the BBB consists of microvascular endothelial cells (ECs) lining the brain microvessels together with closely associated astrocytic end-feet processes and pericytes (5-8). These associated cells play a major role in EC differentiation and acquisition of morphological and functional characteristics unique to the BBB. At the cellular level, the brain microcapillary endothelium is characterized by the presence of tight junctions (TJ), lack of fenestrations, and minimal pinocytotic vesicles (9;10). In particular, TJs between the cerebral endothelial cells form a diffusion barrier, which selectively excludes most of the blood-borne substances from entering the brain, protecting it from systemic influences mediated by substances which are primarily polar in nature (such as electrolytes). Transport of nutrients (as well as other biologically important substances) from the peripheral circulation into brain parenchyma requires translocation through the capillary endothelium by specialized carrier-mediated transport systems. On the other hand, potentially harmful substances that are lipid soluble are discharged back into the cerebral circulation. This is mediated by specialized active efflux systems belonging to the ATP-binding cassette transporters (ABC-transporter) superfamily (such as P-glycoprotein (P-gp) and Multidrug resistance Protein (MRP)) (11;12). These efflux pumps rely heavily on adenosine triphosphate (ATP) as fuel source. Apart from these ATP dependent pumps, energy independent transporters such as organic ion carriers add to the complexity of BBB transport functions (13). Simultaneously, intake of essential nutrients such as glucose, amino acids, peptides, choline occurs through carrier mediated mechanisms (13-17). Topographic membrane localization of these transporters is indicative of the polarity of the endothelial functions and differentiation that sets apart the BBB endothelium from other vascular beds. The BBB endothelial cytoplasm is richly endowed with enzymes(18) including adenosine tri-phosphatase, acid and alkaline phosphatases, Na^+/K^+/ATPase, monoamine oxidase, cytochrome p450s and various dehydrogenases (19-22). The BBB ECs are also characterized by very high density of mitochondria denoting high metabolic activity (23) to support all the specialized cellular activities bestowed upon these highly specialized cells. In addition, previous work from our group has shown that blood flow can modulate the bioenergetic behavior of the BBB endothelial cells favoring the expression of the key metabolic enzyme pyruvate dehydrogenase (switch controller from anaerobic to aerobic pathway) (19). Contrarily, the RNA level of lactate dehydrogenase (switch controller from aerobic to anaerobic pathway) showed decreased expression. In parallel, TCA dehydrogenases such as acotinase, isocitrate dehydrogenase, succinate dehydrogenase were upregulated. These results clearly emphasize how altered rheological conditions (e.g., hypoperfusion and ischemia) may impact the BBB bioenergetic metabolism and how the BBB is well equipped to respond to such changes (see **Figure 2**).

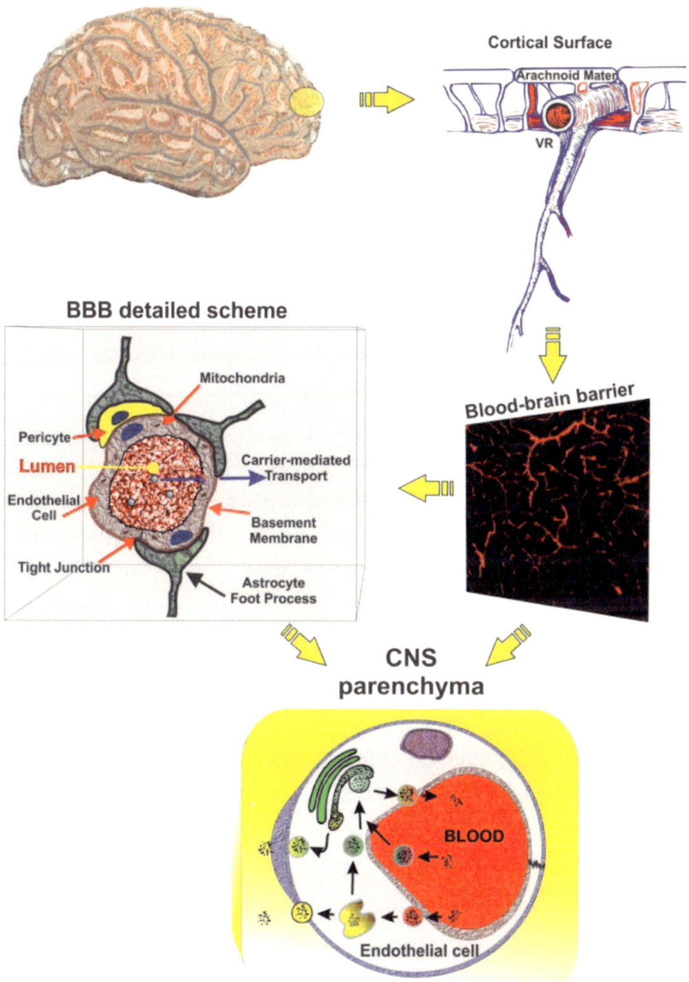

Figure 1. Schematic representation of the brain microvasculature in relation to the brain. Note how the blood vessels start branching in small capillaries while the pia disappears and the endothelium acquires the peculiar characteristics of a tight barrier that regulate the exchange of substances between blood and brain. TJ between adjacent endothelial cells form a diffusion barrier that selectively excludes most blood-borne and xenobiotic substances from entering the brain. In contrast to lipid soluble substances including alcohols, anesthetics and barbiturates, the BBB is highly impermeant to polar molecules or water soluble electrolytes. However, the passage of certain water soluble, but biologically important substances, such as D-glucose or phenylamine are regulated by a variety of specific carrier-mediated transport systems. By contrast, larger vessels (arterioles, small arteries and venous) differ from capillaries by the presence of smooth muscle cells in their walls and a less stringent vascular bed.

A

pI & MW	Gene Match	Fold Change (Flow/No Flow)	±Se	Protein family
pI: 5.63, MW: 56930.99	Acetyl-Coenzyme A acetyltransferase	2.00	0.203	Glucose Metabolism
pI: 5.73, MW: 65949.15	Malate dehydrogenase	2.80	0.184	
pI: 4.92, MW: 82342.21	Succinate dehydrogenase	2.19	0.133	
pI: 5.70, MW: 85916.69	Lactate dehydrogenase A	-2.08	0.141	
pI: 6.23, MW: 76342.28	Glyceraldehyde-3-phosphate dehydrogenase	2.68	0.178	
pI: 5.10, MW: 52396.28	Pyruvate dehydrogenase	2.29	0.152	

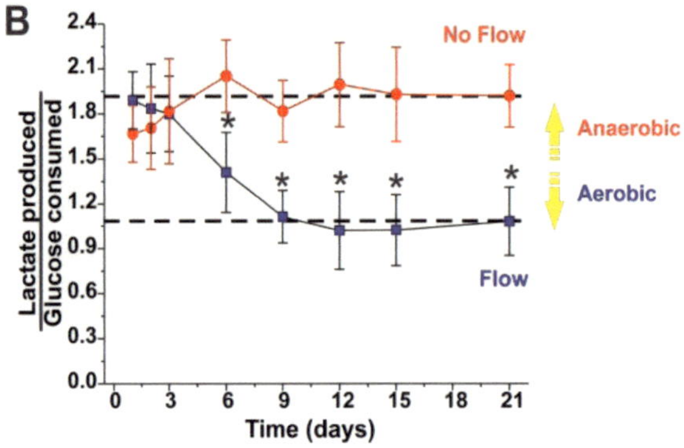

Figure 2. Effect of flow on BBB glucose metabolism. Comparative analysis of the expression level of key enzymes regulating the glycolytic and TCA pathways strongly supported the gene array data (Panel A). Note that the lactate production/glucose consumption ratio measured in the flow-exposed in vitro BBB modules was ≈ 1. Complete anaerobic metabolism would produce 2 lactates/glucose (ratio = 2) thus, indicating that at least 50% of the glucose consumed underwent aerobic metabolism (Panel **B**).

The BBB's ability in maintaining an optimal bioenergetics level at all time is thus crucial to meet the energy demand required by its multiple proprietary functions under normal as well as pathological conditions (e.g., cerebral ischemia). Energetic pathways (such as glycolysis and the tricarboxylic acid cycle -TCA cycle) work in an integrated yet independent manner in the BBB endothelium to respond to physiological (e.g., increased CNS demand in response to changes in neuronal activity) or pathological events that require a prompt BBB response.

The brain on the other hand possesses an incredibly more complex physiology than the BBB vasculature. It is the control center of neuronal as well as hormonal signaling. It is crucial for various functions such as homeostasis, behavior, perception and processing of information, motor control and memory formation. From a physiological stand point, brain functions depend on the ability of neurons to transmit and respond to electrochemical signals. This complex crosstalk is controlled by a wide variety of biochemical and metabolic processes which involve interactions between neurotransmitters and receptors that take place at the synapses. Crucial to this function is the necessity to sustain the bioenergetic demand required to maintain optimal neuronal activity (e.g., generation of action potentials, release of neurotransmitters, restoration of the membrane polarization following an action potential etc). In this respect, the BBB endothelium and glial cells play a major role in brain metabolism, by controlling the influx and distribution of nutrients (e.g., glucose and lactate shuttles as described later) as well as the chemical composition of the extracellular fluids surrounding the neurons.

3. Pathways related to energy metabolism in BBB and brain: Lactate shuttle: NALS or ALNS?

Before we begin describing the role of DHOs in energy metabolism, it is imperative to understand the various metabolic pathways in the brain involved in energy production. Glucose is one of the primary fuel sources available to the brain. ≅ 25% of the total glucose intake is utilized by the brain despite accounting for ≅ 2% of the total body mass (24).

Glucose enters the glycolytic pathway to produce pyruvate along with net production of 2ATP and 2NADH (reduced form of nicotinamide adenine dinucleotide) (see **Figure 3**). Ten intermediate reactions occur until pyruvate is formed in the last step. ATP is first used up in the first part of glycolysis (until formation of glyceraldehyde phosphate-GAP) and is produced later during the second half. The total net ATP gain for each glycolytic cycle is 2 ATP molecules. Another important step of glycolysis involving glyceraldehyde phosphate dehydrogenase (GAPDH), is the conversion of GAP to 1,3-bisphosphate glycerate (1,3- BPG) along with NADH formation. Part of the 1,3-BPG thus formed can be further converted into 2,3-biphosphate glycerate (2,3-BPG) which can bind to hemoglobin enhancing its deoxygenation. The remaining 1,3-BPG undergoes further conversion along the glycolytic pathways and is finally converted into pyruvate.

Pyruvate thus formed can either enter the citric acid chain (or Kreb's cycle or Tricyclic acid cycle-TCA) or get converted to lactate which represents the end product of glycolysis. Pyruvate to lactate conversion is the last step of anaerobic form of respiration which occurs via lactate dehydrogenase and results in 2 molecules of ATP production.

Conversely, pyruvate can be further converted into acetyl co-A and then enter the TCA cycle. This intermediate reaction is crucial in linking glycolysis to the TCA. In this step, pyruvate dehydrogenase (PDH) decarboxylates pyruvate to its acetyl form along with addition of co-enzyme A. Acetyl co-A then combines with oxaloacetate (4C) in presence of water molecule to form citrate (6C). The citrate thus formed cycles through TCA forming oxaloacetate in the last step, which can re-enter the cycle reacting with a new acetyl co-A (see **Figure 3**). The various steps in the TCA cycle involve various oxidation reactions involving various dehydrogenases such as isocitrate dehydrogenase, α-ketoglutarate dehydrogenase, succinate dehydrogenase and malate dehydrogenase. Each of these reactions lead to formation of NADH (except for FADH2 i.e. reduced flavin adenine dinucleotide formation during conversion of succinate to fumarate). These reducing equivalents can then enter the electron transport chain and result in ATP formation or can be used by the cell to counteract the oxidative stress caused by reactive oxygen species (ROS) and free radicals. A complete cycling of two molecules of acetyl co-A (from a single molecule of glucose) results in the production of approximately 10 NADH and 2 FADH2. The complete metabolic conversion of a glucose molecule into water and CO_2 (glycolysis and TCA cycles combined) results in the approximate production of 38 ATP molecules (including ATP formed indirectly through NADH and FADH2 which produce 3 and 2 ATP molecules respectively).

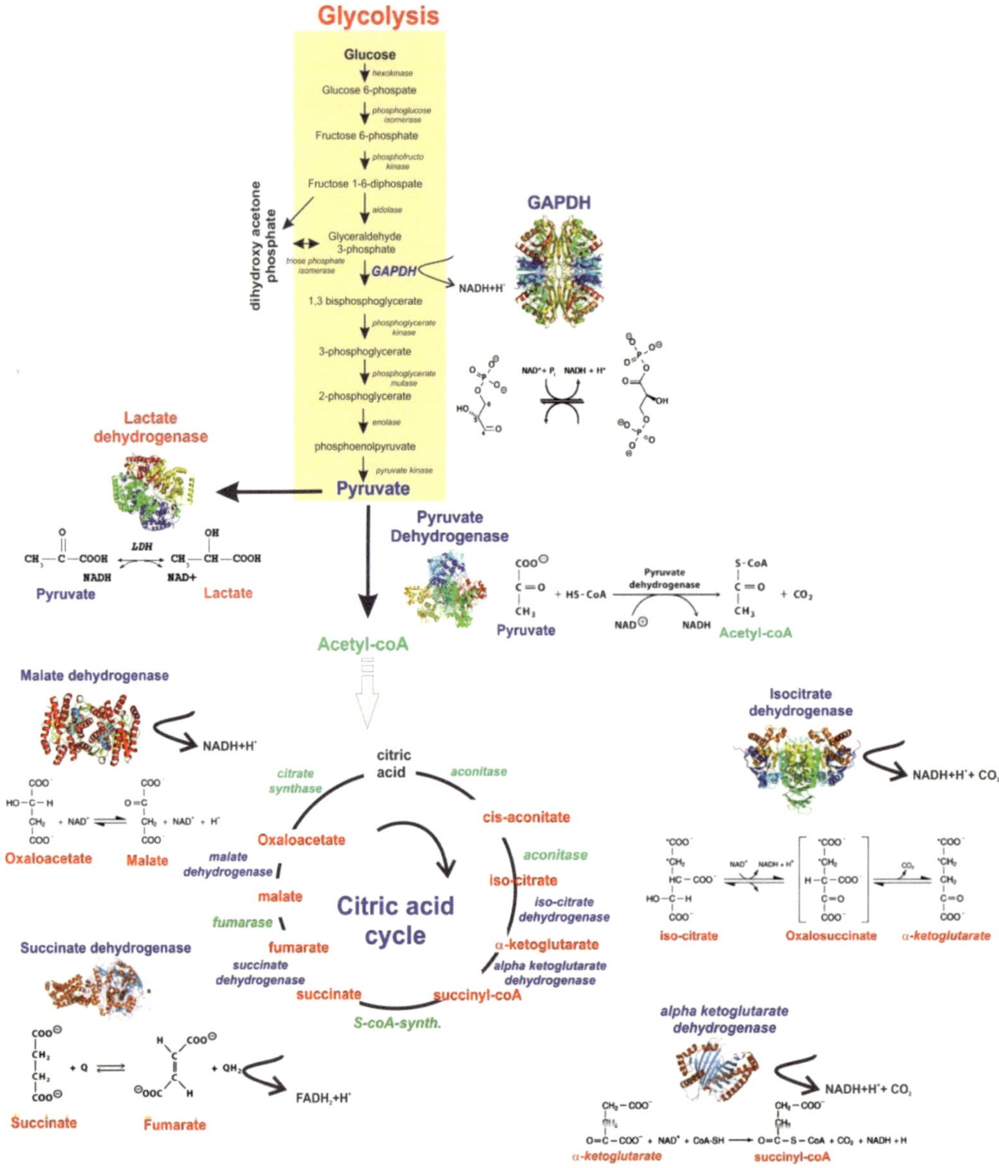

Figure 3. Enzymatic pathways of glycolysis and Kreb's cycle. Note the DHOs involved in the different metabolic pathways. (glycolysis and citric acid cycles).

It has been an age old concept that glucose is the major energy source to maintain brain functions and that lactate does not provide further metabolic use for the neuronal activities (25-28). However, it is now accepted that lactate can be produced in the brain under aerobic conditions, this is known as aerobic glycolysis. Aerobic glycolysis under normal conditions contributes to about 10% of the total energy production in the brain, which can increase under ischemic conditions(29). Lactate thus formed can then be shuttled between the various cell types in the brain, be converted back into pyruvate and be fully reutilized through complete aerobic respiration. Although this concept of lactate shuttle was proposed long ago, it received lot of resistance from the scientific community especially from those who believed that glucose is the sole substrate in the brain (30;31). However now, it is getting accepted that lactose is also used as a main substrate for energy production under normal conditions (32-34). The direction of lactate flow between various cell types in the brain as well as its relative contribution with respect to glucose to the overall energy production is still debatable and not yet fully understood. As far as the lactate flow between the different cell types in the brain there are two current schools of thoughts. One supporting the Astrocyte-lactate-neuron shuttle (ALNS) (33;35-37) and the second advocating a Neuron-Lactate-Astrocyte-shuttle (NLAS) (38-40).

Classically, glucose is transported from the cerebral blood flow into the neurons or astrocytes via glucose transporters such as Glut-1 and Glut-3 respectively (see **Figure 4**). This glucose is then metabolized to produce ATP through complete aerobic respiration (glycolysis and citric acid pathways). In the astrocytes, part of the glucose also gets converted into glycogen, as an energy reserve to be used under critical conditions of low oxygen supply.

Astrocytes can withstand low oxygen tension for a longer period of time than neurons and have proven to be more resilient to hypoxic insults (7;41-43). Based on the concept of lactate shuttle, at the astrocytic level (under resting conditions), glucose can be converted to lactate. Lactate thus produced can shuttle first into the interstitial fluid (through monocarboxylate transporter –MCT-1 & 4). From there lactate can be influxed into neurons by the neuronal specific MCT-2 transporter and converted back into its pyruvate form ready to undertake complete aerobic metabolism (36;44;45). This route is known as the Astrocyte-lactate-neuron shuttle (ALNS shuttle).

Continuous glutamatergic activation of neurons results in a more exhaustive energy expenditure. As pyruvate utilization during the TCA cycle increases and its cytoplasmic levels decrease correspondingly, the condition becomes favorable for increasing both glucose and lactate utilization. By the late phase of activation, glutamate released is taken up by the astrocytes for recycling. Aerobic glycolysis is enhanced in the astrocytes with increased lactate production. The lactate thus formed helps sustain the energy demands of astrocytes as well as replenishing the neurons. During intense and prolonged neuronal stimulation which may occur under certain conditions, energy replenishment becomes crucially important. This is because the continuous glutamate reuptake by the Na^+, glutamate co-transporter in astrocytes (GLutamate ASpartate Transporter – GLAST; glutamate transporter 1 –GLT1) must be paired with an equivalent intense activity of the Na^+, K^+ATPase to efflux the Na^+ back in the extracellular space thus continuing the cycle. When the extracellular glucose levels become insufficient to sustain this level of activity then glycogen stored in the astrocyte is mobilized to provide the extra glycosyl units necessary to support the cellular activity. Thus sustained activation of the neurons results in conversion of the stored energy substrate glycogen to glucose and further lactate production for shuttling to the neurons. This concept of ALNS shuttle during activation has received opposition and few groups suggest that lactate is primarily produced by the neurons and is then transported from the neurons to the astrocytes (38;40).

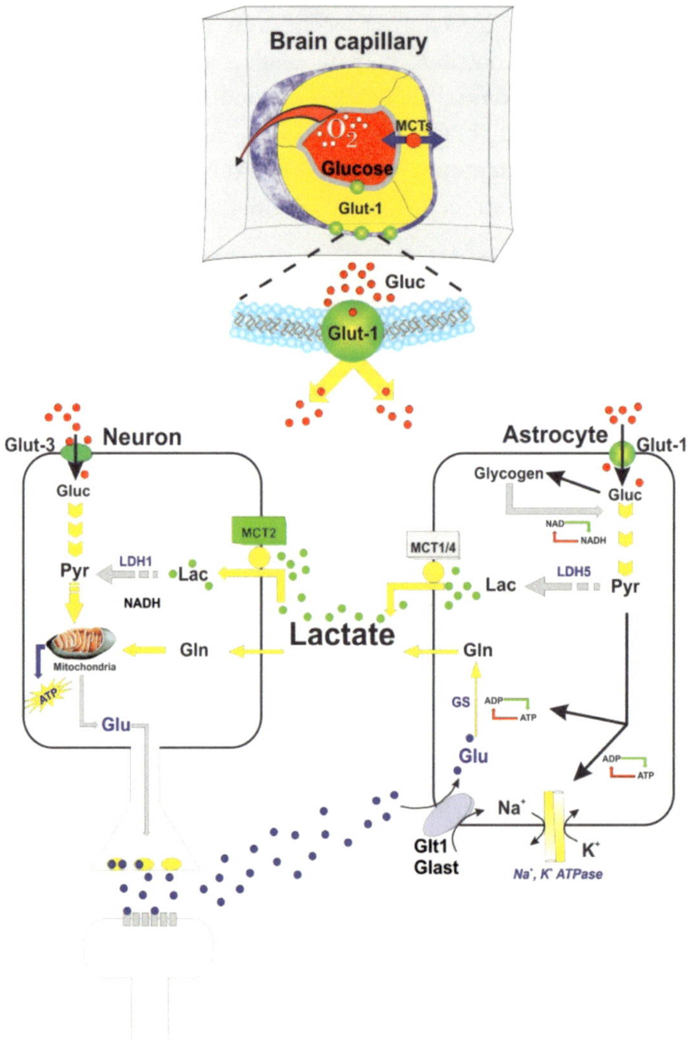

Figure 4. Schematic of neuroenergetics pathways: The astrocyte – neuron lactate shuttle.

In summary whether lactate flux moves from astrocytes to neurons or vice-versa the important point is that glucose is not the sole substrate utilized in the brain. Lactate plays an equally important role especially during activation, neurotransmission and pathological conditions such as under ischemic insults. This further emphasizes the still dismissed importance of lactate dehydrogenase which represents the key switch in the metabolic pathway of glucose in the generation of lactate.

In summary at the microcapillary level, the BBB acts as a functional interface which is charged with the critical task to fuel the brain with energy sources. Whether this is glucose or glucose-derived lactate the BBB is the main fuel distribution system to the brain and a safe for energy storage to which the brain can avail when normal fuel supplies are short.

4. IEM

Mutation in single gene leads to enzyme deficiency that results in type of genetic disorders termed as inborn error in metabolism (IEM) (46-50). IEM related deficiencies are generally autosomal recessive or X-linked. IEM linked enzyme deficiencies may be directly related to metabolic pathways of energy metabolism, purines/pyrimidine synthesis or degradation, amino acid synthesis, fatty acid oxidation, etc. Metabolic errors lead to accumulation of toxic or absence of essential products in the brain with neurological implications such as ataxia-motor control, encephalopathy, mental deficits, learning disabilities and mental retardation with structural anomalies. Some examples of DHOs involved include the pyruvate dehydrogenase complex.

4.1. Pyruvate dehydrogenase (PDH) deficiency

Pyruvate dehydrogenase is a multi-enzyme complex which catalyzes the conversion of pyruvate (the end product of glycolysis) into acetyl-coA- a substrate that can enter citric acid cycle (for production of ATP and energy equivalents). PDH is a six subunit complex composed of E1-pyruvate dehydrogenase, E2-dihydrolipoyl transacetylase and E3-dihydrolipoyl dehydrogenase, E3BP- E3 binding protein and two regulatory subunits - pyruvate dehydrogenase kinase and pyruvate dehydrogenase phosphatase. Although several mutations in the PDH complex deficiency (such as point mutations, deletions, duplications) have been reported so far; deficiency in PDH E1-alpha subunit (abbreviated as PDHA1) is the most common type (51-53). PDH deficiencies due to mutations in other subunits of the PDH complex are comparatively rare. All mutations leading to PDH deficiency are X linked except, the one in regulatory units, which are autosomally recessive (54-57). Since PDH results in acetyl co-A formation, the most common clinical manifestation of PDH deficiency is severe lactic acidosis. Defects in energy metabolism can cause neurological deficits such as mental retardation, developmental delay as well as psychomotor retardation. Hypertonia/hypotonia, ataxia, motor dysfunction like spasticity are the more common symptoms observed (58). Structural anomalies (such as microcephaly, facial dysmorphism) and epilepsy (focal or generalized seizures- both have been reported) may develop *in utero*. Optic atrophy, nystagmus and strabismus are observed at ocular level whereas peripheral neuropathies such as in nerve conduction have also been reported.

Based on clinical case studies, PDH deficiencies have been classified into four typical neurological patterns (58-60):

i. Neonatal encephalopathic pattern with facial dimorphic features and cerebral developmental defects, prenatal brain lesions, affecting the females
ii. Leigh syndrome like presentation with symmetric necrotic lesions of basal ganglia, more common in males

iii. A chronic relapsing ataxia with prolonged survival
iv. Static encephalopathy, cerebral palsy like motor deficits associated with paroxysmal dystonia.

Alkali (such as sodium bicarbonate) administration to neutralize severe lactic acidosis and provide immediate temporary relief has been reported to treat acute episodes of severe acidosis. Chronic treatment strategies for PDH deficiency on the other hand, include incorporation of ketogenic diet consisting of high fat, low carbohydrate and low protein. High doses of thiamine can be beneficial in treating thiamine responsive PDH deficiency. Dichloroacetate (DCA) can reduce the inhibition of PDHc phosphorylation and thus can be used to treat severe lactic acidosis to some extent.

4.2. Branched Chain Alpha Ketoacid dehydrogenase (BCKDH) complex Deficiency - Maple syrup urine disease (MSUD)

Another important dehydrogenase deficiency leading to an inborn error in metabolism is that in the branched chain alpha ketoacid dehydrogenase (BCKDH) complex (61;62). This complex is similar to PDH complex, and autosomal recessive mutations in the different subunits of the complex have been reported for this disease. In this disorder, accumulation of branched chain amino acids (BCAAs like isoleucine, leucine and valine) and branched chain alpha ketoacids (BCKAs) (with maple syrup odor to the urine) is observed along with neurological deficits and developmental disorders. Based on the characteristic manifestations and level of neurological complications, this disease is classified into the following five forms:

i. Classical MSUD: This is the most common type of MSUD with early symptoms in neonates. Neonates born normal, within 3-7 days of birth show symptoms such as lethargy, weight loss, metabolic dearrangement, encephalopathy with hypotonia and hypertonia. It is characterized by seizures, coma and even death if not treated.
ii. Intermediate MSUD: This type shows mild but persistent ketoacidosis and developmental delay with 3-30% less dehydrogenase activity.
iii. Intermittent MSUD: Episodic ataxia, semicoma, elevated BCAAs and BCKAs occur after episodes of infection or acute illness. Cognitive functions may be affected only in case of repeated episodes of acute illness.
iv. Thiamine responsive MSUD: Mutation in E2 protein of the BCKDH complex results in reduced affinity of cofactor thiamine pyrophosphate (TPP), thus this form of disorder can be treated with thiamine substitution. Reduced activity of the complex results in hyperaminoacidemia.
v. E3-deficient MSUD: Combined deficiency of E3 subunit (common component of all three mitochondrial multi enzyme complexes - BCKDH, pyruvate dehydrogenase complex and alpha ketoglutarate dehydrogenase) results in elevated lactate, pyruvate and alpha ketoglutarate along with BCAAs and BCKAs.

Thus, all these forms of IEM are caused due to varying degree of deficiency in the enzyme activity leading to varying levels of neurological complication. Treatment is initiated by high calories leucine free diet rich with BCAA-free formulas and an optimum supplementation of isoleucine and valine. Hemodialysis or hemofiltration may be used to remove deposited BCAAs and BCKAs from the body. During acute MSUD, brain edema and hyponatremia can also occur but can be immediately treated by administration of mannitol or diuretic drugs. In adolescents and adults, it can lead to depression, anxiety. However, the burden of these pathologies can be decreased by treatment with appropriate standard drugs such as psychostimulants or antianxiety drugs. Although no direct drug is used to treat MSUD, recent studies have shown the role of phenylbutyrate in increasing BCKDH activity, reducing levels of BCAAs and BCKAs and causing relief in MSUD patients (63). However, careful monitoring and routine biochemical testing is key in appropriate treatment in MSUD affected patients.

4.3. Succinic semialdehyde dehydrogenase (SSD) deficiency

SSD deficiency is an autosomal recessive disorder of γ-hydroxybutyric acid (GABA) metabolism. In human brain, GABA is the most important inhibitory neurotransmitter. Oxidative conversion of succinate semialdehyde to succinic acid is impaired in this deficiency. This leads to production of γ-hydroxybutyrate (GHB) (see **Figure 5**). ALDH5A1 is the only gene associated with this deficiency. Mild developmental delay, psychomotor retardation, hypotonia, ataxia are observed along with extrapyramidal symptoms such as dystonia, choreoathetosis and myoclonus. More than 50% of affected individuals develop seizures (64-67). Neuroimaging screening generally reveals hyper intensities in globus pallidus, sub cortical white matter, cerebellar dentate nucleus and brain stem (68). Accumulation of γ-amino butyric acid (GABA) and GHB are considered positive indicators for this disease which can be confirmed by testing of SSD enzyme activity in leucocytes.

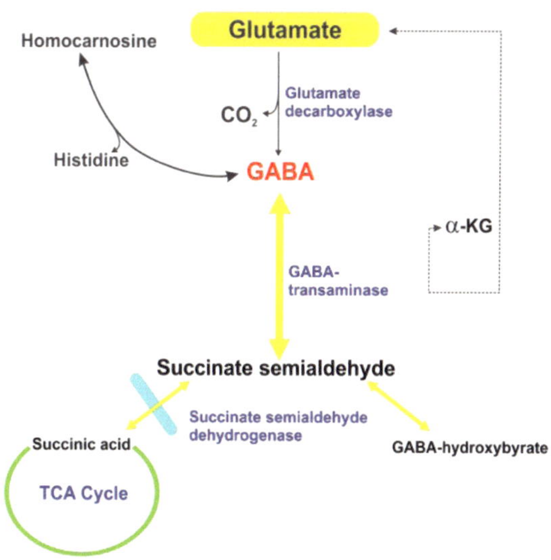

Figure 5. SSD deficiency: In the absence of SSD, transamination of γ-aminobutyric acid (GABA) to succinic semialdehyde is followed by reduction to 4-hydroxybutyric acid (γ-hydroxybutyrate [GHB]). SSADH deficiency leads to significant accumulation of GHB and GABA.

Current therapies are mostly symptomatic, directed at seizure treatment and amelioration of neurobehavioral symptoms. Antiepileptic drugs such as carbamazepine and anti-anxiety drugs may be administered in conjunction with physical and occupational therapy. Early attempts to use Vigabtrin (GABA transaminase inhibitor) did not meet the therapeutic expectations (69).

4.4. IEM related to fatty acid oxidation

Fatty acids are a major source of energy in heart as well as muscle. Fatty acid oxidation (FAO) is a series of four reactions occurring in mitochondria. The first step is catalyzed by four straight chain acyl coA dehydrogenases such as:

- Short chain acyl coA dehydrogenase (C4-C6 fatty acyl coAs)
- Medium Chain acyl coA dehydrogenase (C6-C10 fatty acyl coAs)
- Long Chain acyl coA dehydrogenase (C10-C14 fatty acyl coAs)
- Very Long Chain acyl coA dehydrogenase (C14-C20 fatty acyl coAs)

Medium chain acyl-coA dehydrogenase (MCAD) deficiency is the most common fatty acid oxidation-related disorder (1:10000 to 1:30000 in US) which is inherited in an autosomal recessive fashion (70-72). MCAD is an enzyme that catalyzes breakdown of fatty acids for energy production during long periods of prolonged fasting. Accumulation of octanoylcarnitine with Reye-like syndrome is typical clinical manifestation of this disorder (73;74). Children can exhibit severe hypoglycemia in mild illnesses. It can also lead to sudden infant like death syndrome (75). Symptoms may appear from 2 days to 6.5 years of age, however the patient can also remain asymptomatic for long time. When left undiagnosed MCAD deficiency has a mortality of 20% and 10-15% are severely handicapped. A case study of diagnosis as late as 30 years of age is reported in literature. The 30 year old man exhibited rhabdomyolysis, muscle weakness, acute encephalopathy after exertion in cold and fasting. Urine detection of carnitine led to the diagnosis of MCAD deficiency. Point mutation at position 985 in the coding region of MCAD gene was detected. 449_452 deletion mutation is also studied. During acute episodes, symptomatic relief to overcome hypoglycemia cerebral edema, seizures or metabolic acidosis is the main line of treatment. Avoiding long periods of fasting is the best preventive measure that can be employed in cases with MCAD deficiency.

Short chain acyl coenzyme A dehydrogenase deficiency (SCAD) is another autosomal recessive disorder in mitochondrial fatty acid oxidation. It is characterized by increased C4-carnitine in plasma and ethylmalonic acid in urine. Clinical symptoms which appear early in life include developmental delay, hypotonia, epilepsy and behavioral disorders along with hypoglycemia and myopathy (76;77). Unlike MCAD deficiency, if neonatally screened and followed up it is found to remain asymptomatic, thus the clinical disease outcome of SCAD deficiency is questionable. Thus, need for treatment is not clear. Avoidance of fasting for longer hours with age appropriate diet is the only recommendation for prevention of primary manifestation. Annual checkup for growth, development is generally suggested.

5. Aging: The role of dehydrogenases in metabolic and mitochondrial dysfunction

Aging or growing old is defined as a time related loss and decline in certain morphological, anatomical and functional features of body in comparison to its previous state. Beginning as a maturation process from childhood to young adulthood, it assumes the characteristic of decline through middle and late ages. Accumulation of molecular, cellular, or organ level damage leads to higher vulnerability of disease and eventually death. There have been numerous theories and hypothesis for causes of aging but it is still under investigation and discussion. "The Free Radical Theory of aging (ROS generation), shortening of telomerase, DNA methylation and epigenetics are few main ones. Important to us is the "The Free Radical Theory of aging since it is closely associated with mitochondria, and linked DHOs (78-81).

Broadly, both genetic as well as external environmental factors can be responsible for promoting the age associated decline in functionalities (82). Oxidative stress and dietary restrictions can influence the genes externally. "Oxygen derived species", "Reactive nitrogen species" and "Reactive aldehydic species" can cause changes at cellular level ensuing damage to our natural defense mechanisms affecting repair and elimination processes in the body. In totality, irregularities in function, oxidative changes and the piled up cellular damages can lead to homoeostatic imbalance which finally result in aging as well as age-related diseases. "The Free Radical Theory of aging" suggests generation of superoxide radical, hydrogen peroxide and hydroxyl radical as a side reaction to the electron transport chain at mitochondrial membrane (83). These free radicals can cause enzyme inactivation to different extent with different mechanisms (see **Figure 6**). Studies show that mitochondrial enzymes are resistant to hydrogen peroxide free radical but are fairly affected by hydroxyl free radical. On the other hand oxygen free radical by itself can cause significant oxidative damage with respect to inactivation of mitochondrial enzymes like NADH dehydrogenase, succinate dehydrogenase, NADH oxidase, succinate oxidase and ATPase 2.

Link between aging and various dehydrogenase enzymes is based on the energy demand of our body which involves the participation of different dehydrogenases for production of ATP at cellular level (as elaborated in the earlier section of energy metabolism). Several dehydrogenases involved in energy metabolism can exhibit altered activity or complete inactivation with aging. This can result in hampering energy production as well as accumulation of toxic metabolites in the body.

6. Metabolic dysfunction and its link to Alzheimer's disease: The role of dehydrogenases

Alzheimer's disease is the most common form of dementia characterized by loss of memory, cognitive decline and change in perception and behavior. Pathological hallmarks include accumulation of amyloid beta protein (Aβ) and resulting plaque formation (a cleavage protein of amyloid precursor protein-APP) and formation of neurofibrillary tangles (due to hyper phosphorylation of microtubule associated protein of neurons in the brain). Genetic mutations in APP protein or ApoE protein (a protein linked to lipid and cholesterol in the body) as well as Down's syndrome are some examples of genetic predispositions that increase the propensity to develop Alzheimer's disease. However, the etiology of the disease is still questionable. Both vascular as well as metabolic dysfunction have been accredited as major factors prodromic to the pathogenesis and progression of Alzheimer's disease (84). Vascular dysfunction includes reduction in cerebral blood flow, reduced glucose uptake, and reduced amyloid beta clearance with cerebral amyloid angiopathy. The patients with atherosclerosis have shown to have an increased risk to develop AD. Vascular structural anomalies in cerebral vessels (like increased pinocytic activity as well as swelling) have been seen in early AD further signifying the importance of vascular dysfunction in AD pathogenesis. Defective glucose utilization has been observed earlier than reduced cerebral blood flow (CBF), indicating the role of glucose metabolism in AD development. What is of primary interest to us is that it is slowly becoming well accepted that metabolic dysfunction, related oxidative stress and mitochondrial deficits can precede AD development (84-86). In one such recent study, triple transgenic mice 3xTg-AD were developed (having mutations in human APP_{SWE}, Tau_{P301L}, $PS1_{M146V}$ genes linked with AD) along with their respective controls. Decreased mitochondrial respiration was observed along with reduced PDH activity. High levels of oxidative stress (via measurement of hydrogen peroxide production and related lipid peroxidation), high Aβ levels with high levels of Aβ binding alcohol dehydrogenase (ABAD) was observed. Decreased respiration was also observed in embryonic neurons, which continued till senescence leading to AD pathogenesis. Thus these studies clearly emphasized how mitochondrial dysfunction and resulting metabolic respiration preceded AD development. Defective enzyme function(of dehydrogenases) in pathways of energy metabolism such as glycolysis, tricarboxylic acid pathway and electron transport chain have been well studied in AD progression. Defective PDH, KGDHC, cytochrome oxidase along with reduced activity of hexokinase, phosphofructokinase are the major enzymes reported in AD so far (87). In the paragraph below we will be further discuss some of these DHOs and their mechanism of metabolic dysfunction in AD. We will also look at other DHOs apart from those directly involved in the bioenergetics which have an important role in pathogenesis of AD.

One such DHO as mentioned earlier is Aβ binding alcohol dehydrogenase (ABAD). ABAD is a short chain alcohol dehydrogenase which is also called type II hydroacylcoA dehydrogenase, 17β-hydrosteroid dehydrogenase type 10 and 2-methyl-3-hydroxybutyrylcoA dehydrogenase. ABAD acts on various substrates such as branched chain fatty acids, alcohols, amino acid catabolites and steroids. In the brain, it is primarily localized in mitochondria of neurons. Previous studies have shown high ABAD expression in the temporal lobe and hippocampus of AD affected patients (88;89).

Animals with transgenic (Tg) APP mice have also demonstrated higher expression of ABAD (88). As the name suggests, ABAD directly binds to Aβ protein which is highly expressed in Alzheimer's patients. These links suggest role of ABAD dehydrogenase in the pathophysiology of Alzheimer's disease. Another crucial factor in causing AD as discussed earlier is mitochondrial dysfunction, related oxidative stress and hypometabolism. In recent studies, it was hypothesized that ABAD can act as a crucial link between increased Aβ production and mitochondrial dysfunction in Alzheimer's disease progression (90-92).To test this hypothesis, double transgenic mice with increased levels of ABAD and Aβ were developed (along with Tg mAP, Tg ABAD, and not Tg littermate controls)(93). Neuron cultures derived from these Tg mice showed increased ROS, oxidative stress and relative decrease in ATP production. Further studies indicated defective activity of mitochondrial Complex IV as the source of the ROS species, also such effect was not observed in single Tg mice with increased ABAD alone (suggesting Aβ acting as a crucial element linking the two). Lactate dehydrogenase (LDH) was higher in the Tg mAP/ABAD mice as compared to other groups suggesting the reversal to lactate metabolism. Cell apoptosis via caspase 3 activity was observed in *in vitro* studies. Data from the Tg mice also suggested reduced ATP production at 9 months of age along with reduced Complex IV activity. Overall, ABAD acts a crucial enzyme that can lead to mitochondrial dysfunction and disease progression in AD.

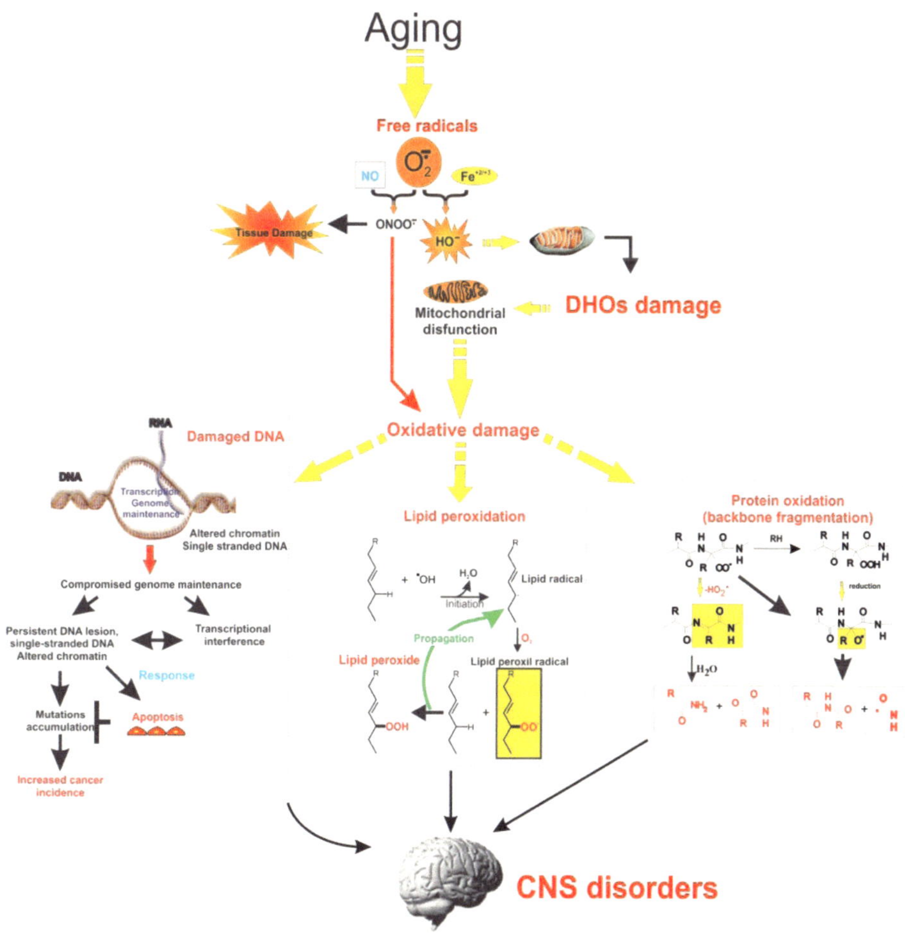

Figure 6. : Correlation between aging free radicals, DHOs and the onset of CNS disorders.

In another study, ABAD-decoy peptide (ABAD-DP) was introduced in the Tg mAPP mice, which prevented the interaction of ABAD with Aβ. As expected, reduction in ABAD-Aβ complex formation accompanied with attenuated oxidative stress, increased oxygen consumption, increased activity of enzymes associated with mitochondrial respiratory chain, improvement in energy metabolism, and increased spatial memory (89). Thus based on these studies, inhibitors of ABAD-Aβ hold promise as potential targets for the treatment of AD.

Another dehydrogenase that is implicated in AD progression is aldehyde dehydrogenase (ALDH) (89;94). Aldehyde dehydrogenase is observed as a key enzyme in the brain involved in metabolism and degradation of biogenic aldehydes, monoamine neurotransmitters such as norepinephrine, dopamine, diamines and GABA. Recent studies have also shown that patients with Down's syndrome have reduced activity of ALDH enzyme (95). Two dimensional analysis of proteins extracted from brain samples of nine aged patients with Down's syndrome and nine controls showed that ALDH was down regulated in the patients with Down's syndrome. This resulted in accumulation of aldehydes and further formation of tangles and plaques as observed in aged patients with Down's syndrome.

Oxidative stress and generation of ROS species has been implicated in Alzheimer's disease as elaborated earlier. These oxygen species modify proteins, nucleic acids as well as lead to lipid peroxidation. Lipid peroxidation produces toxic aldehydes such as 4-hydroxy-2-nonenal (HNE) in several disorders such as Alzheimer's as well as Parkinson's disease. In the brain, normally ALDH2- an isoform of aldehyde dehydrogenase oxidizes and degrades end product of lipid peroxidation such as HNE. The role of ALDH in oxidative stress and age dependent memory loss and decline in cognitive function was studied using a transgenic mouse model with defective ALDH2 (96). A dominant negative form of ALDH2 mice was produced and its effect on the metabolic pathways as well as accumulation of toxic products was tested. As expected HNE accumulation was observed in such transgenic mice compared to controls. Further testing of cognitive capability was performed using object recognition and water maze test. Decreased cognitive function in the transgenic mice was observed along with accumulation of tau phosphorylation (a typical pathological sign of Alzheimer's disease).

A dominant negative form of ALDH2 mice was produced and its effect on the metabolic pathways as well as accumulation of toxic products was tested. As expected HNE accumulation was observed in such transgenic mice compared to controls. Further testing of cognitive capability was performed using object recognition and water maze test. Decreased cognitive function in the transgenic mice was observed along with accumulation of tau phosphorylation (a typical pathological sign of Alzheimer's disease).

7. Metabolic dysfunction and its link to Parkinson's disease (PD): The role of dehydrogenases

PD is a neurological disorder characterized by typical motor features such as tremor bradykinesia, rigidity, slowness of movement and postural instability. Reduction in number of dopaminergic (DA) neurons located in substantia nigra pars compacta is the pathological cause of PD. It is also characterized by accumulation of α-synuclein into inclusions called Lewy bodies. 60% of DA neurons are dead and 70% responsiveness of DA is lost. Mostly PD is idiopathic, however specific genetic mutations have shown to increase the risk to develop PD. Mutations is genes such as α-synuclein, Parkin, PINK1 have been reported so far. After diagnosis of PD based on its classical symptoms and neuroimaging, treatment is usually done using levodopa (L-DOPA). L-DOPA is converted to dopamine in the brain and can temporarily alleviate the motor symptoms. Dopamine receptor agonists as well as selective monoamine oxidase-B (MAO-B) inhibitors are also administered along with L-DOPA (97;98). Treatment thus helps to partially reduce the symptoms of PD, since the actual underlying cause of this disease is still unknown. Altered enzyme activity and mitochondrial dysfunction has been linked to PD as well.

Aldehyde dehydrogenase plays an important role in detoxifying aldehydes in brain. Reduced expression of isoforms of ALDH such as ALDH1A1 and ALDH2 is reported in PD patients. In addition impaired Complex I activity is documented in PD which can reduce the availability of NAD^+ cofactor required by ALDHs to remove toxic biogenic aldehydes. Thus decreased ALDH function could be the underlying factor preceding the development of PD. Using transgenic mice null for both ALDH1A1 and ALDH2, the risk to develop PD was tested (99). Such mice exhibited deficits in motor performance typical of PD. Loss of DA with increased accumulation of biogenic aldehydes such as HNE was observed. L-DOPA administration alleviated the motor deficits suggesting a role of ALDHs in the pathophysiology of PD.

Another DHO implicated in PD is glutamate dehydrogenase (GDH). GDH is a key enzyme involved in interconversion of glutamate to alpha-ketoglutarate and ammonia using NADP(H) and NAD(H) as co factors. It plays an important role in homeostasis by interconnecting amino acid and carbohydrate metabolism pathways. Present in two isoforms in humans, the GDH isoform 2 (hGDH2) is overexpressed in the brain astrocytes and the sertoli cells in testis. ADP levels act as positive regulators for this enzyme and unlike the other isoform it is not inhibited by GTP. Important in recycling glutamate in the brain astrocytes, this enzyme works in concert with glutamine synthetase (GS) providing ammonia as well as ATP for GS activity. Two parallel studies have shown that increased levels of glutamate prepones the onset of the disease by 6 to 13 years (100). Hemizygous individuals with a rare variation in hGDH2 (substitution of Ala for Ser445) was detected in these individuals. GDH deficiencies have also been linked to the onset of epilepsy. All together these results highlight the role of hGDH2 in the maintenance of brain homeostasis.

ABAD associated with Alzheimer's disease has also shown to play some role in PD disease. ABAD expression is seen to be downregulated in PD patients (101). In mouse models of PD generated by administration of neurotoxin 1-methyl-4-phenyl-1,2,3,6-tetrahydropyridine (MPTP) ABAD expression is significantly reduced. By contrast, overexpression of ABAD in transgenic mice is shown to attenuate MPTP-induced dopaminergic neurogeneration. This strongly suggests that ABAD may contribute to the fate of DA neurons during the onset of PD.

8. Conclusion

The brain as well as the BBB have complex structural and functional physiology which demands a continuous supply of high energy. Bioenergetic pathways in the brain utilize multiple pathways (such as glycolytic metabolism, TCA cycle etc) to ensure that the energy requirements of the different cell types in the brain are fulfilled at all time. The BBB acts as a critical interface to buffer and influx energy substrates into the brain. Shuttling of multiple substrates such as glucose, lactose as well as glycogen derived lactate/glucose commonly occurs between the neurons and the astrocytes. Various DHOs are a critical part of these bioenergetic pathways and occurrence of DHO defect can lead to inborn errors in the metabolism followed by strong neurological complications. PDH is an imporant IEM which is directly linked to bioenergectic pathways such as TCA cycle and aerobic respiration. Apart from energy metabolism, BCKDH and SSD are IEMs that correlate to other pathways in the brain such as amino acid metabolism and neurotransmitter degradation. DHOs (such as ALDHs) also play an important role to further degrade the biogenic aldehydes derived from the degradation pathways of neurotransmitters such as for epinephrine, norepinephrine and GABA which are commonly synthesized in the brain. Furthermore, DHOs play an important role in oxidation of fatty acids as an energy supply. Although this does not occur in the brain, IEMs affecting these dehydrogenases have shown to correlate with at least one reported neurological complication (such as Reye-like syndrome).

Aging naturally promotes alterations and/or reduction in DHOs' activity which can alter mitochondrial functions leading to hypometabolism other metabolic dysfunction. This can ultimately facilitate the onset and progression of various neurological disorders such as Alzheimer's disease and PD. Specifically, altered expression/function of ABAD and ALDH2 have been associated with the pathogenesis of Alzheimer's disease whereas alteration of ALDH1A1, ALDH2, GDH2 and ABAD have been linked to PD.

In summary, DHOs play a critical role in supporting neuronal and BBB functions. They constitute an integral part of various metabolic pathways in the brain associated with energy metabolism, as well as synthesis and degradation of neurotransmitters. Their optimal functioning facilitates neuronal signaling and homeostasis. In born as well as acquired defects in DHOs have been shown to correlate with various CNS and BBB pathophysiologies.

Author details

P. Naik and S. Prasad
Department of Pharmaceutical Sciences, Texas Tech University Health Sciences Center, TX, USA

L. Cucullo
Vascular Drug Research Center, Texas Tech University Health Sciences Center, TX, USA
Department of Pharmaceutical Sciences, Texas Tech University Health Sciences Center, TX, USA

9. References

[1] Holman DW, Klein RS, Ransohoff RM. The blood-brain barrier, chemokines and multiple sclerosis. Biochim Biophys Acta 2011 Feb;1812(2):220-30.

[2] Oby E, Janigro D. The blood-brain barrier and epilepsy. Epilepsia 2006 Nov;47(11):1761-74.

[3] Pardridge WM. Alzheimer's disease drug development and the problem of the blood-brain barrier. Alzheimers Dement 2009 Sep;5(5):427-32.

[4] Shapira Y, Setton D, Artru AA, Shohami E. Blood-brain barrier permeability, cerebral edema, and neurologic function after closed head injury in rats. Anesth Analg 1993 Jul;77(1):141-8.

[5] Abbott NJ, Ronnback L, Hansson E. Astrocyte-endothelial interactions at the blood-brain barrier. Nat Rev Neurosci 2006 Jan;7(1):41-53.

[6] Abbott NJ, Dolman DE, Drndarski S, Fredriksson SM. An improved in vitro blood-brain barrier model: rat brain endothelial cells co-cultured with astrocytes. Methods Mol Biol 2012;814:415-30.

[7] Al AA, Taboada CB, Gassmann M, Ogunshola OO. Astrocytes and pericytes differentially modulate blood-brain barrier characteristics during development and hypoxic insult. J Cereb Blood Flow Metab 2011 Feb;31(2):693-705.

[8] Bell RD, Winkler EA, Sagare AP, Singh I, LaRue B, Deane R, et al. Pericytes control key neurovascular functions and neuronal phenotype in the adult brain and during brain aging. Neuron 2010 Nov 4;68(3):409-27.

[9] Abbott NJ, Patabendige AA, Dolman DE, Yusof SR, Begley DJ. Structure and function of the blood-brain barrier. Neurobiol Dis 2010 Jan;37(1):13-25.

[10] Abbott NJ, Revest PA. Control of brain endothelial permeability. Cerebrovasc Brain Metab Rev 1991;3(1):39-72.

[11] Miller DS. Regulation of P-glycoprotein and other ABC drug transporters at the blood-brain barrier. Trends Pharmacol Sci 2010 Jun;31(6):246-54.

[12] Hermann DM, Kilic E, Spudich A, Kramer SD, Wunderli-Allenspach H, Bassetti CL. Role of drug efflux carriers in the healthy and diseased brain. Ann Neurol 2006 Nov;60(5):489-98.

[13] Pardridge WM. Blood-brain barrier delivery. Drug Discov Today 2007 Jan;12(1-2):54-61.

[14] Devraj K, Klinger ME, Myers RL, Mokashi A, Hawkins RA, Simpson IA. GLUT-1 glucose transporters in the blood-brain barrier: differential phosphorylation. J Neurosci Res 2011 Dec;89(12):1913-25.

[15] Hawkins RA, O'Kane RL, Simpson IA, Vina JR. Structure of the blood-brain barrier and its role in the transport of amino acids. J Nutr 2006 Jan;136(1 Suppl):218S-26S.

[16] Lockman PR, Allen DD. The transport of choline. Drug Dev Ind Pharm 2002 Aug;28(7):749-71.

[17] Pardridge WM. Blood-brain barrier carrier-mediated transport and brain metabolism of amino acids. Neurochem Res 1998 May;23(5):635-44.

[18] Cornford EM, Hyman S. Localization of brain endothelial luminal and abluminal transporters with immunogold electron microscopy. NeuroRx 2005 Jan;2(1):27-43.

[19] Cucullo L, Hossain M, Puvenna V, Marchi N, Janigro D. The role of shear stress in Blood-Brain Barrier endothelial physiology. BMC Neurosci 2011;12:40.

[20] Johnson MD, Anderson BD. Localization of purine metabolizing enzymes in bovine brain microvessel endothelial cells: an enzymatic blood-brain barrier for dideoxynucleosides? Pharm Res 1996 Dec;13(12):1881-6.

[21] Meyer J, Mischeck U, Veyhl M, Henzel K, Galla HJ. Blood-brain barrier characteristic enzymatic properties in cultured brain capillary endothelial cells. Brain Res 1990 Apr 30;514(2):305-9.

[22] Sanchez del Pino MM, Hawkins RA, Peterson DR. Biochemical discrimination between luminal and abluminal enzyme and transport activities of the blood-brain barrier. J Biol Chem 1995 Jun 23;270(25):14907-12.

[23] Cornford EM, Varesi JB, Hyman S, Damian RT, Raleigh MJ. Mitochondrial content of choroid plexus epithelium. Exp Brain Res 1997 Oct;116(3):399-405.

[24] Belanger M, Allaman I, Magistretti PJ. Brain energy metabolism: focus on astrocyte-neuron metabolic cooperation. Cell Metab 2011 Dec 7;14(6):724-38.

[25] Hilgier W, Benveniste H, Diemer NH, Albrecht J. Decreased glucose utilization in discrete brain regions of rat in thioacetamide-induced hepatic encephalopathy as measured with [3H]-deoxyglucose. Acta Neurol Scand 1991 Jun;83(6):353-5.

[26] Hoyer S. Brain glucose and energy metabolism during normal aging. Aging (Milano) 1990 Sep;2(3):245-58.

[27] Hoyer S. The effect of age on glucose and energy metabolism in brain cortex of rats. Arch Gerontol Geriatr 1985 Oct;4(3):193-203.

[28] Gardiner M, Smith ML, Kagstrom E, Shohami E, Siesjo BK. Influence of blood glucose concentration on brain lactate accumulation during severe hypoxia and subsequent recovery of brain energy metabolism. J Cereb Blood Flow Metab 1982 Dec;2(4):429-38.

[29] Vaishnavi SN, Vlassenko AG, Rundle MM, Snyder AZ, Mintun MA, Raichle ME. Regional aerobic glycolysis in the human brain. Proc Natl Acad Sci U S A 2010 Oct 12;107(41):17757-62.

[30] Pellerin L, Pellegri G, Bittar PG, Charnay Y, Bouras C, Martin JL, et al. Evidence supporting the existence of an activity-dependent astrocyte-neuron lactate shuttle. Dev Neurosci 1998;20(4-5):291-9.

[31] Brooks GA. The lactate shuttle during exercise and recovery. Med Sci Sports Exerc 1986 Jun;18(3):360-8.

[32] Serres S, Bouyer JJ, Bezancon E, Canioni P, Merle M. Involvement of brain lactate in neuronal metabolism. NMR Biomed 2003 Oct;16(6-7):430-9.

[33] Pellerin L. Lactate as a pivotal element in neuron-glia metabolic cooperation. Neurochem Int 2003 Sep;43(4-5):331-8.

[34] Bergersen LH. Is lactate food for neurons? Comparison of monocarboxylate transporter subtypes in brain and muscle. Neuroscience 2007 Mar 2;145(1):11-9.

[35] Pellerin L, Pellegri G, Bittar PG, Charnay Y, Bouras C, Martin JL, et al. Evidence supporting the existence of an activity-dependent astrocyte-neuron lactate shuttle. Dev Neurosci 1998;20(4-5):291-9.

[36] Pellerin L, Bouzier-Sore AK, Aubert A, Serres S, Merle M, Costalat R, et al. Activity-dependent regulation of energy metabolism by astrocytes: an update. Glia 2007 Sep;55(12):1251-62.
[37] Pellerin L. Brain energetics (thought needs food). Curr Opin Clin Nutr Metab Care 2008 Nov;11(6):701-5.
[38] Simpson IA, Carruthers A, Vannucci SJ. Supply and demand in cerebral energy metabolism: the role of nutrient transporters. J Cereb Blood Flow Metab 2007 Nov;27(11):1766-91.
[39] Mangia S, Tkac I, Gruetter R, Van de Moortele PF, Maraviglia B, Ugurbil K. Sustained neuronal activation raises oxidative metabolism to a new steady-state level: evidence from 1H NMR spectroscopy in the human visual cortex. J Cereb Blood Flow Metab 2007 May;27(5):1055-63.
[40] Mangia S, Simpson IA, Vannucci SJ, Carruthers A. The in vivo neuron-to-astrocyte lactate shuttle in human brain: evidence from modeling of measured lactate levels during visual stimulation. J Neurochem 2009 May;109 Suppl 1:55-62.
[41] Rossi DJ, Brady JD, Mohr C. Astrocyte metabolism and signaling during brain ischemia. Nat Neurosci 2007 Nov;10(11):1377-86.
[42] Takano T, Oberheim N, Cotrina ML, Nedergaard M. Astrocytes and ischemic injury. Stroke 2009 Mar;40(3 Suppl):S8-12.
[43] Ziu M, Fletcher L, Rana S, Jimenez DF, Digicaylioglu M. Temporal differences in microRNA expression patterns in astrocytes and neurons after ischemic injury. PLoS One 2011;6(2):e14724.
[44] Turner DA, Adamson DC. Neuronal-astrocyte metabolic interactions: understanding the transition into abnormal astrocytoma metabolism. J Neuropathol Exp Neurol 2011 Mar;70(3):167-76.
[45] Figley CR. Lactate transport and metabolism in the human brain: implications for the astrocyte-neuron lactate shuttle hypothesis. J Neurosci 2011 Mar 30;31(13):4768-70.
[46] Fingerhut R, Olgemoller B. Newborn screening for inborn errors of metabolism and endocrinopathies: an update. Anal Bioanal Chem 2009 Mar;393(5):1481-97.
[47] Garcia-Cazorla A, Wolf NI, Serrano M, Perez-Duenas B, Pineda M, Campistol J, et al. Inborn errors of metabolism and motor disturbances in children. J Inherit Metab Dis 2009 Oct;32(5):618-29.
[48] Jurecka A. Inborn errors of purine and pyrimidine metabolism. J Inherit Metab Dis 2009 Apr;32(2):247-63.
[49] Raghuveer TS, Garg U, Graf WD. Inborn errors of metabolism in infancy and early childhood: an update. Am Fam Physician 2006 Jun 1;73(11):1981-90.
[50] Wolf NI, Garcia-Cazorla A, Hoffmann GF. Epilepsy and inborn errors of metabolism in children. J Inherit Metab Dis 2009 Oct;32(5):609-17.
[51] Dahl HH, Maragos C, Brown RM, Hansen LL, Brown GK. Pyruvate dehydrogenase deficiency caused by deletion of a 7-bp repeat sequence in the E1 alpha gene. Am J Hum Genet 1990 Aug;47(2):286-93.
[52] Dahl HH, Brown GK. Pyruvate dehydrogenase deficiency in a male caused by a point mutation (F205L) in the E1 alpha subunit. Hum Mutat 1994;3(2):152-5.

[53] Hansen LL, Brown GK, Brown RM, Dahl HH. Pyruvate dehydrogenase deficiency caused by a 5 base pair duplication in the E1 alpha subunit. Hum Mol Genet 1993 Jun;2(6):805-7.
[54] Brown RM, Head RA, Brown GK. Pyruvate dehydrogenase E3 binding protein deficiency. Hum Genet 2002 Feb;110(2):187-91.
[55] Cerna L, Wenchich L, Hansikova H, Kmoch S, Peskova K, Chrastina P, et al. Novel mutations in a boy with dihydrolipoamide dehydrogenase deficiency. Med Sci Monit 2001 Nov;7(6):1319-25.
[56] Prasad C, Rupar T, Prasad AN. Pyruvate dehydrogenase deficiency and epilepsy. Brain Dev 2011 Nov;33(10):856-65.
[57] Robinson BH, MacKay N, Petrova-Benedict R, Ozalp I, Coskun T, Stacpoole PW. Defects in the E2 lipoyl transacetylase and the X-lipoyl containing component of the pyruvate dehydrogenase complex in patients with lactic acidemia. J Clin Invest 1990 Jun;85(6):1821-4.
[58] Patel KP, O'Brien TW, Subramony SH, Shuster J, Stacpoole PW. The spectrum of pyruvate dehydrogenase complex deficiency: clinical, biochemical and genetic features in 371 patients. Mol Genet Metab 2012 Jan;105(1):34-43.
[59] Barnerias C, Saudubray JM, Touati G, de LP, Dulac O, Ponsot G, et al. Pyruvate dehydrogenase complex deficiency: four neurological phenotypes with differing pathogenesis. Dev Med Child Neurol 2010 Feb;52(2):e1-e9.
[60] Robinson BH, MacMillan H, Petrova-Benedict R, Sherwood WG. Variable clinical presentation in patients with defective E1 component of pyruvate dehydrogenase complex. J Pediatr 1987 Oct;111(4):525-33.
[61] Mitsubuchi H, Owada M, Endo F. Markers associated with inborn errors of metabolism of branched-chain amino acids and their relevance to upper levels of intake in healthy people: an implication from clinical and molecular investigations on maple syrup urine disease. J Nutr 2005 Jun;135(6 Suppl):1565S-70S.
[62] Strauss KA, Puffenberger EG, Morton DH. Maple Syrup Urine Disease. 1993.
[63] Brunetti-Pierri N, Lanpher B, Erez A, Ananieva EA, Islam M, Marini JC, et al. Phenylbutyrate therapy for maple syrup urine disease. Hum Mol Genet 2011 Feb 15;20(4):631-40.
[64] Pearl PL, Reehal T, Drillings I, Gibson KM. Succinic Semialdehyde Dehydrogenase Deficiency. 1993.
[65] Pearl PL, Gibson KM, Cortez MA, Wu Y, Carter SO, III, Knerr I, et al. Succinic semialdehyde dehydrogenase deficiency: lessons from mice and men. J Inherit Metab Dis 2009 Jun;32(3):343-52.
[66] Pearl PL, Novotny EJ, Acosta MT, Jakobs C, Gibson KM. Succinic semialdehyde dehydrogenase deficiency in children and adults. Ann Neurol 2003;54 Suppl 6:S73-S80.
[67] Buzzi A, Wu Y, Frantseva MV, Perez Velazquez JL, Cortez MA, Liu CC, et al. Succinic semialdehyde dehydrogenase deficiency: GABAB receptor-mediated function. Brain Res 2006 May 23;1090(1):15-22.
[68] Acosta MT, Munasinghe J, Pearl PL, Gupta M, Finegersh A, Gibson KM, et al. Cerebellar atrophy in human and murine succinic semialdehyde dehydrogenase deficiency. J Child Neurol 2010 Dec;25(12):1457-61.

[69] Gropman A. Vigabatrin and newer interventions in succinic semialdehyde dehydrogenase deficiency. Ann Neurol 2003;54 Suppl 6:S66-S72.
[70] Iafolla AK, Thompson RJ, Jr., Roe CR. Medium-chain acyl-coenzyme A dehydrogenase deficiency: clinical course in 120 affected children. J Pediatr 1994 Mar;124(3):409-15.
[71] Matern D, Rinaldo P. Medium-Chain Acyl-Coenzyme A Dehydrogenase Deficiency. 1993.
[72] Rice G, Brazelton T, III, Maginot K, Srinivasan S, Hollman G, Wolff JA. Medium chain acyl-coenzyme A dehydrogenase deficiency in a neonate. N Engl J Med 2007 Oct 25;357(17):1781.
[73] Cyriac J, Venkatesh V, Gupta C. A fatal neonatal presentation of medium-chain acyl coenzyme a dehydrogenase deficiency. J Int Med Res 2008 May;36(3):609-10.
[74] Gosalakkal JA, Kamoji V. Reye syndrome and reye-like syndrome. Pediatr Neurol 2008 Sep;39(3):198-200.
[75] Yusupov R, Finegold DN, Naylor EW, Sahai I, Waisbren S, Levy HL. Sudden death in medium chain acyl-coenzyme a dehydrogenase deficiency (MCADD) despite newborn screening. Mol Genet Metab 2010 Sep;101(1):33-9.
[76] van Maldegem BT, Wanders RJ, Wijburg FA. Clinical aspects of short-chain acyl-CoA dehydrogenase deficiency. J Inherit Metab Dis 2010 Oct;33(5):507-11.
[77] Wolfe L, Jethva R, Oglesbee D, Vockley J. Short-Chain Acyl-CoA Dehydrogenase Deficiency. 1993.
[78] Harman D. Origin and evolution of the free radical theory of aging: a brief personal history, 1954-2009. Biogerontology 2009 Dec;10(6):773-81.
[79] Biesalski HK. Free radical theory of aging. Curr Opin Clin Nutr Metab Care 2002 Jan;5(1):5-10.
[80] Ashok BT, Ali R. The aging paradox: free radical theory of aging. Exp Gerontol 1999 Jun;34(3):293-303.
[81] Harman D. Free radical theory of aging: history. EXS 1992;62:1-10.
[82] Yu BP. Aging and oxidative stress: modulation by dietary restriction. Free Radic Biol Med 1996;21(5):651-68.
[83] Cadenas E, Davies KJ. Mitochondrial free radical generation, oxidative stress, and aging. Free Radic Biol Med 2000 Aug;29(3-4):222-30.
[84] Cai H, Cong WN, Ji S, Rothman S, Maudsley S, Martin B. Metabolic dysfunction in Alzheimer's disease and related neurodegenerative disorders. Curr Alzheimer Res 2012 Jan;9(1):5-17.
[85] Swerdlow RH, Khan SM. The Alzheimer's disease mitochondrial cascade hypothesis: an update. Exp Neurol 2009 Aug;218(2):308-15.
[86] Swerdlow RH, Burns JM, Khan SM. The Alzheimer's disease mitochondrial cascade hypothesis. J Alzheimers Dis 2010;20 Suppl 2:S265-S279.
[87] Blass JP, Sheu RK, Gibson GE. Inherent abnormalities in energy metabolism in Alzheimer disease. Interaction with cerebrovascular compromise. Ann N Y Acad Sci 2000 Apr;903:204-21.

[88] Yao J, Irwin RW, Zhao L, Nilsen J, Hamilton RT, Brinton RD. Mitochondrial bioenergetic deficit precedes Alzheimer's pathology in female mouse model of Alzheimer's disease. Proc Natl Acad Sci U S A 2009 Aug 25;106(34):14670-5.

[89] Yao J, Du H, Yan S, Fang F, Wang C, Lue LF, et al. Inhibition of amyloid-beta (Abeta) peptide-binding alcohol dehydrogenase-Abeta interaction reduces Abeta accumulation and improves mitochondrial function in a mouse model of Alzheimer's disease. J Neurosci 2011 Feb 9;31(6):2313-20.

[90] Marques AT, Fernandes PA, Ramos MJ. ABAD: a potential therapeutic target for Abeta-induced mitochondrial dysfunction in Alzheimer's disease. Mini Rev Med Chem 2009 Jul;9(8):1002-8.

[91] Takuma K, Yao J, Huang J, Xu H, Chen X, Luddy J, et al. ABAD enhances Abeta-induced cell stress via mitochondrial dysfunction. FASEB J 2005 Apr;19(6):597-8.

[92] Yan SD, Stern DM. Mitochondrial dysfunction and Alzheimer's disease: role of amyloid-beta peptide alcohol dehydrogenase (ABAD). Int J Exp Pathol 2005 Jun;86(3):161-71.

[93] Yao J, Irwin RW, Zhao L, Nilsen J, Hamilton RT, Brinton RD. Mitochondrial bioenergetic deficit precedes Alzheimer's pathology in female mouse model of Alzheimer's disease. Proc Natl Acad Sci U S A 2009 Aug 25;106(34):14670-5.

[94] Petersen DR. Aldehyde dehydrogenase and aldehyde reductase in isolated bovine brain microvessels. Alcohol 1985 Jan;2(1):79-83.

[95] Lubec G, Labudova O, Cairns N, Berndt P, Langen H, Fountoulakis M. Reduced aldehyde dehydrogenase levels in the brain of patients with Down syndrome. J Neural Transm Suppl 1999;57:21-40.

[96] Ohsawa I, Nishimaki K, Murakami Y, Suzuki Y, Ishikawa M, Ohta S. Age-dependent neurodegeneration accompanying memory loss in transgenic mice defective in mitochondrial aldehyde dehydrogenase 2 activity. J Neurosci 2008 Jun 11;28(24):6239-49.

[97] Kalinderi K, Fidani L, Katsarou Z, Bostantjopoulou S. Pharmacological treatment and the prospect of pharmacogenetics in Parkinson's disease. Int J Clin Pract 2011 Dec;65(12):1289-94.

[98] Riederer P, Laux G. MAO-inhibitors in Parkinson's Disease. Exp Neurobiol 2011 Mar;20(1):1-17.

[99] Wey MC, Fernandez E, Martinez PA, Sullivan P, Goldstein DS, Strong R. Neurodegeneration and motor dysfunction in mice lacking cytosolic and mitochondrial aldehyde dehydrogenases: implications for Parkinson's disease. PLoS One 2012;7(2):e31522.

[100] Plaitakis A, Latsoudis H, Kanavouras K, Ritz B, Bronstein JM, Skoula I, et al. Gain-of-function variant in GLUD2 glutamate dehydrogenase modifies Parkinson's disease onset. Eur J Hum Genet 2010 Mar;18(3):336-41.

[101] Tieu K, Perier C, Vila M, Caspersen C, Zhang HP, Teismann P, et al. L-3-hydroxyacyl-CoA dehydrogenase II protects in a model of Parkinson's disease. Ann Neurol 2004 Jul;56(1):51-60.

Chapter 6

Hydrohysteroid Dehydrogenases – Biological Role and Clinical Importance – Review

Nina Atanassova and Yvetta Koeva

Additional information is available at the end of the chapter

http://dx.doi.org/10.5772/54149

1. Introduction

Hydroxysteroid dehydrogenases (HSDs) belong to the NADPH/NAD+-dependent oxidoreductases, which interconvert ketones and the corresponding secondary alcohols. As their names imply, they catalyze the oxidoreduction in different positions of steroidal substrates (3α-, 3β-, 11β-, 17β-, 20α- and 20β-position). The steroid-converting HSDs play central roles in the biosynthesis and inactivation of steroid hormones, but some of them are also involved in the metabolism of diverse non-steroidal compounds [1]. The HSDs are integral parts of systemic (endocrine) and local (intracrine) mechanisms. In target tissues they convert inactive steroid hormones to their corresponding active forms and viceversa, thus modulating the transactivation of steroid hormone receptors or other elements of the non-genomic signal transduction pathways. Therefore, HSDs act as molecular switches allowing pre-receptor modulation of steroid hormone action [2].

It is also well recognized that human and certain other primates are unique among animal species in having adrenals that secrete large amounts of inactive steroid precursors including dehydroepiandrosterone (DHEA). These steroids do not bind to the androgen receptor but exert either estrogenic or androgenic action after their conversion into active estrogens and/or androgens in target tissues [3]. Imbalanced action of sex steroid hormones, i.e. androgens and estrogens, is involved in the pathogenesis of various severe diseases in human. Hormone-dependent cancers are commonly lethal both in women and in men, with breast cancer being the most prevalent cancer in women and prostate cancer in men in several Western countries [4]. In addition, there are various other common hormone-dependent diseases, such as polycystic ovary syndrome (PCOS) and endometriosis, having poorly understood aetiology and lacking efficient pharmacological treatment [5, 6]. However, changes in circulating hormone concentrations do not explain all pathophysiological processes occured in hormone-dependent tissues. A more inclusive explanation is provided by paracrine and intracrine action of sex steroids, namely the regulation of intratissue hormone concentrations by expression of steroidogenic enzymes. The modulation of local sex steroid production using pharmaceutical compounds is also a valuable treatment option for developing of novel therapies against hormonal diseases [7]. In the view of successful practice of inhibiting of non-HSD enzymes (aromatase and 5α-reductase) [8, 9], recent attempt are made for development of HSD inhibitors as therapeutic strategy. Several of HSD enzymes are also considered as promising drug targets and inhibitors, for example most of the isoforms of 17β-HSD enzyme [10].

In this review, we summarise the data from the literatute and our own data on the main HSDs (11β-HSD, 3β-HSD 17β-HSD) focusing our attention on the localization/tissue distribution and regulation of the enzyme isoforms and their role in normal and pathological processes as revealed by experimental models and clinical observations. The review would provide better understanding on multifunctionality of HSDs and their relevance to the clinic and that would be helpful for scientists and clinicians, working in a new challenging area of development of HSD-inhibitors as new drugs for hormone-related deceases.

2. Steroid hormones and role of hydroxysteroid dehydrogenases in steroidogenesis: steroidogenic pathways and general regulatory mechanisms

Steroid hormones are produced by the gonads, adrenal gland and placenta and they play vital role in physiological and reproductive processes. Structurally, steroids have a basic or common nucleus called the cyclopentanoperhydrophenanthrene, consisting of three, six-membered fully hydrogenated (perhydro) phenanthrene rings designated A, B and C, and one five-membered cyclopentane ring designated D (Fig 1, right top). In 1967, the International Union of Pure and Applied Chemistry (IUPAC) established rules for the number of carbons in a steroid and thus its biological action can be predicted. For instance, 21-carbon steroids have progestogenic or corticoid activity, 19-carbon steroids have androgenic activity and 18-carbon steroids have estrogenic activity. Cholesterol is a 27-carbon steroid that gives rise pregnenolone (21-carbon) after cleavage of its side chain. Pregnenolone is subsequently converted to progesterone, which in turn give rise androgens or corticoids. Androgens are subjected to aromatization of ring A thus giving rise estrogens [11]. The pathways of steroidogenesis differ between species, but the pathways of human steroidogenesis are shown in the Figure 1. [12]. Cholesterol is the precursor of the steroid hormones, providing backbone of the steroid molecule. The enzymes involved in the synthesis of steroid hormones can be divided into two major classes of proteins: the cytochrome P450 heme-containing proteins (CYP) and the hydroxysteroid dehydrogenases (HSD) [13, 14]. These enzymes are primarily expressed in the gonads, adrenal and placenta. Interestingly, some of these enzyme activities have been demonstrated in non-endocrine tissues, where they may be involved in important paracrine and autocrine actions. This is particularly the case in the human fetus where steroid precursors circulates at high levels and could be metabolized within tissues to produce active steroid hormones. The first class of steroidogenic enzymes, CYP proteins called hydroxylases catalyze reaction of xydroxylation (introduction of hydroxyl group –OH into organic compound) and cleavage of the steroid substrate utilizing molecular oxygen and nicotinamide adenine dinucleotide phosphate (NADPH, reduced) as the source of reductive potential. Several enzymes are included: cytochrome P450 cholesterol side-chain cleavage enzyme (P450scc, *CYP11A1)*, cytochrome P450 17α-hydroxylase (P450c17, 17α-hydroxylase, 17-20 lyase, *CYP17A1*), P450 aromatase (aromatase, *CYP19A1*), 21α-hydroxylase (*CYP21A*), 11β-hydroxylase (*CYP11B1)* and aldosterone synthase (*CYP11B2*). The second class of steroidogenic enzymes, HSD enzymes called alcohol oxydoreductases catalyze the dehydrogenation of hydroxysteroids. Acting as oxydoreductases, HSD enzymes require nicotinamide adenine dinucleotide (NAD, oxidized) and/or NADPH as electron acceptor/donor. HSD enzymes include: 3β-hydroxysteroid dehydrogenase (3β-HSD),11β-hydroxysteroid dehydrogenase (11β-HSD) and 17β-hydroxysteroid dehydrogenase (17β-HSD). While each P450 enzyme is the product of a single gene, the HSD enzymes have several isoforms that are products of distinct genes [15]. There are four types, classified by the number of the carbon acted upon.

In all species, the first and rate-limiting step in steroidogenesis, in particular androgen biosynthesis, is conversion of the C27 cholesterol to the C21 steroid, pregnenolone (Figure 1). This reaction is catalyzed by cytochrome P450scc enzyme located in the inner mitochondrial membrane. Pregnenolone diffuses across the mitochondrial membrane and it is further metabolized by enzymes associated with the smooth endoplasmic reticulum. These enzymes are: 1) cytochrome P450c17, which catalyzes the conversion of the C21 steroids pregnenolone or progesterone to the C19 steroids dehydroepiandrosterone or androstenedione, respectively; 2) 3β-HSD (Δ5-Δ4 isomerase), which catalyzes the conversion of the Δ5 hydroxysteroids - pregnenolone or dehydroepiandrosterone to the Δ4 ketosteroids - progesterone or androstenedione, respectively; 3) 17β-HSD (17-ketosteroid reductase), which catalyzes the final step in the biosynthesis of testosterone [16].

Corticosteroids (mineralocorticoids and glucocorticoids, C-21 cabons) derive from progestagens (progesterone and 17α-OH progesterone) after hydroxylation of carbon-21 by the enzyme 21α-hydroxylase. So, aldosterone and corticosterone share the first part of their biosynthetic pathway. The last part is mediated either by aldosterone synthase (for aldosterone) or by 11β-hydroxylase (for corticosterone). These enzymes are nearly identical (they share 11β-hydroxylation and 18-hydroxylation functions). Aldosterone synthase is also able to perform 18-oxidation. 11β-hydroxysteroid dehydrogenase (11β-HSD) catalyzes the conversion of active cortisol to inert 11 keto-products (cortisone), or vice versa, thus regulating the access of glucocorticoids to the steroid receptors.

The steroidogenic pathays/steroid output are controlled by complex regulatory mechanisms that involved wide range of factors like pituitary trophic hormones, growth factors, cytokines and steroids. The major factors, expressed since early fetal life, are steroidogenic acute regulatory protein (StAR) and Steroidogenic Factor-1 (SF-1). StAR actively transports cholesterol from the outer to the inner mitochondrial membrane and allows *CYP11A* (located in the inner membrane) access to cholesterol [17]. Cell specific expression of StAR and P450 enzymes are regulated by Steroidogenic Factor-1 (SF-1), which binds to promoter region of StAR gene and of all CYP genes, activating their expression [18, 19]. The most compelling evidence for the essential requirement for StAR in steroidogenesis is provided by StAR-specific knockout mice and human mutations that caused the potentially lethal condition known as congenital lipoid adrenal hyperplasia. It is not surprising that 46XY individuals with mutated SF1 have XY sex reversal, indicative of disrupted fetal testosterone biosynthesis and masculinization. In mice with Leydig cell-specific knockout of SF-1 gene there is lack of *CYP11A* and StAR expression resulting in adrenal and gonadal agenesis [20-23]. The activity of P450scc enzyme is regulated by mitochondrial environment [24] and the vital role of this enzyme is demonstrated by homozygous mutation of *CYP11A* gene that is lethal due to inability of placenta to produce progesterone [25]. Consequently, 46XY genetic males with partial inactivation of *CYP11A* exhibit major deficiencies in masculinization [26, 27].

The combined enzymatic actions of 3β-HSD and P450c17 catalyze the overall conversion of pregnenolone to androstenedione, the precursor of testosterone. This conversion can occur via one of two main pathways, either via Δ4 or Δ5 pathway and the preferred route is both species- and age-dependent. [14] (Figure 2.).

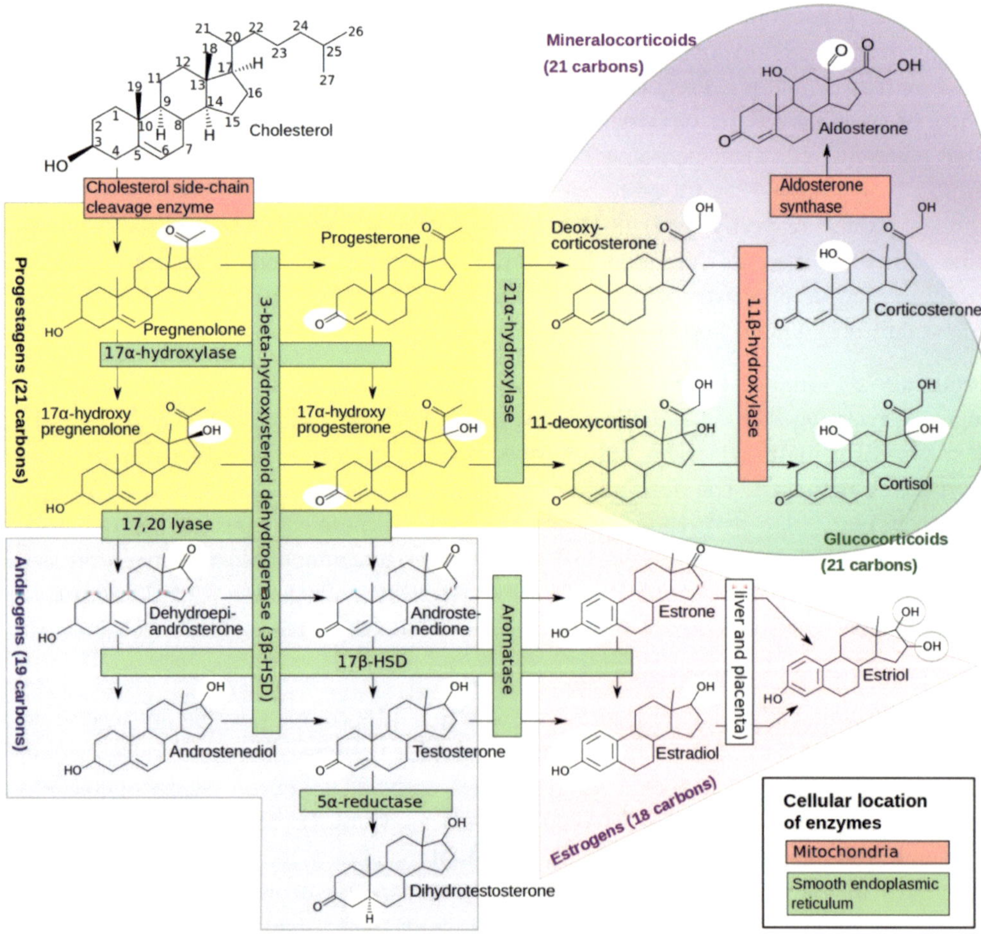

Figure 1. Pathways of human steroidogenesis [12].

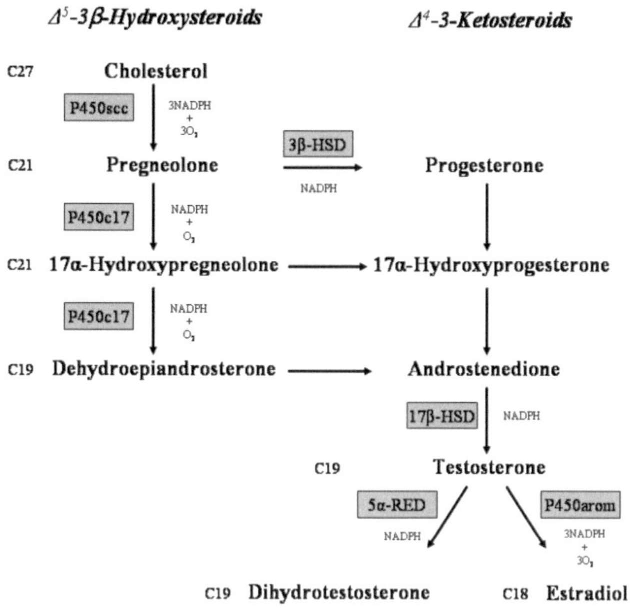

Figure 2. Steroid biosynthetic pathways as adapted to Payne, 2007 [16]

The Δ4 pathway (pregnenolone, progesterone, androstenedione, testosterone) was the first indentified route in rat testis and subsequently shown to be preferred one. In the human and higher primates, as well as in pig and rabbit the Δ5 pathway predominates in the adult and fetal testis because human P450c17 enzyme readily converts 17α-hydroxypregnenolone to dehydroepiandrosterone (DHEA), but has little emzyme activity when 17α-hydroxyprogesterone is the substrate. In the rat, P450c17 readily cleaves both the Δ4 and Δ5 C21 steroids, but in contrast to the human, it has a preference for the Δ4 pathway. In the mouse the Δ4 pathway dominates before puberty but in adult animals the Δ5 pathway may also contribute to overall testosterone production. Therefore, differences in preferred pathways between species are likely to depend upon relative substrate affinity of P450c17 enzyme [6, 14].

The clinical importance of P450c17 enzyme is demonstrated by numerous reports on *CYP17A* gene mutations [28, 29, 30]. Both male and female patients are hypertensive because overproduction of mineralocorticoids as well as impaired production of cortisol. Affected females exhibit abnormal sexual development resulting in primary amenorrhea. Male patients are phenotypic females due to the deficiency of testosterone production.

3. 3β-HSD gene family – function, tissues distribution, regulation and clinical importance

The 3β-HSD was described in 1951 and later characterized as bifunctional dimeric enzyme required for the biosynthesis of all classes of steroid hormones (glucocorticoids, mineralocorticoids, progestagens, androgens, and estrogens). Therefore the 3β-HSD controls the critical steroidogenic reactions in the adrenal cortex, gonads, placenta, and peripheral target tissues [31]. The 3β-HSD isoforms catalyze the conversion of the Δ5-3β-hydoxysteroids - pregnenolone, 17α-hydroxypregnenolone, and DHEA, to the Δ4-3-ketosteroids - progesterone, 17α-hydroxyprogesterone, and androstenedione, respectively. Two sequential reactions are involved in the conversion of the Δ5-3β-hydroxysteroid to a Δ4-3 ketosteroid. The first reaction is the dehydrogenation of the 3β-hydroxysteroid, requiring the coenzyme NAD^+, yielding the Δ5-3-keto intermediate, and reduced NADH. The reduced NADH, activates the isomerization of the Δ5-3- keto steroid to yield the Δ4-3-ketosteroid (Figure 2.). Stopped-flow spectroscopy studies show that NADH activates the isomerase activity by inducing a time-dependant conformational change in the enzyme [15, 32]. Using histochemical and imunohistochemical techniques 3β-HSD activity was detected to the smooth endoplasmic reticulum and mitochondrial cristae and later in the microsomal fraction suggesting that 3β-HSD is a membrane-associated enzyme [16]. Submitochondrial fractionation studies showed that 3β-HSD is in a functional steroidogenic complex with P450scc located in the inner mitochondrial membrane [33, 34], that provides the enzyme with immediate substrate metabolized from cholesterol. However, 3β-HSD activity could be preferentially distributed to the mitochondria under certain physiological conditions [35, 36].

Isoforms: Structural studies of 3β-HSD family characterized several isoforms, products of distint genes. The number of isozymes varies in different species. The isoenzymes differ in tissue distribution, catalytic activity (whether they function predominantly as dehydrogenases or reductases), in substrate and cofactor specificity, and in subcellular distribution [6]. So far, two isoforms were reported in human (h) 3β-HSD, six in mouse, four in rat and three in hamster. Multiple 3β-HSD isoenzymes have been cloned from several other species, further illustrating that the 3β-HSD gene family is conserved in vertebrate species The human type I 3β-HSD gene (*HSD3B1*) encodes an enzyme of 372 amino acids predominantly expressed in the placenta and peripheral tissues (skin, mammary gland, prostate, and several other normal and tumor tissues) [37, 38]. In comparison, the type II gene (*HSD3B2*), which encodes a protein of 371 amino acids, shares 93.5% identity with the type I and it is almost exclusively expressed in the adrenals ovaries and testes. It is most homologous to the type I gene expressed in mice, rats and other species [39, 40]. The structure of h*HSD3B1* and h*HSD3B2* genes consists of four exons which are included within a DNA fragment of 7.8 kb and genes are assigned to chromosome 1p13.1 [41].

The rat type I and II 3β-HSD proteins are expressed in the adrenals, gonads, kidney, placenta, adipose tissue, and uterus and share 93.8% identity. The type III protein shares 80% identity with the type I and II proteins but, in contrast to other types, it is a specific 3-ketosteroid reductase (KSR) [42, 43]. The type III gene is exclusively expressed in male liver, and there is marked sexual dimorphic expression, which results in pituitary hormone-induced gene repression in the female rat liver [44]. The rat type IV protein shares 90.9%, 87.9%, and 78.8% identity with types I, II, and III proteins, respectively. Furthermore, types I and IV possess a 17β-HSD activity specific to 5α-androstane-17β-ol steroids, thus suggesting a key role in controlling the bioavailibility of the active androgen dihydrotestosterone DHT

[45, 46, 47]. Concerning to an enzyme having dual activity, such secondary activity could be explained by binding of the steroid in the inverted substrate orientation, in this case C-17 rather than C-3 possition. [47].

To date, six distinct cDNAs encoding murine members of the 3β-HSD family have been cloned and all of them are highly homologous and encode a protein of 372 amino acids. Functionally, the different forms fall into two distinct classes of enzymes - 3β-HSD types I, II and III function as dehydrogenase/isomerases, and are essential for the biosynthesis of active steroid hormones whereas 3β-HSD type IV and type V (analogous to rat type III) function as 3-KSRs and they are involved in the inactivation of active steroid hormones [48, 49]. In the adult mouse 3β HSD I is expressed in gonads and adrenal gland, whereas 3β-HSD II and III are expressed in liver and kidney. The type V isoenzyme is expressed only in the liver of the male mouse and the expression starts in late puberty. The type VI isoenzyme is the earliest isoform expressed during the first half of pregnancy in cells of embryonic origin and in uterine tissue suggesting that this isoenzyme may be involved in the local production of progesterone, required for the successful implantation and/or maintenance of pregnancy [50]. In the adult mouse, 3β-HSD type VI appears to be the only isoenzyme expressed in skin. The aminoacid sequences among the different isoforms and between mouse and human isoforms show a high degree of identity. Mouse 3β-HSD I has 84% identity to mouse VI, and 71% identity to human II [31, 50].

Tissue distribution: As 3β-HSD gene family is widely expressed within the steroidogenic organs (adrenal, ovary and testis) as well as in peripheral tissues, the distribution and local regulation will be described separately for each organ.

Adrenal: The onset of 3β-HSD expression in the fetal primate adrenal cortex correlates with the ability of the definitive zone to synthesize aldosterone and also allows cortisol production by transitional zone cells. Although 3β-HSD is not expressed to a high degree in the fetal cortex, P450c17 is expressed, thereby directing the steroidogenic pathway toward Δ5-hydroxysteroid (*i.e.*, DHEA) production. There is zone-specific steroid secretion pattern dependent on the relative expression levels of 3β-HSD, P450c17 and P450 21α-hydroxylase (P450c21) that serve as molecular markers of the adrenocortical developmental state [51, 52]. After birth, the coexpression of 3β-HSD and P450c21 leads to aldosterone production, whereas the coexpression of 3β-HSD and P450c17 results in production of cortisol. The expression of P450c17 along with low levels of 3β-HSD expression leads to synthesis of DHEA. The differential expression of the enzymes required for zonal-specific steroid production in the adrenal is under the control of multiple factors as Adrenocorticotropic hormone (ACTH), Epidermal Growth Factor (EGF), Fibroblast Growth Factors (FGFs), Insulin-like Growth Factors (IGFs), thyroid hormone (T3), Transforming Growth Factor-β (TGFβ) [31, 53, 54]. Therefore, there appears to be a complex interplay of factors controlling adrenal development, and combinations of these factors could be involved in the regulation of 3β-HSD and other steroidogenic enzymes *in vivo*.

Ovary: Ontological studies for 3β-HSD have shown that fetal human ovaries are steroidogenically quiescent except for a window late in gestation [55], so most of the estrogens seen by the primate fetus are of placental origin [56]. 3β-HSD is not expressed in mouse and rat ovary until first week after birth.This is in contrast to testicular expression because androgen production by the male embryo is critical for male sexual development [57]. PCOS is an ovarian disorder associated with hyperthecosis of the ovary and elevated serum LH, insulin, and androgen levels. Several studies provide evidence of aberrant 3β-HSD regulation in polycystic thecal cells although the mechanisms are unclear [58].

Preantral/antral *follicular* expression studies show 3β-HSD mRNA and protein expression in the human ovary initially in the theca and then in the granulosa layer as folliculogenesis continues [59]. In nonprimate species, 3β-HSD has been shown to have different expression patterns. In the rat, preantral, antral, and preovulatory rat follicles showed 3β-HSD expression in the theca, but no expression was seen in the granulosa layer [60]. In contrast to rodents, pigs, and primates, 3β-HSD expression in the cow was seen in all the stages of the preovulatory follicle in both theca and granulosa layers [61]. Pituitary hormones are the primary means of the regulation of the steroidogenesis in the ovary.The gonadotropins, FSH and LH cause an increase in 3β-HSD expression concomitantly with other steroidogenic enzymes. The role of prolactin (PRL) on primate 3β-HSD is unclear, although PRL was shown to be inhibitory. Interestingly in postmenopausal women 30% of circulating Δ4-DIONE is of ovarian origin [62]. These studies suggest that ovarian steroid production in postmenopausal women continues, but the decline in pituitary control dramatically changes the steroid profile. After ovulation, *Corpus Luteum (CL)* is developed to secrete large amount of progesterone that is controlled in part by the amount of 3β-HSD. The enzyme is considered as a marker for progesterone production of the CL [63]. In primates, LH/hCG action through LH receptor provides a primary mean of luteotropic support [64, 65]. In addition, FSH increased 3β-HSD protein and mRNA levels in human granulosa-lutal cells, and this effect could be enhanced by insulin [66]. Although the direct control of 3β-HSD by PRL in humans has yet to be demonstrated, PRL has been shown to up-regulate 3β-HSD transcriptional activity *in vitro* [67]. During regression of CL (luteolysis) the expression of 3β-HSD dramatically decreased and there is evidence that LH is mainly involved in induction of luteolysis [68].

Testis: Testis is the major place for production of androgens, mainly testosterone although local conversion/reduction of testosterone to dihydrotestosterone (DHT) by 5α-Reductase (5α-Red) occurred in the following part of reproductive system (epididymis and prostate). Within the testis, the Leydig cells (LC) are primary place for steroidogenesis as they are only cell type in the male that expressed all of the enzymes essential for the conversion of cholesterol to testosterone [16]. During development two distinct population of LCs arise sequentially, namely fetal and adult LC population, being differentially regulated [20]. Immunohistochemical studies have revealed that human Leydig cells express 3β-HSD as early as 18 wk of gestation. During gestation in human, 3β-HSD expression is an indicator of testicular androgen production. Adult Leydig cells arise postnatally and encompass three developmental stages: progenitor, immature and adult Leydig cells [69]. Rat testes of postnatal day 15 showed 3β-HSD localization to the smooth ER in precursor Leydig cells and that points the beginning of differentiation of adult LC population. At this time point

LC expressed P450scc and P450c17, as well. Therefore an antibody against 3β-HSD is highly applicable as a marker for visualization both, fetal and adult LC. The expression of 3β-HSD protein overlapped with expression of other steroidogenic enzymes, P450scc and P450c17, clearly demonstrated on Figure 3 and that was confirmed by other authors [70]. Development of triple co-localization immunohistochemical technique allows distinguishing of presumptive progenitors cells form adult or fetal LC that is very helpful to study kinetic and differentiation pattern of LCs (Figure 4) [71]. Application of IHC for 3β-HSD is widely used by many authors in quantification studies of LC under normal and experimental/pathological conditions, especially those of hormonal manipulations [72]. 3β-HSD immunohistochemistry is also useful tool for validation of EDS (ethane dimethanesulfonate) model for selective ablation of adult LC and thus testosterone withdrawal. The major regulator of postnatal testicular expression of 3β-HSD in rodents and human is the LH, acting via LH receptor located in LCs. That is in contrast to the fetal testis where an independent mechanism is suggested [73]. Steroids and growth factors (EGF, TGFβ, FGFs, Activin A) are also suggested to control the expression of 3β-HSD [31].

Peripheral tissues; Expression of 3β-HSD in peripheral tissues such breast, prostate, placenta, liver, blain and skin will be briefly described in relation to clinical importance. Sex steroids are well recognized to play a predominant role in the regulation of cell growth and differentiation of normal mammary gland as well as in hormone-sensitive breast carcinomas. Estrogens stimulate cell growth of hormonesensitive breast cancer cells, whereas androgens exert an antiproliferative action in breast cancer cells [74]. Stage II/III infiltrating ductal primary breast tumors demonstrated 3β-HSD activity [75], and 3β-HSD protein was seen in 36% of breast carcinoma samples tested [76, 77]. The 3β-HSD expressed in human placenta is the peripheral isoform, type I 3β-HSD, and it is under differential regulatory control than the adrenal/gonadal isoform, type II 3β-HSD [78, 79]. In the prostate epithelium 3β-HSD expression was colocalized with 17β-HSD type V in normal conditions. 3β-HSD was found in human hyperplastic prostates suggesting the capacity of the human prostate for local androgen production, that increase the hypertrophic potential of the organ [80, 81]. Hepatic 3β-HSD expression is presumed to be important in the metabolism and inactivation of steroids. 3β-HSD activity in human liver microsomes was shown to be three times higher for the reduction of DHT to 3Δ-DIOL than the reverse reaction [82]. The circulating levels of steroids might affect regulation of 3β-HSD activity in the liver, principally through altering Growth Hotmone (GH) and PRL levels, and thereby resulting in feedback on steroid degradation [83]. In skin 3β-HSD was confined to keretinocytes, co-expressed with 17β-HSD. Aberrant expression of these enzymes results in increased scalp DHT levels and possibly acceleration of the balding process in genetically predisposed men and women [84, 85]. 3β-HSD expression was reported in the central nervous system (CNS) and peripheral neurons demonstrating the importance of steroid hormones for growth maturation and differentiation of nerve tissue. For instance, 3β-HSD together with P450scc are expressed in the hippocampus, dentate gyrus, cerebellum, olfactory bulb, and Purkinje cells of the rat brain with highest levels in cerebellum [86] as well as in cultured neuronal cells [87]

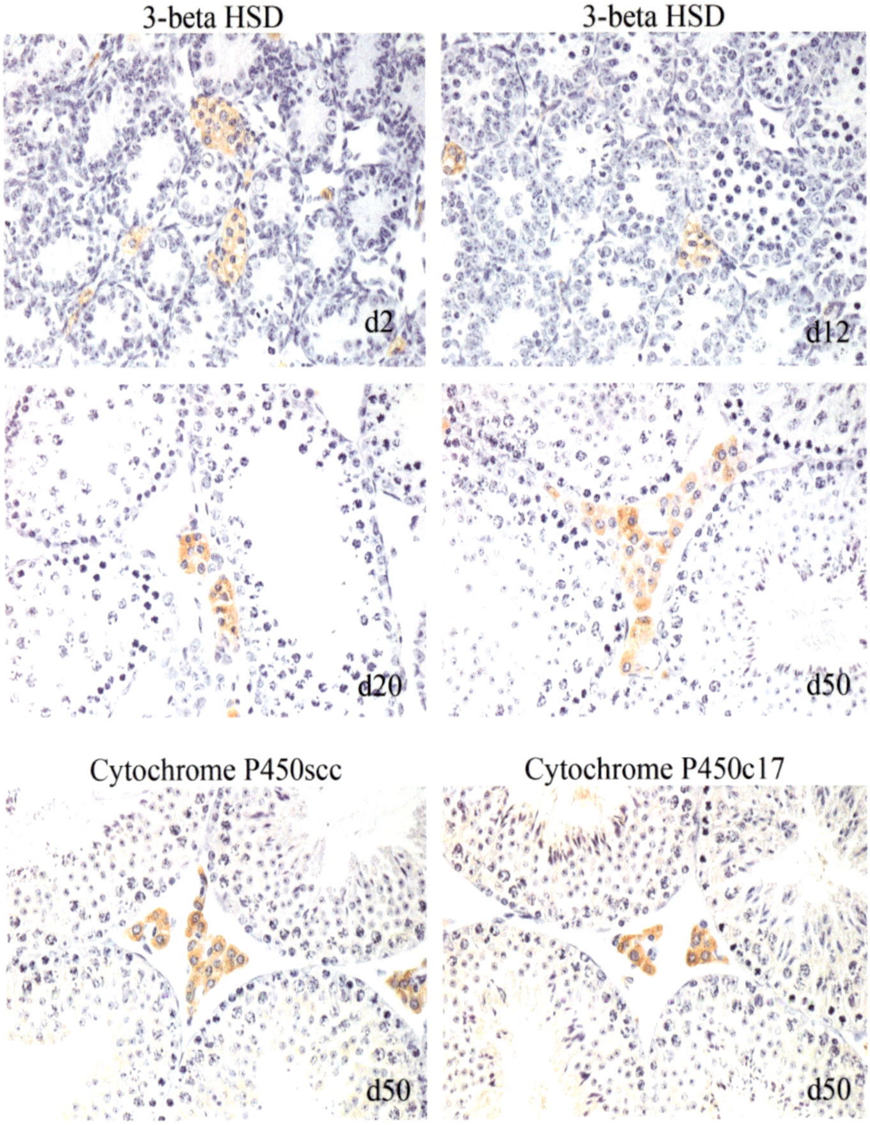

Figure 3. Immunoexpression of steroideogenic ezymes (3β-HSD, cytochrome P450scc and cytochrome P450c17 in the Leydig cells (DAB-brown) of postnatal mouse testis after birth to sexual maturity (d2-neonatal, d12-prepubertal, d20-pubertal, d50-adult) x400.

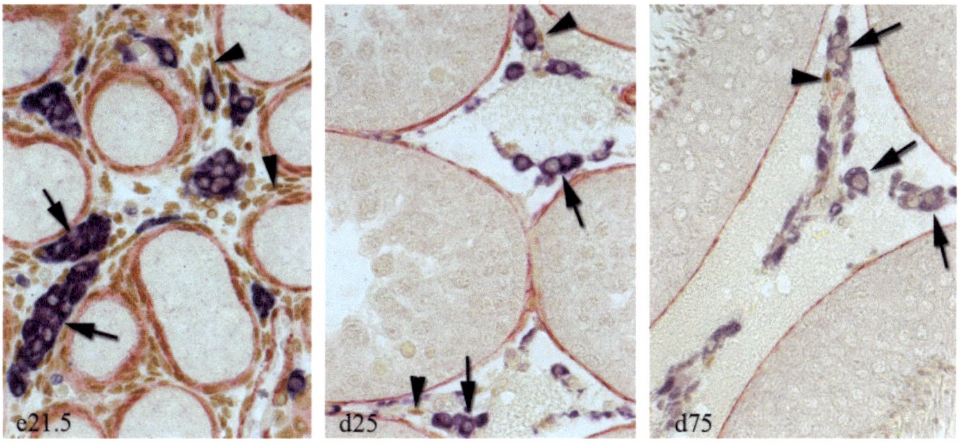

Figure 4. Triple immunostaining for 3β-HSD (blue), α-smooth muscle actin (red) and COUP TFII (brown) in fetal (embryonal day 21.5) and postnatal (pubertal-d25 and adult-d75) rat testes. Fetal and adult LCs (arrows) are clrearly distinguishable from presumptive ptogenitors cells (arrowheads) x400.

Regulation: The regulation of 3β-HSD gene family is quite complex process involving multiple signal transduction pathways that are activated by growth factors, steroids and cytokines and they are differentially dependent on ontogeny and tissue distribution. Initial studies investigating the transcriptional regulation of the human *HSD3B2* gene are primarily focused on the trophic hormones, including ACTH in the adrenal cortex, LH/human chorionic gonadotropin (hCG) in theca cells and corpus luteum, as well as LH in testicular Leydig cells. cAMP is well known intracellular mediator of trophic hormone stimulation of 3β-HSD expression but mechanisms by which cAMP stimulate transcription of the *HSD3B2* gene are not clear yet. [31].

Gonadal expression of human 3β-HSD II and mouse 3β-HSD I is dependent on SF-1 as described for the gonadal-specific expression of the P450 steroidogenic enzymes [88]. Studies on mouse *Hsd3b1* promoter identified three potential SF-1 consensus binding sites [89]. The regulation of *HSD3B2* human gene expression involved the transcription factors of Stat family (signal transducers and activators of transcription) [90]. Interestingly, the Stat5 knockout mice displays luteal failure [91]. DAX-1 (dosage-sensitive sex reversal adrenal hypoplasia congenita critical region on X chromosome gene-1) was originally isolated by positional cloning from patients with DAX-mutation exhibiting adrenal congenita hypoplasia associated with hypogonadotropic hypogonadism. The studies examining the effects of DAX-1 overexpression on adrenal cell showed suppression of steroidogenesis associated with inhibition of the expression of StAR, P450scc, and 3β-HSD [92]. The exact mechanisms by which DAX-1 overexpression affects 3β-HSD expression remain unclear. Interestingly, transcription factors belonging to the GATA family are emerging as novel regulators of steroidogenesis. In fetal and adult adrenals and gonads several target genes for GATA protein were identified such as StAR, *CYP11A, CYP17A, CYP19A, HSD17B1,* human *HSD3B1* and *HSD3B2* [93]. Moreover, deregulation of GATA expression and/or activity might be relevant to pathological processes associated with aberrant *HSD3B2* expression such as adrenal insufficiency, male pseudohermaphroditism and polycystic ovary syndrome (PCOS) [31]. Immune cell populations in the ovary undergo changes during the reproductive cycle and cytokines from these immune cells (Interleukin-4, IL-4) have been shown to affect steroidogenesis, mediated by Stat [94]. Some growth factors like members of the TGFβ family and nerves growth factor have been shown to regulate *HSD3B2* gene expression [95-97]. There is growing evidence in the literature that steroid hormones modulate type II 3β-HSD expression. For example, glucocorticoids stimulate the expression of 3β-HSD in adrenal cells [98], whereas androgens inhibit 3β-HSD expression in the adrenal cortical cells and in testicular Leydig cells [99, 100]. There are number of questions concerning the mechanisms of steroids and the action of their receptors. In relation to structure-function aspects the question is what is the influence of known steroid agonists and antagonists on the efficacy of activation? What is the effect of other nonsteroid factors, which are known to activate other intracellular signaling pathways on steroid-regulated transcription?

Clinical importance of 3β-HSD genetic deficiency:

Homozygous mutations in *HSD3B1* are lethal in human due to interruption of pregnancy before the end of the first trimester because 3β-HSD I protein is required for progesterone synthesis in the placenta (as described above for *CYP11A).* Many mutations in the *HSD3B2* gene have been identified and are summarized in a review by Simard et al. 2005 [31]. The classical 3β-HSD deficiency results from mutations in the *HSD3B2* gene (the *HSD3B1* gene in these patients is normal) and it can be divided, depending upon the severity of the salt-wasting (salt-wasting or non-salt-wasting forms). The classical 3β-HSD deficiency is a rare form of congenital adrenal hyperplasia (CAH) accounting for about 1–10% of cases of CAH. The salt-losing forms of CAH are a group of life-threatening diseases that require prompt recognition and treatment. Indeed, the autosomal recessive mutations in the *CYP21, CYP17, CYP11B1,* and *HSD3B2* genes encoding steroidogenic enzymes can cause CAH, each resulting in different biochemical consequences and clinical features. In these cases the cortisol secretion is impaired resulting in compensatory hypersecretion of ACTH and consequent hyperplasia of the adrenal cortex. However, only deficiencies in 21-hydroxylase (*CYP21*) and 11β-hydroxylase (*CYP11B1*) predominantly result in virilizing disorders. Indeed, in patients with the classical form of these two defects, the most noticeable abnormality in the sexual phenotype is the masculinization of the female fetus due to oversynthesis of adrenal DHEA. Male individuals suffering from classical 3β-HSD deficiency present hypospadias. On the other hand, the complete or partial inhibition of 3β-HSD activity in the adrenals and ovaries was not accompanied by a noticeable alteration in the differentiation of the external genitalia of female patients. The reason for this striking difference in phenotype between the male and female individuals is that the deficiency of 3β-HSD in the fetal testis results in lowering of the T levels below the levels required for the normal development of male external genitalia.

The basal plasma levels of Δ5-3β-hydroxy steroids such as pregnenolone (PREG), 17OH-PREG, and DHEA are elevated in affected individuals. An elevated ratio of Δ5/Δ4-steroids is considered to be the best biological parameter for the diagnosis of 3β- HSD deficiency. The best criteria for the correct diagnosis of this disorder now appears to be a plasma level of 17OH-PREG but 17OH Progesterone (17OH-PROG) also should be measured for correct diagnosis of 3β-HSD deficiency. It is well recognized that plasma levels of 17OH-PROG and Δ4-DIONE and other Δ4-steroids are frequently elevated in 3β-HSD-deficient patients. Such observations are consistent with a functional type I 3β-HSD enzyme that is expressed in peripheral tissues. Moreover, the peripheral type I 3β-HSD activity could explain why certain patients were initially misdiagnosed as suffering from 21-hydroxylase deficiency, in view of elevated levels of 17OH-PROG and mild virilization seen in girls at birth. Therefore, measurement of the levels of 17OH-PREG should be performed when an elevated level of 17OH-PROG has been observed in a female neonate without ambiguity of external genitalia or if the patient is a male pseudohermaphrodite [31].

4. 11β-hydroxysteroid dehydrogenase – biological role in the regulation of glucocorticoid metabolisms and cortisol levels

The glucocorticosteroids exert diverse actions throughout the body and many of them have important implications in the reproduction and metabolite syndrome. It was recognized that within potential target cells, the actions of glucocorticoids are modulated by 11β-hydroxysteroid dehydrogenases (11β-HSD) which catalyse the reversible inactivation of cortisol and corticosterone to their inert 11-ketosteroid metabolites, cortisone and 11-dehydrocorticosterone, respectively [101]. The actions of physiological glucocorticoids (cortisol and corticosterone) are modulated by isoforms of the enzyme 11β-HSD (Figure 5, [108]). To date, two isoforms of 11β-HSD have been identified: 1) 11β-HSD1 acts predominantly as an NADP(H)-dependent reductase that converts inactive circulating 11-ketosteroids, into active glucocorticoids generating active cortisol or corticosterone; 2) 11β-HSD2 is a high affinity NAD+-dependent enzyme that catalyses the inactivation of glucocorticoids [102-107]. Although the biochemistry of 11β-HSD is well established, the physiological significance of glucocorticoid metabolism by these enzymes is still not fully

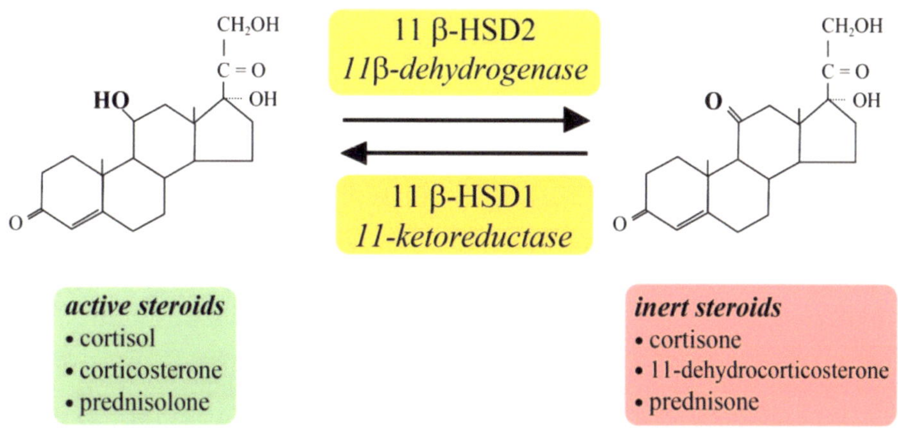

Figure 5. 11β-hydroxysteroid dehydrogenase (11β-HSD) (Adopted by Seckl et al., 2004 [108])

understood. The enzymatic inactivation of cortisol and corticosterone by 11β-HSD enzymes appears to be of central importance for protection of gonadal steroidogenesis, prevention of intra-uterine growth retardation and metabolite syndrome.

This review focuses on the importance of 11β-HSD isoenzymes in the developing and aging testis, ovary, adrenal gland, placenta and adipose tissue. The current work aims to provide recent understanding of the biological roles played by 11β-HSD in different processes and diseases including reproduction, adrenal gland function, cystic ovarian disease, and the metabolite syndrome. In addition, this review summarizes recent knowledge based on human data and genetic models on the clinical importance of 11β-HSD in relation to metabolite syndrome.

5. 11β-hydroxysteroid dehydrogenase in developing testis- marker for differentiation of the Leydig cells

The enzyme 11β-hydroxysteroid dehydrogenase (11β-HSD) is hypothesized to modulate LCs steroidogenesis by controlling the intracellular concentration of glucocorticoids. By doing so, 11β-HSD can protect the LCs against the suppressive effect of glucocorticoids [109-112]. Glucocorticoids have been found to directly inhibit the transcription of genes encoding the key enzymes of testosterone biosynthesis [113,114]. Excessive glucocorticoid exposure suppress androgen synthesis and thus decrease serum testosterone (T) levels by inducing LC apoptosis and reducing the number of LCs per testis [115,116]. The effects of glucocorticoids on LCs are not only associated with the classic glucocorticoid receptor-mediated mechanism but possibly through the plasma membrane receptor or prereceptor-mediated action by the glucocorticoid metabolizing enzyme 11β-HSD1 [117]. Both isoforms of 11β-HSD are localized in testicular LCs [118-121]. Recent studies showed that reductase activity predominates in both human and rat type 1 11β-HSD [109]. In contrast, the other 11β-HSD isoform, type 2, has been found to be exclusively oxidative [118,110,131]. Predominance of oxidative activity results in glucocorticoid inactivation, whereas the reductive activity of the enzyme has an opposite effect [109]. Hu et al. [122] postulated that inhibition of 11β-HSD1 in rats *in vivo*, increases intracellular active glucocorticoid concentration and thereby affects serum T concentration and steroidogenic enzyme expression in the LCs. The above mentioned data suggest an important role of 11β-HSD1 in modulating intracellular corticosterone concentrations and, in turn, for a direct effect of glucocorticoids on LCs. On the other hand, 11β-HSD type 1 mRNA and its activity was decreased corticosterone deficiency, and it seems that LCs need to maintain their intracellular concentration of corticosterone for normal function [123].

Several authors have demonstrated that 11β-HSD in LCs is predominantly an oxidase [109-111] and the enzyme has been suggested as a marker for the functional maturity of rat adult LCs [111,112,124,125]. The appearance of 11β-HSD correlates with the postnatal increase in testicular weight, LCs number, total surface area of the intracellular membranes and T production by LCs [112]. Neumann *et al.* [126] reported a temporal coincidence of the first appearance of elongated spermatids in the seminiferous epithelium and the first histochemical demonstration of 11β-HSD in the rat LCs on 35 pnd. The developmental pathway of ALCs population is accompanied with an increase in the 11β-HSD activity and thus the enzyme can be used as a marker for steroidogenic differentiation of LCs [112,124,126,127]. Examination of 11β-HSD in the LCs revealed that both oxidative and reductive activities were barely detectable in the progenitors (PLCs), intermediate in immature type (ILCs), and highest in ALCs. The ratio of the two activities favored reduction in PLCs and ILCs and oxidation in ALCs [109]. Clear recognizable oxidative activity of 11β-HSD is present from 31 pnd onward, first in single ALCs and later in majority of these cells [127]. ALCs population expresses high levels of 11β-HSD oxidative activity [109,125] and enzymatic behavior of 11β-HDS in LCs is not consistent with the presence of type 1 alone [127,128]. Developmental analysis of 11β-HSD in rat LCs revealed that 11β-HSD reductive activity predominated in LCs precursors, whereas in adult LCs, the enzyme was primarily oxidative [118]. This switch, observed in the predominant direction of catalysis of 11β-HSD from reduction to oxidation in adult LCs, may protect this cell type from glucocorticoid-mediated inhibition of steroidogenesis. It was demonstrated that the adult LCs expressed not only 11β-HSD type 1, an oxidoreductase, but also type 2, an unidirectional oxidase [129, 130]. Due to its high affinity for glucocorticoid substrates and exclusively oxidative activity, 11β-HSD type 2 may also play a protective role in blunting the suppressive effects of glucocorticoids on LCs steroidogenesis. The inhibition of 11β-HSD1 predominantly lowered reductase activity whereas by inhibition of 11β-HSD2 alone, the oxidase activity was more prominently suppressed [131]. Recently, it has been reported that products such 7α-hydroxytestosterone significantly switched 11β-HSD1 oxidoreductase activities toward reductase in developing rat testis and thus regulates the direction of 11β-HSD1 activity in LCs [132]. It seems that the switch of 11β-HDS activity from reduction to oxidation during the transition from PLCs to ALCs [109] can be associated with the presence of 11β-HSD2.

As mentioned above the main function of glucocorticoids in adult LCs is inhibition of T biosynthesis [111]. Glucocorticoids directly regulate T production in LCs through glucocorticoid receptor (GR)-mediated repression of the genes that encode T biosynthetic enzymes [143,109]. The response of LCs to glucocorticoids depends not only on the number of GR and the circulating concentration of glucocorticoids, but also on the ratio of 11β-HSD oxidative and reductive activities [144]. When oxidation predominates over reduction, 11β-HSD decreases the intracellular availability to active glucocorticoid, attenuating GR-mediated responses [118]. In this way, T production is maintained in the presence of normal serum concentrations of corticosterone and it is inhibited only if 11β-HSD oxidative capacity in LCs is reduced.

By using experimental model for treatment with ethane-dimethnesulphonate (EDS) of mature rats our studies provided new data about expression pattern of 11β-HSD during renewal of LCs population [133]. The quantitative immunohistochemical analysis of 11β HSD2 pattern after EDS treatment revealed progressive increases in the reaction intensity during postnatal development (on d 21after EDS) and reached a maximum on d35 and that is a turning point in the development from immature to mature LCs [133]. These changes in 11β-HSD2 expression are consistent with previous data about structural and functional maturation of the new population of LCs after EDS [134,135]. Therefore, 11β-HSD2 can be a useful marker for ALCs differentiation and the reaction intensity might be associated with increased 11β-HSD oxidative activity that occurred during the transition from PLCs to ALCs in postnatal rat testis [109,127]. Moreover, the gene profiling of rat PLCs, immature LCs and ALCs showed increased expression of 11β-HSD2 gene that is in parallel with enhanced 11β-HSD2 enzyme activity during postnatal development [136]. Together with previous studies [126] the data from EDS model suggest the relationship between 11β-HSD and kinetics of spermatid differentiation and restoration of T production by new LCs population.

6. 11β-hydroxysteroid dehydrogenase in aging testis- role in the response of Leydig cells to the glucocorticoids

It has been established that circulating levels of testosterone decrease with age in both male rodents and men [137]. It was demonstrated by analyzing cohorts of healthy men and rodents that the decline in androgen levels result from specific age-related changes in the male reproductive system and not secondarily from increased disease frequency associated with the aging process, [138,139]. Data indicated that the hypothalamic-pituitary axis in the aging individuals is still intact [140]. Indeed, it is unlikely that the deficiency in the hypothalamic-pituitary axis are primarily responsible for age-related changes in steroidogenesis. The reduced ability of aging LCs to produce T might be caused by events occurring outside these cells that impinge upon them or by events that occur within LCs themselves [141]. It seems that functional changes in LCs themselves rather than their loss cause reduced steroidogenesis during aging [142].

Our data demonstrated that aging affects T production not only through the direct suppression of 3β-HSD, a key marker for LCs steroidogenic activity but also through the inhibition of 11β-HSD type 2 and insulin-like 3 (INSL3) factor that are involved in functional maturation of the adult LCs [146]. These data suggest that increasing functional hypogonadism in aging male rats is likely caused by dedifferentiation of the LCs themselves. Our findings for reduced 11β-HSD type 2 expression in aging LCs provide new evidence for the functional properties of this enzyme in rat testis and bring an additional elucidation of the intracellular mechanisms underlying the decrease in T production accompanying aging. Significant diminished expression of 11β-HSD type 2 in LCs with aging implies suppression in 11β-HSD oxidative capacity resulting in elevated inhibitory potency of corticosterone on T production [136]. The reduced expression of 11β-HSD type 2 in aging rat LCs is also suggestive for decline in LCs protection ability as opposed to adverse effect of glucocorticoids on T production [146]. Inhibition of 11β-HSD 2 oxidative activity by treatment with 11β-HSD 2 antisense oligomer results in excess of glucocorticoids due to lowering the rate of their inactivation [136]. On the other hand, the elevated levels of corticosterone caused decline in oxidative activity of 11β-HSD leading to impaired LCs steroidogenesis [147]. Therefore, the reduction of 11β-HSD type 2 oxidase occurred during LC aging [146] appears to be a key event that leads to down-stream deficits in the response of LCs to prevent glucocorticoid-mediated suppression of steroidogenesis. (Figure.6)

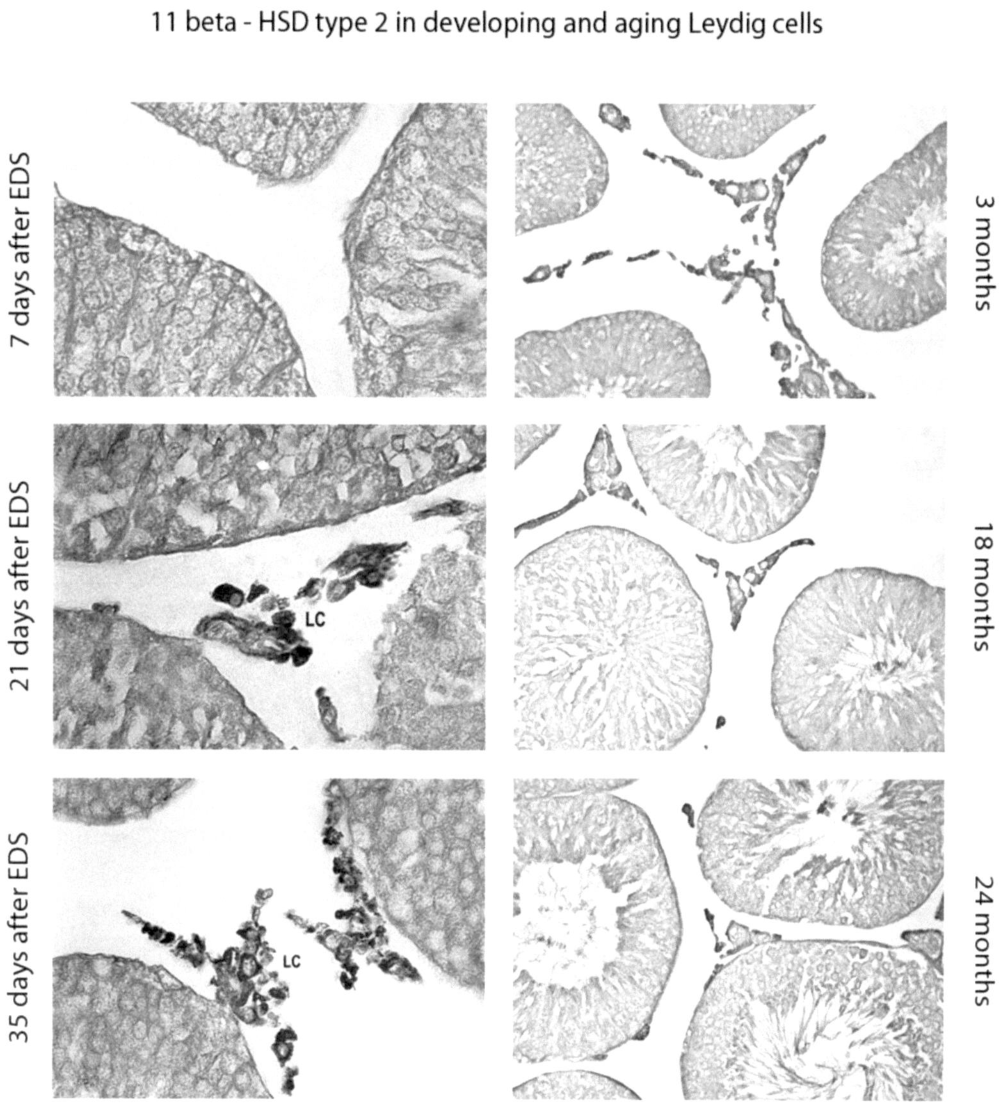

Figure 6. 11β-HSD type 2 in developing Leydig cells (LC)- 7, 21 and 35 days after EDS; and aging Leydig cells- 3, 18 and 24-months of age. x 400.

7. 11β-hydroxysteroid dehydrogenase in the adrenal gland - expression profile under conditions of testosterone withdrawal

As mentioned above, the enzyme11β-HSD catalyzes the interconversion of glucocorticoids to inert metabolites in man and rodents and plays a crucial role in regulating the action of corticosteroids. Inhibition of 11β-HSD allows access of cortisol or corticosterone to the mineralcorticoid receptors where they act as mineralcorticoids [148]. Northern blot analyses revealed expression of mRNAs encoding both 11β-HSD1 and 11β-HSD2 in the whole rat adrenal gland. *In situ* hybridization of rat adrenal cortex and medulla demonstrated specific localization of 11β HSD1 mRNA predominantly to the cells at the corticomedullary junction, within the inner cortex, suggesting that the oxoreductase enzyme may serve to maintain high medullary glucocorticoid concentrations required for catecholamine biosynthesis. In contrast, 11β-HSD2 mRNA was more uniformly distributed in the cortex and was low/absent in the medulla [149, 150]. The expression of 11β-HSD2 has been demonstrated in rat adrenal gland by immunohistochemical and molecular analyses and the 11β-HSD2 antigen was confined to the zona fasciculata and zona reticularis, but not in the zona glomerulosa or medulla [149-151]. The ubiquitous presence of 11β-HSD2 in sodium-transporting epithelia revealed that mineralcorticosteroid action is facilitated by this enzyme which metabolizes glucocorticoids and allows aldosterone to bind to the nonselective mineralcorticoid receptor [151].

Using EDS experimental model in adult rats [152] we found that the dynamic of 11β-HSD2 expression correlated with the changes of serum T levels following the exposure after EDS [153]. The lowest 11β-HSD2 staining intensity was found 7 days after EDS followed by progressive increase in the immunoreactivity on day 14 and 21 after EDS [152]. Moreover, the restoration of 11β-HSD2 activity on day 14 after EDS corresponded with unchanged glandular and serum corticosterone levels in treated rats on day 15 reported by Plecas et al. [154]. Enzymatic assays on tissue homogenates showed extensive conversion of corticosterone to its 11β-dehydro product in an NAD+-dependent manner in adrenal gland [151]. Using enzymehistochemistry a strong reduction was found in the activity of $NADH_2$-cytochrome-C-reductase that is involved in NAD+-synthesis as a cofactor in the adrenal gland after EDS treatment of adult rats [155]. Immunohistochemical analysis revealed that the 11β-HSD2 expression pattern in adrenal gland of EDS treated rats [152] is very similar to the enzymehistochemical profile of $NADH_2$- cytochrome-C-reductase [155], supporting the view that 11β-HSD2 acts as high-affinity NAD+-dependent dehydrogenase in the rat adrenal gland [151]. On the other hand, the increase in the expression of 11β-HSD2 in rat adrenal gland on day 14 after EDS treatment [152] coincided with the appearance of the repopulation of testosterone-producing Leydig cells in the testis [135]. These data suggested a possible role of the gonadal steroids, especially of testosterone, as modulators of the adrenal gland functional activity and they are consistent with previously reported results related to the direct impact of testosterone on the key steps in the adrenal gland steroidogenesis [156]. The above mentioned findings characterized 11β-HSD2 (high-affinity NAD+- dependent unidirectional dehydrogenase) as a potential target of testosterone action in rat adrenal cortex. Our data from EDS experimental model provided new evidence for expression of 11β-HSD2 in the adrenal gland under conditions of testosterone withdrawal. The EDS results bring additional elucidation on the functional significance of 11β-HSD system in rat adrenal gland and the regulatory role of testosterone in its activity [152]. Together with our previous studies [135,153], these data suggested the relationship between 11β-HSD2 expression in adrenal gland and kinetics of restoration of testosterone production during renewal of testicular adult LCs population after EDS treatment. (Figure 7)

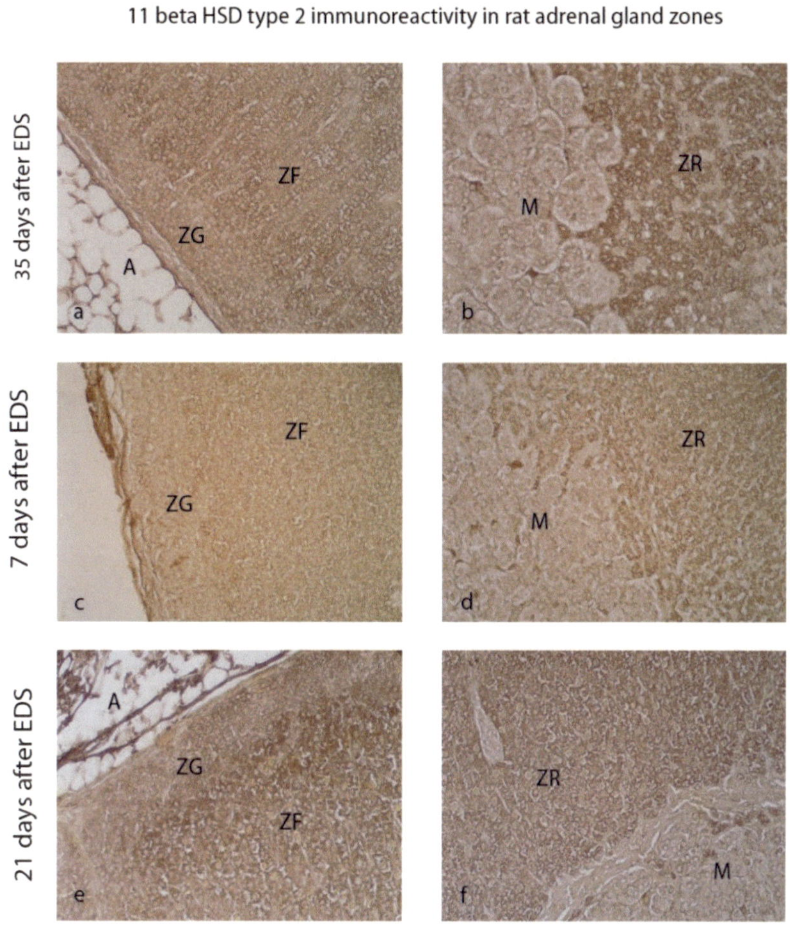

Figure 7. 11β-HSD2 immunoreactivity in rat adrenal gland zones. 35 days after EDS (a, b); 7 days after EDS (c, d); 21 days after EDS (e, f). 11β-HSD2- immunoreactivity in the zona fasciculata (ZF) and zona reticularis (ZR), and the adipocytes of adrenal capsula adipose (A). Less sensitive were the adrenocorticocytes of zona glomerulosa (ZG). No positive signals in the medulla (M). x 200.

8. 11β –hydroxysteroid dehydrogenase in the ovary – cellular localization/distribution and relation to Polycistic Ovaries Syndrome and obesity in women

Glucocorticoids exert their effects in all parts of the body and they are involved in a number of physiological processes, including female reproduction. The ovary is also affected by the glucocorticoids and it is well known that the reproductive function may be impaired in cases of adrenal hyperactivity. The ovaries express glucocorticoid receptors and one of the prominent glucocorticoids affecting ovarian function is the cortisol [157]. Ovaries lack the necessary enzymes for cortisol synthesis and cortisol is not produced *de novo* [158] but it was delivered by the circulation. The 11β-HSD enzymes play a crucial role in controlling the tissue concentration of cortisol. The two types of 11β-HSD (1 and 2) with opposite action modifies cortisol exposure by interconversion between active and inactive glucocorticoids [159,160].

In the human ovary expression of 11β-HSD types 1 and 2 is well documented. 11β-HSD type 2 expression is most prominent during the luteal phase in the corpus luteum and in non-luteinized granulosa cells from follicles before the mid-cycle surge of gonadotrophins. In contrast 11β-HSD type 1 is only seen in granulosa cells from preovulatory follicles [161]). As a result, developmentally regulated pattern of 11β-HSD types 1 and 2 promotes high levels of cortisol during the mid-cycle surge of gonadotrophins, immediately prior to ovulation, whereas reduced levels are maintained throughout the rest of the menstrual cycle [162,163]. Therefore the high levels of local free cortisol are suggested to act as anti-inflammatory agent that limited the tissue damage occurring in connection with follicular rupture [163,164]. This considerations suggest that the regulation of concentration of biologically active cortisol in the ovary may be an important physiological mechanism by which glucocorticoids affect female reproductive organs.

The polycystic ovary syndrome (PCOS) is a common endocrine and metabolic disorder among premenopausal women. The symptoms include the consequences of excessive androgen production (hyperandrogenemia), anovulation and infertility. The hallmark of PCOS is follicular maturation arrest and hyperandrogenemia that is believed to be a critical component of the syndrome [165, 66]. Studies regarding the pathophysiology of PCOS focus attention to primary defects in the hypothalamic–pituitary axis, ovarian function, insulin secretion and action but none of these hypotheses can fully elucidate the multiple clinical phenotypes of PCOS [167-169]. Insulin resistance and the associated compensatory hyperinsulinemia and centripetal obesity, perhaps reflect an association and linkage of the insulin gene with PCOS [170]. PCOS is of unknown etiology, but several lines of evidence suggest that there is an underlying genetic cause for PCOS. Ovarian androgen production occurs primarily in the theca cells and examination of the metabolism of radiolabeled steroid hormone precursors and steady-state levels of mRNAs, encoding steroidogenic enzymes, revealed that there are multiple alterations in the steroidogenic machinery of PCOS theca cells [171-173]. These observations are consistent with the notion that dysregulation of androgen biosynthesis is intrinsic property of PCOS theca cells and that PCOS may develop as a consequence of a primary genetic abnormality in ovarian androgen production [174]. Elevated adrenal androgen levels are common in PCOS, but the underlying pathogenetic mechanisms are poorly understood. One proposed contributing mechanism is altered cortisol metabolism. Moreover, PCOS and obesity are independently associated with increased expression of 11β-HSD1 mRNA in subcutaneous abdominal tissue from lean and obese women with and without PCOS. Decreased peripheral insulin sensitivity and central obesity were associated with increased expression of 11β-HSD1 but not of 11β-HSD2 mRNA expression [175]. Previous studies have described an increased 5alpha-reduction of cortisol and impaired regeneration of cortisol from cortisone by 11β-HSD1 in PCOS, supporting the concept of an altered cortisol metabolism in POCS [176].

In the rare syndrome of cortisone reductase deficiency, impaired ability of 11β-HSD1 to convert cortisone to cortisol, results in compensatory activation of ACTH secretion and adrenal hyperandrogenism [177,178]. This syndrome has been associated with the polymorphisms in the *HSD11B1* gene, which encodes 11β-HSD1, and female patients affected by cortisol reductase deficiency exhibited hyperandrogenism and a phenotype resembling PCOS [179,180]. Lower ratios of cortisol/cortisone metabolites in urine in patients with PCOS were found compared to controls, suggesting a reduced 11β-HSD1 activity [179]. Gambineri et al., [180] reported that polymorphism, predicting lower peripheral regeneration of cortisol by 11β-HSD1, is related to PCOS status and it is associated with increased adrenal hyperandrogenism in lean PCOS. These data strongly support a role for the *HSD11B1* gene in the pathogenesis of PCOS. According to Gambineri et al. [180], the association of the *HSD11B1* genotype with PCOS was mainly attributable to lean rather than obese PCOS patients, suggesting that in obese PCOS women adrenal hyperandrogenism must have a different pathogenetic mechanism as hyperinsulinemia [181] or increased cortisol clearance [182]. The above mentioned findings differ from studies by San Milla'n et al. [183] and White [184] where no association between *HSD11B1* genotype and PCOS was found. This fact suggests that *HSD11B1* polymorphisms may be relevant only in some subgroups of patients and that the pathogenesis of PCOS is different among the different phenotypes of the syndrome [180]. Recently, the functional consequences in these polymorphisms in *HSD11B1* gene were examined and the results confirm previous reports that the variant in *HSD11B1* confer increased 11β-HSD1 expression and activity, that are associated with the metabolic syndrome [183, 185] but are not associated with the prevalence of PCOS [186]. These findings are confirmed by study by Mlinar et al. [187], reporting that PCOS is not associated with increased *HSD11B1* expression. The elevated expression of this gene correlates with markers of adiposity and predicts insulin resistance and an unfavorable metabolic profile, independently of PCOS.

9. 11β –hydroxysteroid dehydrogenase in adipose tissue – relation to obesity and metabolic syndrome

The metabolic syndrome describes a cluster of risk factors like insulin resistance, type 2 diabetes, dyslipidemia, hypertension [188] and co-occurrence of visceral (abdominal, central) obesity. There are strong morphological and metabolic similarities between the Cushing's syndrome of endogenous or exogenous glucocorticoid excess and the metabolic syndrome [189]. Glucocorticoid excess exerts opposing effects on adipose tissue, with an increase in central fat deposition through stimulation of preadipocyte differentiation, gluconeogenesis and triglyceride synthesis, while peripheral fat is reduced as a result from increased lipolysis and lipoprotein lipase downregulation [108]. Glucocorticoid-induced obesity has been investigated in animal models and in humans. It has been shown that cortisol levels are modestly elevated in patients with the metabolic syndrome and tend to be normal or even reduced in simple obesity [189].

The preponderance of data suggest that the intracellular glucocorticoid reactivation was elevated in adipose tissue of obese rodent models and humans [108, 190]. The enzyme that mediates this activation, locally within tissues, is 11β-HSD1 that converts inactive metabolite cortisone to active cortisol, thereby amplifying local glucocorticoid action [104]. 11β-HSD1 expression in adipose tissue was first reported by Monder and White [144] and it is thought to be a dehydrogenase. Studies in leptin-resistant obese rats revealed that obesity was associated with an increase in 11β-HSD1 in abdomenal adipose tissue [191]. In human subcutaneous abdominal adipose tissue, 11β-HSD1 activity is increased both *in vivo* and *in vitro* and the enhanced 11β-HSD1 activity in biopsies is accompanied by elevated 11β- HSD1 mRNA levels [108]). It is interesting to note, that increased subcutaneous adipose 11β-HSD1 is associated with insulin resistance in obesity, but it is not linked specifically with visceral fat accumulation or hypertension [192]. The mechanisms underlying the increase in adipose 11β-HSD1 activity in obesity and metabolic syndrome are still not fully inderstood. 11β-HSD1 transcription is regulated by many factors like cytokines, sex steroids, growth hormone, insulin and induced weight loss [193-195].

The key question is whether increased 11β-HSD1 in adipose tissue is a cause or a consequence of obesity and it is associated with metabolic syndrome. In order to determine this, mice over-expressing 11β-HSD1 selectively in adipose tissue have been generated, using the adipocyte fatty acid binding protein (aP2) promoter [196, 197]. The adipose-selective 11β-HSD1 transgenic mice exhibited elevated intra-adipose, but not systemic corticosterone levels, as well as the major features of the metabolic syndrome-abdominal obesity, hyperglycaemia, insulin resistance, dyslipidaemia and hypertension. Conversely, transgenic mice with overexpression of 11β-HSD1 in liver showed an attenuated metabolic syndrome with modest insulin resistance and hypertriglyceridemia, hypertension and fatty liver, but with normal body weight [198]. 11β-HSD1-knock-out mice fed on a high-fat diet are protected from obesity and metabolic complications [199-201]. Recently, polymorphisms in *HSD11B1*, the gene encoding 11β-HSD1, have been associated with components of the metabolic syndrome [186, 202-205]. Moreover, subjects with single nucleotide polymorphisms (SNPs) in *HSD11B1* gene exhibit increased adipose 11β-HSD1 expression and increased whole-body 11β-HSD1 activity, associated with increased prevalence of the metabolic syndrome. These findings strengthen the view that variations in 11β-HSD1 activity influence the metabolic profile and provide a new evidence that *HSD11B1* gene influence enzyme activity *in vivo* [186].

10. 11β-HSD and metabolite syndrome - clinical importance

Based on human data and genetic models, 11β-HSD1 seems to be cause and promising pharmaceutical target for the treatment of metabolic disease. In mice, the increased enzyme activity in adipose tissue enhances local glucocorticoid levels and produces a metabolic syndrome [196], whereas the decreased enzyme activity protects against obesity and the metabolic syndrome [200, 201]. In human, 11β-HSD1 expression is elevated in adipose tissue in obesity [206], whereas inhibition of 11β -HSD1 enhances insulin sensitivity and provides a new approach to treat type 2 diabetes [207-209]. Polymorphisms in the HSD11B1 gene that encodes 11β-HSD1 have been associated with type 2 diabetes [203] and hypertension [204, 205]. On the other hand, a polymorphism that predicts 11β-HSD1 deficiency may protect against obesity and its metabolic consequences because of impaired regeneration of cortisol in adipose tissue [180]. 11β-HSD1 inhibition is a tempting target for treatment of the metabolic syndrome and its complications. Selective 11beta-HSD1 inhibitors in rodents cause weight loss, improve insulin sensitivity and delay progression of cardiovascular disease [210-212]. Pharmacological inhibition of 11b-HSD1 with the anti-ulcer drug carbenoxolone has provided evidence that cortisol regeneration influences insulin sensitivity, particularly glycogen turnover in healthy human subjects and in patients with type 2 diabetes [207, 208]. This corroborated the notion that the enzyme may be an attractive option to treat the metabolic disease [108, 190, 202, 212, 213]. Moreover, 11β-HSD1 gene knock-out (11β-HSD1-/-) mice exhibited cardioprotective phenotype with improved glucose tolerance and lipid profile, reduced weight and visceral fat accumulation in condition of chronic high-fat feeding [190, 200, 201, 214]. These data support the beneficial effects of 11β-HSD1 inhibitors to lower intracellular glucocorticoid levels and to treat both obesity and its metabolic complications.

11. 11β –hydroxysteroid dehydrogenase and pregnancy – role of 11b-HSD type 2 as a protective barrier for fetus to overexposure to glucocorticoids; implication in intrauterine growth retardation

In mammals, glucocorticoids are important for fetal growth, tissue development and maturation of various organs (surfactant production by the fetal lung, gut enzymes activation and development of the brain and liver). However, supraphysiological levels of glucocorticoids have been shown to cause fetal growth retardation in mammalian models and in human. A number of studies in animal models have examined the effects of prenatal exposure to synthetic glucocorticoids on the fetal development and offspring biology. Maternal glucocorticosteroid treatment reduces birth weight of the offspring and adults exhibit hypertension, hyperinsulinemia, increased hypothalamic–pituitary–adrenal (HPA) axis activity and altered affective behavior [215, 216]). Moreover, human intrauterine growth retardation is associated with high maternal and fetal concentrations of glucocorticoids [217]. Normally, fetal physiological glucocorticoid levels are much lower than maternal levels [218]. The physiological fetoplacental barrier to glucocorticoid exposure is placental 11β-HSD2 that catalyses the rapid conversation of active cortisol and corticosterone to physiologically inert cortisone and corticosterone [219]. 11β-HSD2 acts as a protective barrier to glucocorticoids but a small proportion of maternal glucocorticoid passes through the placenta [220] thus, maternal stress elevates fetal glucocorticoid levels [221]. Different factors are involved in the regulation of placental 11β-HSD2 expression - progesterone, estrogen, hypoxia, infection and proinflammatory cytokines reduce placental 11β-HSD2 activity. Conversely, placental 11β-HSD2 activity is stimulated by glucocorticoids, retinoids and leptin [221]. Studies in rats and human indicate that the deficiency in placental 11β-HSD2 activity results in high fetal exposure to maternal glucocorticoids, with subsequent effects on fetal development and birth weight and offspring biology - high plasma cortisol levels, permanent hypertension, hyperglycemia and increased HPA axis activity was present through the adult life [222-224]. Moreover, individuals homozygous for deleterious mutations of *HSD11B2* gene encoding 11β-HSD have low birth weight. Intrauterine growth retardation in human is associated with increased fetal cortisol levels and reduced placental 11β-HSD2 activity [217]. Studies on prenatal exposure to 11β-HSD inhibitors such as glycyrrhetinic acid and carbenoxolone have indicated that these agents cause fetal growth retardation and adult offspring changes that are very similar to those that are caused by prenatal exposure to glucocorticoids such as dexamethasone (readily crosses the placenta) [221]). Mice that are homozygous for disrupted alleles of *HSD11B2* (i.e. 11β-HSD2–/– mice) also have lower birth weight and the offspring display anxiety-related behaviors in adulthood. It seems that the conditions of increased fetal glucocorticoid levels, in response to different maternal restrictions, sometimes have persistent effects in the offspring - so-called concept of developmental physiological programming and that placental 11β-HSD2 is a key player in fetal programming [215, 216, 221].

12. 17β-HSD dehydrogenase and multifunstional izoforms: localization, function and relevance to clinical therapeutic strategies

17βHydroxysteroid dehydrogenases (17β-HSDs, 17HSD/KSRs) are NAD(H)- and/or NADP(H)-dependent enzymes that catalyze the oxidation and reduction of active 17β-hydroxy- and low active/inactive 17-ketosteroids, respectively. In the presence of substantial excess of a suitable cofactor and/or in the absence of a preferred cofactor, 17HSD/ KSRs can be compelled to catalyze both oxidative and reductive reactions. Depending on their reductive or oxidative activities, they modulate the intracellular concentration of inactive and active steroids. Acting as oxidoreductases at the 17-position of the steroid, they play a key role in estrogen/androgen steroid metabolism by catalyzing the final steps of steroid biosynthesis. Both estrogens and androgens have the highest afnity for their receptors in the 17β-hydroxy form and hence, 17HSD/KSR enzymes regulate the biological activity of the sex hormones.17KSR activities are essential for estradiol and testosterone biosynthesis in the gonads, but they are also present in certain extragonadal tissues and can convert low-activity precursors to their more potent forms in peripheral tissues. Instead, 17HSD activities tend to decrease the potency of estrogens and androgens and consequently may protect tissues from excessive hormone action [10, 225].

Up to now, 14 different subtypes have been identified in mammals and they differ in tissue distribution, sub-cellular localization, function and catalytic preference (oxidation or reduction using the cofactor NAD(H) and NADP(H), respectively) (Table 1). In fact, 17β-HSDs have diverse substrate specificities in vivo as they also catalyze the conversions of other substrates than steroids as for example lipids or retinoids. Until recently, besides 17β-HSD3 and 17β-HSD14, 17β-HSD1 and 2 were thought to be exclusively converting sex steroids. However, the participation of the two latter enzymes (17β-HSD1 and 2) in retinoic acid metabolism recently was suggested. Other 17β-HSD types were already known to be multifunctional and some of them play important roles in different metabolic pathways.

17β-HSD7 is mainly involved in cholesterol synthesis, 17β-HSD4 is implicated in β-oxidation of fatty acids, 17β-HSD5 participates in both prostaglandin and steroid metabolism, and 17β-HSD12 is required in fatty acid elongation. 17β-HSD10 catalyzes the oxidation of short chain fatty acids. 17β-HSD6 and 9 play a role in retinoid conversion. For some 17β-HSDs, the physiological function is not yet clear. For several types of 17β-HSDs participation in the pathophysiology of human diseases has been postulated [225]. The specificity of each 17β-HSD subtype for a preferred substrate together with distinct tissue localization, suggests that these proteins are promising therapeutic targets for diseases like breast cancer, endometriosis, osteoporosis, and prostate cancer. For some of them, their

Type	Gene	Function	Disease-associations	References
1	HSD17B1	Steroid (estrogen) synthesis	Breast and prostate cancer, endometriosis	[226, 227]
2	HSD17B2	Steroid (estrogen, androgen, progestin) inactivation	Breast and prostate cancer, endometriosis Abnormal eye develpment	[10,226, 227]
3	HSD17B3	Steroid (androgen) synthesis	Pseudohermaphroditism in males associated with obesity, prostate cancer	[10,228]
4	HSD17B4	Fatty acid β-oxidation, steroid (estrogen, androgen) inactivation	D-specific bifunctional protein-deficiency, prostate cancer	[229]
5	HSD17B5	Steroid (androgen, estrogen, prostaglandin) synthesis	Breast and prostate cancer	[230,231]
6	HSD17B6	Retinoid metabolism, 3α-3β-epimerase, steroid (androgen) inactivation?		[232]
7	HSD17B7	Cholesterol biosynthesis, steroid(estrogen) synthesis	Breast cancer	[233, 234]
8	HSD17B8	Fatty acid elongation, steroid inactivation, estrogens, androgens	Polycystic kidney disease	[235, 236]
9	HSD17B9	Retinoid metabolism		[237]
10	HSD17B10	Isoleucine, fatty acid, bile acid metabolism, steroid (estrogen, androgen) inactivation	X-linked mental retardation MHBD deficiency Alzheimer's disease	[238] [239]
11	HSD17B11	Steroid (estrogen, androgen) inactivation, lipid metabolism?		[240]
12	HSD17B12	Fatty acid elongation, steroid(estrogen) synthesis		[241, 242]
13	HSD17B13	Not demonstrated		[243]
14	HSD17B14	Steroid (estrogen, androgen?) inactivation, fatty acid metabolism	Breast cancer, prognostic marker	[244, 245]

Table 1. Human 17β-Hydroxysteroid dehydrogenases

expression level can be used as prognostic marker in breast or prostate cancer. The selective inhibition of the concerned enzymes might provide an effective treatment and a good alternative for treatment of steroid dependent diseases [246]. Having in mind multifunctionaloty 17β-HSD enzymes, the biological and clinical aspects of each isoform will be described separately.

17β-HSD type1: 17β-HSD1 catalyzes the activation of estrone (E1) to the most potent estrogen estradiol (E2), predominantly considered as an ezyme of estradiol biosynthesis. It is abundantly expressed in granulosa cells of developing follicles and variable amounts of the enzyme are also expressed in human breast epithelial cells. The enzyme is known to have a crucial role in the development of estrogen-dependent diseases. Based on the in vitro studies, human (h) 17β-HSD1 has been considered as highly estrogen specific, with markedly lower catalytic efficacy towards androgenic substrates. There is a clear difference in the substrate specificity between human and rodent 17β-HSD1 enzymes; the catalytic efficacy of rodent enzyme in vitro is similar for both androgens and estrogens. According a recent review by Saloniemi et al. [10], the h17β-HSD1 is not fully estrogenen-specific but it possesses significant androgenic sctivity. The enzyme catalyses both oxidative (17-hydroxy to 17-keto) and reductive (17-keto to 17-hydroxy) 17b-HSD activity with a proper cofactor added in vitro. However, in cultured cells, the h17β-HSD1 has been shown to catalyse predominantly the reductive reaction [247]. Although h17β-HSD1 expression in various peripheral tissues is low, its catalytic efficacy is markedly higher than those measured for 17β-HSD7 and 17β-HSD12 [248, 242], suggesting an important role for 17β-HSD1 in peripheral E2 formation. Data from animal models further demonstrated the ability of h17β-HSD1 to enhance estrogen action in target tissues and its decrease after treating the mice with 17β-HSD1 inhibitors [10]. These data suggest that 17β-HSD1 plays a major role in determining the gradient between the E2 concentrations in serum and peripheral tissues. An increased E2/E1 ratio by the 17β-HSD1 point out the pivotal role of 17β-HSD1 in breast cancer, ovarian tumor, endometriosis, endometrial hyperplasia and uterine leiomyoma [249, 250]. Consequently, inhibition of 17β-HSD1 is considered as a valuable therapeutic approach for treatment of these deseases. In vivo evaluation of 17β-HSD1 inhibitors is complicated by the fact that the rodent enzymes only show moderate homology/identity to the human one. Due to these species differences, there is a high probability that inhibitors optimized for activity toward rodent 17β-HSD1 do not inhibit the human enzyme. In addition, rodents and humans vary considerably in enzyme distribution in the different tissues. Attempts to overcome these problems include xenograft models using nude mice.

Recently generated mouse genetic model for overexpression of17β-HSD1 (HSD17B1-TG mice) by Saloniemi et al [10] provided valuable data about common female reproductive disorders like Polycystic Ovarian Syndrome (PCOS), ovarian carcinogenesis and endometiosis. Overexpression of hHSD17B1 leads to increased androgen exposure during embryonic development that caused androgen-dependent phenotypic alterations in female, such as increased anogenital distance, lack of vaginal opening and combination of vagina with urethra. These alterations observed in the HSD17B1-TG females were effectively rescued by prenatal anti-androgen (flutamide) treatment, further confirming the dependence of these phenotypes on androgens. Interestingly, the androgen exposure during pregnancy in the HSD17B1-TG mice resulted in benign ovarian serous cystadenomas in adulthood. As ovarian serous borderline tumours are positively associated with a history of PCOS, thus with a history of (foetal) hyperandrogenism, 17β-HSD1 may promote ovarian carcinogenesis via increased estrogen concentration, but also via enhanced androgen production. Endometrial hyperplasia in HSD17B1-TG mice closely resembled human disease and it was efficiently reversed by 17β-HSD1 inhibitor treatment. The data concerning the expression of 17β-HSD1 in normal and diseased human endometrium are not fully conclusive. However, in most of the studies, the 17β-HSD1 expression is detected in normal endometrium, endometriosis specimens and endometriotic cancer. Other 17β-HSD enzymes including 17β-HSD2, 17β-HSD5, 17β-HSD7 and 17β-HSD12 have also been detected in the endometrium under different pathological conditions like endometriosis and PCOS [10]. Collectively, the data suggest that 17β-HSD1 inhibition is one of the several possible approaches to reduce estrogen production both in eutopic and in ectopic endometrial tissue.

17β-HSD type-2: 17-HSD/KSR2 converts 17β-hydroxy forms of estrogens and androgens (estradiol, testosterone and 5α-dihydrotestosterone) to their less active 17-keto forms (estrone, androstenedione and 5α-androstanedione).The enzyme also possesses 20α-HSD activity, thereby activating 20α-hydroxyprogesterone to progesterone. The 17β-HSD2 enzyme is widely and abundantly expressed in both adult and fetal tissues such as placenta, uterus, liver, the gastrointestinal and urinary tracts. Due to its expression pattern and enzymatic characteristics, it has been suggested that the 17β-HSD2 enzyme protects tissues from excessive steroid action [251]. 17β-HSD2 is localised in the endoplasmic reticulum, and it is widely expressed in various estrogen and androgen target tissues both in human and in rodents including breast endometrium, placenta and prostate. Furthermore, the 17β-HSD2 expression in the placenta and in foetal liver and intestine, together with the observed oxidative 17β-HSD2activity, are the basis for the hypothesis, suggesting a role for the enzyme in lowering the sex steroid exposure of the foetus.

Phylogenic analyses have indicated that 17β-HSD2 is a close homologue of retinoid-converting enzymes and has a high sequence similarity to retinol dehydrogenase type 1. In addition, studies have shown that retinoic acid (RA) induces expression of 17β-HSD2 in a dose- and time-dependent manner in human endometrial epithelial and placental cells [10]. Recent data from transgenic mice (HADS17B2-TG) provide evidence for importance of 17β-HSD2 for prenatal eye morphogenesis and eye development [10]. These TG mice overexpressing human 17β-HSD2 showed growth retardation, disrupted spermayogenesis, female masculinization, delayed eye opening, squint appearance of the eyes and some of these defects closely resembeled those identified in retinoid receptor mutant mice. The most notable changes in the HSD17B1TG mice are well explained by alterations in sex steroid action, whereas in the HSD17B2-TG mice the connection to sex steroids is weaker. The opposite mouse model of deficiency of 17β-HSD2 provide evidence for the essential role of 17β-HSD2. Embryonic death in the HSD17B-KO mice is reported, related to lack of action of 17β-HSD2 enzyme in placenta. Furthermore, the treatment of pregnant female mice with an anti-estrogen or with progesterone did not prevent the foetal loss of the HSD17B2-KO mice, thus indicating that embryonic deaths is likely not due to the lack of progesterone or due to an increased action of estrogens.

Osteoporosis is well known to occurs in elderly people when the level of active sex steroids decreases. Estrogen replacement therapy is beneficial for the treatment of osteoporosis but it is no longer recommended because of adverse effects (breast, endometrial and ovarian cancers, stroke, thromboembolism). Since 17β-HSD2 oxidizes E2 into E1, decreasing the amount of E2 in bone cells, inhibition of this enzyme is a promising approach for the treatment of this disease [225]. Ovariectomized cynomolgus monkeys were used as an osteoporosis model to evaluate the efficacy of 17β-HSD2 inhibitors. Decrease in bone resorption and maintenance of bone formation was achieved in this experimental model.

17β-HSD type-3: 17-HSD/KSR3 17β-HSD3 converts Δ4-androstenedione into testosterone and it is essential for testosterone biosynthesis. The enzyme is present exclusively in the testis and the deficiency of the active enzyme results in male pseudohermaphroditism [252]. In addition to the conversion of androstenedione to testosterone, the enzyme is capable of catalyzing conversion of 5α-androstanedione to 5α- dihydrotestosterone as well as estrone to estradiol [108]. Messender RNA for 17β-HSD3 are over-expressed in prostate cancer tissues. As T is known to be responsible for cell proliferation in androgen dependent diseases, 17β-HSD3 inhibitors (exerting effects equivalent of chemical castration) could be therapeutics for the treatment of such diseases [225]. Day et al. [253] developed the first xenograft model in castrated mice to evaluate 17β-HSD3 inhibitors and strong suppression of tumor growth by 81% was found, suggesting that 17β-HSD3 inhibition might be an efficient strategy for the treatment of hormone dependent prostate cancer.

There are only few observations in human male deficient in 17β-HSD as rare mutation associated with 46XY disorder of sexual development [254]. Patients with 17β-HSD deficiency are usually classified as female at birth (although abdominal testes) but developed secondary male features at pubery with diminished virilization [255].

17β-HSD type-4: Among 17-HSD/KSRs, type 4 is an unique multifunctional enzyme consisting of 17-HSD/KSR-, hydratase- and sterol carrier 2-like domains. 17β-HSD4 is ubiquitously expressed, but in some tissues it shows cell-specific expression. In the brain it is present only in Purkinje cells, in the lung only in bronchial epithelium and in the uterus in luminal and glandular epithelium. The deficiency of 17β-HSD4 leads to disease known as Zellweger syndrome [251].

17β-HSD type-5: 17-HSD/KSR5 is also known as type 2 3α-HSD, and diferently from other 17-HSD/KSRs it belongs to the AKR (aldo-keto reductase) family. With other members of the AKR family (type 1 3α-HSD, type 3 3α-HSD and 20α-HSD), 17β-HSD5 shares 84%, 86% and 88% identity, respectively. Both human and mouse 17β-HSD5 catalyze the conversion of androstenedione to testosterone, and additionally possess 3α-HSD activity. Human 17β-HSD5 has been previously identified predominantly as 3α-HSD. Human, but not mouse, 17β-HSD5 also converts progesterone to 20α-dihydroprogesterone effectively. 17β-HSD5 appears to be involved in the formation of androgens in the testis and several peripheral tissues. Using specific probes and antibodies, human 17β-HSD5 has been localized in liver, adrenal, testis, basal cells of the prostate, and in prostatic carcinoma cell lines [251]. Recently, up-regulation of 17β-HSD5 was found in breast and prostate cancer [256].

17β-HSD type-6: 17-HSD/KSR6 is part of the catabolic cascade of 5α-dihydrotestosterone (DHT). The 17β-HSD6 shows low dehydrogenase activity with DHT, testosterone and estradiol and possesses a weak oxidative 3α-HSD activity. The 17β-HSD6 enzyme shares 65% sequence identity with retinol dehydrogenase type 1 and it is most abundantly expressed in liver and prostate, at least in rodent tissues [251].

17β-HSD type-7: 17β-HSD7 is expressed in the developing follicles and in luteinized cells, being the enzyme of ovarian estradiol biosynthesis. Both rodent and human 17β-HSD7 catalyze exclusively the conversion of estrone to estradiol. The 17β-HSD7 is abundantly expressed in corpus luteum during pregnancy and the enzyme is considered to be important in E2 production, especially during pregnancy. In addition, 17β-HSD7 mRNA has been detected in placental, mammary gland and kidney samples [251]. The 17β-HSD7 enzyme was first characterised as a prolactin receptor-associated protein in the rat corpus luteum, although its role in prolactin signalling has remained unknown.

A role for mouse 17β-HSD7 in cholesterol biosynthesis was also suggested by the studies, showing a similar expression pattern of 17β-HSD7 and cholesterogenic enzymes during mouse embryonic development. Data from HSD17B7-KO mouse embryos evidently showed the essential role of 17β-HSD7 for cholesterol biosynthesis in vivo. The lack of 17β-HSD7 resulted in a marked blockage in foetal de novo cholesterol synthesis. Histological analysis revealed that the 17β-HSD7 deficiency results in defects in the development of nerve system, vasculature, heart, associated with defect in cholesterol synthesis. HSD17B-KO deficient mice exhibit embryonic lethal phenotypes Tese data suggest a possible role of 17β-HSD7 in cholesterol biosynthesis in mice, while its role in E2 production in vivo needs further clarification [10].

17β-HSD type-8: The *Ke 6* gene product has been characterized as a protein whose abnormal regulation is linked to the development of recessive polycystic kidney disease in mice and later it was discovered to be a 17βHSD8. In *in vitro* conditions, 17β-HSD8 converts most eficiently estradiol to estrone and, to some extent, it also catalyses oxidative reactions of androgens and the reduction from estrone to estradiol. The 17β-HSD8 is abundand in kidney, liver and gonads. Interestingly, in the ovary, 17β-HSD8 is present in cumulus cells and not in granulosa or luteal cells like 17βHSD1 and 7, respectively [251].

17β-HSD type-10: The 17β- -HSD10 has a very broad substrate profile. Interestingly, it has been proposed that this enzyme plays an important role in the pathological processes of Alzheimer's disease (AD), mainly because 17β-HSD10 binds to amyloid-β peptide and appears to be up-regulated in patients suffering from this disease [225]. The mechanism by which 17β-HSD10 contributes to the pathology of AD is still not completely understood. The protein-protein interaction of 17β-HSD10 with amyloid-β appears to inhibit the enzymatic activity of 17β-HSD10. In vitro studies with a potent 17β-HSD10 inhibitor [257] have shown that inhibition of this enzyme can prevent its interaction with the amyloid-β peptide, suggesting 17β HSD10 as a potential target for the treatment of AD.Transgenic mice over-expressing human 17β-HSD10 suggesting that inhibition of 17β-HSD10 could protect from cerebral infarction and ischemia [258].

17β-HSD type-12: The mammalian 17β-HSD12 was initially characterised as a 3-ketoacyl-CoA reductase, involved in the long-chain fatty acid synthesis, particularly essential for brain arachidonic acid synthesis. Both the human and the mouse 17β-HSD12 share 40% sequence similarity with 17β-HSD3, and the data indicate that 17β-HSD12 is an ancestor of 17β-HSD3. In human and rodents, 17β-HSD12 is expressed universally and the highest expression of 17β-HSD12 is detected in tissues involved in the lipid metabolism, including the liver, kidney hearth, and skeletal muscle. In mice, the expression has also been detected in brown and white adipose tissue. 17β-HSD12 expression is also regulated by sterol regulatory element binding proteins, identically to that shown to be involved in fatty acid and cholesterol biosynthesis. Interestingly, a reduced expression of 17β-HSD12 in cultured breast cancer cells results in significant inhibition of cell proliferation that is fully recovered by supplementation of arachidonic acid. In addition to its putative role in fatty acid synthesis, human 17β-HSD12 has been shown to catalyse the conversion of E1 to E2 in cultured cells, and the enzyme was suggested to be a major enzyme converting E1 to E2 in postmenopausal women [10]. Analysis of the HSD17B12-KO embryos indicated that the embryos initiated gastrulation but further organogenesis was severely disrupted. The mutant embryos exhibited severe defects in the neuronal development (ectoderm-derives), they failed to grow several mesoderm-derived structures. Therefore, the embryos at the age of E8.5–E9.5 were avoid of all normal embryonic structures that caused their death.

13. Conclusion

HSD enzymes are broadly expressed in all steroidogenic organs as different isoforms with differential localization and function. HSD are key enzymes involved in growth and reproduction and they are considered as suitable targets to modulate the concentration of the potent steroids in case of steroid-dependent diseases. As they could act selectively in an intracrine manner, inhibitors of these enzymes might be superior to the existing endocrine therapies regarding the off-target effects. Although commont mechanisms operate in regulation of steroidogenesis, there are some differences/specificities between rodent and human, in particular the susceptibility of fetal testicular stereoidogenesis to environmental chemicals with estrogenic/antiandrogenic activity. As the latter appeared to be devoid of effect on fetal human testis, this should be taken into account when dial with risk assessment of endocrine disruptors for human reproductive health. Species specific diffences in steroiodogenesis cause real obstacles in investigation of HSD inhibitors. Some of the most active and selective inhibitors were investigated in vivo in animal disease-oriented models. They showed efficacy, but none of them reached the clinical trial stage. One reason for this might be the difficulty to identify an appropriate species to conduct the functional assays, as very potent inhibitors of the human enzyme show little activity toward HSD of other species (rodents). In this respect, experiments by using xenograft approach (human tissue xenografting in immunocompromised nude mice) would enable us to develop our studies for better understanding of regulatory mechanisms of the expression of HSD enzymes. Elucidation of molecular events involved in transcription control of HSD is of great importance for molecular desigh of new HSD inhibitors and development of new strategies for appropriate treatment of steroid-dependent deceases without use of invasive techniques.

Author details

Nina Atanassova
Inst. Experimental Morphology, Pathology and Anthropology with Museum, Bulgarian Academy of Sciences, Sofia, Bulgaria

Yvetta Koeva
Dept. Anatomy and Histology, Medical University, Plovdiv, Bulgaria

Acknowledgement

The authors thank to Professor Richard Sharpe for providing samples from experimental models for hormonal manipulations, Chris McKinnel for technical expertise in immunohistochemistry. We are also grateful to Professor Michail Davidoff and Assoc. Professor Mariana Bakalska for studies on EDS experimental model. Authors' work was supported in part by Grant DEER # 212844 funded by FP7-ENV-CP and Grant # DO 02/113 funded by NF "Scientific Research" of Ministry of Education Youth and Science in Bulgaria.

14. References

[1] Hoffmann F, Maser E. Carbonyl reductases and pluripotent hydroxysteroid dehydrogenases of the short-chain dehydrogenase/reductase superfamily, Drug Metabolism Reviews 2007; 39: 87 144.
[2] Duax WL., Ghosh D, Pletnev V. Steroid dehydrogenase structures, mechanism of action, and disease. Vitamines & Hormones 2000; 58: 121–148.
[3] Labrie F, Luu-The V, Labrie C, Simard J. DHEA and its transformation into androgens and estrogens in peripheral target tissues: intracrinology. Frontiers in Neuroendocrinol 2001; 22: 185–212.
[4] Cancer Research:http://info.cancerresearchuk.org/cancerstats/
[5] Giudice L. Clinical practice. Endometriosis. New England Journal of Medicine 2010; 362: 2389–2398.
[6] Goodarzi MO, Dumesic DA, Chazenbalk G & Azziz R. Polycystic ovary syndrome: etiology, pathogenesis and diagnosis. Nature Reviews Endocrinology 2010; 7: 219–331.
[7] Labrie F. Drug insight: breast cancer prevention and tissue-targeted hormone replacement therapy. Nature Clinical Practice in Endocrinology & Metabolism 2007; 3: 584–593.

[8] Schuster D, Laggner C, Steindl TM, Palusczak A, Hartmann RW, Langer T. Pharmacophore modeling and in silico screening fornewP450 19 (aromatase) inhibitors, Journal of Chemical Information & Modeling 2006; 46: 1301–1311.

[9] Aggarwal S, Thareja S, Verma A, Bhardwaj TR, Kumar K, An overview on 5alpha-reductase inhibitors. Steroids 2010; 75: 109–153.

[10] Saloniemi T, Jokela H, Strauss L, Pakarinen P and Poutanen M. The diversity of sex steroid action: novel functions of hydroxysteroid (17*b*) dehydrogenases as revealed by genetically modified mouse models (Thematic Review). Journal of Endocrinology 2012: 212, 27-40.

[11] Hafez ESE. Hormones, Growth Factors, and Reproduction. In: Hafez ESE.(ed.) Reproduction in Farm Animals. Philadelphia: Lea & Febiger; 1993. p59-93.

[12] Barrett E. Section VIII The Endocrine System. In: Boron WF, Boulpaep EL (eds) Medical Physiology. A Cellular And Molecular Approach. Philadelphia, PA: Elsevier/Saunders; 2003 (1st edition) p1009-1110.

[13] Pezzi V, Mathis JM, Rainey WE, Carr BG. Profiling transcript levels for steroidogenic enzymes in fetal tissues. Journal of Steroid Biochemistry and Molecular Biology 2003; 87: 181-189.

[14] Scott HM, Mason JI, Sharpe RM. Steroidogenesis in the fetal Testis and its Susceptibility to Disruption by Exogenous Compounds. Endocrine Review 2009; 30: 883-925.

[15] Payne AH, Hales DB. Overview of steroidogenic enzymes in the pathway from cholesterol to active steroid hormones. Endocrine Review 2004; 25: 947-970.

[16] Payne AH. Steroidogenic Enzymes in Leydig Cells. In: Payne AH. & Hardy MP. (eds) The Leydig Cell in Health and Disease. Totowa, NJ: Human Press Inc; 2007, p157-171.

[17] Arakane F, Kallen CB, Watari H, Foster JA, Sepuri NB, Pain D, Stayrook SE, Lewis M, Gerton GL, Strauss 3rd JF. The mechanism of action of steroidogenic acute regulatory protein (StAR). StAR acts on the outside of mitochondria to stimulate steroidogenesis. J Biological Chemistry 1998; 273: 16339–16345.

[18] Dube′ C, Bergeron F, Vaillant MJ, Robert NM, Brousseau C, Tremblay JJ. The nuclear receptors SF1 and LRH1 are expressed in endometrial cancer cells and regulate steroidogenic gene transcription by cooperating with AP-1 factors. Cancer Letters 2009; 275: 127–138.

[19] Parker KL, Schimmer BP. Transcriptional regulation of the genes encoding the cytochrome P-450 steroid hydroxylases. Vitamins and Hormones 1995; 51: p339–370.

[20] O'Shaughnessy PJ, Johnston H, Baker PJ. Development of Leydig Cell Steroidogenesis. In: Payne AH. & Hardy MP. (eds) The Leydig Cell in Health and Disease. Totowa, NJ: Human Press Inc; 2007, p173-179.

[21] Achermann JC, Ozisik G, Ito M, Orun UA, Harmanci K, Gurakan B, Jameson JL. Gonadal determination and adrenal development are regulated by the orphan nuclear receptor steroidogenic factor-1, in a dose-dependent manner. Journal of Clinical Endocrinology and Metabolism 2002; 87: 1829–1833.

[22] Jeyasuria P, Ikeda Y, Jamin SP, Zhao L, De Rooij DG, Themmen AP, Behringer RR, Parker KL. Cell-specific knockout of steroidogenic factor 1 reveals its essential roles in gonadal function. Molecular Endocrinology 2004;1: 1610–1619.

[23] Stocco DM.The Role of StAR in Leydig Cell Steroidogenesis. In: Payne AH. & Hardy MP. (eds) The Leydig Cell in Health and Disease. Totowa, NJ: Human Press Inc; 2007, p149-155
[24] Black SM, Harikrishna JA, Szklarz GD, Miller WL. The mitochondrial environment is required for activity of the cholesterol side-chain cleavage enzyme, cytochrome P450scc. Proceedings of the National Acadamy of Science USA 1994; 91: 7247–7251.
[25] Miller WL. Why nobody has P450scc(20,22 desmolase deficiency). Journal of Clinical Endocrinology and Metabolism 1998;83: 1399–1400.
[26] Kim CJ, Lin L, Huang N, Quigley CA, AvRuskin TW, Achermann JC, Miller WL. Severe combined adrenal and gonadal deficiency caused by novel mutations in the cholesterol side chain cleavage enzyme, P450scc. Journal of Clincal Endocrinology and Metabolism 2008; 93: 696-702.
[27] Pang S, Yang X, Wang M, Tissot R, Nino M, Manaligod J, Bullock LP, Mason JI. Inherited congenital adrenal hyperplasia in the rabbit: absent cholesterol side-chain cleavage cytochrome P450 gene expression. Endocrinology 1992;131, 181–186.
[28] Auchus RJ. The genetics, pathophysiology, and management of human deficiencies of P450c17. Endocrinology, Metabolism Clinics of North America 2001; 30, 101–119.
[29] Van Den Akker EL, Koper JW, Boehmer AL, Themmen AP, Verhoef-Post M, Timmerman MA, Otten BJ, Drop SL, De Jong FH. Differential inhibition of 17 α-hydroxylase and 17,20-lyase activities by three novel missense CYP17 mutations identified in patients with P450c17 deficiency. Journal of Clinical Endocrinology & Metabolism 2002; 87: 5714–5721.
[30] Martin RM, Lin CJ, Costa EM, de Oliveira ML, Carrilho A, Villar H, Longui CA, Mendonca BB. P450c17 deficiency in Brazilian patients: biochemical diagnosis through progesterone levels confirmed by CYP17 genotyping. Journal of Clinical Endocrinology & Metabolism 2003; 88: 5739–5746.
[31] Simard J, Ricketts ML, Gingras S, Soucy P, Feltus FA, Melner MH. Molecular biology of the 3beta-hydroxysteroid dehydrogenase/delta5-delta4 isomerase gene family. Endocrine Review 2005; 26: 525-582.
[32] mas JL, Duax WL, Addlagatta A, Brandt S, Fuller RR, Norris W. Structure/function relationships responsible for coenzyme specificity and the isomerase activity of human type-1 3β-hydroxysteroid dehydrogenase/isomerase. Journal of Biological Chemistry 2003; 278: 483–490.
[33] Cherradi N, Defaye G, Chambaz EM. Dual subcellular localization of the 3β-hydroxysteroid dehydrogenase isomerase: characterization of the mitochondrial enzyme in the bovine adrenal cortex. Journal Steroid Biochemistry & Molecular Biology 1993; 46, 773–779.
[34] Cherradi N, Chambaz EM, Defaye G. Organization of 3β-hydroxysteroid dehydrogenase/isomerase and cytochrome P450scc into a catalytically active molecular complex in bovine adrenocortical mitochondria. Journal Steroid Biochemistry & Molecular Biology 1995; 55: 507–514.
[35] Pelletier G, Li S, Luu-The V, Tremblay Y, Belanger A, Labrie F. Immunoelectron microscopic localization of three key steroidogenic enzymes (cytochrome P450(scc), 3β-

hydroxysteroid dehydrogenase and cytochrome P450(c17)) in rat adrenal cortex and gonads. Journal of Endocrinology 200; 171: 373–383.

[36] Chapman JC, Waterhouse TB, Michael SD. Changes in mitochondrial and microsomal 3_-hydroxysteroid dehydrogenase activity in mouse ovary over the course of the estrous cycle. Biology of Reproduction 1992; 4: 992–997.

[37] Luu-The V, Lachance Y, Labrie C, Leblanc G, Thomas JL, Strickler RC, Labrie F. Full length cDNA structure and deduced amino acid sequence of human 3β-hydroxy-5-ene steroid dehydrogenase. Molecular Endocrinology 1989; 3: 1310–1312.

[38] Rheaume E, Lachance Y, Zhao HF, Breton N, Dumont M, de Launoit Y, Trudel C, Luu-The V, Simard J, Labrie F. Structure and expression of a new complementary DNA encoding the almost exclusive 3β-hydroxysteroid dehydrogenase/ Δ5-Δ4-isomerase in human adrenals and gonads. Molecular Endocrinology 1991; 5: 1147–1157.

[39] Dumont M, Luu-The V, Dupont E, Pelletier G, Labrie F. Characterization, expression, and immunohistochemical localization of 3β-hydroxysteroid dehydrogenase/Δ5-Δ4 isomerase in human skin. Journal of Investigave Dermatology 1992; 99: 415–421.

[40] Lachance Y, Luu-The V, Labrie C, Simard J, Dumont M, de Launoit Y, Guerin S, Leblanc G, Labrie F. Characterization of human 3β-hydroxysteroid dehydrogenase/Δ5-Δ4 isomerase gene and its expression in mammalian cells. Journal of Biological Chemistry 1992; 267: 3551

[41] Morissette J, Rheaume E, Leblanc JF, Luu-The V, Labrie F, Simard J 1995 Genetic linkage mapping of HSD3B1 and HSD3B2 encoding human types I and II 3β-hydroxysteroid dehydrogenase/Δ5-Δ4- isomerase close to D1S514 and the centromeric D1Z5 locus. Cytogenetics & Cell Genetics 1995; 69: 59–62.

[42] Zhao HF, Labrie C, Simard J, de Launoit Y, Trudel C, Martel C, Rheaume E, Dupont E, Luu-The V, Pelletier G. Characterization of rat 3β-hydroxysteroid dehydrogenase/Δ5-Δ4 isomerase cDNAs and differential tissue-specific expression of the corresponding mRNAs in steroidogenic and peripheral tissues. Journal of Biological Chemistry 1991; 266: 583–593.

[43] Simard J, Couet J, Durocher F, Labrie Y, Sanchez R, Breton N, Turgeon C, Labrie F. Structure and tissue-specific expression of a novel member of the rat 3β-hydroxysteroid dehydrogenase/ Δ5-Δ4 isomerase (3β-HSD) family. The exclusive 3β-HSD gene expression in the skin. Journal of Biological Chemistry1993; 268: 19659–19668.

[44] Couet J, Simard J, Martel C, Trudel C, Labrie Y, Labrie F. Regulation of 3-ketosteroid reductase messenger ribonucleic acid levels and 3β-hydroxysteroid dehydrogenase/Δ5-Δ4 isomerase activity in rat liver by sex steroids and pituitary hormones. Endocrinology 1992; 131: 3034–3044.

[45] Sanchez R, de Launoit Y, Durocher F, Belanger A, Labrie F, Simard J. Formation and degradation of dihydrotestosterone by recombinant members of the rat 3β-hydroxysteroid dehydrogenase/ Δ5-Δ4 isomerase family. Molecular & Cellular Endocrinolology 1994; 103: 29–38.

[46] de Launoit Y, Simard J, Durocher F, Labrie F. Androgenic 17β-hydroxysteroid dehydrogenase activity of expressed rat type I 3β-hydroxysteroid dehydrogenase/Δ5-Δ4 isomerase. Endocrinology 1992; 130: 553–555.

[47] Mason JI, Howe BE, Howie AF, Morley SD, Nicol MR, Payne AH. Promiscuous 3β-hydroxysteroid dehydrogenases: testosterone 17β-hydroxysteroid dehydrogenase activities of mouse type I and VI 3β-hydroxysteroid dehydrogenases. Endocrine Research 2004; 30: 709–714.
[48] Payne AH, Clarke TR, Bain PA. The murine 3β-hydroxysteroid dehydrogenase multigene family: structure, function and tissue-specific expression. Journal Steroid Biochemistry & Molecular Biology 1995; 53: 111–118.
[49] Payne AH, Abbaszade IG, Clarke TR, Bain PA, Park CH. The multiple murine 3β hydroxysteroid dehydrogenase isoforms: structure, function, and tissue- and developmentally specific expression. Steroids 1997; 62: 169–175.
[50] Abbaszade IG, Arensburg J, Park CH, Kasa-Vubu JZ, Orly J, Payne AH 1997 Isolation of a new mouse 3β-hydroxysteroid dehydrogenase isoform, 3β-HSD VI, expressed during early pregnancy. Endocrinology 1998; 138: 1392–1399.
[51] Belanger B, Belanger A, Labrie F, Dupont A, Cusan L, Monfette G. 1989 Comparison of residual C-19 steroids in plasma and prostatic tissue of human, rat and guinea pig after castration: unique importance of extratesticular androgens in men. Journal of Steroid Biochemestry 1989; 32: 695–698.
[52] Dupont E, Luu-The V, Labrie F, Pelletier G. Ontogeny of 3β-hydroxysteroid dehydrogenase/Δ5-Δ4 isomerase (3β-HSD) in human adrenal gland performed by immunocytochemistry. Molecular & Cellular Endocrinology 1990; 74: R7–R10.
[53] Simonian MH. ACTH and thyroid hormone regulation of 3β-hydroxysteroid dehydrogenase activity in human fetal adrenocortical cells. Journal Steroid Biochem 1986; 25: 1001–1006.
[54] Lo MJ, Kau MM, Chen YH, Tsai SC, Chiao YC, Chen JJ, Liaw C, Lu CC, Lee BP, Chen SC, Fang VS, Ho LT, Wang PS. Acute effects of thyroid hormones on the production of adrenal cAMP and corticosterone in male rats. American Journal of Physiology 1998; 274: E238–E245.
[55] Dupont E, Labrie F, Luu-The V, Pelletier G. Immunocytochemical localization of 3β-hydroxysteroid dehydrogenase/Δ5-Δ4 isomerase in human ovary. Journal of Clinical Endocrinology & Metabolism 1992; 74: 994–998.
[56] Kaplan S, Grumach M. Pituitary and placental gonadotropin and sex steroids in the human and sub-human primate fetus. Journal of Clinical Endocrinology & Metabolism 1978; 7: 487–511.
[57] Grumbach M, Conte F. Disorders of sex differentitiation. In: Wilson JD & Foster EW (eds) Williams textbook of endocrinology. Philadelphia: W. B. Saunders; 1999: 1303–1425
[58] Nelson VL, Legro RS, Strauss 3rd JF, McAllister JM. Augmented androgen production is a stable steroidogenic phenotype of propagated theca cells from polycystic ovaries. Molecular Endocrinology 1999; 13: 946–957.
[59] Doody KJ, Lorence MC, Mason JI, Simpson ER. Expression of messenger ribonucleic acid species encoding steroidogenic enzymes in human follicles and corpora lutea throughout the menstrual cycle. Journal of Clinical Endocrinology & Metabolism 1990; 70: 1041–1045.

[60] Teerds KJ, Dorrington JH. Immunohistochemical localization of 3β-hydroxysteroid dehydrogenase in the rat ovary during follicular development and atresia. Biology of Reprod 1993; 49: 989–996.
[61] Voss AK, Fortune JE. Levels of messenger ribonucleic acid for cholesterol side-chain cleavage cytochrome P-450 and 3β-hydroxysteroid dehydrogenase in bovine preovulatory follicles decrease after the luteinizing hormone surge. Endocrinology 1993; 132: 888–894.
[62] Labrie F, Belanger A, Cusan L, Gomez JL, Candas B. Marked decline in serum concentrations of adrenal C19 sex steroid precursors and conjugated androgen metabolites during aging. Journal of Clinical Endocrinology & Metabolism 1997; 82: 2396–2402.
[63] Sasano H, Suzuki T. Localization of steroidogenesis and steroid receptors in human corpus luteum. Classification of human corpus luteum (CL) into estrogen- producing degenerating CL, and nonsteroid-producing degenerating CL. Seminars of Reproductive Endocrinology 1997; 15: 345–351.
[64] Duncan WC, Cowen GM, Illingworth PJ. Steroidogenic enzyme expression in human corpora lutea in the presence and absence of exogenous human chorionic gonadotrophin (HCG). Molecular & Human Reproduction 1999; 5: 291–298.
[65] Benyo DF, Little-Ihrig L, Zeleznik AJ. Noncoordinated expression of luteal cell messenger ribonucleic acids during human chorionic gonadotropin stimulation of the primate corpus luteum. Endocrinology 1993; 133: 699–704,
[66] McGee E, Sawetawan C, Bird I, Rainey WE, Carr BR. The effects of insulin on 3β-hydroxysteroid dehydrogenase expression in human luteinized granulosa cells. Journal of the Society for Gynecologic Investigations 1995; 2:535–541.
[67] Feltus FA, Groner B, Melner MH. Stat5-mediated regulation of the human type II 3β-hydroxysteroid dehydrogenase/Δ5-Δ4 isomerase gene: activation by prolactin. Molecular Endocrinology 1999; 13: 1084–1093.
[68] Stocco CO, Deis RP. Participation of intraluteal progesterone and prostaglandin F2α in LH-induced luteolysis in pregnant rat. J Endocrinol 1998; 156: 253–259.
[69] Mendis-Handagama SMLC, Ariyaratne HBS. Differentiation of adult Leydig cell proliferation in the postnatal testis. Biology of Reproduction 2001; 65: 660-671.
[70] Davidoff MS, Middendorff R, Enikolopov G, Riethmacher D, Holstein AF, Muller D. Progenitor cells of the testosterone-producing Leydig cells revealed. Journal of Cell Biology 2005; 167: 935-944.
[71] Kilcoyne K, Sharpe RM, McKinnell C, van den Driesche S, Smith LB, Atanossova N. Putative adult Leydig progenitor cells in the rat are reduced in number following DBP-induced suppression of fetal intratesticular testosterone. Proceedings of 17th European Testis Workshop on Molecilar and Cellular Endocrinology of the Testis. April 20-24 2012, Stockholm, Sweeden.
[72] Atanassova N. Morpho-functional aspect of androgen-estrogen regulation of mammalian testis and male reproductive tract. DSci Thesis, Bulgarian Academy of Sciences, Sofia, 2007.
[73] Habert R, Lejeune H, Saez JM. Origin, differentiation and regulation of fetal and adult Leydig cells. Molecular and Cellular Endocrinology 2001; 179: 47-74.

[74] Bernstein L, Ross RK 1993 Endogenous hormones and breast cancer risk. Epidemiology Review 1993; 15 :48–65.
[75] Gunasegaram R, Peh KL, Loganath A, Ratnam SS. Expression of 3β-hydroxysteroid dehydrogenase-5,4-ene isomerase activity by infiltrating ductal human breast carcinoma in vitro. Breast Cancer Research & Treatment 1998; 50: 117–123.
[76] Reed MJ, Purohit A. Breast cancer and the role of cytokines in regulating estrogen synthesis: an emerging hypothesis. Endocine Review 1997; 18: 701–715.
[77] Turgeon C, Gingras S, Carriere MC, Blais Y, Labrie F, Simard J. Regulation of sex steroid formation by interleukin-4 and interleukin-6 in breast cancer cells. Journal of Steroid Biochemistry & Molecular Biology 1998; 65: 151–162,
[78] Riley SC, Dupont E, Walton JC, Luu-The V, Labrie F, Pelletier G,Challis JR. Immunohistochemical localization of 3β-hydroxy- 5-ene-steroid dehydrogenase/Δ5-Δ4 isomerase in human placenta and fetal membranes throughout gestation. Journal of Clinical Endocrinology & Metabolism 1992; 75: 956–961.
[79] Morrish DW, Linetsky E, Bhardwaj D, Li H, Dakour J, Marsh RG, Paterson MC, Godbout R. Identification by subtractive hybridization of a spectrum of novel and unexpected genes associated with in vitro differentiation of human cytotrophoblast cells. Placenta 1996; 17: 431–441
[80] Amet Y, Simon B, Quemener E, Mangin P, Floch HH, Abalain JH. Partial purification of 3α- and 3β-hydroxysteroid dehydrogenases from human hyperplastic prostate. Comparison between the two enzymes. Journal of Steroid Biochemistry & Molecular Biology 1992; 41: 689–692.
[81] Labrie F, Belanger A, Cusan L, Labrie C, Simard J. History of LHRH agonist and combination therapy in prostate cancer. Endocrine Related Cancer 1996; 3, 243–278
[82] Pirog EC, Collins DC 1999 Metabolism of dihydrotestosterone in human liver: importance of 3α- and 3β-hydroxysteroid dehydrogenase. Journal of Clinical Endocrinology & Metabolism 1999; 84: 3217–3221.
[83] Keeney DS, Murry BA, Bartke A, Wagner TE, Mason JI 1993 Growth hormone transgenes regulate the expression of sex-specificisoforms of 3β-hydroxysteroid dehydrogenase/Δ5-Δ4 isomerase in mouse liver and gonads. Endocrinology 1993; 133: 1131–1138.
[84] Anderson D. Steroidogenic enzymes in skin. Journal of Dermatology 2001; 11, 293-295.
[85] Asada H, Linton J, Katz SI 1997 Cytokine gene expression during the elicitation phase of contact sensitivity: regulation by endogenous IL-4. Journal of Investigarive Dermatology 1997; 108: 406–411.
[86] Mensah-Nyagan AG, Do-Rego JL, Beaujean D, Luu-The V, Pelletier G, Vaudry H 1999 Neurosteroids: expression of steroidogenic enzymes and regulation of steroid biosynthesis in the central nervous system. Pharmacology Review 1999; 51: 63–81.
[87] Zwain IH, Yen SS 1999 Neurosteroidogenesis in astrocytes, oligodendrocytes, and neurons of cerebral cortex of rat brain. Endocrinology 1999; 140: 3843–3852.
[88] Baker PJ, Johnston H, Abel M, HM C, O'Shaughnessy PJ. Differentiation of adult-type Leydig cells occurs in gonadotropin-deficient mice. Reproductive Biology & Endocrinology 2003; 1: 1–9.

[89] Martin LJ, Taniguchi H, Robert NM, Simard J, Tremblay JJ, Viger RS. GATA Factors and the Nuclear Receptors, Steroidogenic Factor 1/Liver Receptor Homolog 1, Are Key Mutual Partners in the Regulation of the Human 3β-Hydroxysteroid Dehydrogenase Type 2 Promoter. Molecular Endocrinology 2005; 19: 2358–2370.

[90] Darnell Jr JE. STATs and gene regulation. Science1997; 277: 1630–1635.

[91] und S, McKay C, Schuetz E, van Deursen JM, Stravopodis D, Wang D, Brown M, Bodner S, Grosveld G, Ihle JN. Stat5a and Stat5b proteins have essential and nonessential, or redundant, roles in cytokine responses. Cell 1998; 93: 841–850.

[92] Lalli E, Bardoni B, Zazopoulos E, Wurtz JM, Strom TM, Moras D, Sassone-Corsi P. A transcriptional silencing domain in DAX-1 whose mutation causes adrenal hypoplasia congenita. Molecular Endocrinology 1997; 11: 1950–1960.

[93] Tremblay JJ, Viger RS. Novel roles for GATA transcription factors in the regulation of steroidogenesis. Journal of Steroid Biochemestry & Molecular Biology 2003; 85: 291–298.

[94] Cote S, Feltus AF, Gingras S, Freeman M, Melner MH, Simard J 2000 IL-4 stimulation of ovarian 3β-hydroxysteroid dehydrogenase/ Δ5-Δ4-isomerase type II gene expression: mechanisms of activation. Proceedings of the 82nd Annual Meeting of The Endocrine Society, 2000, Toronto, Ontario, p 313 (Abstract 1295).

[95] Rainey WE, Naville D, Mason JI. Regulation of 3β-hydroxysteroid dehydrogenase in adrenocortical cells: effects of angiotensin- II and transforming growth factor β. Endocrine Research 1991; 17: 281–296

[96] Havelock JC, Smith AL, Seely JB, Dooley CA, Rodgers RJ, Rainey WE, Carr BR. The NGFI-B family of transcription factors regulates expression of 3β-hydroxysteroid dehydrogenase type 2 in the human ovary. Molecular & Human Reproduction 2005; 11: 79–85

[97] Martin LJ, Tremblay J. The human 3β-hydroxysteroid dehydrogenase/Δ5-Δ4 isomerase type 2 promoter is a novel target for the immediate early orphan nuclear receptor Nur77 in steroidogenic cells. Endocrinology 2005;146: 861–869

[98] Feltus FA, Cote S, Simard J, Gingras S, Kovacs WJ, Nicholson WE, Clark BJ, Melner MH. Glucocorticoids enhance activation of the human type II 3β-hydroxysteroid dehydrogenase/Δ5-Δ4 isomerase gene. Journal of Steroid Biochemistry & Molecular Biology 2002; 82: 55–63.

[99] Perry JE, Stalvey JR. Gonadal steroids modulate adrenal fasciculata 3β-hydroxysteroid dehydrogenase isomerase activity in mice. Biology of Reproduction 1992; 46: 73–82.

[100] Stalvey JR, Clavey SM. Evidence that testosterone regulates Leydig cell 3β-hydroxysteroid dehydrogenase-isomerase activity by a trans-acting factor distal to the androgen receptor. Journal of Andrology 1992; 13: 93–99.

[101] Bush IE, Hunter SA and Meigs RA. Metabolism of 11-oxygenated steroids Biochemical Journal 1968; 107: 239–258.

[102] Lakshmi V and Monder C. Purification and characterization of the corticosteroid 11β-dehydrogenase component of the rat liver 11β- hydroxysteroid dehydrogenase complex Endocrinology 1988; 123: 2390–2398.

[103] Jamieson PM, Chapman KE, Walker BR, Seckl JR. 11β -hydroxysteroid dehydrogenase type 1 is a predominant 11-reductase in the intact perfused rat liver. Journal of Endocrinol 2000; 165: 685–692.

[104] Seckl JR, Walker BR. Minireview: 11ß-hydroxysteroid dehydrogenase type 1- a tissue-specific amplifier of glucocorticoid action. Endocrinology 2001; 142: 1371-1376.
[105] Albiston AL, Obeyesekere VR, Smith RE and Krozowski ZS. Cloning and tissue distribution of the human 11β -hydroxysteroid dehydrogenase type 2 enzyme Molecular and Cellular Endocrinology 1994; 105: R11–R17.
[106] MercerWR and Krozowski ZS. Localization of an 11_-hydroxysteroid dehydrogenase activity to the distal nephron. Evidence for the existence of two species of dehydrogenase in the rat kidney.Endocrinology1992; 130: 540–543.
[107] Brown RW, Chapman KE, Edwards CRW and Seckl JR. Human placental 11β-hydroxysteroid dehydrogenase: evidence for and partial purification of a distinct NAD-dependent iosform. Endocrinology 1993; 132: 2614–2621.
[108] Seckl JR, Morton NM, Chapman KE, Walker BR. Glucocorticoids and 11beta-hydroxysteroid dehydrogenase in adipose tissue. Recent Progess in Hormone Research 2004; 59: 359-93.
[109] Ge RS, Hardy DO, Catterall JE, Hardy MP. Developmental changes in glucocorticoid receptor and 11ß–hydroxysteroid dehydrogenase oxidative and reductive activities in rat Leydig cells. Endocrinology. 1997;138: 5089-95.
[110] Ge RS, Hardy MP. Initial predominance of the oxidative activity of type 11ß-hydroxysteroid dehydrogenase in primary rat Leydig cells and transfected cell lines. Journal of Andrology 2000; 21: 303-310.
[111] Monder C, Miroff Y, Marandici A, Hardy MP. 11ß–dehydrogenase alleviates glucocorticoid-mediated inhibition of steroidogenesis in rat Leydig cells. Endocrinology 1994; 134: 1199-1204.
[112] Phillips DM, Lakshmi V, Moder C. Corticosteroid 11ß–hydroxysteroid dehydrogenase in rat testis. Endocrinology 1989;125: 209-216.
[113] Hales DB, Payne AH. Glucocorticoid-mediated repression of P450scc mRNA and de novo synthesis in cultured Leydig cells. Endocrinology 1989; 124: 2099–2104.
[114] Payne AH, Sha LL. Multiple mechanisms for regulation of 3b-hydroxysteroid dehydrogenase/D5-D4-isomerase, 17a-hydroxylase/C17-20 lyase cytochrome P450, and cholesterol side-chain cleavage cytochrome P450 messenger ribonucleic acid levels in primary cultures of mouse Leydig cells. Endocrinology 1991; 129: 1429–1435.
[115] Gao HB, Tong MH, Hu YQ, Guo QS, Ge R, Hardy MP. Glucocorticoid induces apoptosis in rat Leydig cells. Endocrinology 2002; 143: 130–138.
[116] Gao HB, Tong MH, Hu YQ, You HY, Guo QS, Ge RS, Hardy MP. Mechanisms of glucocorticoidinduced Leydig cell apoptosis. Molecular & Cellular Endocrinology 2003; 199: 153–163.
[117] Guo-Xin Hu, Qing-Quan Lian, Han Lin, Syed A. Latif, David J. Morris, Matthew P. Hardy, and Ren-Shan Ge. Rapid mechanisms of glucocorticoid signaling in the Leydig cell. Steroids 2008; 73: 1018–1024.
[118] Gao HB, Ge RS, Lakshmi A, Hardy MP. Hormonal regulation of oxidative and reductive activities of 11ß–hydroxysteroid dehydrogenase in rat Leydig cells. Endocrinology 1997; 138: 156-161.

[119] Jamieson PM, Walker BR, Hapman KE, Andrew R, Rossiter S, Seckl JR. 11 beta-hydroxysteroid dehydrogenase type 1 is a predominant 11 beta-reductase in the perfused rat liver. Journal of Endocrinology 2000; 165: 685-692.
[120] Seckl JR, Walker BR. Minireview: 11ß-hydroxysteroid dehydrogenase type 1- a tissue-specific amplifier of glucocorticoid action. Endocrinology 2001; 142: 1371-1376.
[121] Latif SA, Shen M, Ge RS, Sottas CM, Hardy MP, Morris DJ. Role of 11β-OH-C(19) and C(21) steroids in the coupling of 11β-HSD1 and 17β-HSD3 in regulation of testosterone biosynthesis in rat Leydig cells. Steroids 2011; 76: 682-689.
[122] Hu GX, Lin H, Sottas CM, Morris DJ, Hardy MP, Ge RS. Inhibition of 11beta-hydroxysteroid dehydrogenase enzymatic activities by glycyrrhetinic acid in vivo supports direct glucocorticoid-mediated suppression of steroidogenesis in Leydig cells. Journal of Andrology 2008; 29: 345-51.
[123] Parthasarathy C, Yuvaraj S, Ilangovan R, Janani P, Kanagaraj P, Balaganesh M, Natarajan B, Sittadjody S, Balasubramanian K. Differential response of Leydig cells in expressing 11beta-HSD type I and cytochrome P450 aromatase in male rats subjected to corticosterone deficiency. Molecular & Cellular Endocrinology 2009; 311:18-23.
[124] Haider SG, Passia D, Rommert FFG. Histochemical demonstration of 11ß–hydroxysteroid dehydrogenase as a marker for Leydig cell maturation in rat. Acta Histochemica (Suppl) 1990; 38: 203-207.
[125] Monder C, Hardy MP, Blanchard RJ, Blanchard DC. Comparative aspects of 11ß-hydroxysteroid dehydrogenase: development of a model for the mediation of Leydig cell function by corticosteroids. Steroids 1994; 59: 69-73.
[126] Neumann A, Haider SG, Hilscher B. Temporal coincidence of the appearance of elongated spermatids and of histochemical reaction of 11ß–hydroxysteroid dehydrogenase in rat Leydig cells. Andrologia 1993; 25: 263-269.
[127] Schafers BA, Schlutius BG, Haider SG. Ontogenesis of oxidative reaction of 17β-hydroxysteroid dehydrogenese and 11ß–hydroxysteroid dehydrogenase in rat Leydig cells, a histochemical study. The Histochemical Journal 2001; 33: 585-595.
[128] Ge RS, Gao HB, Nacharaju VL, Gunsalus GL, Hardy MP. Identification of a kinetically distinct activity of 11ß–hydroxysteroid dehydrogenase in rat Leydig cells. Endocrinology 1997;138: 2435-2442.
[129] Brereton PS, Van Driel RR, Suhaimi FB, Koyama K, Dilley R, Krozowski Z. Light and electron microscopy localization of the 11ß-hydroxysteroid dehydrogenase type I enzyme in the rat. Endocrinology 2001; 142: 1644-1651.
[130] Hardy MP, Gao HB, Dong Q, Ge R, Wang Q, Chai WR, Feng X, Sottas C. Stress hormone and male reproductive function.Cell & Tissue Research 2005; 322:147-53.
[131] Ge RS, Dong Q, Sottas CM, Chen H, Zirkin BR, Hardy MP. Gene expression in rat Leydig cells during development from the progenitor to adult stage: a cluster analysis. Biology of Reproduction 2005; 72: 1405-1415.
[132] Hu GX, Lian QQ, Chen BB, Prasad PV, Kumar N, Zheng ZQ, Ge RS. 7alpha-hydroxytestosterone affects 1 beta-hydroxysteroid dehydrogenase 1 direction in rat Leydig cells. Endocrinology 2010; 151: 748-54.
[133] Koeva Y, Bakalska M, Atanassova N, Georgieva K, Davidoff M. 11β hydroxysteroid dehydrogenase type 2 expression in the newly formed Leydig cells after ethane

dimethanesulphonate treatment of adult rats. Folia Histochemica & Cytobiologica 2007; 45: 381-6.

[134] Bakalska M, Atanassova N, Angelova P, Koeva I, Nikolov B, Davidoff M. Degeneration and restoration of spermatogenesis in relation to the changes in Leydig cell population following ethane dimethanesulfonate treatment in adult rats. Endocrine Regulations 2001; 35: 211-217.

[135] Bakalska M, Koeva I, Atanassova N, Angelova P, Nikolov B, Davidoff M. Steroidogenic and structural differentiation of new Leydig cell population following exposure of adult rats to ethane dimethanesulphonate. Folia Biologica (Praha) 2002; 48: 205-209.

[136] Ge RS, Dong Q, Niu EM, Sottas CM, Hardy DO, Catterall JF, Latif SA, Morris DJ, Hardy MP. 11 beta-hydroxysteroid dehydrogenase 2 in rat Leydig cells: its role in blunting glucocorticoid action at physiological levels of substrate. Endocrinology 2005; 146: 2657-2664.

[137] Hardy M and Schlegel P. Testosterone production in the aging male: Where does the slowdown occur? Endocrinology 2004; 145: 4439-40.

[138] Harman SM, Metter EJ, Tobin JD, Pearson J, Blackman MR. Longitudinal effects of aging on serum total and free testosterone levels in healthy men. Baltimore Longitudinal Study of Aging. Journal of Clinical Endocrinology & Metabolism 2001; 86: 724-31.

[139] Wang C, Hikim AS, Ferrini M, Bonavera JJ, Vemet D, Leung A, Lue YH, Gonzalez-Cadavid NF, Schwerdlloff RS. Male reproductive ageing: using the brown Norway rat as a model for man. Novartis Found Symposium. 2002; 242: 82-95.

[140] Chen H, Huhtaniemi I, Zirkin BR. Depletion and repopulation of Leydig cells in the testes in aging Brown Norway rats. Endocrinology 1996; 137: 3447-52.

[141] Zirkin BR and Chen H. Regulation of Leydig cell steroidogenic function during aging. Biology of Reproduction 2000;63: 977-81.

[142] Chen H, Luo L, Zirkin BR. Leydig cell structure and function during aging. In: Payne AH, Hardy MP, Russell LD, eds. The Leydig cell. Cache River Press, Vienna, IL; 1996: p221-30.

[143] Schultz R, Isola J, Parvinen M, Honkaniemi J, Wikstrom AC, Gustafsson JA, Pelto-Huikko M. Localization of the glucocorticoid receptor in testis and accessory sexual organs of male rat. Molecular & Cellular Endocrinology 1993; 95:115-20.

[144] Monder C, White PC. 11β hydroxysteroid dehydrogenase. Vitamines & Hormones 1993; 47: 187-271.

[145] Walker BR, Connacher AA, Lindsay RM, Webb DJ, Edwards CR. Carbenoxolone increases hepatic insulin sensitivity in man: a novel role for 11-oxosteroid reductase in enhancing glucocorticoid receptor activation. Journal of Clinical Endocrinology & Metabolism 1995; 80 :3155-59.

[146] Koeva Y, Bakalska M, Atanassova N, Georgieva K, Davidoff M. Age-related changes in the expression of 11beta-hydroxysteroid dehydrogenase type 2 in rat Leydig cells. Folia Histochemica & Cytobiologica 2009; 47: 281-287.

[147] Sankar BR, Maran RR, Sudha S, Govindarajulu P, Balasubramanian K. Chronic corticosterone treatment impairs Leydig cell 11beta-hydroxysteroid dehydrogenase

activity and LH-stimulated testosterone production. Hormone & Metabolism Research 2000; 32: 142-146.

[148] Morita H, Cozza EN, Zhou MY, Gomez-Sanchez EP, Romero DG, Gomez-Sanchez CE. Regulation of the 11 beta-hydroxysteroid dehydrogenase in the rat adrenal. Decrease enzymatic activity induced by ACTH.Endocrine 1997; 7: 331-5.

[149] Shimojo M, Condon J, Whorwood CB, Stewart PM. Adrenal 11 beta-hydroxysteroid dehydrogenase. Endocrine Research 1996; 22:771-80.

[150] Shimojo M, Whorwood CB, Stewart PM. 11 beta-hydroxysteroid dehydrogenase in the rat adrenal. Journal of Molecular Endocrinology 1996; 17: 121-30.

[151] Smith RE, Li KX, Andrews RK, Krozowski Z. Immunohistochemical and molecular characterization of the rat 11 beta-hydroxysteroid dehydrogenase type II enzyme. Endocrinology 1997; 138: 540-547.

[152] Koeva YA, Bakalska MV, Petrova EI, Atanassova NN. 11beta hydroxysteroid dehydrogenase type 2 in the adrenal gland by testosterone withdrawal of adult rats. Folia Medica (Plovdiv) 2010; 52: 38-42.

[153] Atanassova N, Koeva Y, Bakalska M, Pavlova E, Nikolov B, Davidoff M. Loss and recovery of androgen receptor protein expression in the adult rat testis following androgen withdrawal by ethane dimethanesulfonate. Folia Histochemica et Cytobiologica 2006; 44: 81-86.

[154] Plecas B, Pesic VP, Mirkovic D, Majkic-Singh N, hristic M, Solarovic T. Opposite effects of dexamethasone and ACTH on the adrenal cortex response to ethane dimethanesulphonate (EDS). Experimenatal Toxicology & Pathology 2001; 53: 31-34.

[155] Petrova E, Koeva Y, Bakalska M, Atanassova N, Davidoff M. Morphofunctional characteristics of rat adrenocorticocytes after treatment with ethane dimethanesulphonate. Jubilee Scientific Session of Medical University, Plovdiv, 2005, abstract book; 164.

[156] Stalvey JR. Inhibition of 3 beta-hydroxysteroid dehydrogenase- isomerase in mouse adrenal cells: a different effect of testosterone. Steroids 2002; 67: 721-31.

[157] Michael AE & Cooke BA. A working hypothesis for the regulation of steroidogenesis and germ cell development in the gonads by glucocorticoids and 11β-hydroxysteroid dehydrogenase (11β-HSD). Molecular and Cellular Endocrinology 1994; 100: 55–63.

[158] Omura T & Morohashi K. Gene regulation of steroidogenesis. Journal of Steroid Biochemistry and Molecular Biology 1995; 53: 19–25.

[159] Monder C & Lakshmi V. Evidence for kinetically distinct forms of corticosteroid 11β-hydroxysteroid dehydrogenase in rat liver microsomes. Journal of Steroid Biochemistry 1989; 32: 77–83.

[160] Mercer W, Obeyeskere V, Smith R & Krozowski Z. Characterization of 11βHSD1B gene expression and enzyme activity. Molecular and Cellular Endocrinology 1993; 92: 247–251.

[161] Yding Andersen C. Possible new mechanism of cortisol action in female reproductive organs: physiological implications of the free hormone hypothesis. Journal of Endocrinology 2002; 173: 211–217.

[162] Yding Andersen C, Morineau G, Fukuda M, Westergaard LG, Ingerslev HJ, Fiet J & Byskov AG. Assessment of the follicular cortisol:cortisone ratio. Human Reproduction 1999; 14: 1563–1568.
[163] Yong PYK, Thong KJ, Andrew R, Walker BR & Hillier SG. Development-related increase in cortisol biosynthesis by human granulosa cells. Journal of Clinical Endocrinology and Metabolism 2000; 85: 4728–4733.
[164] Hillier SG & Tetsuka M. An anti-inflammatory role for glucocorticoids in the ovaries? Journal of Reproductive Immunology 1998; 39: 21–27.
[165] Knochenhauer ES, Key TJ, Kahsar-Miller M, Waggoner W, Boots LR, Azziz R. Prevalence of the polycystic ovary syndrome in unselected black and white women of the Southeastern United States: a prospective study. Jornal of Clinical Endocrinology & Metabolism 1998; 83: 3078–3082.
[166] Legro, R.S. & J.F. Strauss III. Molecular progress in infertility: polycystic ovary syndrome. Fertility & Sterility 2002; 78: 569–576.
[167] Milutinović DV, Macut D, Božić I, Nestorov J, Damjanović S, Matić G. Hypothalamic-pituitary-adrenocortical axis hypersensitivity and glucocorticoid receptor expression and function in women with polycystic ovary syndrome. Experimental Clinical Endocrinology & Diabetes. 2011; 119: 636-43.
[168] Diamanti-Kandarakis E, Xyrafis X, Boutzios G, Christakou C. Pancreatic beta-cells dysfunction in polycystic ovary syndrome. Panminerva Medicine 2008; 50: 315-25.
[169] Goodarzi MO, Dumesic DA, Chazenbalk G, Azziz R. Polycystic ovary syndrome: etiology, pathogenesis and diagnosis. Nature Reviews of Endocrinology 2011;7 :219-31.
[170] Waterworth DM, Bennett ST, Gharani N, McCarthy MI, Hague S, Batty S, Conway GS, White D, Todd JA, Franks S, Williamson R. Linkage and association of insulin gene VNTR regulatory polymorphism with polycystic ovary syndrome. Lancet 1997; 349: 986–990.
[171] Nelson VL, Legro RS, Strauss JF 3rd, McAllister JM.. Augmented androgen production is a stable phenotype of propagated theca cells from polycystic ovaries. Molecular Endocrinology 1999; 13: 946–957.
[172] Nelson VL, Qin KN, Rosenfield RL, Wood JR, Penning TM, Legro RS, Strauss JF 3rd, McAllister JM.The biochemical basis for increased testosterone production in theca cells propagated from patients with polycystic ovary syndrome. Journal of Clinical Endocrinology and Metabolism 2001; 86: 5925–5933.
[173] Wickenheisser JK, Quinn PG, Nelson VL, Legro RS, Strauss JF 3rd, McAllister JM. Differential activity of the cytochrome P450 17 α-hydroxylase and steroidogenic acute regulatory protein gene promoters in normal and polycystic ovary syndrome theca cells. Journal of Clinical Endocrinology and Metabolism 2000; 85: 2304–2311.
[174] J.F.Strauss III. Some New Thoughts on the Pathophysiology and Genetics of Polycystic Ovary Syndrome. Annals of New York Academy of Sciences 2003; 997: 42–48.
[175] Svendsen PF, Madsbad S, Nilas L, Paulsen SK, Pedersen SB. Expression of 11beta-hydroxysteroid dehydrogenase 1 and 2 in subcutaneous adipose tissue of lean and obese women with and without polycystic ovary syndrome. International Journal of Obesity (Lond) 2009; 33:1249-56.
[176] Tsilchorozidou T, Honour JW, Conway GS. Altered cortisol metabolism in polycystic ovary syndrome: insulin enhances 5alpha-reduction but not the elevated adrenal

steroid production rates. Journal of Clinical Endocrinology and Metabolism 2003; 88: 5907-13.

[177] Draper N, Walker EA, Bujalska IJ, Tomlinson JW, Chalder SM, Arlt W, Lavery GG, Bedendo O, Ray DW, Laing I, Malunowicz E, White PC, Hewison M, Mason PJ, Connell JM, Shackleton CHL, Stewart PM. Mutations in the gene encoding 11β-hydroxysteroid dehydrogenase type 1 and hexose- 6-phosphate dehydrogenase interact to cause cortisone reductase deficiency. Nature Genetics 2003; 34: 434–439.

[178] Phillipov G, Palermo M, Shackleton CH. Apparent cortisone reductase deficiency: a unique form of hypercortisolism. J Clin Endocrinol Metab, 1996; 81:3855–3860.

[179] Rodin A, Thakkar H, Taylor NJ, Clayton R. Hyperandrogenism in polycystic ovary syndrome: evidence of dysregulation of 11β-hydroxysteroid dehydrogenase. New English Journal of Medicine 1994; 330: 460–465.

[180] Gambineri A, Vicennati V, Genghini S, Tomassoni F, Pagotto U, Pasquali R, Walker BR. Genetic variation in 11beta-hydroxysteroid dehydrogenase type 1 predicts adrenal hyperandrogenism among lean women with polycystic ovary syndrome. Journal of Clinical Endocrinology and Metabolism 2006; 91: 2295-302.

[181] Hines GA, Smith ER, Azziz R. Influence of insulin and testosterone on adrenocortical steroidogenesis in vitro: preliminary studies. Fertility & Sterilility 2001; 76: 730–735.

[182] Andrew R, Phillips DIW, Walker BR. Obesity and gender influence cortisol secretion and metabolism in man. Journal of Clinical Endocrinology and Metabolism 1998; 83: 1806–1809.

[183] San Milla´n JL, Botella-Carretero JI, Alvarez-Blasco F, Luque-Ramı´rez M, Sancho J, Moghetti P, Escobar-Morreale HF. A study of the hexose-6-phosphate dehydrogenase gene R453Q and 11β-hydroxysteroid dehydrogenase type 1 gene 83557insA polymorphisms in the polycystic ovary syndrome. Journal of Clinical Endocrinology and Metabolism 2005; 90: 4157–4162.

[184] White PC. Genotypes at 11β-hydroxysteroid dehydrogenase type 11B1 and hexose-6-phosphate dehydrogenase loci are not risk factors for apparent cortisone reductase deficiency in a large population-based sample. Journal of Clinical Endocrinology and Metabolism 2005; 90: 5880–5883.

[185] Draper N, Powell BL, Franks S, Conway GS, Stewart PM & McCarthy MI. Variants implicated in cortisone reductase deficiency do not contribute to susceptibility to common forms of polycystic ovary syndrome. Clinical Endocrinology 2006; 65: 64– 70.

[186] Gambineri A, Tomassoni F, Munarini A, Stimson RH, Mioni R, Pagotto U, Chapman KE, Andrew R, Mantovani V, Pasquali R, Walker BR. A combination of polymorphisms in HSD11B1 associates with in vivo 11{beta}-HSD1 activity and metabolic syndrome in women with and without polycystic ovary syndrome. European Journal of Endocrinology 2011; 165: 283-92.

[187] Mlinar B, Marc J, Jensterle M, Bokal EV, Jerin A, Pfeifer M. Expression of 11βhydroxysteroid dehydrogenase type 1 in visceral andsubcutaneous adipose tissues of patients with polycystic ovary syndrome is associated with adiposity. Journal of Steroid Biochemistry & Molecular Biology 2011; 123: 127-32.

[188] Reaven G. Metabolic syndrome — pathophysiology and implications for management of cardiovascular disease. Circulation 2002; 106: 286–288.

[189] Walker B, Seckl J. Cortisol metabolism. In: Bjo¨rntorp P, ed. International Textbook of Obesity. 2001, Chichester, UK: John Wiley and Sons; 241–268.
[190] Morton NM, Seckl JR. 11beta-hydroxysteroid dehydrogenase type 1 and obesity. Frontiers in Hormone Research 2008; 36: 146-64.
[191] Livingstone DEW, Jones G, Smith K, Jamieson PM, Andrew R, Kenyon CJ, Walker BR. Understanding the role of glucocorticoids in obesity: tissue-specific alterations of corticosterone metabolism in obese Zucker rats. Endocrinology 2000; 141: 560–563.
[192] Westerbacka J, Yki-Ja¨rvinen H, Vehkavaara S, Ha¨kkinen A, Andrew R, Wake D, Seckl J, Walker B. Body fat distribution and cortisol metabolism in healthy men: enhanced 5-reductase and lower cortisol/cortisone metabolite ratios in men with fatty liver. Journal of Clinical Endocrinology and Metabolism 2003; 88: 4924–4931.
[193] Livingstone DEW, Kenyon CJ, Walker BR. Mechanisms of dysregulation of 11 beta hydroxysteroid dehydrogenase type 1 in obese Zucker rats. Journal of Endocrinol 2000; 167: 533–539.
[194] Mattsson C, Olsson T. Estrogens and glucocorticoid hormones in adipose tissue metabolism. Current Medicinal Chemistry 2007; 14:2918-24.
[195] Andersson T, Söderström I, Simonytė K, Olsson T. Estrogen reduces 11beta-hydroxysteroid dehydrogenase type 1 in liver and visceral, but not subcutaneous, adipose tissue in rats. Obesity (Silver Spring). 2010; 18: 470-5.
[196] Masuzaki H, Paterson J, Shinyama H, Morton NM, Mullins JJ, Seckl JR, Flier JS. A transgenic model of visceral obesity and the metabolic syndrome. Science 2001; 294: 2166–2170.
[197] Masuzaki H, Yamamoto H, Kenyon CJ, Elmquist JK, Morton NM, Paterson JM, Shinyama H, Sharp MGF, Fleming S. Transgenic amplification of glucocorticoid action in adipose tissue causes high blood pressure in mice. Journal of Clinical Investigations, 2003; 112: 83–90.
[198] Paterson JM, Morton NM, Fievet C, Kenyon CJ, Holmes MC, Staels B, Seckl JR & Mullins JJ. Metabolic syndrome without obesity: hepatic overexpression of 11b-hydroxysteroid dehydrogenase type 1 in transgenic mice. Proceedings of the National Academy of Sciences USA 2004; 101: 7088–7093.
[199] Kotelevtsev YV, Holmes MC, Burchell A, Houston PM, Scholl D, Jamieson PM, Best R, Brown RW, Edwards CRW, Seckl JR & Mullins. 11b-Hydroxysteroid dehydrogenase type 1 knockout mice show attenuated glucocorticoid inducible responses and resist hyperglycaemia on obesity and stress. Proceedings of the National Academy of Sciences USA,1997; 94: 14924–14929.
[200] Morton NM, Holmes MC, Fievet C, Staels B, Tailleux A, Mullins JJ & Seckl JR. Improved lipid and lipoprotein profile, hepatic insulin sensitivity, and glucose tolerance in 11b hydroxysteroid dehydrogenase type 1 null mice. Journal of Biological Chemistry, 2001; 276: 41 293–300.
[201] Morton NM, Paterson JM, Masuzaki H, Holmes MC, Staels B, Fievet C, Walker BR, Flier JS, Mullins JJ & Seckl JR. Novel adipose tissue-mediated resistance to diet-induced visceral obesity in 11b-hydroxysteroid dehydrogenase type 1 deficient mice. Diabetes 2004; 53: 931–938.

[202] Stimson RH, Walker BR. Glucocorticoids and 11beta-hydroxysteroid dehydrogenase type 1 in obesity and the metabolic syndrome. Minerva Endocrinology 2007; 32: 141-159.

[203] Nair S, Lee YH, Lindsay RS, Walker BR, Tataranni PA, Bogardus C, Baier LJ & Permana PA. 11Beta-hydroxysteroid dehydrogenase type 1: genetic polymorphisms are associated with type 2 diabetes in Pima Indians independently of obesity and expression in adipocyte and muscle. Diabetologia 2004; 47: 1088–1095.

[204] Franks PW, Knowler WC, Nair S, Koska J, Lee YH, Lindsay RS, Walker BR, Looker HC, Permana PA, Tatarani PA, Hanson RL. Interaction between an 11bHSD1 gene variant and birth era modifies risk of hypertension in Pima Indians. Hypertension 2004; 44: 681–688.

[205] Morales MA, Carvajal CA, Ortiz E, Mosso LM, Artigas RA, Owen GI & Fardella CE. Possible pathogenetic role of 11 beta-hydroxysteroid dehydrogenase type 1 (11beta HSD1) gene polymorphisms in arterial hypertension. Revista Me´dica de Chile, 2008; 136: 701–710.

[206] Rask E, Walker BR, Soderberg S, Livingstone DE, Eliasson M, Johnson O, Andrew R, Olsson T. Tissue-specific changes in peripheral cortisol metabolism in obese women: increased adipose 11beta-hydroxysteroid dehydrogenase type 1 activity. Journal of Clinical Endocrinology & Metabolism 2002; 87: 3330–3336.

[207] Walker BR, Connacher AA, Lindsay RM, Webb DJ, Edwards CRW. Carbenoxolone increases hepatic insulin sensitivity in man: a novel role for 11-oxosteroid reductase in enhancing glucocorticoid receptor activation. Journal of Clinical Endocrinology & Metabolism 1995; 80: 3155–3159.

[208] Andrews RC, Rooyackers O, Walker BR. Effects of the 11beta-hydroxysteroid dehydrogenase inhibitor carbenoxolone on insulin sensitivity in men with type 2 diabetes. Journal of Clinical Endocrinology & Metabolism 2003; 88: 285–291.

[209] Sandeep TC, Andrew R, Homer NZ, Andrews RC, Smith K, Walker BR. Increased in vivo regeneration of cortisol in adipose tissue in human obesity and effects of the 11beta-hydroxysteroid dehydrogenase type 1 inhibitor carbenoxolone. Diabetes 2005; 54: 872–879.

[210] Alberts P, Nilsson C, Selen G, Engblom LO, Edling NH, Norling S, Klingström G, Larsson C, Forsgren M, Ashkzari M, Nilsson CE, Fiedler M, Bergqvist E, Ohman B, Björkstrand E, Abrahmsen LB. Selective inhibition of 11bhydroxysteroid dehydrogenase type 1 improves hepatic insulin sensitivity in hyperglycaemic mice strains. Endocrinology 2003; 144: 4755–4762.

[211] Hermanowski-Vosatka A, Balkovec JM, Cheng K, Chen HY, Hernandez M, Koo GC, Le Grand CB, Li Z, Metzger JM, Mundt SS, Noonan H, Nunes CN, Olson SH, Pikounis B, Ren N, Robertson N, Schaeffer JM, Shah K, Springer MS, Strack AM, Strowski M, Wu K, Wu T, Xiao J, Zhang BB, Wright SD, Thieringer. 11b-HSD1 inhibition ameliorates metabolic syndrome and prevents progression of atherosclerosis in mice. Journal of Experimental Medicine 2005; 202: 517–527.

[212] Wamil M, Seckl JR. Inhibition of 11beta-hydroxysteroid dehydrogenase type 1 as a promising therapeutic target. Drug Discovery Today 2007; 12) :504-520.

[213] Morton NM. Obesity and corticosteroids: 11beta-hydroxysteroid type 1 as a cause and therapeutic target in metabolic disease. Molecular & Cellular Endocrinology 2010 25; 316: 154-64.

[214] Morton NM, Ramage L & Seckl JR. Down-regulation of adipose 11b-hydroxysteroid dehydrogenase type 1 by high-fat feeding in mice: a potential adaptive mechanism counteracting metabolic disease. Endocrinology 2004;145: 2707–2712.

[215] Harris A, Seckl J. Glucocorticoids, prenatal stress and the programming of disease. Hormones & Behavior 201; 59: 279-89.

[216] Marciniak B, Patro-Małysza J, Poniedziałek-Czajkowska E, Kimber-Trojnar Z, Leszczyńska-Gorzelak B, Oleszczuk J. Glucocorticoids in pregnancy. Current Pharmacology & Biotechnology 201; 12:750-757.

[217] McTernan CL, Draper N, Nicholson H, Chalder SM, Driver P, Hewison M, Kilby MD, Stewart PM. Reduced placental 1β-hydroxysteroid dehydrogenase type 2 mRNAlevels in human pregnancies complicated by intrauterine growth restriction: an analysis of possible mechanisms. Journal of Clinical Endocrinology & Metabolism 2001; 86: 4979–4983.

[218] Beitens IZ, Bayard F, Ances IG, Kowarski A, Migeon CJ. The metabolic clearance rate, blood production, interconversion and transplacental passage of cortisol and cortisone in pregnancy near term. Pediatric Research1997; 37: 509–519.

[219] Brown RW, Diaz R, Robson AC, Kotelevtsev Y, Mullins JJ, Kaufman MH, Seckl JR. Isolation and cloning of human placental 11β hydroxysteroid dehydrogenase-2 cDNA. Biochemical Journal 1996; 313: 1007–1017.

[220] Benediktsson R, Calder AA, Edwards CRW, Seckl JR. Placental 11β-hydroxysteroid dehydrogenase type 2 is the placental barrier to maternal glucocorticoids: ex vivo studies. Clinical Endocrinology 1997; 46: 161–166.

[221] Seckl JR, Holmes MC. Mechanisms of disease: glucocorticoids, their placental metabolism and fetal 'programming' of adult pathophysiology. Nat Clinical Practice of Endocrinology and Metabolism 2007; 3:479-488.

[222] Edwards CRW, Benediktsson R, Lindsay R, Seckl JR. Dysfunction of the placental glucocorticoid barrier: a link between the foetal environment and adult hypertension? Lancet 1993; 341: 355–357.

[223] Stewart PM, Rogerson FM, Mason JI. Type 2 11β-hydroxysteroid dehydrogenase messenger RNA and activity in human placenta and fetal membranes: its relationship to birth weight and putative role in fetal steroidogenesis. Journal of Clinical Endocrinology & Metabolism 1995; 80: 885–890.

[224] Murphy VE, Zakar T, Smith R, Giles WB, Gibson PG, Clifton VL. Reduced 11β-hydroxysteroid dehydrogenase type 2 activity is associated with decreased birth weight centile in pregnancies complicated by asthma. Journal of Clinical Endocrinology & Metabolism 2002; 87: 1660–1668.

[225] Marchais-Oberwinkler S, Henn C, Möller G, Klein T, Negri M, Oster A, Spadaro A, Werth R, Wetzel M, Xu K, Frotscher M, Hartmann RW, Adamski J. 17β-Hydroxysteroid dehydrogenases (17β-HSDs) as therapeutic targets: protein structures, functions, and recent progress in inhibitor development. Journal of Steroid Biochemistry & Molecular Biology 2011; 125: 66-82.

[226] Vihko P, Herrala A, Harkonen P, Isomaa V, Kaija H, Kurkela R, Pulkka A. Controlof cell proliferation by steroids: the role of 17HSDs. Molecular & Cellular Endocrinology 2006; 248: 141–148.

[227] Vihko P, Herrala A, Harkonen P, Isomaa V, Kaija H, Kurkela R, Li Y, Patrikainen L, Pulkka A, Soronen P, Torn S. Enzymes as modulators in malignant transformation. Journal of Steroid Biochemistry & Molecular Biology 2005; 93: 277–283.

[228] Geissler W, Davis D, Wu L, Bradshaw K, Patel S, Mendonca B, Elliston K, Wilson J, Russell D, Andersson S. Male pseudohermaphroditism caused bymutations of testicular 17β-hydroxysteroid dehydrogenase 3, Nature Genetics 1994, 7: 34–39.

[229] Rasiah KK, Gardiner-Garden M, Padilla FJ, Moller G, Kench JG, Alles MC, Eggleton SA, Stricker PD, Adamski J, Sutherland RL, Henshall SM, Hayes VM. HSD17B4 overexpression, an independent biomarker of poor patient outcome in prostate cancer. Molecular & Cellular Endocrinology 2009; 301: 89–96.

[230] Jin Y, Penning TM. Aldo–keto reductases and bioactivation/detoxication, Annual Review of Pharmacology & Toxicology 2006; 47: 263–292.

[231] Stanbrough M, Bubley GJ, Ros K, Golub TR, Rubi MA, Penning TM, Febbo PG, Balk. SP. Increased expression of genes converting adrenal androgens to testosterone in androgen independent prostate cancer. Cancer Research 2006; 66: 2815–2825.

[232] Biswas MG, Russell DW. Expression cloning and characterization of oxidative 17beta- and 3alpha-hydroxysteroid dehydrogenases from rat and human prostate. Journal of Bioogical Chemistry 1997; 272: 15959–15966.

[233] Prehn C, Moller G, Adamski J. Recent advances in 17beta-hydroxysteroid Dehydrogenases. Journal of Steroid Biochemistry & Molecular Biology 2009; 114, 72–77.

[234] Haynes BP, Straume AH, Geisler J, A'Hern R, Helle H, Smith IE, Lønning PE, Dowsett M. Intratumoral estrogen disposition in breast cancer, Clinical Cancer Research 2010; 16: 1790–1801.

[235] Fomitcheva J, Baker ME, Anderson E, Lee GY, Aziz N. Characterization of Ke 6, a new17beta-hydroxysteroid dehydrogenase, and its expression in gonadal tissues. Journal of Biological Chemistry 1998; 273: 22664–22671.

[236] Maxwell MM, Nearing J, Aziz N. Ke 6 gene. Sequence and organization and aberrant regulation in murine polycystic kidney disease, Journal of Biological Chemistry 1995; 270: 25213–25219.

[237] Su J, Lin M, Napoli JL. Complementary deoxyribonucleic acid cloning and enzymatic characterization of a novel 17beta/3alpha-hydroxysteroid/retinoid short chain dehydrogenase/reductase. Endocrinology 1999; 140: 5275–5284.

[238] Froyen G, Corbett M, Vandewalle J, Jarvela I, Lawrence O, Meldrum C, Bauters M, Govaerts K, Vandeleur L, Esch H, Chelly J, Sanlaville D, Bokhoven H, Ropers HH, Laumonnier F, Ranieri E, Schwartz CE, Abidi F, Tarpey PS, Futreal PA, Whibley A, Raymond FL, Stratton MR, Fryns JP, Scott R, Peippo M, Sipponen M, Partington M, Mowat D, Field M, Hackett A, Marynen P, Turner G, Gecz J. Submicroscopic duplications of the hydroxysteroid dehydrogenase HSD17B10 and the E3 ubiquitin ligase HUWE1 are associated with mental retardation. American Journal of Human Genetetics 2008; 82: 432–443.

[239] Yang SY, He XY , Miller D. HSD17B10: a gene involved in cognitive function through metabolism of isoleucine and neuroactive steroids, Molecular Genetics & Metabolism 2007; 92: 36–42.
[240] Brereton P, Suzuki T, Sasano H, Li K, Duarte C, Obeyesekere V, Haeseleer F, Palczewski K, Smith I, Komesaroff P, Krozowski Z. Pan1b (17betaHSD11)- enzymatic activity and distribution in the lung, Molecular & Cellular Endocrinology 2001; 171: 111–117.
[241] Day JM, Foster PA, Tutill HJ, Parsons MF, Newman SP, Chander SK, Allan GM, Lawrence HR, Vicker N, Potter BV, Reed MJ, Purohit A. 17betahydroxysteroid dehydrogenase Type 1, and not Type 12, is a target for endocrine therapy of hormone-dependent breast cancer. International Journal of Cancer 2008; 122: 1931–1940.
[242] Luu-The V, Tremblay P, Labrie F. Characterization of type 12 17betahydroxysteroid dehydrogenase (17beta-HSD12), an isoform of type 3 17beta-hydroxysteroid dehydrogenase responsible for estradiol formation in women. Molecular Endocrinology 2006; 20: 437–443.
[243] Horiguchi Y, Araki M, Motojima K.17beta-hydroxysteroid dehydrogenase type 13 is a liver-specific lipid droplet-associated protein. Biochemical Biophysical Research Communications 2008; 370: 35–238.
[244] Lukacik P, Keller B, Bunkoczi G, Kavanagh KL, Lee WK, Adamsk Ji, U. Oppermann. U Structural and biochemical characterization of human orphan DHRS10 reveals a novel cytosolic enzyme with steroid dehydrogenase activity. Biochemical Journal 2007; 402: 419–427.
[245] Jansson AK, Gunnarsson C, Cohen M, Sivik T, Stal O. 17beta-hydroxysteroid dehydrogenase 14 affects estradiol levels in breast cancer cells and is a prognostic marker in estrogen receptor-positive breast cancer. Cancer Research 2006; 66: 11471–11477.
[246] Day JM, Tutill HJ, Purohit A & Reed MJ. Design and validation of specific inhibitors of 17beta-hydroxysteroid dehydrogenases for therapeutic application in breast and prostate cancer, and in endometriosis. Endocrine-Related Cancer 2008; 15: 665–692.
[247] Day JM, Foster PA, Tutill HJ, Parsons MF, Newman SP, Chander SK, Allan GM, Lawrence HR, Vicker N, Potter BV. 7Betahydroxysteroid dehydrogenase type 1, and not type 12, is a target for endocrine therapy of hormone-dependent breast cancer. International Journal of Cancer; 2008 122: 1931–1940.
[248] Yang SY, He XY, Schulz H. Multiple functions of type 10 17betahydroxysteroid Dehydrogenase. Trends in Endocrinoogy & Metabolism 2005; 6: 167–175.
[249] Vihko P, Herrala A, Harkonen P, Isomaa V, Kaija H, Kurkela R, Pulkka A. Control of cell proliferation by steroids: the role of 17HSDs. Molecular & Celular Endocrinoogy 2006; 248: 141–148.
[250] Vihko P, Herrala A, Harkonen P, Isomaa V, Kaija H, Kurkela R, Li Y, Patrikainen L, Pulkka A, Soronen P, Torn S. Enzymes as modulators in malignant transformation. Journal of Steroid Biochemistry & Molecular Biology 2005; 93: 277–283.
[251] Peltoket H, Luu-The V, Simard J, Adamski J. 17β-Hydroxysteroid dehydrogenase (HSD)/17-ketosteroidreductase (KSR) family; nomenclature and maincharacteristics of the 17HSD/KSR enzymes. Journal of Molecular Endocrinology 1999; 23: 1-11.

[252] Geissler W, Davis D, Wu L, Bradshaw K, Patel S, Mendonca B, Elliston K, Wilson J, Russell D. Andersson S. Male pseudohermaphroditism caused by mutations of testicular 17β hydroxysteroid dehydrogenase 3. Nature Genetic 1994; 7: 34–39.

[253] Day MJ, Tutill HJ, Foster PA, Bailey HV, Heaton WB, Sharland CM, Vicker N, Potter BV, Purohit A, Reed MJ. Development of hormonedependent prostate cancer models for the evaluation of inhibitors of 17beta-hydroxysteroid dehydrogenase type 3, Molecular & Cellular Endocrinology 2009; 301: 251–258.

[254] Neocleous V, Sismani C, Shammas C, Efstathiou E, Alexandrou A, Ioannides M, Argyrou M, Patsalis PC, Phylactou LA, Skordis N. Duplication of exons 3-10 of the HSD17B3 gene: A novel type of genetic defect underlying 17β-HSD-3 deficiency. Gene 2012; 499: 250-255.

[255] Faienza MF, Giordani L, Delvecchio M, Cavallo L. Clinical, endocrine, and molecular findings in 17beta-hydroxysteroid dehydrogenase type 3 deficiency. Journal of Endocrinological Investigations 2008; 31: 85-91.

[256] Penning TM, Byrns MC. Steroid hormone transforming aldo-keto reductases and cancer. Annals of New York Academy of Sciences 2009; 1155: 33–42.

[257] Kissinger CR, Rejto PA, Pelletier LA, Thomson JA, Showalter RE, Abreo MA, Agree CS, Margosiak S, Meng JJ, Aust RM, Vanderpool D, Li B, Tempczyk-Russell A, Villafranca JE. Crystal structure of human ABAD/HSD10 with a bound inhibitor: implications for design of Alzheimer's disease therapeutics. Journal of Molecular Biology 2004; 342; 943–952.

[258] Du Yan S, Zhu Y, Stern ED, Hwang YC, Hori O, Ogawa S, Frosch MP, Connolly ES Jr., Taggert RMc, Pinsky DJ, Clarke S, Stern DM, Ramasamy R. Amyloid beta-peptide-binding alcohol dehydrogenase is a component of the cellular response to nutritional stress. Journal of Biological Chemistry 2000; 275: 27100–27109.

Chapter 7

Functions of Dehydrogenases in Health and Disease

Nwaoguikpe Reginald Nwazue

Additional information is available at the end of the chapter

http://dx.doi.org/10.5772/48278

1. Introduction

Dehydrogenases are a group of biological catalysts (enzymes) that mediate in biochemical reactions removing hydrogen atoms [H] instead of oxygen [O] in its oxido-reduction reactions. It is a versatile enzyme in the respiratory chain pathway or the electron transfer chain. T. Turnberg discovered this group of enzymes between1900-1922. Several dehydrogenases are present in tissues of humans, plants and micro-organisms having enormous biochemical interests. As a result of the polymorphic nature of this enzyme, it is pertinent therefore to limit our interest on the different functions of Lactate dehydrogenase in the diagnosis and treatment of malaria. Lactate dehydrogenase, an oxidoreductase [EC 1.1.1.27] exists in different forms in different tissues possessing different subunits as a multi-enzyme complex called isoenzyme. It is the last enzyme of the glycolytic sequence or pathway essential for ATP generation. The enzyme, 17β-Hydroxysteroid dehydrogenase exists in at least fourteen isoforms in tissues involved in the biosynthesis of estrogenic and androgenic steroids. Lactate dehydrogenase is a tetrameric enzyme, but only two distinct subunits have been found; those designated H for heart (myocardium) and M for muscle. These two subunits are combined in five different ways. The lactate dehydrogenase isoenzymes, subunit compositions and major locations are shown below.

Following myocardial infarction (MI), the serum levels of LDH rise within 24-28 hrs, reaching a peak by 2-3 days and return to normal in 5-10 days. Especially diagnostic is a comparison of the LDH1/LDH2 ratio. Normally, the ratio is less than 1. A reversal of the ratio is referred to as "flipped LDH". Following an acute myocardial infarction, the flipped ratio will appear in 12-24 hours and it is definitely present by 48 hours in over 80% of cases. Also important, is the fact that, persons suffering from chest pain due to angina only, will not likely have LDH altered levels.

Type	composition	Location
LDH1	HHHH	Found in the heart and red blood cells and Is 17%-27% of the normal serum level.
LDH2	HHHM	Found in the heart and red blood cells and 27%-37% of the normal serum level
LDH3	HHMM	Found in a variety of organs and is 18%-25% of the normal serum level.
LDH4	HMMM	Found in a variety of organs and is 3%-8% of the normal serum level.
LDH5	MMMM	Found in liver and skeletal muscles and is 0%-5% Of the normal serum level.

Table 1. Subunit compositions and major locations of Lactate Dehydrogease Isoforms

2. Measurement of parasite lactate dehydrogenase (pLDH) activity of plasmodium falciparum

Malaria is the most lethal parasitic disease in the world, annually affecting approximately 500 million people and resulting in 800,000 deaths, mostly in Africa and Sub-Saharan countries [1]. Some countries, for example Brazil registered more than 306,000 deaths in 2009, most of which were in the Amazonian region [2]. In Africa, the figures may be higher due to endemic nature of the infection. Transmission occurs through the bite of the female anopheles mosquitoes infected with the parasite of which there are five species affecting humans. Plasmodium falciparum is the most pathogenic species and may cause severe malaria and death in non-immune individuals, especially children under five years [3].Drug resistant malaria parasites have emerged and has resulted in treatment failures. This resistance might be as a result of mutation at the active sites of drug targets or from biochemical changes in the drug receptors [4].

Plasmodium falciparum is a significant cause of morbidity and mortality in travelers to areas where the parasite is endemic. Non –specific manifestations may equally result in failure to recognize malaria until autopsy, when it is often too late to obtain blood for microscopic evaluation, which has been in use for years for the diagnosis of malarial parasitemia. The Plasmodium falciparum lactate enzyme (pLDH) has been considered, a potential molecular target for antimalarials. This is an enzyme assay for the detection of Plasmodium falciparum, employed in the assessment of malarial parasitism. The enzyme assay is based on the observation that Lactate dehydrogenase(LDH) enzyme of *P. falciparum* has the ability to rapidly use 3-acetylpyridine dinucleotide(APAD) as a co-enzyme instead of NAD in the reaction leading to the formation of pyruvate from lactate. Human red blood cells' LDH carries out this reaction at very low rate in the presence of APAD. The measured development of APADH leads to the formation of a product that could establish the basis of an assay that detects the presence of *P. falciparum* from *in vitro* cultures at parasitemic levels of 0.02 %. Lactate dehydrogenase is the most abundant enzyme expressed by *P. falciparum.*

A correlation between levels of parasitemia and the activity of parasite LDH from patients with malaria is worthwhile. The serum assay for pLDH is followed up to monitor the level of pLDH in a patient with cerebral malaria prior to antimalarial therapy and also during recovery period. It is evident that measurement of pLDH has a strong correlation with malarial parasitemia and can follow a method that can be developed into a simple test for the detection of *Plasmodium falciparum* as an assessment of plasmodium parasitemia. In malarial falciparum parasitemia, LDH does not persist in blood, but clears about the time as the parasite, following successful treatment. The lack of antigen persistence after treatment makes the pLDH test useful in predicting treatment failure. In this respect, pLDH is similar to pGluDH. LDH for *P.vivax*, *P.ovale*, and *P. malariae* exhibit 90-92% identity to pLDH *from P. falciparum.*

3. Functions of inhibitors of parasite lactate dehydrogenase (pLDH) as potential antimalarial agents

The *plasmodium falciparum* lactate dehydrogenase enzyme (Pf LDH), has been considered a potential molecular target to antimalarials due to this parasite dependence on glycolysis for energy production by catalyzing the reduction of pyruvate to lactate. It has been a routine activity among drug designers for malarial infestation to embark on the screening of analogs to NADH(an essential cofactor) to pLDH.The continued search for new molecular targets for drug design is an endless search, since the introduction of the quinolones in malaria therapy. Chloroquine interacts specifically with PfLDH in the NADH binding pocket, occupying a position similar to that of the adenyl ring cofactor, hence acting as a competitive inhibitor for this critical glycolytic enzyme [5,6,7,8]. Analogs of NADH have been identified as new potential inhibitors to PfLDH[9]. Computational studies have been undertaken to recognize the potential binding of selected compounds to the pLDH active site. This was analyzed using Motegro Virtual Docker Software. The researchers selected fifty (50) compounds based on their similarity to NADH. The compounds with the best bonding energies included: itraconazole, atorvastatin and posaconazole. These were tested against P. falciparum , chloroquine resistant blood parasites. All these compounds proved to be active in two immunoenzymatic assays performed in parallel using monoclonals specific to pfLDH or a histidine rich protein 2 (HRP 2). The IC_{50} values for each drug in both tests are similar; values were lowest for posaconazole (< 5μM) and were 40-and 100-fold less active than chloroquine. The compounds so tested reduced *P. berghi* parasitemia in treated mice, in comparison to untreated control. The drug itraconazole is the least active compound. Posaconazole is an inhibitor of ergosterol biosynthesis [10]. In this study, it was the most active drug against *P. falciparum*. It is also, the most effective compound against murine malaria caused by *P. berghei* and was the most promising agent in vitro and in vivo. Itraconazole is normally acquired as a tablet, causes a strong inhibition of P. falciparum growth in vitro and is partially active against *P. berghei*. The results of these trials according to the authors proved that molecular docking studies are important in the strategy for discovering new antimalarial drugs. This approach is more practical and less expensive than discovering novel compounds that require studies on human toxicology. The parasite enzyme, lactate dehydrogenase has recently received a great deal of attention, since it may constitute a valid therapeutic target for diseases such as malaria and cancer. Because the LDH enzymes found in *P.vivax*, *P. ovale* and *P.malariae*, all exhibit 90% identity to PfLDH; It would be desirable to have new anti-pLDH drugs particularly, ones that are effective against *P. falciparum*, the most virulent species of human malaria. Most invasive tumor phenotypes show a metabolic switch (Warburg effect) from oxidative phosporylation to an increased anerobic glycolysis by promoting an up-regulation of the human isoform-5 of lactate dehydrogenase (hLDH-5 or LDH-A), which is normally present in muscles and liver. Hence, inhibition of hLDH-5 may constitute an efficient way to interfere with tumor growth and invasiveness.

4. New enzymatic assay using parasite ldh in diagnosis of malaria in Kenya

The biochemical basis for this assay is on the fact that human red blood cells do not utilize APAD in the metabolism of glucose. The study subjects were of three different categories: the healthy non-infected individuals staying out of malaria endemic area(controlled GP-1).The non-symptomatic and parasitemic healthy individuals living in endemic regions (both field study –GP 2).The non-parasitemic and parasitemic symptomatic individuals living in endemic region (both clinical study Group 3).In the clinical studies, thin smear microscopy gave the highest sensitivity as 75.6 % for plasma, while the highest specificity was 71.4 %. For red blood cells, the highest sensitivity was 78.4% while specificity was 80 %. In field trials, the highest sensitivity was 89 %) using thin smear microscopy, where as the specificity was 45% for the plasma cut off, using thick smear. For red blood cells, the highest sensitivity was 79% while the specificity was 66.7%. The variations in sensitivity and specificity of this assay in comparison to microscopy is a strong indication that pLDH may even be measuring sequestered parasites that cannot be visualized by microscopy. The results of the study validates the use of pLDH as an alternative objective test for malarial diagnosis against microscopy.

5. Malaria antigen detection tests

There are currently over twenty (20) such tests commercially available (WHO product testing, 2008) and these consist of a group of commercially available tests that allow rapid diagnosis of malaria by unskilled laboratory traditional techniques. The first malaria antigen suitable as target for Rapid Diagnostic Tests (RDTs) was a soluble glycolytic enzyme, Glutamate dehydrogenase. None of the current tests is as sensitive as a thick blood film. A major drawback in the use of all dipstick methods is that the result is essentially qualitative. In many endemic areas of tropical Africa; however, the quantitative assessment of parasitemia is important as a large percentage of the population will test positive. An accurate diagnosis of malaria is becoming more and more important in view of the increasing resistance of *Plasmodium falciparum* and the high price of alternatives to chloroquine. The enzyme pGluDH (parasite Glutamate Dehydrogenase) does not occur in the host red blood cells and was regarded as a marker enzyme for Plasmodium species. The malaria marker enzyme test is suitable for routine work and it's now a standard in most nations afflicted with malaria. The presence of pGluDH is known to represent parasite viability, and a rapid diagnostic test using pGluDH as antigen would have the ability to differentiate live and dead organisms. It is possible to note that a complete RDT with pGluDH as antigen has been developed in China. Glutamate dehydrogenases are ubiquitous enzymes that occupy an important branch point between carbon and nitrogen metabolisms. Both nicotinamide adenine dinucleotide[(NAD)EC 1.4.1.2] and nicotinamide adenine dinucleotide phosphate (NADP) dependent GluDH [EC1.4.1.4] enzymes are present in plasmodia. The NAD-dependent GluDH is relatively unstable and not useful for diagnostic purposes. Glutamate dehydrogenase provides an oxidizable carbon source used in production of energy as well as a reduced electron carrier, NADH. Glutamate is a principal donor to other amino acids in subsequent transamination reactions. The multiple roles of glutamate in nitrogen balance, make it a gateway between free ammonia and the amino groups of most amino acids. The GluDH activity in *P. vivax, P.ovale* and *P. malariae* has been tested, but given the importance of GluDH as a branch point enzyme, every cell must have a high concentration of GluDH. It is well known that enzymes with high molecular weights like GluDH have many isoenzymes which allow strain differentiation (given the right monoclonal antibody).The host produces antibodies against the parasite enzyme, indicating a low sequence identity.

6. New therapeutic approaches for treatment of plasmodium falciparum

The antimalarial treatment so far recommended for P. falciparum consists of drug combinations containing Artemisinin derivatives (ACT) known as artemisinin combination therapy with other antimalarials, including quinolone compounds such as Amodiaquine and Mefloquine. The mechanism of action of the quinolones involve the inhibition of hematin polymerization, thus intoxicating the parasite with the ferriprotoporphyrin groups generated by hemoglobin degradation. Other antimalarials used in the ACT therapy include-Pyrimethamine and proguanil, which inhibit the tetrahydrofolic acid cycle (tetrahydropterate reductase, limiting the formation of folic acid, an important cofactor in DNA biosynthesis. Despite the arsenal of drugs available for malaria treatment, the disease remains a worldwide public health problem. P. falciparum develops resistance under selected drug pressure. Plasmodium vivax is the most prevalent human malaria parasite world over and has been shown to be resistant to chloroquine, including in Brazil and other countries where malaria is endemic. Various efforts have been made to develop new drugs (antimalarials), but resistance to drugs has limited the search. The continued search for new molecular targets for drug design has broadened the therapeutic arsenal and strategies to fight drug resistance in human malarial infestation.

7. Three new parasite ldh (pan-pLDH) tests for diagnosis of uncomplicated malaria

Since Charles Laveran first visualized the malaria parasite in 1880; the mainstay of malaria diagnosis has been the microscopic examination of blood smear. It is the main economic, most preferred and reliable diagnosis using two types of blood films, which is amenable to the four species of malaria parasite. These two types of films are (a) the thin blood film which is similar to the usual blood films, which allows species to be identified because they can be visualized and the appearance of the parasites are much more distorted.(b)the thick film from which an experienced microscopist can detect parasite levels or parasitemia down to levels as low as 0.0000001% of red blood cells. Diagnosis of species can be difficult because, the early trophozoites ("ring form") of all four species look identical and never possible to diagnose the species on the basis of simple ring form. The success of the method above requires well trained staff, quality equipment and supervision. The scarcity of these facilities within malaria endemic areas becomes limiting. In sub-Saharan Africa and some other areas, clinicians often have to rely on clinical signs and symptoms for diagnosis and in some areas where increasing emphasis is laid on home based management , malaria diagnosis is often equated[11] with fever. It is to be noted that such presumptive treatment without laboratory confirmation could contribute to the development of drug resistance[11]. Today, an alternative method to the blood film diagnosis approach is the rapid diagnostic test (RDTs), recommended by WHO, where reliable microscopy is not reliable or available. Rapid diagnostic tests (RDTs) are antigen detection tests, which are simple to use and interpret, although the tests also use peripheral blood. The most commonly used RDTs, is the histidine-rich protein 2 (HRP 2), produced by trophozoites and young gametocytes of plasmodium falciparum. HRP 2 test has been the most widely evaluated to date test, and has shown consistently high sensitivity. The limitation of this test is that RDTs detect P. falciparum only and can remain positive for several weeks after antimalarial treatment. Besides these, a study to assess the diagnostic capabilities of three parasite lactate dehydrogenase (pan- LDH)-,Vistapan® ,Carestat™ and Parabank® were conducted in Uganda. Similarly, a histidine- rich protein 2 (HPR 2) test, Paracheck-Pf® and a Geimsa-stained blood film were performed with pfLDH tests for outpatients. A total of 460 patients were recruited for the exercise, 248 had positive blood films and 212 with negative blood films. *Plasmodium falciparum* was present in 95% of infections. Sensitivity of the tests above 90% was shown by two pLDH tests-Carestart (96.5%) and Vistapan (91.9%) and sensitivity above 90% by Parabank (94.3%) and Carestart (91.5%). The benefits of these tests when compared with the previous gold standard for laboratory confirmation of malaria diagnosis which is a peripheral blood film examined microscopically shows the high specificity and validity of the tests.

An alternative diagnostic method to the rapid diagnostic test (RTD), recommended by WHO, where reliable microscopy is not available. RTDs are antigen detection tests, which are simple to use and interpret and also use peripheral blood. The most commonly used RTD detects histidine -rich protein 2 (HRP2), produced by trophozoites and young gametocytes of *Plasmodium falciparum*. HRP 2 tests have been the most widely evaluated to date and show high sensitivity. However, they are limited in that they detect P. falciparum only and can remain positive for several weeks after successful treatment[12,13].

The second type of RTD detects the malaria antigen parasite lactate dehydrogenase (pLDH), an enzyme produced in the glycolytic cycle of the asexual stage of all species of Plasmodium. Parasite lactate dehydrogenase(pLDH) are produced only by viable parasites, thus being cleared from the blood stream more rapidly after treatment, resulting in test becoming negative more quickly. There is no doubt that these characteristics suggest that pLDH tests could be used with more confidence for malaria diagnosis at the peripheral level. The development of several new pLDH monoclonal antibodies by Flow Inc. has enabled the production of a new generation of pLDH tests. These characteristics suggest that pLDH tests could be used with more confidence for malaria diagnosis at the peripheral level [14]. For confident diagnosis of malaria in routine outpatient department conditions, a sensitivity of more than 90% is crucial and this has been achieved by both Carestart and Vistapan. The pLDH tests have also demonstrated desirable qualities that could reduce the possibility of patients without malaria being given antimalarials, which may therefore reduce drug pressure, a major concern at a time when Artemesinin combination therapy (ACTs) are being introduced in Africa. The validation of these tests for malaria diagnosis involves many stages from the selection of site, enrolment of patients, sample size, study procedures, laboratory procedures, study outcomes, analysis and results. The validity of the tests after all these procedures must be rated from 90% and above. From the validity tests carried out by some researchers, Carestart had estimates for all validity parameters greater than 90%. Vistapan and Carestart were also sensitive as Paracheck Pf ($p=0.14$ and $p=0.38$) respectively. Parabank was less sensitive than all other tests ($P<0.001$ for each comparison). There was no significant difference between the three pLDH tests, but Parabank had a higher specificity compared with Paracheck Pf ($P=0.02$) for P. falciparum detection. In the study, the ages of patients were taken into consideration. Sensitivity decreased with older age for both Vistapan (97.4%) for the under fives versus 85.7%, $P<0.011$ and Parabank (95.4%) for the under fives versus 73.1%, $P<0.001$. Three tables were used to summarize the results of a study conducted in Mbarara Regional Hospital in Uganda, in a mesoendemic area of malaria transmission. These results are shown in tables 2, 3 and 4 respectively.

7.1. Materials and methods

The first approach is to select a site, which should be a highly malarial infested zone.

7.2. Enrolment for the study.

Patients from the outpatient department were systematically screened for symptoms suspected to be malaria and referred to the research clinic. Inclusion criteria were a clinical suspicion of malaria; weight$\geq$ 5kg; resident in Mbarara Municipality available for two weeks follow up period; and signed informed consent from the study subjects or their legal guardians. Exclusion criteria were signs of severe or complicated malaria [15b], signs of severe disease; and women with visible pregnancy or suspicion of pregnancy based on the assessment of the last normal menstrual period.

7.3. Sample size

The required number of patients with positive blood film was calculated using an estimated sensitivity of the RDTs of 90%, an alpha error of 0.05 and a precision of 6%. This number (n-

96) was doubled to permit a stratified analysis by age group (0--4 and ≥5 years. The same parameters were used to calculate the required number of patients with a negative blood film, thus giving a final minimum sample size of 200 blood-film-positive and 200 blood –film –negative patients.

7.4. Study procedures

On the day of inclusion, demographic and clinical information were recorded, and a thick/thin blood film and the four rapid tests (Vistapan, Carestart, Parabank , and Paracheck-Pf) were performed. Women with positive pregnancy test and hyperparasitemic patients (P. falciparum›250,000 parasites/µl) were given quinine and excluded from further follow up. All other patients with positive blood film received an artemether-lumefantrine(Coartem®, Norvatis Pharma AG,Bassel,Switzerland), six –dose regimen under directly observed therapy. This treatment modality have been shown to be very efficacious with a prompt reversion to a negative [16] blood film after treatment. Patients receiving Coartem were asked to return to the Clinic on the third, seventh and 14th day after inclusion to repeat the blood film and all RDTs

7.5. Laboratory procedures

Blood films and rapid tests were performed from the same finger-prick blood. Blood films were dried, thin films fixed in methanol, and both films stained with 3% Giemsa for 45 minutes. Smears were read by experienced technicians, counting parasites against 200 or 500 white blood cells (WBC) or 200 high power fields before declaring a blood slide negative. The parasite density per micro-liter was calculated by multiplying the asexual parasite count by 8000 and dividing by the number of WBC counted [17].Plasmodium species were confirmed on the thin film and slides with mixed infections had only P.falciparum monoinfection, had the asexual density per microliter calculated as for P.falciparum. Gametophytes were recorded with species identification where possible. All inclusion slides were blinded and double read, with a third reading performed in case of discordance, ie; positive/negative discordance for asexual stages; asexualdensity discordance (difference in parasitemia≥50%);positive/negative gametocyte discordance. Twenty percent of the follow-up visit slides were also blinded and double read. External quality control of 290 inclusion slides was performed by Shokia Malaria Research Unit, Thailand, giving Mbarara laboratory, a sensitivity of 95.5% and a specificity of 100%. All RDTs were performed and interpreted according to the manufacture's instructions. Each test result was interpreted by two independent health care providers blind to the result of the blood film and reading according to a rota to avoid observer bias. The first reading was performed at the time specified by the manufacturer (15 min after preparation for Paracheck Pf and Parabank and 20 mins for Carestart and Vistapan. The second reading was performed within 15 min of the first one. Discordant results were read by the laboratory supervisor for a definitive result. Each reader also classified the test as either invalid or doubtful. A doubtful test was defined as a test for which the reader was not sure if there was any indication of a line present. At the end of the study, two test readers and two laboratory technicians involved in preparing the tests completed a questionnaire concerning the ease of use and interpretation of each test.

Analysis. All data were either recorded directly or transcribed from source data forms to an individually numbered case report form (CRF). Data were double entered and validated using EpiData version 3.1(EpiData Association,Odense, Denmark)and analysed using Stata 9.1 (Stata Corp. college station,TX,USA).The study profile and base-line characteristics were summarized, including comparative tests between age groups (x 2 test ,Ma-Whitey U test). The validity for each test was calculated overall and then stratified by age group, level of parasitemia(parasites/μl 1---99,≥100,≥200,≥500), presence/absence of fever, duration of illness(0—2vs.3 days and above) and a history of taking antimalarials, using comparative tests(x 2 test, Mann-Whitney U test) to compare differences between groups. Kappa statistics were calculated for iter-reader reliability for each test on the day of diagnosis. A test was considered as reliable if $k \geq 0.8$. Univariable and multivariate analyses were performed to investigate the association between explanatory factors and the test remaining positive at each follow- up visit.

Results. Demographical and Parasitological characteristics of study subjects.

Between 26 April and 27th July, 2005, 485 patients from the out-patient department were screened. Nine were ineligible (three had severe illness, five were non-residents and one was not in the appropriate age group after completion of recruitment in the under fives. Sixteen patients did not consent to participate in the study; 239 under fives and 221 aged 5 years and above. The mean age was 12 years(SD 13years;Table 2). There were 248 positive blood films with P. falciparum monoinfections(93.6%), P. malariae monoinfections(2.4%), P. falciparum+ P. malariae mixed infections(0.8%) and P. falciparum+ P. vivax mixed infections(0.8%). Of the 212 negative films, nine had gametocytes present. Parasitological characteristics of positive subjects are given in Table 2. Slides positive with P. falciparum had higher parasite densities than those of the other two species.

Validity of RDTs.

Only Carestart had estimates for all validity parameters greater than 90% (Table 3). Vistapan and Carestart were as sensitive as Paracheck-Pf(P=0.14 and P=0.38 respectively. Parabank was less sensitive than all other tests (P‹0.001 for each comparison). There was no significant difference in specificity between the three pLDH tests, but Parabank had a higher specificity compared with Paracheck.Pf (P= 0.02) for P. falciparum detection. Sensitivity decreased with older age for both Vistapan [97.7% (under fives) vs 85.7%, P‹0.01] and Parabank[95.4%(under fives)vs73.1%, P‹0.001]. Sensitivity increased with axillary temperature≥ 37.5 ^{0}C at inclusion for Paracheck.Pf (98.8 vs91.4%, P=0.04), Vistapan(97.6 vs 89.0%, P=0.03) and Parabank(91.8vs81.0%, P=0.04) compared with patients with axillary temperature ‹37.5 ^{0}C. Although, the small number of non-falciparum monoinfections does not permit reliable calculations of validity of non-falciparum mutants, all tests detected 100 % (n=6) of the P. malariae monoinfections. Plasmodium vivax was detected in 4/6 infections by Carestart, 2/6 by Vistapan and 1/6 by Parabank.

Parameters	Group A (<5yrs)	Group B (≥5yrs)	Overall	p-value
Baseline characteristics	n=239	n=221	n=460	
Gender ratio (M:F)	0.98 (118:121)	0.52 (76:145)	0.72 (194:266)	0.001(X 2)
Mean age(SD)	2yrs (14 months)	22yrs (12yrs)	12yrs (13 yrs)	N/A
Median duration of illness in days (range)	3(1-14)	3(1-30)	3(1-30)	0.2(Kruskai-Wallis)
Previously taken antimalarial (n,%)	81(33.9)	60(27.3)	141(30.7)	0.13(X 2)
Fever on presentation (axillary temp≥ 37.5 ^{0}C)	99(41.4)	31(14.0)	130(28.3)	<0.001(X 2)
Parasitological characteristic	n=129	n=119	n=248	-----------
Asexual parasitemia range (parasites/μl)	16-703411	16-233241	16-703411	0.001(Kruskal-Wallis)
Geometric mean of asexual parasitemia(95%Cl)	7433(4869-11346)	1524(975-2384)	3475(2521-4790)	0.001(t test)
Interquartile range (Interquartile value)	1682-45748(44066)	166-11070(10904)	641-23827(23186)	----------
Gametocyte carriage(n,%)	36(27.9)	22(18.5)	58(23.4)	0.11(X 2)

Table 2. Baseline characteristics of all study subjects and Parasitological characteristics of Slide-positive subjects attending Mbarara Regional Referal Hospital, Outpatient department, South -Western Uganda

Reliability

The k statistic for the inter-reader reliability for all tests was above 0.90(very good agreement) [Carestart,k=0.96(95.0%, Cl 0.94—0.99); Vistapan, k=0.94(95% Cl 0.91- t 0.97); Parabank, k=0.96(95% Cl 0.94-0.99); Paracheck.Pf, k=0.97(95% Cl 0.95-1.0)]

Time to Negativity of RDTs.

There were no positive blood films on follow up visits, and therefore, every positive RDT result on day 3,7 or 14 was considered a false positive result (Table 4). All three RDTs tested had significantly fewer false positive results on every day of follow-up compared with Paracheck.Pf(P‹0.001 for all tests on day 3,7 and 14). There was o difference between the pLDH tests by day 14, with the percentage of positive tests ranging from 4.6 to 9.5%.

Younger age group and higher parasite level at inclusion were related to positive Paracheck. Pf on all follow-up days (logistic regression, P‹0.01), for all. Age group, fever at diagnosis and presence of gametocytes on day 3 were all related to a positive pLDH test on day 3 (except age group for Parabank)(age group:Vistapan P=0.026, Carestart P‹0.001.

Parameters	Carestart % [95 Cl]	Vistapan% [95%Cl)	Parabank% [95% cl)	Paracheck =Pf%(95%cl)
Sensitivity	95.6(237/248) [90.2-96.6]	91.9(228/248) [87.8-95]	84.7(210/248) [79.6-88.9]	94(233/248) [90.2-96.6]
Specificity	91.5(194/212) [86.9-94.9]	89.6(190/212) [84.7-93.4]	94.3(200/212) [90.3-97.0]	87.3(185/212) [82.0-91.4]
PPV	92.9(237/255) [89.1-95.8]	91.2(228/250) [87-94.4]	94.6(210/222) [90.7-97.2]	89.6(233/260) [85.3-93]
NPV	94.6(194/205) [90.6-97.3]	90.5(190/210) [85.7-94.1]	84.0(200/238) [78.7-88.4]	92.5(185/200) [87.9-95.7]

PPV: Positive Predictive Value; NPV: Negative Predictive Value

Table 3. Validity of four rapid diagnostic tests for the detection P. falciparum species in patients attending Mbarara Regional Hospital, out-patient department, southwestern Uganda

RDT	Day 0 n[a]	Day 3% [95% Cl]	Day 7 %[95Cl]	Day 14 % [95% Cl]
Paracheck - Pf	226	86.2(193/224) [81.7-90.7]	80.8(181/224) [75.6-86.0]	69.7(152/218) [63.1-75.7]
Vistapan	221	36.1(79/219)[29.7-42.5]	3.4(51/218)[17.8-29.0]	8.9(19/213) [5.1-12.7]
Carestart	230	42.5(97/228) [36.1-48.9]	27.6(63/228) [21.8-33.4]	9.5(21/221) [5.6-13.4]
Parabank	204	17.8(36/202) [12.5-23.1]	8.9(18/202) [5.0-12.8]	4.6(9/196) [1.7-7.5]

a n is the number of positive tests for each RDT on day 0 in patients who were followed up.

Table 4. Percentage of positive tests on each follow up visit in patients attending Mbarara Regional Hospital, Outpatient department, SouthWestern Uganda.

In overall, there are not large differences between the tests in terms of ease of use. Some tests have small advantages or disadvantages over others. For example, Vistapan has individual buffer sachets, considered to be an advantage, where as Carestart has a delay time of over 60 seconds between blood application and buffer application, considered to be a disadvantage. All tests results are stable for a minimum of 24hr.The number of invalid tests was ‹0.5% for Parabank and between 0.5 and 2% for Carestart and Vistapan. No test had items requiring refrigeration and all tests have undergone temperature stability studies up to 30 ^{0}C.

This study appears to be a pioneer study [17] to evaluate a new generation of pLDH tests for malaria diagnosis, performed in a mesoendemic African setting with a predominance of P. falciparum infections. The authors showed that several of these tests were validated and should be of great use in malaria endemic countries, where microscopy is not available and well trained microscopists are lacking. For confident diagnosis of malaria in routine outpatient departments, a sensitivity of more than 90% is crucial and this was achieved for Carestart and Vistapan.

The pLDH tests also demonstrated desirable qualities that could reduce the possibility of patients without malaria being treated or given antimalarial drugs, thereby reducing drug pressure and resistance, a major concern at a time when artemisinin combination therapy (ACTs) are being introduced throughout Africa. Their high specificity would reduce the number of patients with false positive results. Secondly, the great reduction in the number of tests remaining positive for a long time after treatment had been effected. This is reminiscent of the test carried out by HRP 2. Thirdly, the ability to detect both P. falciparum and P. vivax, would increase confidence on a negative test result. A variety of factors may contribute towards differing sensitivities of the test, such as patients age and level of parasitemia, which will vary according to endemic nature of P. falciparum in the locality. Lower test sensitivity may be related to low parasitemia in adults in an area of stable transmission. This may be a limitation of the tests; although, such patients are less at risk from severe clinical episodes to perpetuate parasite transmission.

8. Comparison of two rapid field immunochromatographic tests to expert microscopy in malaria diagnosis.

The vast majority of malarial tests adopted and used in the past depended on light microscopy and expertise in attaining results. Currently, the vast majority of malaria cases in the world are diagnosed by the century old standard of light microscopy and stained blood smears. Although the technique is sensitive and very inexpensive, there are several disadvantages. The disadvantages include practical issues such as electrical requirement, the need for experienced staff as well as difficulties in accurate species identification. Efforts have been made to develop malaria rapid diagnostic devices (MRDDs) to facilitate field diagnosis [18]. While the first generation tests diagnosed only Plasmodium falciparum, newer devices were designed to recognize both Plasmodium falciparum and P. falciparum specific antigen as well as Plasmodium genus specific antigen.

The ICT malaria Pf/P.v test is a rapid immunochromatographic assay for the detection of P. falciparum-specific Pf HRP 2 and a pan-malarial antigen, manufactures in test card form. The name is misleading as the pan malarial antigen may be produced by Plasmodium ovale, and Plasmodium malariae as well. The test has been shown to be specific and sensitive. Malaria P.f/P.v test has been shown to be 96% sensitive and 90% specific for P. falciparum and 75% sensitive and 95% specific for P. vivax [19,20].

The OptiMAL dipstick test detects two forms of plasmodium lactate dehydrogenase parasites pLDH1; one P. falciparum-specific and one common to the four plasmodium species which infect humans. Many researchers reported a 95% sensitivity and 100% specificity for P. falciparum, and a 96 % sensitivity and 100% specificity for P.vivax. Sensitivity for P. ovale and P.malariae were significantly lower (57% and 47%) respectively. Using a batch OptiMAL dipstick procedure in 1999, a large scale field evaluation showed the following result: 91% sensitivity and 95% specificity for P.falciparum and 83% sensitivity and 100 % specificity for P.vivax [21].

Methodology. The methodology involved the use of patients both symptomatic and asymptomatic , who are villagers referred to the study and informed consent obtained. For each patient, a finger prick was made and the following were obtained. Fifty microliters of blood in a pre-heparinized Eppendorf tube for dipstick assays and two thick/ thin smears, one for on –site microscopic diagnosis by Acridine orange(AO) technique (for immediate treatment purposes, and the other, for reference Geimsa-microscopist.OptiMAL reader and ICT reader, were all blinded to each other's diagnosis. Patients positive for malaria were treated. Thick and thin blood smears were prepared and stained with Geimsa according to standard procedures. To declare a sample negative, thick smears were read for 200 microscopic fields (1000 X) without finding a parasite.If found positive, the number of asexual malaria parasites were counted per 500 WBC separately for each species. If there are more than 250 parasites/500 WBC, parasites were counted on the corresponding thin film per 10,000RBC. Density calculations were based on approximations of7500 WBC/μl and 5X10 6 RBC/μl. ICT malaria P.f/P.v test kits were used as per manufacturers instructions. Ten microliters (10μl) of whol blood was transferred to a sample pad. A buffer reagent was added to induce cell lysis and allow PfHRP 2 and pan malarial antigens to bind to colloidal gold-labeled antibodies. Additional buffer caused the blood and immune complexes to migrate up the test strip and cross monoclonal (mAb) lines. Finally, more buffer was added to clear blood from the membrane and facilitate reading.

Tests are counted as valid, if control lines are observed. They were counted P. falciparum positive, if Pf HRP 2 specific and pan-malarial antigen lines were visible or if only PfHRP 2-specific lines were seen only. If the control and pan-malarial antigen lines were observed, the sample was counted as positive for a malaria parasite other than P. falciparum. The test result was assigned a value of +0, if no line was seen; +1, if test line intensity was less than control line intensity; +2,if it was equal and +3, if it was greater in intensity. Other factors can exacerbate drug resistance. The second type of RDT detects the malarial antigen parasite-lactate dehydrogenase (pLDH), an enzyme produced in the glycolytic pathway or cycle of the asexual stage of all species of plasmodium. Parasite lactate dehydrogenase is produced only by viable parasites, being cleared from the blood stream more quickly after treatment, resulting in the test becoming negative more quickly [22].

9. Clinical and epidemological findings based on histopathology and immunohistochromatographic detection of p.falciparum antigens

This new technique for determination of P.falciparum is not adopted during autopsy when actual diagnosis of P.falciparum infestation is missed mostly during improper diagnosis. Many organs and tissues manifest the severity of this type of malaria before death. The heart, lung, liver tissues are always available for post mortem analysis. Some other tissues used include: spleen, kidney, and brain. The following tissues were equally used in a case of five travelers suspected to have died from other chronic diseases not related to malaria. These other tissues include: tongue, trachea, thyroid and adrenal glands, gall bladder and testis. Viral and or bacterial hemorrhagic fever pathogens were suspected at death [23, 24,

25]. In the study, three novel IHC assays targeting HRP 2, aldolase, and pLDH were developed and confirmed on severe P. falciparum infection in five travelers whose deaths were wrongly suspected.

Malaria is the most common cause of fever in travelers returning to industrialized cities or countries from malaria endemic countries. However, the clinical features of the disease are not specific and in some areas where P.falciparum is not endemic, fatal malaria is often not suspected. The pathological features of malaria resemble many other viral, rikettsial and bacterial infections. An unequivocal diagnosis can be made only by laboratory testing. IHC assays and histopathologic review confirmed P. falciparum infections in many study cases as being responsible for unsuspected deaths. Abundant hemozoin pigment, a by-product of parasite metabolism was distributed diffusely throughout peripheral tissues and in the blood vessels in the central nervous system. The findings are consistent with reports of hemozoin localization. The density of hemozoin increases in proportion to the duration of falciparum infestation and decreases with adequate and appropriate therapy. It has been discovered that the hallmark of P.falciparum infection is sequestration, characterized by the adherence of mature stage falciparum pRBCs(trophozoites and schizonts) to endothelial cells of capillaries and venules.

Many studies in humans and other animals have described sequestration of trophozoites and schizonts in a variety of tissues, including the brain, heart, lung, skeletal muscles and subcutaneous tissues. As a result of sequestration, peripheral blood parasitemia, traditionally evaluated using Geimsa bood smear, may not give substantive correlative result in the pathogenesis of the severity of P. falciparum, hence,the severity of the infection may be underestimated. Plasmodium falciparum infection caused respiratory symptoms that resemble influenza like-illness in correlation with results from studies. There is pulmonary edema with intra-aveolar hyaline membranes and proteinaceous debri, associated with malarial antigens. Pulmonary edema is associated with high parasitemias and often leads to respiratory distress syndrome. Rust tinged urine described for several patients afflicted with this malaria, is associated with hyperbilirubinemia caused by erythrocyte destruction. In many cases of P. falciparum, malarial antigens can be detected in tubular epithelial cells in association with erythrocyte casts. The HRP 2 antibody used in this study was specific for P.falciparum, whereas the aldolase and the pLDH antibodies reacted with both P. falciparum and P. vivax.

In conclusion, the current approach in the diagnosis and development of new drugs for the treatment of Plasmodium falciparum infections is quite novel and holds promise for the future. The ubiquitous nature of the enzyme, Lactate dehydrogenase especially, parasite lactate dehydrogease (pLDH) as malaria antigen, is indicative of the vital biochemical process of metabolism of pyruvate and lactate in microbial cells. The endemic nature of malaria in Africa and some Mediterranean countries poses a great challenge to humanity. There should be a more radical approach especially in Africa and other countries afflicted to tackle this problem which tend to decimate world population. Other frontline drugs which are designed to inhibit the enzyme should be developed to add to the success of this protocol. In as much as mosquitoes have developed resistance against chloroquine, the drug in some places remains the only option for radical cure of malaria.

Author details

Nwaoguikpe Reginald Nwazue
Department of Biochemistry, Federal, University of Technology, Owerri, Imo State, Nigeria

10. References

[1] Kawada M, Anaka M, Kato H, Shibasaki S, Hikosaka K, Mizuno H, Masuda Y, Inamatsu F (2011).Evaluation of the a simultaneous detection kit for the Glutamate dehydrogenase(pan-LDH) tests for diagnosis of uncomplicated malaria. Journal Infect. chemotherapy , 17(6):807-811

[2] Carol F, Roggers T, Vincent B et al. (2008).Assessment of three new parasite Lactate dehydrogease(pan-LDH)tests for the diagnosis of uncomplicated malaria. Transactions of the Royal Society of Medicine and Hygiene, 102: 25-31

[3] Daniel PM, Fumihiko K, KhIn I, Anintita L, Chainsuda W (2002).A comparison of two rapid field immuochromatographic tests for expert microscopy in the diagnosis of malaria. Acta Tropica, 82:51-59

[4] Gillian L, Genrich JG, Christopher D, Paddock WS, Patricia WG, John WB, Sherif FZ (2007). Fatal malaria infection in travelers. Immuochemical Assays for the detection of Plasmodium falciparum in tissues and Implication for Pathogenesis. American Journal of Tropical Medicine and Hygiene, 76(2):251-259

[5] Sharon W, Majorie M, Christopher TVB, Rachel C, Suzanne P, Lauret F, Henri V (2010). Reliability of Antimalarial sensitivity depends on drug mechanisms of action. Journal of Clinical Microbiology, 48(5): 1651-1660

[6] Odhiambo RA, Odulaja A (2005). New Enzymatic assay parasite lactate dehydrogenase in diagnosis of malaria in Kaya. East African Medical Journal, 82(3): 111-117

[7] Perilla T, Gashaw M, Eva NK (2007). Potential and Utilization of thermophiles and thermostable enzymes in biorefining. Microbial Cell Factories, 6:9-41

[8] Marker MT, Hinnchs DJ (1993).Measurement of Lactate dehydrogenase activity of Plasmodium falciparum as an assessment of parasitemia. American Journal of Tropical Medicine, 48(2): 205-210

[9] Sibylie Gersti, Sophie Dukley, Ahmed Mukhtar, Martin De Smet, Samuel Baker, Jacob Maker (2010). Assessment of two malaria rapid diagnostic tests in children under five years of age, with follow up of false-positive pLDH test results, in a hyperendemic falciparum malaria area, Seirra Leone. Malaria Journal, 9: 28-39

[10] Pores Xsoler, J Panes, X Pares et al. (1987). Hepatic alcohol and aldehyde dehydrogenases in liver diseases. Alcohol and Alcoholism, Oxfordshire Supplement, 1:513-517

[11] Meter M, Moller G, Adamski J (2009).Perspectives in understanding the role of human 17β-hydroxysteroid dehydrogenase in health and disease. American New York Academy of Science, 1155:15-24

[12] WHO 2000a. New Perspective for Malaria Diagnosis.Report of a joint WHO/USID informal consultation, 25-27 October, 1999 World Health Orgaizatio, Geneva, WHO/MAL/2000, 1091

[13] Beadle, C., Long, G.W., Weiss W.R., McElroy , P.D., Maret S.M., Oloo A.J., Hoffman S.L. (1994).Diagnosis of Malaria by detection of Plasmodium falciparum HRP-2 antigen with a rapid dipstick antigen capture assay. Lancet, 343:564-568
[14] Humar, A., Ohrt, C., Harrigto, M.A., Pillal, D., Kain, K.C., (1997). ParaSight® F test compared with the Polymerase chain reaction and microscopy for the diagnosis of Plasmodium falciparum malaria in travelers. Am. J. Trop. Med. Hyg., 56:44-48
[15] Pattaasin, S., Proux, S., Chompasuk, D., Luwiradaj, K., Jacquier P, Looareesuwan, S., Nosten, F. (2003). Evaluation of a new Plasmodium lactate dehydrogenase assay (OptiMAL-IT®) for the detection of malaria. Trans. R. Soc. Trop. Med Hyg., 97: 672-674
[16] WHO 2000b. Management of severe malaria, a Practical Handbook, second ed.World Health Organization, Geneva.
[17] Piola, P. Fogg C., Bajuninwe, F., Biraro, S., Grandesso, F., Ruzagira, E., Babigumira, J., Kigozi et al. (2005). Supervised versus Unsupervised intake of six-dose artemeter-lumefantrine for treatment of acute, uncomplicated Plasmodium falciparum malaria in Mbarara, Uganda: a randomized trial. Lancet 365, 1467-1473
[18] WHO 1991. Basic Malaria Microscopy, World Health Organization, Geneva.
[19] Wongarichanalai, C. Chuanak, N. Tulyayon S., Thanoos-ingha, N., Laoboonchai, A., Thimasarn, K., Brewer, T.G., Heppner, D.G. (1999). Comparison of a repid field immunochromatographic test to expert microscopy for the detection of Plasmodium falciparum asexual parasitemia in Thailand. Acta Trop., 73:263-273
[20] Tjiltra E., Suprianto, S., Dyer, M., Currie, B.J., Anstey, N.M. (1999). Field evaluation of the ICT Malaria P.f/P.v immunochromatographic test for the detection of Plasmodium falciparum and Plasmodium vivax in patients with presumptive clinical diagnosis of malaria in Eastern Indonesia. Journal Clinical Microbiology, 37:2412-2417
[21] Mason D.P., Wogsrichanalai C., Lin, K., Miller, R.S., Kawamoto, F. (2000). The pan-malarial antigen detected by the ICT Malaria P.f/P.v. Immunochromatograpic test is expressed by Plasmodium malariae. Journal Clinical Microbiology, 39:2035.
[22] Gasser, R.A., Forney, J.R., Magili A.J., Sirichasinthop, J., Bautista C., Wongsrichanalai, C. (2000). Continuing progress in rapid diagnostic technology for malaria field trial performance of a revised immunochromatographic assay (OptiMAL®) detecting Plasmodium- specific lactate dehydrogenase. Malaria Diagnosis Symposium, xvth International Congress for Tropical Medicine and Malaria, Cartagen, a, Columbia, 20-25 August, 2000.
[23] Connor, D.H., Neafie RC, Hockmeyer, WT. (1976). MALARIA. Binford CH, Connor, DH, eds. Path ology of Tropical and Extraordinary Disease Washington, DC.Armed Forces Institute of Pathology, 273-283
[24] Pongponratn E, Turner GD, Day NP, Phu, NH.Simpson JA, Stepniewska K, Mai NT, Viriyavejakul P., Looareesuwa S, Hein TT, Ferguson DJ, White NJ (2003). An Ultrastructural study of the brain in fatal Plasmodium falciparum malaria. Am. J. Trop. Med. Hyg.69:345-359
[25] Nakazawa S, Looareesuwan S, Fujioka H, Pongponratn E, Luc KD, Rabbege J. Aikawa M. (1995). A correlation between sequestered parasitized erythrocytes in subcutaneous tissue and Cerebral malaria, American Journal TROP. Med.Hyg., 53:544-546

Section 3

Dehydrogenases and Physiological Role

Chapter 8

Dehydrogenase Activity in the Soil Environment

Agnieszka Wolińska and Zofia Stępniewska

Additional information is available at the end of the chapter

http://dx.doi.org/10.5772/48294

1. Introduction

The main purpose of the chapter is clarify description of the role of intracellular enzyme-dehydrogenase in the soil environment, as well as presentation of soil factors, influencing an enzymatic activity, by either stimulation or inhibition effect on soil dehydrogenase activity (DHA).

The most common laboratory procedure used for DHA determination is the method developed by Casida et al. (1964). According this method, specific dyes such as the triphenyltetrazolium chloride (TTC), that can specify the flow of electrons are useful indicators of electron transport system (ETS) activity. By the reduction of colorless, water soluble substrate (TTC) by dehydrogenases present in the soil environment, an insoluble product with red color (triphenylformazan-TPF) is formed. TPF can be easily quantified calorimetrically at the range of visible light (485 nm). This test however, reflected positive answer only at neutral range of pH and in presence of calcium carbonate for buffering soil system. Briefly, if the red colors of soil samples prepared for spectrophotometer analyses are more intensive, the measured level of DHA is higher. Consequently, soil samples without red colors or those with light red colors are characterized by lower DHA values.

Determination of DHA in the soil samples gives us large amount of information about biological characteristic of the soil. It was confirmed that although oxygen and other electron acceptors can be utilized by dehydrogenases, most of the enzyme is produced by anaerobic microorganisms. In other words, soil DHA strongly increases under anaerobic conditions.

Several environmental factors, including soil moisture, oxygen availability, oxidation-reduction potential, pH, organic matter content, depth of the soil profile, temperature, season of the year, heavy metal contamination and soil fertilization or pesticide use can affect significantly DHA in the soil environment. In the current chapter we would like to concentrate on precise description of mentioned factors effect on soil DHA level. Presented results of laboratory experiments were conducted on different soil types, representing dominant types of arable soils in Poland, in order to demonstrate changeability and variability of DHA at diverse soil environment.

2. Role of dehydrogenase activity in the soil environment

There are lots of enzymes in soil the environment, such as Oxidoreductases, Hydrolases, Isomerases, Lyases and Ligases. Each of them play key biochemical functions in the overall process of material and energy conversion (Gu et al., 2009).

Soil dehydrogenases (EC 1.1.1.) are the major representatives of the Oxidoreductase enzymes class (Gu et al., 2009). Among all enzymes in the soil environment, dehydrogenases are one of the most important, and are used as an indicator of overall soil microbial activity (Quilchano & Marañon, 2002; Gu et al., 2009; Salazar et al., 2011), because they occur intracellular in all living microbial cells (Moeskops et al., 2010; Zhao et al., 2010; Yuan & Yue, 2012). Moreover, they are tightly linked with microbial oxidoreduction processes (Moeskops et al., 2010). What is important dehydrogenases do not accumulate extracellular in the soil.

Dehydrogenases play a significant role in the biological oxidation of soil organic matter (OM) by transferring hydrogen from organic substrates to inorganic acceptors (Zhang et al., 2010). Many specific dehydrogenases transfer hydrogen to either nicotinamide adenine dinucleotide or nicotinamide adenine dinucleotide phosphate (Subhani et al., 2001). Throughout mentioned co-enzymes hydrogen atoms are involved in the reductive processes of biosynthesis. Due to this fact, the overall DHA of a soil depends on the activities of various dehydrogenases, which are fundamental part of the enzyme system of all living microorganisms, like enzymes of the respiratory metabolism, the citrate cycle, and N metabolism (Subhani et al., 2011). Thus, DHA serves as an indicator of the microbiological redox-systems and could be considered a good and adequate measure of microbial oxidative activities in soil.

Brzezińska et al. (2001) found that active dehydrogenases can utilize both O_2 and other compounds as terminal electron acceptors, although anaerobic microorganisms produce most dehydrogenases. Therefore, DHA reflects metabolic ability of the soil and its activity is considered to be proportional to the biomass of the microorganisms in soil. However, the relationships between an individual biochemical property of soil DHA and the total microbial activity is not always obvious, especially in the case of complex systems like soils, where the microorganisms and processes involved in the degradation of the organic compounds are highly diverse (Salazar et al., 2011).

3. Soil factors stimulating dehydrogenase activity

Among different environmental factors with special emphasis on enzymatic activities in the soil environment it is possible to screen some, which have positive impact on DHA. The most important soil factors stimulating soil DHA are described below.

3.1. Soil moisture

Life in the soil environment, as well as land use is related to alternate cycles of humidification and drainage (Wolińska & Bennicelli, 2010). Water availability strongly affects on soil microbial activity, community composition (Geisseler et al., 2011), and consequently on soil enzymatic activities. As soils dry, the water potential increases, and as well microbial activity as intracellular enzyme activity slows down (Geisseler et al., 2011). In the case of wet soils, increased moisture could bring into soil solution soluble OM, what might be responsible for increase of bacterial population number (Subhani et al., 2001). What is important, we should have consciousness that any compound, which alters the number or activity of microorganisms, could on the other hand affect on soil biochemical properties, and ultimately also on soil fertility and plant growth (Subhani et al., 2001).

A basic hydrophysical characteristic of soil is water retention, that can be described as a dependence between soil water content and soil water potential. Soil water content in the function of the soil water tension is described by pF curve, which provides information about the ability for water retaining by the soil pores at any given water tension, or conversely, how tightly a water is held between soil aggregates (Wolińska & Bennicelli, 2010).

The Figure 1 demonstrates diminishing trend for DHA behaviour at different soil moisture, described as water potential values. During this experiment gig set of soils (n=315), including all representatives among the most typical Polish mineral soils (*Eutric Cambisol, Eutric Histosol, Eutric Fluvisol, Mollic Gleysol, Orthic Podzol, Rendzina Leptosol, Haplic Phaeozem*) were investigated. However, each of soil unites displayed DHA reducing trend with increase of soil pF value, what means that maximum values of DHA in the soil profiles are indirectly connected with maximum soil moisture (pF 0).

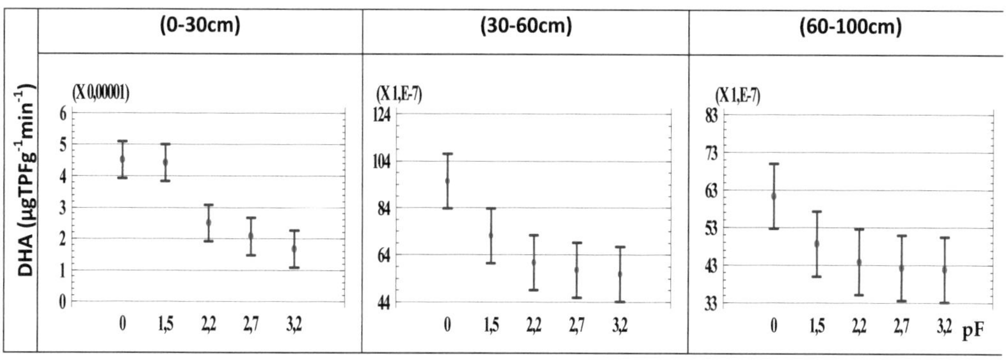

Figure 1. DHA ($\mu gTPFg^{-1}min^{-1}$) dependence from water potential (pF) at different mineral Polish soil types, during reoxidation (n=315), according to Wolińska (2010)

Statistical relationships between DHA and soil water content, described as pF value in the range of pF0 – pF3.2, determined by Wolińska & Bennicelli (2010) are presented in Table 1.

Founded significant negative relationships between DHA and pF are confirmed by our above mentioned observations, that DHA is strongly affected by soil moisture. These strong correlations are undoubtedly connected with the fact that the metabolism and the survival of soil microorganisms are also strongly impacted by the availability of water (Uhlirova et al., 2005), what is essential for microbial survival and activity. Consequently, low water availability can inhibit microbial activity by lowering intracellular water potential, and thus by reducing of hydration and enzymes activity (Wall & Heiskanen, 2003). Periods of moisture limitation may affect microbial communities through starvation. Thus, the most common environmental stress for soil microorganisms is perhaps drought (Wolińska & Stępniewska, 2011).

It was shown in many studies that DHA is significantly influenced by water content and dropped with the decrease of soil humidity. For example, Gu et al. (2009) observed higher DHA level (even by 90%) in flooded soil, rather than in non-flooded conditions. The higher DHA values in flooded conditions agreed also with results presented by Zhao et al. (2010) and Weaver et al (2012).

DHA response	Depth (cm)	pF
Rendzina Leptosols	0-20 50-60	-0.98*** -0.95**
Eutric Fluvisol	0-20 50-60	-0.97*** -0.22 n.s.

*, **, *** - indicate significance at the 5, 1 and 0.1% level, respectively, n.s. – not significant differences

Table 1. Statistical significance of differences between DHA and pF described by correlation coefficient (R) (95% LSD method, n=15), according to Wolińska & Bennicelli, 2010

The decline of DHA with an increase of pF value, could be also explained by the fact, that flooding of soil with water significantly increased the electron transport system (Wolińska & Stępniewska, 2011). Dehydrogenases however, are responsible for electron transport in the soil environment. It was also reported that DHA is higher in flooded, anaerobically soils, than aerobically incubated soils (Trevors, 1984; Subhani et al., 2001).

3.2. Soil aeration state (redox potential and oxygen diffusion rate)

Oxygen diffusion rate (ODR) is usually considered to be the most critical proximal regulator of microbial activities (Hutchinson, 1995). Moreover, it is often assumed that a decrease of soil water content (higher value of pF), cause a significant (P<0.001) increase of ODR and redox potential (Stępniewski et al., 2000; Wolińska & Bennicelli, 2010). The available literature shows that low ODR level, ranged below its critical values (35 $\mu g\ O_2\ m^{-2}s^{-1}$), is favorable and optimal for DHA (Stępniewski et al., 2000; Brzezińska et al., 2001; Wolińska & Bennicelli, 2010).

We confirmed that dehydrogenases are sensitive enzymes, indirectly depended on the soil aeration status (Wolińska & Bennicelli, 2010). Based on performed measurements we found, that pF constitutes a significant factor, determining ODR in the soil environment, as well as its DHA level (P<0.01). The reoxidation processes, occurring in the direction from pF 0 to pF

3.2, were the reason of DHA inhibition and stimulation of ODR level in the *Rendzina Leptosols* and *Eutric Fluvisol* soil samples (Fig. 2). We also stated that soil DHA at pF 3.2 was lower by about 60.86%, in comparison to the activity estimated at pF 0.

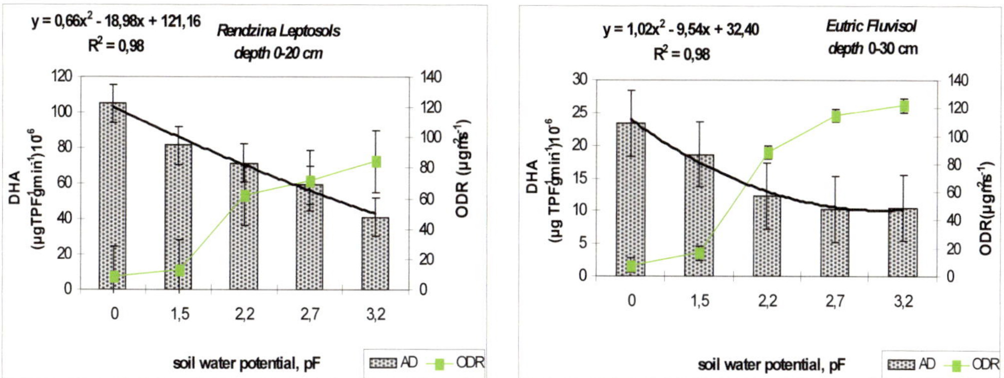

Figure 2. The response of soil DHA to varied aeration factors (pF and ODR), at surface layers of *Rendzina Leptosols* and *Eutric Fluvisol* , during reoxidation process (according to Wolińska & Bennicelli, 2010). Averaged values of three replicates with standard deviations are presented

The Figure 3. demonstrates that low oxygen diffusion rate (2.8-25 µg O_2 m^{-2} s^{-1}) was optimal for DHA, what was also confirmed by correlation coefficient (Wolińska & Bennicelli, 2010; Wolińska, 2010). Our results and founding's are compatible with work of Stępniewski et al. (2000), Brzezińska et al. (2001), and Yang et al. (2005).

Statistical relationships between DHA and ODR, determined for two soil types (*Rendzina Leptosols* and *Eutric Fluvisol*) by Wolińska & Bennicelli (2010) are presented in Table 2. At every case negative correlations DHA-ODR were determined.

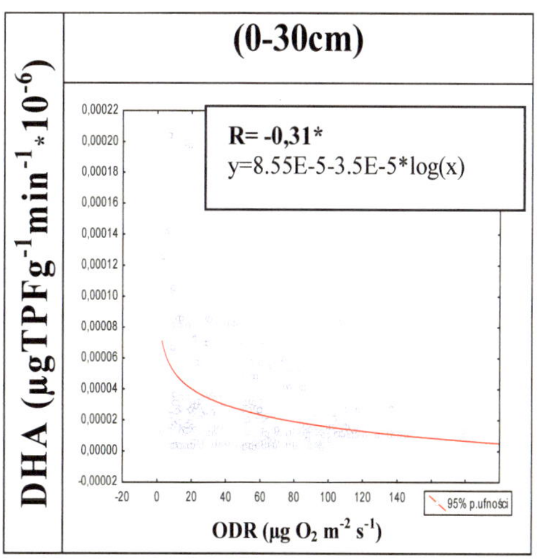

Figure 3. Relationship between DHA (µgTPF$g^{-1}$$min^{-1}$ * 10^{-6}) and ODR (µg m^{-2} s^{-1}), in surface layer of different mineral Polish soil types (n=315, *P*<0.05), according to Wolińska (2010)

DHA response	Depth (cm)	ODR
Rendzina Leptosols	0-20 50-60	-0.90** -0.84*
Eutric Fluvisol	0-20 50-60	-0.96** -0.16 n.s.

*, **, *** - indicate significance at the 5, 1 and 0.1% level, respectively, n.s. – not significant differences

Table 2. Statistical significance of differences between DHA and ODR described by correlation coefficient (R) (95% LSD method, n=15)

Redox potential (Eh) is the next, important, environmental factor, which expresses the tendency of an environment to receive or to supply electrons in solution (Stępniewski et al., 2005). The well-oxygenated soils are characterized by high values of Eh (600-800 mV), in quite well-oxygenated soils Eh ~ 500-600 mV, whereas in anaerobic conditions drop of Eh below 300 mV or even lower values were observed (Pett-Ridge & Firestone, 2005; Stępniewski et al., 2000).

It is well known, that Eh play a crucial role in regulating microbial activity as well as community structure (Pett-Ridge & Firestone, 2005; Song et al., 2008), and affecting on soil enzymatic activity, especially DHA. Brzezińska et al. (1998) indicated that among all aeration parameters, Eh plays the most important role in determining soil DHA level. Similar conclusions were also reported by Włodarczyk et al. (2001) and Menon et al. (2005).

We founded significant negative relationships between DHA and Eh (Fig. 4) at surface layers of *Mollic Gleysols, Eutric Fluvisols, Rendzina Leptosols* and *Haplic Phaeozems*, where determined correlation coefficients equaled as follows: r=-0.91*, r=-0.43*, r=-0.47** and r=-0.48** (Wolińska, 2010).

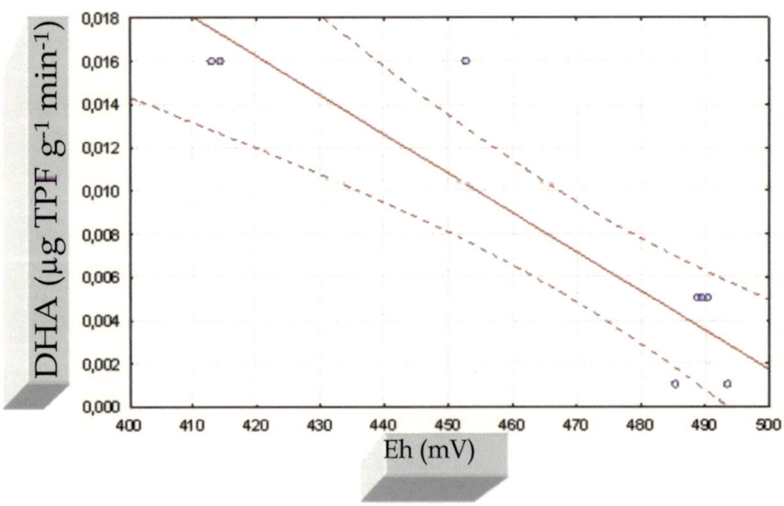

Figure 4. Relationship between soil DHA level and Eh at *Mollic Gleysol* (n=9, r=-0.91*), according to Wolińska (2010)

Negative correlations DHA-Eh were also described by Brzezińska et al. (1998), who determined r=-0.75***, r=-0.83*** and r=-0.87*** for temperature 10, 20 and 30°C, respectively, and by Stępniewski et al. (2000), and Nyak et al. (2007).

Mentioned relationships DHA-Eh have significant negative character, what means that increase of soil DHA level is indirectly connected with decrease of Eh values, as most of microorganisms, which are responsible for DHA prefer rather anaerobic conditions, and belong to obligate anaerobes. What is more, anaerobic conditions are consequence of flooding and decrease of oxygen availability in soil environment. Competition for oxygen limits aerobic processes and the subsequent oxygen deficiency creates local anaerobic microsites, which stimulates growth of anaerobic bacteria (Wolińska & Stępniewska, 2011), and finally DHA. Also, in the absence of oxygen in the soil a decline of Eh and the reduction of oxidized forms (nitrate, Mn^{4+}, Fe^{2+} and SO_4^{2-}) takes place. Bohrerova et al. (2004) reported that the most common ions forming the redox couples of soil include NO_3^-/NO_2^-, Fe^{3+}/Fe^{2+}, and Mn^{4+}/Mn^{2+}. In the literature data, it was also assumed that DHA is strongly affected by both Fe as Mn presence in the soil (Brzezińska et al., 1998; Włodarczyk et al., 2002).

3.3. Organic matter content

Soil organic matter (OM) has important effects not only on soil enzymes activities but first of all on microorganisms activities. Soil OM has been considered as an indicator of soil quality (similarly like dehydrogenases,) because of its character of nutrient sink and source that can enhance soil physical and chemical properties, and also promote biological activity (Salazar et al., 2011). Interestingly, not only amount of OM in the soil is important but most of all its quality, as OM affects the supply of energy for microbial growth and enzyme production (Fontaine et al., 2003).

It is evident that soil enzymatic activity is strongly connected with soil OM content. The higher OM level can provide enough substrate to support higher microbial biomass, hence higher enzyme production (Yuan & Yue, 2012). Several authors reported positive correlation between DHA and OM content (Chodak & Niklińska, 2010; Moeskops et al., 2010; Romero et al., 2010; Zhao et al., 2010; Yuan & Yue, 2012).

Zhang et al. (2010) indicated also that as well DHA and $CaCO_3$ correlated with OM content, and what is more DHA, OM and $CaCO_3$ were correlated with each other in their spatial distribution, suggesting that abundant OM content contributed to the formation of pedogenic calcium carbonate.

Salazar et al. (2011) hypothesized that activities of dehydrogenases in different forest ecosystems are involved in the carbon cycling, and they also reported their positive relationships. Dehydrogenases, are highly associated with microbial biomass (MB), which in turn affects on decomposition of OM and the release of $CaCO_3$ (Zhang et al., 2010).

We also investigated effect posed by total organic carbon (TOC) and response of DHA in the agricultural used *Mollic Gleysol*, taken from Kosiorów village (SE part of Poland). We determined significant (*P*<0.0001) correlation between TOC-DHA (Fig. 5). Mentioned strong relationship was also confirmed by high value of correlation coefficient (r=0.99***). In our laboratory conditions the optimal value of TOC content for reaching maximal values of soil DHA was its level above 25%.

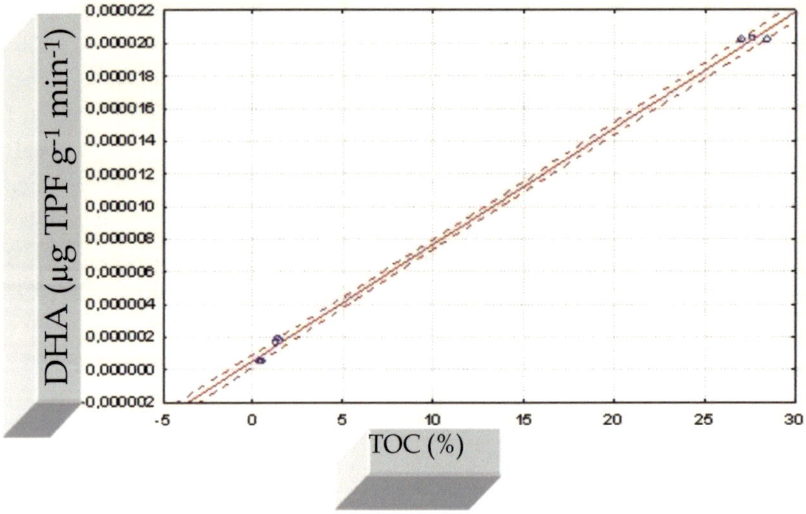

Figure 5. Relationship between DHA and TOC content in the *Mollic Gleysol* (n=9, r=0.99***), according to Wolińska & Stępniewska (unpublished data)

Analogically to our investigations also Koper et al. (2008) found and reported strong significant relationships between DHA and organic carbon content in *Haplic Podzol* soil samples, and they described mentioned correlations by r coefficient ranged between 0.56* and 0.98*.

The study of Kumar et al. (1992) indicated that DHA displayed the close, positive correlations not only with OM content but also with fungal population abundance in four forest stands (two at low and two at higher attitudes).

High correlation coefficient reported for enzymatic activities and TOC level suggested an important role of these enzymes in transformations of basic components of soil OM (Wolińska & Stępniewska, 2011). There is in general agreement with previous results indicated by Pascual et al. (2000), who found that soils characterized with low microbial and biological activity (e.g. low microbial carbon and low respiration rate), also display the lowest values of DHA.

Summarizing, the higher content of OM, the more active the soil microorganisms. Microorganisms accelerate the degradation of OM, which is reflected in soil respiration and release of carbon dioxide from the rizosphere (Zhang et al., 2010), thus DHA is positively correlated with OM content. Similarly, increase of DHA with higher microorganisms number was reported (Fontaine et al., 2003).

3.4. pH

The literature data, currently available, referring to the connections between DHA and soil pH are still ambiguous.

Generally, enzyme activities tend to increase with soil pH (Błońska, 2010; Moeskops et al., 2010) – please put a space before Moeskops. Błońska (2010) determined significant positive correlation (r=0.50***) DHA-$pH_{(water)}$ in the pH range 3.67-5.88.

Fernandez-Calviño et al. (2010) noted significantly positive correlations among soil DHA and pH in the range of 4.1 (pH_{KCl}) and 4.9 (pH_{water}), suggesting that acidity suppressed potential enzyme activity.

Adequately, a study by Levyk et al. (2007) demonstrated that acidic conditions in the pH range between 1.5–4.5 resulted with strong DHA inhibition in relation to alkaline soils, whereas Ghaly & Mahmoud (2006) noted that under acidic conditions with pH less than 6.5, the rate of TTC - specific substrate for DHA, did not decrease.

According to Frankenberger & Johanson (1982), the weakening of enzymatic activity in soil with the increase of soil acidity is the effect of destroying ion and hydrogen bonds in enzyme active centre.

On the other hand, study performed by Włodarczyk et al. (2002) indicated maximum DHA at pH 7.1, similarly to the work of Ros et al. (2003), where optimum for DHA was noted for pH 7.6-7.8. Also Brzezińska et al. (2001) reported that the best pH conditions for DHA ranged between 6.6-7.2.

Natywa & Selwet (2011) noted positive correlation between DHA and pH in soils under maize growth at pH range from 5.17 to 7.27.

Trevors (1984) concluded that very little DHA is observed below pH 6.6. and above pH 9.5. According to Nagatsuka & Furosaka (1980) the optimum range for DHA is contained between 7.4–8.5. However we should realize that many heterogeneous soil types might not be included in mentioned above range.

Our investigations, performed on *Mollic Gleysol* sample (from Kosiorów village) indicated however, that DHA also reached high level at lower pH values–between 5.5-5.73 (Fig. 6). Significant inhibition of DHA (even by 95%) we scarcely noted when soil pH was above 5.75.

It is often assumed that pH may affects soil enzymes level in three different ways (Shuler & Kargi, 2010):

1. by changing in the ionic form of the active sites of the enzymes, which consequently affect the enzyme activity and hence the reaction rate,
2. by altering the three-dimensional shape of enzyme, and
3. by affecting the affinity of the substrate to the enzyme.

Thus, the pH factor is considered to be the best predictor of DHA in the soil environment (Quilchano & Marañon, 2002; Moeskops et al., 2010).

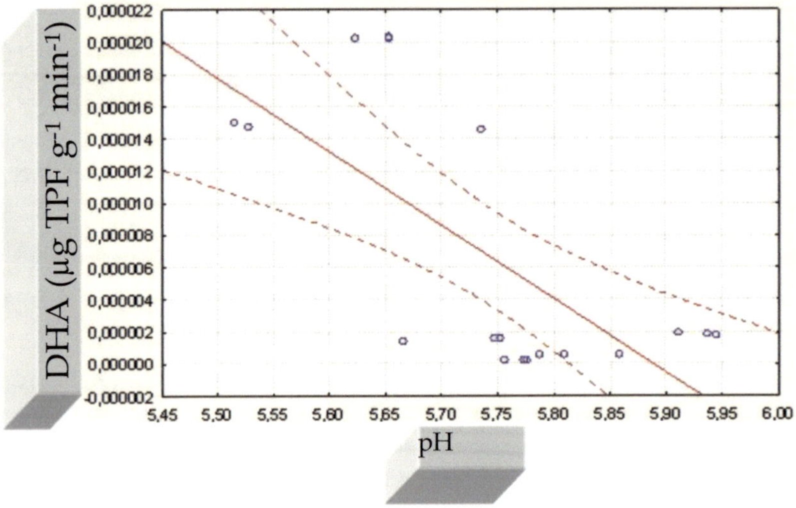

Figure 6. Relationship between DHA and pH values in the *Mollic Gleysol* (n=18, r=-0.70**), according to Wolińska & Stępniewska (unpublished data)

3.5. Temperature

Many researchers have studied effect posed by temperature incubation on soil DHA and/or on soil microorganisms abundance (Subhani et al., 2001; Ghaly & Mahmoud, 2006; Trasar-Cepeda et al., 2007). Taking into account the important fact that DHA is found inside the viable soil microbial cells only, its activity must be the highest at a temperature close to optimum temperature for microorganisms growth and their development (Wolińska & Stępniewska, 2011).

It is known that, the rate of enzyme catalysis generally increases with increase in temperature until the unfavorable temperature, at which enzyme becomes denaturized and hence its activity reduces (Wolińska & Stępniewska, 2011).

Our investigations were concentrated on investigations of DHA changeability at temperature range 5-30ºC, what reflect natural changes of soil temperature during seasons. Surface layer (0-20 cm) of *Mollic Gleysol* was used for experiment. Soil samples were incubated at the following temperatures: 5, 10, 20 and 30ºC. DHA was measured after 30 h incubations at proper temperature and after ethanol extractions. Absorbance was tested at λ=485 nm (UV-1800 Shimadzu). Received results are presented in Fig. 7.

We found growing, linear trend for DHA with increase of temperature at the range from 5 to 30ºC, what we described by R^2=0.97. The differences between DHA level estimated at 5 and 30ºC were significant (P<0.01), analogically like between 5 and 20ºC (P<0.05). The lowest values of DHA at 5ºC were found, where DHA equaled 1.259 (µg TPF g^{-1} $min^{-1} \cdot 10^{-6}$), whereas the same soil sample incubated at 30ºC reached DHA level of 3.149 (µg TPF g^{-1} $min^{-1} \cdot 10^{-6}$), what was by c.a. 60% higher in relation to DHA level from 5ºC. Quite high DHA level (2.741 µg TPF g^{-1} $min^{-1} \cdot 10^{-6}$) was also estimated at 20ºC, where mentioned value was only by 13% lower than maximum DHA, from 30ºC.

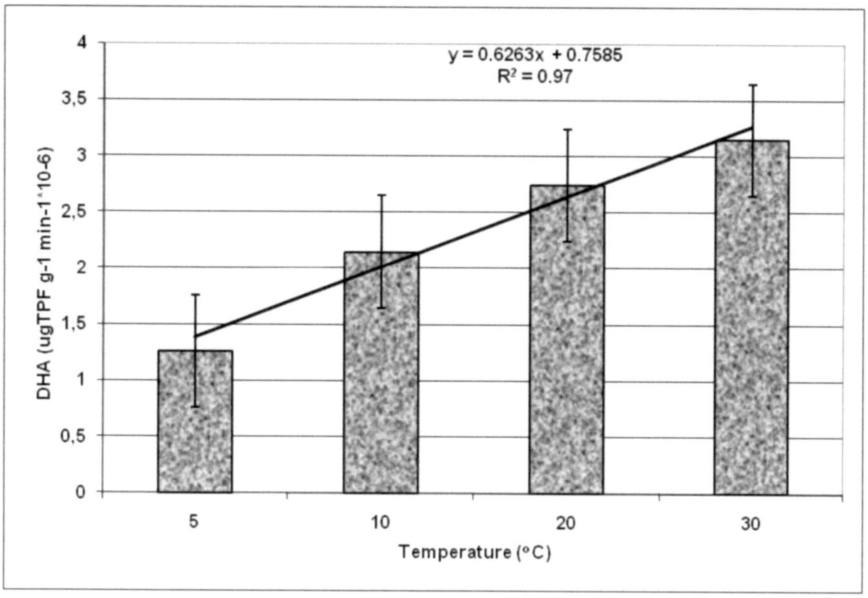

Figure 7. The dependence between DHA and temperature incubation in the *Mollic Gleysol*, according to Wolińska & Stępniewska (unpublished data). Averaged values of three replicates with standard deviations are presented

Casida et al. (1964) indicated that incubation of soil samples at 37ºC increased of soil DHA above the value normally observed at lower temperatures.

Trevors (1984) described positive significant correlation among DHA and temperature in the range from 5 to 70ºC and determined r coefficient on the level of 0.99*. Moreover, study by Trasar-Cepeda et al. (2007) reported that increased temperatures up to 57-70ºC enhanced the product formation in the reaction catalyzed by soil dehydrogenases increased with, explained by the fact that specific substrate (TTC), used for DHA determination, is chemically reduced at high temperatures.

Analogically, Subhani et al. (2001) noted positive correlation in soil samples incubated at 10, 25 and 40ºC (under constant moisture – flooded conditions), what confirmed by r=0.82*.

As suggested by Cirilli et al. (2012) optimum temperature for soil DHA is 30ºC, what is in agreement with our findings. Similarly, Brzezińska et al. (1998) indicated that under laboratory conditions DHA demonstrated the highest activity at 28-30ºC.

3.6. Season of the year

Seasonal variations in both microbial biomass and soil enzymatic activities reflect the combine effects of temperature, moisture, substrate availability and other environmental factors. Dehydrogenases belong to the enzymes displaying strong fluctuations in their activities caused by season of the year, as they are in close relationships with dynamic of microbial activity.

Yuan & Yue (2012) stated the highest DHA level in autumn season and the lowest value of DHA in winter time. The study performed by Piotrowska & Długosz (2012) indicated that DHA level in *Luvisols* revealed significantly higher values in April (by 96%) than in August, probably due to intensive winter wheat growth with an increased secretion of substrates such as polysaccharides, organic acids, which may have affected the growth and activity of microorganisms.

Similarly, our investigations demonstrated the highest level of DHA in *Eutric Fluvisol* sample taken in May (0.0087 μg TPF g^{-1} min^{-1}), than in the same soil type taken in October, where DHA was reduced by 42.5 % (Fig. 8). Quite high level of DHA (lower by 14.9% from its maximum reaching in May) we also noted in July. Moreover, we did not found significant differences (*P*>0.05) in DHA values during autumn season, where DHA remained on similar level equaled 0.000598 (μg TPF g^{-1} min^{-1}) and 0.0005 (μg TPF g^{-1} min^{-1}), for September and October, respectively.

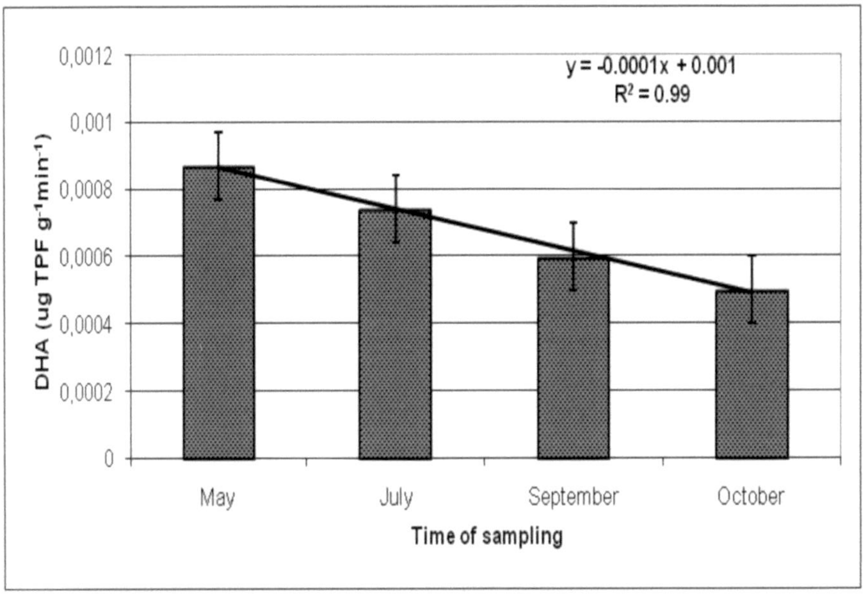

Figure 8. Effect of the season of the year on soil DHA in the *Eutric Fluvisol*, according to Wolińska & Stępniewska (unpublished data). Averaged values of three replicates with standard deviations are presented

Analogical trend like our observations, related to high Oxidoreductases activity at the time form spring to autumn was noted by Januszek (1993). A study by Włodarczyk (2000) performed on *Orthic Luvisol* sample, showed that DHA demonstrated seasonal pattern and reached the highest values in September, whereas the lowest in winter time. Similarly effect noted Tripathi et al. (2007), who indicated maximum DHA in September and its reduction in January.

Spring season is strongly connected with increase in microbial activity, intensification of oxido-reduction reactions and temperature change, what is indirectly impacted with DHA, and is the reason of slight DHA increase during this time. Moreover, taking into account that DHA is present inside viable microbial cells, its activity must be the highest at temperature 20-30ºC (temperature characteristic for summer and early autumn), close to the optimum temperature for microbial growth, activity and development (Wolińska & Stępniewska, 2011).

4. Soil factors inhibiting dehydrogenase activity

Some of environmental factors have ability to affect negatively on DHA, by reducing its activity. In the role of enzyme inhibitors usually different molecules are involved, which by binding to enzymes activation sites are the reason of prohibition the enzymes from catalyzing its reaction, and finally decrease their activity. The most important soil factors inhibiting soil DHA are described below.

4.1. Depth of the soil profile

Depth of the soil profile is one of the most known and popular environmental factor reducing soil DHA level. It is well known that the highest microorganisms abundance is in the surface layer of the soil profile (till to the depth of 30 cm), at the deepest part of the soil the number of microbial cells is limited, and consequently also DHA level display diminishing trend.

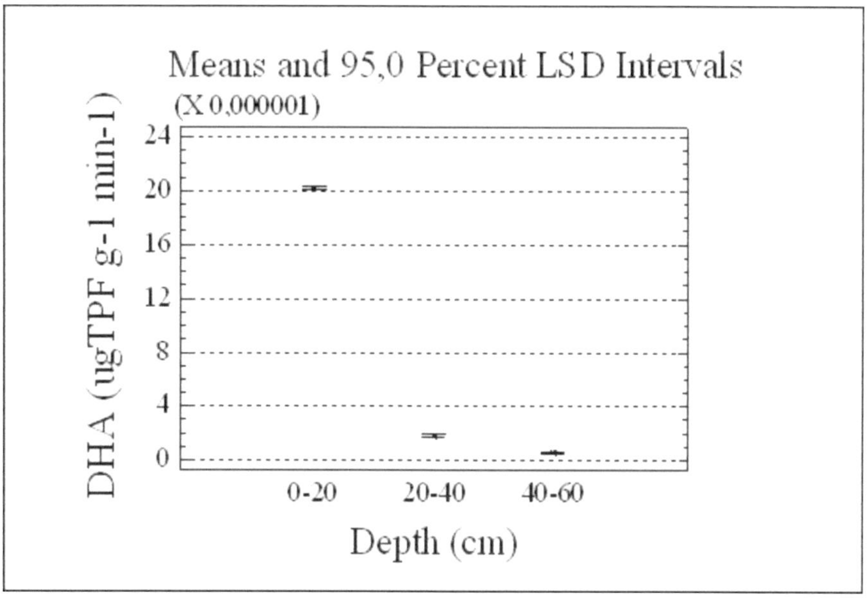

Figure 9. DHA (µgTPFg^{-1}min^{-1}) at different depth of the *Mollic Gleysol* profile (n=18, *P*<0.001), according to Wolińska founding (unpublished data)

The confirmation of the above statement might be the Fig. 9, where effect of depth on DHA in *Mollic Gleysol* is presented. The highest level of DHA we noted in surface layer (0-20 cm), whereas at the deepest part of the soil profile (40-60 cm) DHA was reduced by 95%, in relation to the surface layer. This trend is undoubtedly connected with presented in literature data and mentioned above spatial distribution of soil microorganisms (Agnelli et al., 2004; Levyk et al., 2007; Wolińska, 2010), and its preference to inhabiting the surface layers, where optimum conditions for its growth and development are guaranteed.

Our results are in agreement and might be supported by the findings of Xiang et al. (2008), who observed that DHA was roughly 4-fold higher in surface (till to 5 cm depth), than in subsoil (90-100 cm). It was also suggested by study of Gajda (2008), that values of DHA noted in the anthropogenic soil, till depth to 4 cm were by c.a. 30% higher in relation to the deeper part of the soil profile.

Brzezińska (2006), reported even 9-fold increase of DHA in the surface layer of the soil, than in the subsurface parts and 25-fold higher enzymatic activity in surface than in subsoil.

Generally it is possible to state, that both diversity, abundance as distribution of microorganisms are more even under oxic (surface layers) conditions, relative to anoxic (deeper layers) conditions (Fierer et al., 2003; Wolińska & Stępniewska, 2011).

4.2. Fertilization and pesticide amendment

Organic and inorganic fertilizers are commonly used to increase nutrient availability (Macci et al., 2012). The balanced fertilization of major elements (N, P, K) for plant nutrient could be beneficial for the growth of plant aboveground parts and roots (Chu et al., 2007), and also for improvement of soil structure (Macci et al., 2012).

However, fertilization could affect on the population of soil microorganisms and consequently soil enzymatic activities. It is often assumed, that inorganic fertilizers had relatively less effect on soil enzymes activity than organic fertilizers (Chu et al., 2007; Xie et al., 2009; Romero et al., 2010). Macci et al. (2012) noted, that DHA usually reached higher level in the organic treatments.

As was suggested by Chu et al. (2007) and Xie et al. (2009) long-term balanced fertilization greatly increased DHA level in the soil environment, rather than nutrient-deficiency fertilization. Zhao et al. (2010) indicated, that soils with higher fertility are more capable of maintaining the original biological functions (i.e. have a higher functional stability).

On the other hand, Moeskops et al. (2010) compared the effect of organic and conventional farming practices on soil enzymatic activities. On the organic farms, soil fertility was maintained mainly with composted OM, in contrast to conventional farmers, who combined fresh manure and chemical fertilizers, and typically applied large amounts of pesticides. As a consequence, a strong negative impact of intensive fertilizer and also pesticide use on DHA was demonstrated (Moeskops et al., 2010).

Soil DHA is an indicator of soil quality and microbial activity and also is the most frequently used to determining the influence of various pollutants (like pesticides or excessive fertilization) on the microbiological quality of soils (Xie et al., 2009; Tejada et al., 2010).

Despite the fact that pesticides are important tools in agriculture that help to minimize economic losses caused by weeds, insects and pathogens, they also are recognized as a source of potential adverse environmental impacts (Tejada et al., 2010). It is often assumed that less than 0.3% of the pesticide reaches its target pest, the remaining 99.7% is released to the environment, representing a potential hazard for non-target organisms (Muñoz-Leoz et al., 2011).

Stepniewska et al. (2007) noted the relationship between soil DHA and Fonofos (Stauffer Chemical Co., Westport, USA) concentration in the *Mollic Gleysol*. In the investigated samples influence of pesticide on soil enzymatic activity started to be observed after one week of incubation, but since 14th day to the end of experiment this effect was significant and noticeable (Fig. 10). Generally, 1μg g^{-1} dose of Fonofos was responsible for about 26% inhibition of soil DHA, whereas ten times higher factor reduced activity for 46.6% at 21st day of incubation time, later fall of enzymatic activity ranged from 22.5% to 30% in relation to the control samples was considered.

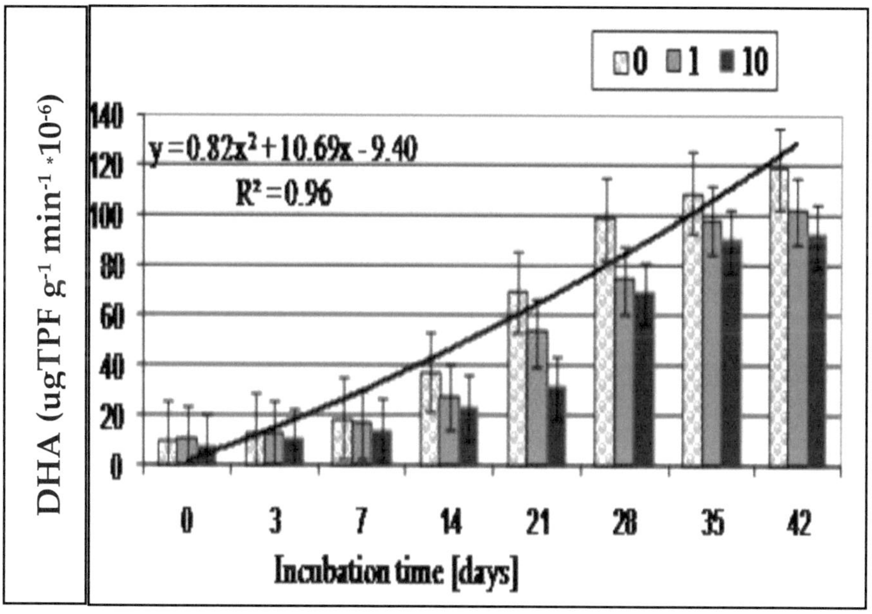

Figure 10. Dynamic of DHA during incubation at 20ºC (0 - control, 1 - 1μg g^{-1} Fonofos supplement; 10 - 10 μg g^{-1} Fonofos supplement), according to Stępniewska et al., 2007. Averaged values of three replicates with standard deviations are presented

Our results suggest a negative effect of Fonofos on soil DHA in the first stage after application (1-7 day), later an initial, almost linear growth of DHA was observed and the final day of incubation resulted in significant extension of DHA, presumably because the process of Fonofos decomposition in the soil environment was almost completely finished.

Tejada et al. (2010) indicated that MCPA herbicide rate of 1.5 l ha^{-1} (manufactures rate recommended) was the reason of 39.3% soil DHA inhibition, what suggest that the MCPA caused toxic effect on soil enzymatic activity. A field half-life of MCPA ranged from 14 days to 1 month, dependently on soil moisture, pH and microorganisms abundance. The most important soil factor in predicting MCPA effect on soil enzyme activities is pH, as at acidic conditions persistence of pesticide may last even 5 years, whereas at alkaline pH only 6 days. Moreover, decrease of soil moisture and reduction of microbial abundance influence on elongation of MCPA decomposition process.

We also studied effect posed by MCPA (Organika Sarzyna, Poland) on soil DHA behavior (Fig. 11). The following MCPA dosage were introduced into the soil samples: 0.165; 0.30 and 3.3 mg MCPA per g of soil. Non-amended with pesticide soil sample was marked as 0 and used as a control. As a result of realized experiment we found linear inhibition of DHA by increasing MCPA doses (R^2=0.99). Decrease of DHA level at 3.3. mg g^{-1} MCPA dose by c.a. 38.5%, in comparison to the control sample, was noted. However, registered inhibition was not significant (*P*>0.05). Our conclusions are comparable with results presented by Tejada et al. (2010).

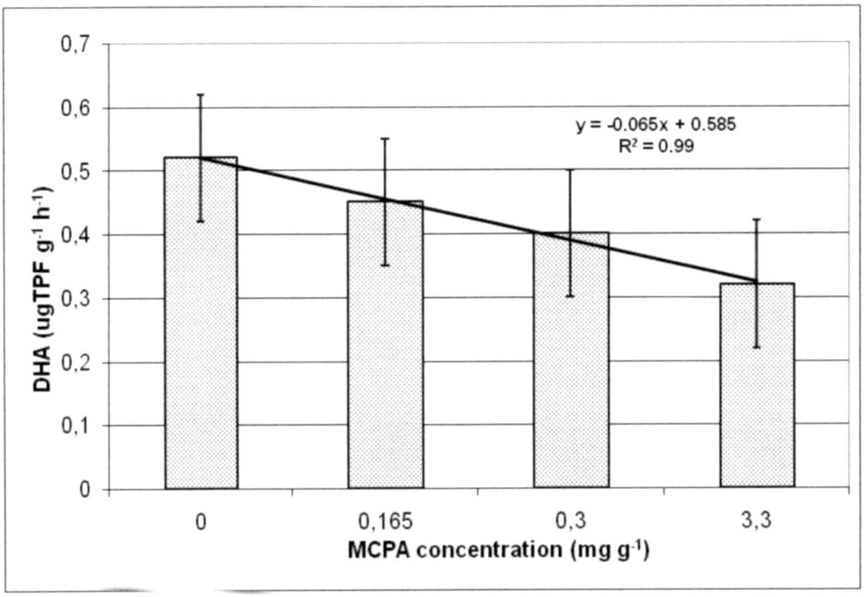

Figure 11. Effect of MCPA pesticide on DHA level in *Eutric Fluvisol*, according to Stępniewska founding (unpublished data). Averaged values of three replicates with standard deviations are presented

Other pesticide, which we take into account in our investigations was Glyphosate – commonly used by Polish farmers (in the form of RUNDUP), a broad spectrum, non-selective, systemic and post-emergence herbicide, widely popular in soil cultivation, forestry, rights-of-way and aquatic systems to prevent grass and weeds competition with plant seedlings (Bennicelli et al., 2009). At low doses it is used as a plant growth regulator.

Glyphosate (Monsanto Co., USA) is a polar substance that is highly soluble in water (12 g l^{-1} at 25°C), and insoluble in most organic solvents. In soil is moderately persistent; its half-life is reported between 1 to 174 days (Bennicelli et al., 2009). Glyphosate in soil is transformed to aminomethylphosphonic acid (AMPA), which is non-persistent metabolite. As a effect of mentioned transformations and in presence of dehydrogenases (microorganisms), Glyphosate give CO_2 and H_2O (Forlani et al., 1999). Glyphosate degradation in soil is mainly the reason of microbial activity, while the chemical decomposition and photolysis play a minor role (Bennicelli et al., 2009). As was reported by Zabaloy et al. (2008), Glyphosate, as an organophosphonate can be used as a source of P, C or N by either gram-positive as gram negative bacteria.

The purpose of our study was to research the influence of Glyphosate on soil DHA in the *Mollic Gleysol* (from Wieprz river valley), *Eutric Fluvisol* (from Vistula river valley) and *Terric Histosol* (from Bystrzyca river valley), taken from surface layer (0-20 cm). Soil samples were enriched with Glyphosate, as follows: with 1 µg (first combination), and 10 µg (next version), and 0 µg (control) of pesticide per 1g of soil. Thus prepared samples were incubated in thermostatic chamber at 20°C. Received results are presented in Fig. 12.

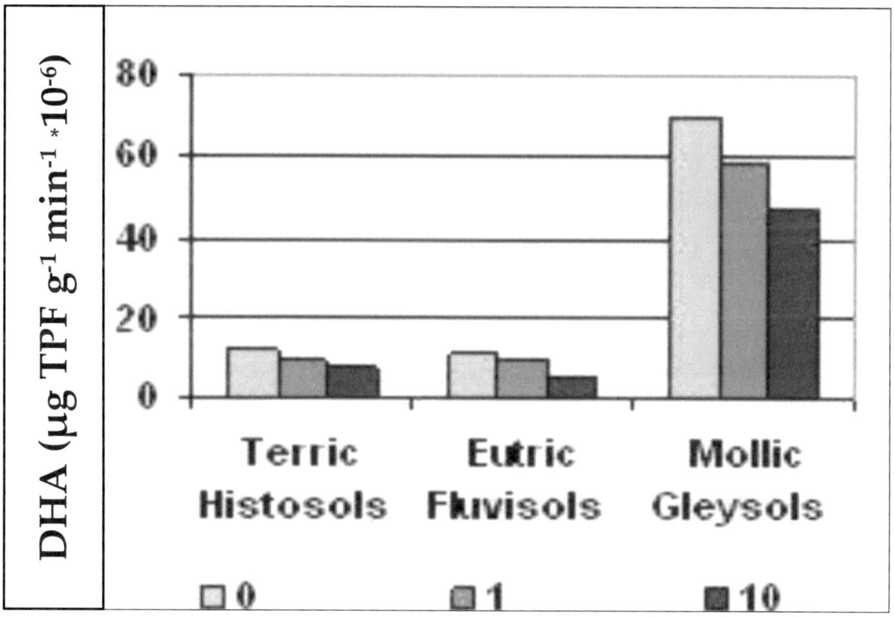

Figure 12. Mean DHA in *Terric Histosols, Eutric Fluvisols* and *Mollic Gleysols,* in three combinations of Glyphosate doses: 0-control; 1-1µg g^{-1}; 10-10µg g^{-1} (according to Bennicelli et al., 2009)

We found that both 1 and 10 µg of Glyphosate additions to soils caused a decrease of DHA, dependently on the pesticide doses. The strongest effect of Glyphosate was observed in *Terric Histosols* and *Eutric Fluvisol* (10 µg g^{-1} of soil),where reduction of DHA by 33-47%, relative to control soils (non-amended with Glyphosate), was noted. The most resistant to

Glyphosate supplement seemed to be *Mollic Gleysol,* in 10 µg g^{-1} of soil dose, where DHA dropped by c.a. 24%.

Suggested by us inhibition effect, may be supported by founding's of Zabaloy et al. (2008), who in typical Ardiudoll from Argentina observed reduction of DHA for about 48%, as an effect of Glyphosate contamination, in comparison to control sample.

Results, suggesting inhibitory pesticide effect on DHA level are also in agreement with those obtained by other plaguicides such as: chlorpyrifos (Kadian et al., 2012), or vermicompost (Romero et al., 2010). Moreover, Muñoz-Leoz et al. (2011) noted that DHA was inhibited by 14%, as a effect of application 5 mg kg^{-1} tebuconazole fungicide dosage.

Conversely, others have found also different results. For example, Tejada et al. (2011) noted insignificant (by 10%) growth of DHA, when the Prochloraz fungicide applied to the soil increased, possibly because the fungicide is commonly used by bacterial communities, as a source of energy and nutrients. Also Andreá et al. (2003), noted that DHA was slight higher after month from Glyphosate application. In that case authors reported, that Glyphosate stimulated DHA, which means that the herbicide might stimulate the soil oxidative processes.

4.3. Heavy metals presence

Heavy metals, even though they are natural constituents of soil, could have long-term hazardous impacts on the health of soil ecosystems, and adverse influences on soil biological processes (Pan & Yu, 2011). Generally, it was assumed that heavy metals can reduce enzyme activity by interacting with the enzyme-substrate complex, denaturing the enzyme protein or interacting with the protein-active groups, they could also affect the synthesis of enzyme microbial cells (Pan & Yu, 2011).

Xie et al. (2009) noted that Cu of 100 mg kg^{-1} could suppress DHA significantly, while Cd of 5 mg kg^{-1} had relative greater influence on soil microbial diversity, what suggest that the effect of each soil pollutant on soil microbes and their enzymatic activities was specific. On the contrary, a study by Fernandez-Calviño et al. (2010) indicated adverse effect of Cu on DHA ($r=-0.24$, $P<0.01$). Threshold Cu concentrations at which changes in the enzyme activities became evident were 150-200 mg total Cu kg^{-1} and 60-80 mg bioavailable Cu kg^{-1}.

A study by Pan & Yu (2011) undertaken with brown soil, showed that DHA was significantly lower by 37.8% and by 51.1% in Cd and Pb treatments, than in control. Moreover, mentioned researchers noted that the effect of Cd and Pb combined on DHA were higher, than Cd or Pb alone.

We also investigated effect posed by Cd (2 and 20 mg kg^{-1}) on soil DHA (Fig. 13). Incubation of soil material with mentioned Cd doses lasted 42 days. After that DHA was determined via Casida et al. (1964) method.

We observed that Cd presence at concentration of 2 mg kg^{-1} had stimulating effect on soil DHA level, and we noted increase of DHA by 8.8%, in comparison with control sample (without Cd contamination). However, 10-fold higher Cd amendment (20 mg kg^{-1})

consequence with strong DHA inhibition, by as follows: 29.4% and 35%, in relation to control and 2 mg kg^{-1} sample, respectively. Observed inhibition effect was probably caused by Cd interaction with enzyme-substrate complex, what resulted with strong decrease of DHA level.

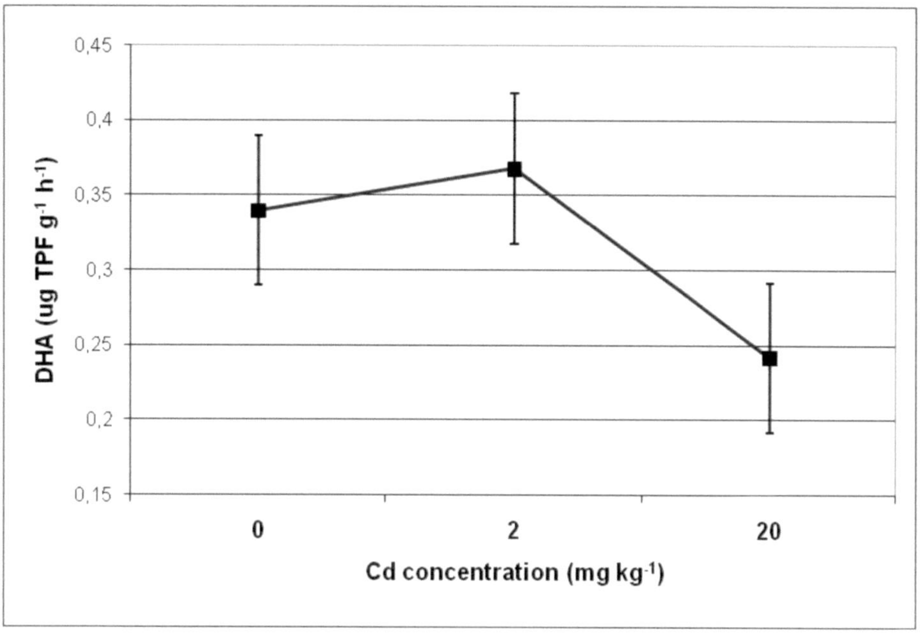

Figure 13. Effect of Cd on DHA in *Eutric Fluvisol* (according to Stępniewska & Wolińska, unpublished data) from Cd introduction into the soil. Averaged values of three replicates with standard deviations are presented

Our results, may be supported by findings of Moreno et al. (2001), who by investigating the influence of Cd on DHA stated, that Cd content strongly affected on DHA, by reducing its activity, and this effect is noticeable even after 3 hours.

Negative effect of heavy metals on DHA was reported also by Kizilkaya et al. (2004), who organized the following order of this inhibition: Cu > Cd > Co. Analogically, strong reduction of DHA by Cd contamination was indicated by Welp (1999), who tested the effect of Co, As, Hg, Cd, Pb and Cu on soil DHA, and demonstrated that the strongest effect was displayed by Hg and Cd.

Stępniewska & Wolińska (2005) found that the application of trivalent and hexavalent chromium compounds had a noticeable negative effect on soil DHA (Fig. 14). The soil sample (*Haplic Luvisol*) was amended with Cr (III), as a $CrCl_3$ and with Cr (VI), as a $K_2Cr_2O_7$ in the concentration range from 0 to 20 mg kg^{-1} and 0-100 μg kg^{-1}, for Cr (III) and Cr (VI), respectively. The differences in Cr (III) and Cr (VI) doses resulted from the fact that Cr (VI) is highly toxic and much mobile form of Cr, and is considered to cause much stronger effect on living organisms, than Cr (III).

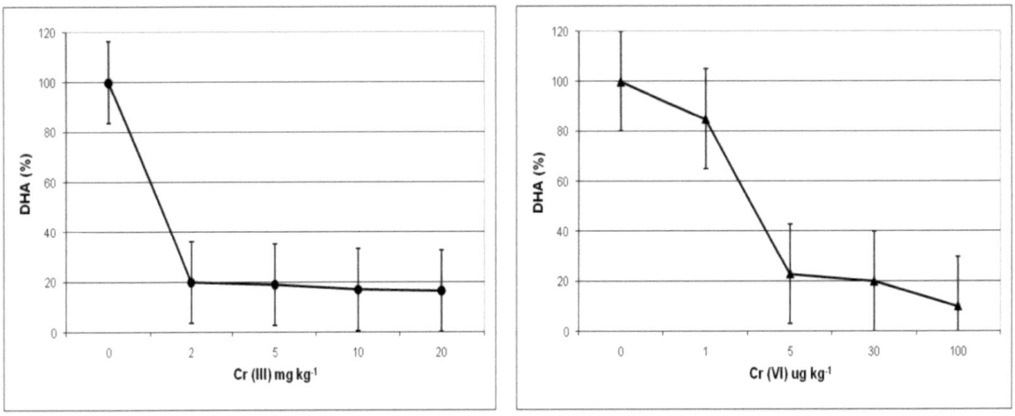

Figure 14. The variations of DHA in *Haplic Luvisol* at different Cr (III) and Cr (VI) concentrations (according to Stępniewska & Wolińska, 2005). Averaged values of three replicates with standard deviations are presented

Non-amended soil samples were used as a control, and their enzymatic activity were estimated as 100%. Effect posed by Cr content was calculated as a decrease of its level, in relation to the control value. We found that the lowest values of DHA were the effect of increasing Cr(III) and Cr(VI) doses. *Haplic Luvisol* seemed to be very sensitive on Cr contamination. DHA was reduced to 18-20% in the samples enriched in Cr (III) forms. Surprisingly, the more dangerous form of Cr (VI) was less harmful for DHA in the *Haplic Luvisol*, because enzymatic activity remained on the level of 84%, with a 1μg kg^{-1} addition and decreased to the value of 14% with the highest supplement of Cr (VI). One possible explanation for this fact is that. the more dangerous form of Cr (VI) was reduced to the less toxic form of Cr (III) by microorganisms, living in the soil (Stępniewska & Wolińska, 2005).

In the same way we investigated effect of Cr forms on *Eutric Cambisol* (Stępniewska & Wolińska, 2004). Received results are shown in Fig. 15. We stated that excess of Cr forms in soil disturb homeostatic metabolism of microbes, what reflect their enzymatic activity. DHA demonstrated a tendency to decrease with increase of Cr concentration. The lowest content of both Cr (III) and Cr (VI), at the level of 2 mg kg^{-1} reduced soil DHA to 51-66%, respectively. But at the same time the highest Cr (III) and Cr (VI) supplement at the level of 20 mg kg^{-1} limited DHA to 6-15%, in relation to the control.

Inhibition of DHA by applied Cr compounds was also reported by Wyszkowska et al. (2001), who noted that decrease of enzymatic activity in soil should be considered as very unfavourable in terms of soil fertility, because soils of good quality and high content of soil OM show high enzymatic activity.

The decrease of soil DHA by several metallic elements (Al, Be, Cu, U) was also discussed by Antunes et al. (2011), whereas a study by Nowak et al. (2002) found that DHA decreased by up to 85% at 5 mM selenic acid (IV) presence.

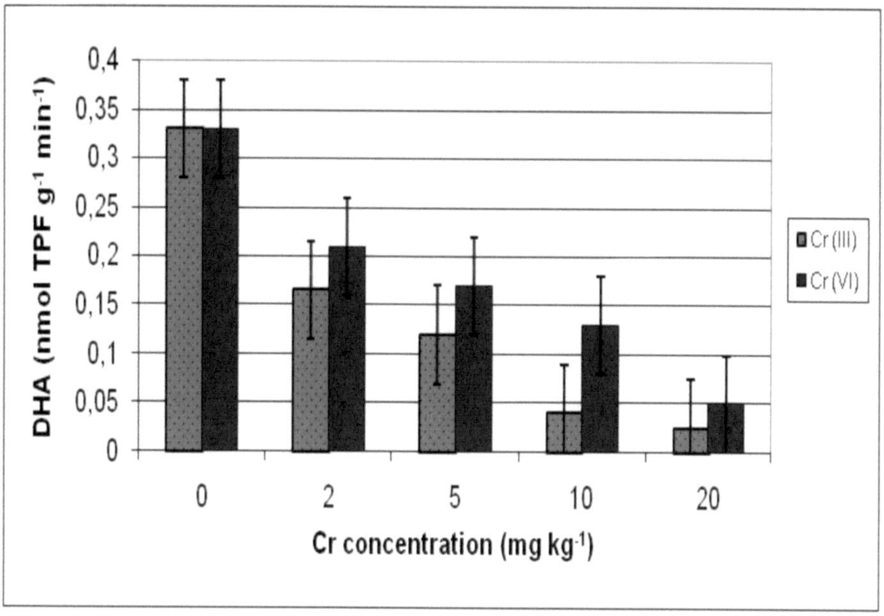

Figure 15. The variations of DHA in *Eutric Cambisol* at different Cr (III) and Cr (VI) concentrations (according to Stępniewska & Wolińska, 2004). Averaged values of three replicates with standard deviations are presented

5. Conclusions

Soil is a part of the terrestrial compartment, and supports all terrestrial life forms. Thus, without proper soil protection policies, numerous problems may arise, like reduction of soil fertility, erosion, groundwater contamination, insufficient water holding capacity and loss of biodiversity. To asses soil quality, it is essential to measure all potential changes in biological soil properties, because they are highly sensitive to any environmental perturbations and stresses. A usual approach to diagnose soil quality, is to use a soil microbial indicators, which are very sensitive and respond quickly to environmental alterations.

Among different soil indicators, DHA is one of the most adequate, important and one of the most sensitive bioindicators, relating to soil quality and fertility. Moreover, their routine measurement is simple and low-cost under laboratory condition. However, we should not remind about limitations, resulting from laboratory conditions, when we are able to measure and estimate only potential DHA, similarly like we are able to cultivate only small percentage of soil microorganisms, on artificial media.

Soil enzymes are strongly associated with microorganisms. Soil enzymatic activity plays an important role in catalyzing reactions indispensable in life processes of soil microorganisms, decomposition of organic residues, circulation of nutrients, as well as forming organic matter and soil structure. Thus, it is possible to say that without proper soil enzymes system, soil life processes will be disturbed.

DHA is related to quantitative changes in microorganisms populations, as only strictly intracellular enzymes can truly reflect microbial activity, because with respect to the degradation processes of extracellular soil enzymes, they will be quickly mineralized by other enzymes (i.e. proteases), unless they are either adsorbed by clays or immobilized by humic molecules.

It should be also remind, that overall soil DHA level depends most of all from the activities of various types of dehydrogenases, which are fundamental part of the enzyme system of all living soil microorganisms, i.e. the respiratory metabolism, the citrate cycle, and N metabolism.

Due to this fact, DHA is proposed as the best indicator of the microbiological redox-systems, and could be considered as good and adequate parameter of microbial oxidative activities in soil. Furthermore, soil DHA is also used as a measure of any soils disruption posed by pesticides, heavy metals, or other soils contaminates and improper management practices.

As DHA is strictly connected with living microbial cells, its activity depends from the same environmental factors, which influence on microorganisms abundance, activity and life processes. Consequently, when entertaining soil DHA behavior in the soil environment, we should be not only limited to DHA, but it is necessity to consider on the most important soils factors and conditions, affecting measuring by us DHA level.

In the presented chapter we described the most important soil parameters, affecting DHA, which poses ability either for stimulation or inhibition its activity.

To sum up the forgoing observations it was demonstrated, that DHA display increasing trend under anaerobic conditions, what suggest that the facultative and anaerobic member of soil microbial community become more important in soil respiration processes. Thus, soil DHA was reported to be negatively correlated with soil water potential, oxygen diffusion rate, and redox potential, what means that DHA reached higher values at lower soil water potential, lower oxygen diffusion rate and lower redox potential conditions. Analogically, negative correlation we also found in the case of soil depth–what was connected with spatial stratification of microorganisms abundance and its preference for inhabiting the surface layers of the soil profiles. Inhibiting effect on DHA level have also pesticides and soil contamination with heavy metals.

Important parameter affecting soil biological activity is pH. Our investigations demonstrated, that optimal pH range for DHA is between 5.5-5.73, what was confirmed by correlation coefficient ($r=-0.70^*$).

Soil DHA depends also from the season of the year, similarly like dynamics of microbial activity, and reached the highest level in May, as spring season is strongly connected with increase in microbial activity, and intensification of oxido-reduction reactions, what is indirectly linked with DHA.

Positive relationships we noted between DHA and two parameters: TOC and temperature, what means that DHA reached higher values at soils with higher TOC content (what is also preferred by soil microorganisms), analogically like increase of temperature to 30°C (temperature close to optimum for microorganisms growth and development) resulted in DHA stimulation.

Presented and discussed above results are based on our several years studies, however additional investigations are needed and recommended to determine the relative contribution of the different environmental effects on soil DHA. However, the discussion highlights the strong interactions between the soil environment, soil enzymes (dehydrogenases especially) and soil microorganisms.

Author details

Agnieszka Wolińska and Zofia Stępniewska
The John Paul II Catholic University of Lublin, Institute of Biotechnology, Department of Biochemistry and Environmental Chemistry, Lublin, Poland

Acknowledgement

The study was partly funded by Polish Ministry of Science and Higher Education (grant N 305 009 32/0514).

6. References

Agnelli, A.; Ascher, J.; Corti, G.; Ceccechrini, M.; Nannipieri, P. & Pietramellara, G. (2004). Distribution of Microbial Communities In a Forest Soil Profile Investigated by Microbial Biomass, Soil Respiration and DGGE of Total and Extracellular DNA. *Soil Biology & Biochemistry*, 36, pp. 859-868

Andrea, M.; Peres, T.; Luchini, L.; Bazarin, S.; Papini, S.; Matallo, M. & Savoy V. (2003). Influence of Repeated Applications of Glyphosate on Its Persistence and Soil Bioactivity. *Pesquisa Agropecuaria Brasileira*, 38, pp. 1329-1335.

Antunes, S.; Pereira, R.; Marques, S.; Castro, B. & Gonçalves, F. (2011). Impaired Microbial Activity Caused by Metal Pollution: A Field Study In a Deactivated Uranium Mining Area. *Science of the Total Environment*, 410, pp. 87-95

Bennicelli, R.; Szafranek-Nakonieczna, A.; Wolińska, A.; Stępniewska, Z. & Bogudzińska, M. (2009). Influence of Pesticide (Glyphosate) on Dehydrogenase Activity, pH, Eh and Gases Production In Soil (Laboratory Conditions). *International Agrophysics*, 23, pp. 117-122

Błońska, E. (2010). Enzyme Activity in Forest Peat Soils. *Folia Forestalia Polonica*, 52, pp.20-25.

Brzezińska, M. (2006). Impact of Treated Wastewater on Biological Activity and Accompanying Processes in Organic Soils. *Acta Agrophysica*, 131, pp. 1-163 (in Polish with English Summary)

Brzezińska, M.; Stępniewska, Z. & Stępniewski, W. (1998). Soil Oxygen Status and Dehydrogenase Activity. *Soil Biology & Biochemistry*, 30, pp. 1783-1790

Brzezińska, M.; Stępniewski, W.; Stępniewska, Z. & Przywara, G. (2001). Effect of Oxygen Deficiency On Soil Dehydrogenase Activity In a Pot Experiment With Triticale CV. Jago Vegetation. *International Agrophysics*, 15, pp. 145-149

Casida, L.; Klein, D. & Santoro, T. (1964). Soil Dehydrogenase Activity. *Soil Science*, 98, pp. 371-376

Chodak, M. & Niklińska, M. (2010). Effect of Texture and Tree Species On Microbial Properties of Mine Soils. *Applied Soil Ecology*, 46, pp. 268-275

Chu, H.; Lin, X.; Fujii, T.; Morimoto, S.; Yagi K.; Hu, J. & Zhang, J. (2007). Soil Microbial Biomass, Dehydrogenase Activity, Bacterial Community Structure In Response To Long-Term Fertilizer Management. *Soil Biology & Biochemistry*, 39, pp. 2971-2976

Cirilli, F.; Bellincontro, A.; De Santis, D.; Botondi, R.; Colao, M.; Muleo, R. & Mencarelli, F. (2012). Temperature and Water Loss Affect ADH Activity and Gene Expression In Grape Berry During Postharvest Dehydration. *Food Chemistry*, 132, pp. 447-454

Fernandez-Calviño, D.; Soler-Rovira, P.; Polo, A.; Diaz-Raviña, M.; Arias-Estevez, M. & Plaza C. (2010). Enzyme Activities In Vineyard Soils Long-Term Treated With Copper-Based Fungicides. *Soil Biology & Biochemistry*, 42, pp. 2119-2127

Fierer, N.; Schimel, J. & Holden P. (2003). Variations In Microbial Community Composition Through Two Soil Depth Profiles. *Soil Biology & Biochemistry*, 35, pp. 167-176

Fontaine, S.; Marotti, A. & Abbadie, L. (2003). The Priming Effect of Organic Matter: A Question of Microbial Competition. *Soil Biology & Biochemistry*, 35, pp. 837-843

Forlani, G.; Mangiagalli, A.; Nielsen, E. & Suardi, M. (1999). Degradation Of the Phosphonate Herbicide Glyphosate In

Soil: Evidence For a Possible Involvement of Unculturable Microorganisms. *Soil Biology & Biochemistry*, 31, pp. 991-997

Frankenberger, W. & Johanson, J. (1982). Effect of pH On Enzyme Stability in Soils. *Soil Biology & Biochemistry*, 14, pp. 433-437

Gajda, A. (2008). Effect of Different Tillage Systems On Some Microbiological Properties of Soils Under Winter Wheat. *International Agrophysics*, 22, pp. 201-208

Geisseler, D.; Horwath, W. & Scow, K. (2011). Soil Moisture and Plant Residue Addition Interact In Their Effect On Extracellular Enzyme Activity. *Pedobiologia*, 54, pp. 71-78

Ghaly, A. & Mahmoud, N. (2006). Optimum Conditions For Measuring Dehydrogenase Activity of *Aspergillus niger* Using TTC. *American Journal of Biochemistry & Biotechnology*, 2, pp. 186-194

Gu, Y.; Wag, P. & Kong, C. (2009). Urease, Invertase, Dehydrogenase and Polyphenoloxidase Activities In Paddy Soils Influenced By Allelophatic Rice variety. *European Journal of Soil Biology*, 45, pp. 436-441

Hutchinson, G. (1995). Biosphere-Atmosphere Exchange of Gaseous N Oxides. In: *Soil and global change*, R. Lal, (Ed.), 219-236, CRC Lewis Publisher, Boca Raton, FL, USA.

Januszek, K. (1993). Seasonal Changes of Enzyme Activity In Mor, Moder and Mull Humus of Selected Forest Soils In The Western Beskid Mountains. *Folia Forestalia Polonica*, 35, pp. 59-75

Kadian, N.; Malik, A.; Sataya, S. & Dureja, P. (2012). Effect of Organic Amendments on Microbial Activity in Chlorpyrifos Contaminated Soil. *Journal of Environmental Management*, 95, pp. 199-202

Kizilkaya, R.; Askin, T.; Bayrakli, B. & Saglam, M. (2004). Microbiological Characteristics of Soils Contaminated With Heavy Metals. *European Journal of Soil Biology*, 40, pp. 95-102

Koper, J.; Piotrowska, A. & Siwik-Ziomek, A. (2008). Dehydrogenase and Inwertase Activities In a Rusty Soil In The Neighborhood of The Włocławek Nitrogen Plant "Anwill". *Proceedings of ECOpole*, 2, pp. 197-202

Kumar, J.; Sharma, G. & Mishra, R. (1992). Soil Microbial Population Number and Enzyme Activities In Relation To Altitude and Forest Degradation. *Soil Biology & Biochemistry*, 24, pp. 761-767

Quilchano, C. & Marañon, T. (2002). Dehydrogenase Activity In Mediterranean Forest Soils. *Biology & Fertility of Soils*, 35, pp. 102-107

Levyk, V.; Maryskevych, O.; Brzezińska, M. & Włodarczyk, T. (2007). Dehydrogenase Activity of Technogenic Soils of Former Sulphur Mines (Yvaoriv and Nemyriv, Ukraine). *International Agrophysics*, 21, pp. 255-260

Macci, C.; Doni, S.; Peruzzi, E.; Masciandro, G.; Mennone, C. & Ceccanti, B. (2012). Almond Tree and Organic Fertilization for Soil Quality Improvement In Southern Italy. *Journal of Environmental Management*, 95, pp. 215-222

Menon, P.; Gopal, M. & Parsad, R. (2005). Effects of Chlorpyrifos and Quinalphos On Dehydrogenase Activities and Reduction of Fe^{3+} In The Soils of Two Semi-Arid Fields of Tropical India. *Agriculture, Ecosystems & Environment*, 108, pp. 73-83

Moeskops, B.; Buchan, D.; Sleutel, S.; Herawaty, L.; Husen, E.; Saraswati, R.; Setyorini, D. & De Neve, S. (2010). Soil Microbial Communities and Activities Under Intensive Organic and Conventional Vegetable Farming In West Java, Indonesia. *Applied Soil Ecology*, 45, pp. 112-120

Moreno, J.; Aliaga, A.; Navarro, S.; Hernendez, T. & Garcia, C. (2007). Effects of Atrazine On Microbial Activity In Semiarid Soil. *Applied Soil Ecology*, 35, pp. 120-127

Muñoz-Leoz, B.; Ruiz-Romera, E.; Antigüedad, I. & Garbisu, C. (2011). Tebuconazole Application Decreases Soil Microbial Biomass and Activity. *Soil Biology & Biochemistry*, 43, pp. 2176-2183

Nagatsuka, T. & Furosaka, C. (1980). Effect of Oxygen Tension On Growth, Respiration and Types of Bacteria Isolated From Soil Suspensions. *Soil Biology & Biochemistry*, 12, pp. 397-403

Natywa, M. & Selwet, M. 2011. Respiratory and Dehydrogenase Activities In The Soils Under Maize Growth In The Conditions of Irrigated and Nonirrigated Fields. *Agricultura*, 10, pp. 93-100

Nayak, D.; Babu, J. & Adhya, T. (2007). Long-Term Application of Compost Influences Microbial Biomass and Enzyme Activities In a Tropical Aeric Endoaquept Planted To Rice Under Flooded Condition. *Soil Biology & Biochemistry*, 39, pp. 1897-1906

Nowak, J.; Kaklewski, K. & Klódka, D. (2002). Influence of Various Concentrations of Selenic Acid (IV) On The Activity of Soil Enzymes. *The Science of Total Environment*, 291, pp. 105-110

Pan, J. & Yu, L. (2011). Effects of Cd or/and Pb On Soil Enzyme Activities and Microbial Community Sructure. *Ecological Engineering*, 37, pp. 1889-1894

Pascual, J.; Garcia, C.; Hernandez,T.; Moreno, J. & Ros, M. (2000). Soil Microbial Activity As a Biomarker of Degradation and Remediation Processes. *Soil Biology & Biochemistry*, 32, pp. 1877-1883

Pett-Ridge, J. & Firestone, M. (2005). Redox Fluctuation Structures Microbial Communities In a Wet-Tropical Soil. *Applied & Environmental Microbiology*, 71, pp. 6998-7007

Romero, E.; Fernandez-Bayo, J.; Diaz, J. & Nogales, R. (2010). Enzyme Activities and Diuron Persistence In Soil Amended With Vermicompost Derived From Spent Grape Marc and Treated With Urea. *Applied Soil Ecology*, 44, pp. 198-204

Ros, M.; Hernandez, M. & Garcia, C. (2003). Soil Microbial Activity After Restoration of a Semiarid Soil By Organic Amendments. *Soil Biology & Biochemistry*, 35, pp. 463-469

Salazar, S.; Sanchez, L.; Alvarez, J.; Valverde, A.; Galindo, P.; Igual, J.; Peix, A. & Santa-Regina, I. (2011). Correlation Among Soil Enzyme Activities Under Different Forest System Management Practices. *Ecological Engineering*, 37, pp. 1123-1131

Shuler, M. & Kargi, F. (2010). Bioprocess Engineering Basic Concepts. Prentice-Hall Incorporation, Englewood Cliffs, New Yersey, USA. ISBN-10: 0130819085

Song, Y.; Deng, S.; Acosta-Martinez, V. & Katsalirou, E. (2008). Characterization of Redox-Related Soil Microbial Communities Along a River Floodplain Continuum By Fatty Acid Methyl Ester (FAME) and 16S rRNA Genes. *Applied Soil Ecology*, 40, pp. 499-509

Stępniewska, Z. & Wolińska, A. (2004). Enzyme Activity In The Soil Contaminated by Chromium (III, VI) Forms. Multi author work: *Modern Physical & Physicochemical Methods & Their Applications in Agroecological Research*. Institute of Agrophysics PAS, Lublin, pp. 201-207

Stępniewska, Z. & Wolińska, A. (2005). Soil Dehydrogenase Activity In The Presence of Chromium (III) and (VI). *International Agrophysics*, 19, pp. 79-83

Stępniewska, Z.; Wolińska, A. & Lipińska, R. (2007). Effect of Fonofos On Soil Dehydrogenase Activity. *International Agrophysics*, 21, pp. 101-105

Stępniewski, W.; Stępniewska, Z.; Gliński, J.; Brzezińska, M., Włodarczyk, T.; Przywara, G.; Varallyay, G. & Rajkai, J. (2000). Dehydrogenase Activity of Some Hungarian Soils as Related To Their Water and Aeration Status. *International Agrophysics*, 14, pp. 341-354

Stępniewski, W.; Stępniewska, Z.; Bennicelli, R. & Gliński, J. (2005). Oxygenology In Outline. Institute of Agrophysics Polish Academy of Sciences, pp. 18-33, Lublin, Poland

Subhani, A.; Changyong, H.; Zhengmiao, Y.; Min, L. & El-ghamry, A. (2001). Impact of Soil Environment and Agronomic Practices On Microbial/Dehydrogenase Enzyme Activity In Soil. A Review. *Pakistan Journal of Biological Sciences*, 4, pp. 333-338

Tejada, M.; Garcia-Martinez, A.; Gomez, I. & Parrado, J. (2010). Application of MCPA Herbicide On Soils Amended With Biostimulants: Short-time Effects On Soil Biological Properties. *Chemosphere*, 80, pp. 1088-1094

Tejada, M.; Gomez, I.; Garcia-Martinez, A.; Osta, P. & Parado, J. (2011). Effects of Prochloraz Fungicide On Soil Enzymatic Activities and Bacterial Communities. *Ecotoxicology & Environmental Safety*, 74, pp. 1708-1714

Trasar-Cepeda, C.; Gil-Sotres, F. & Leiros, M. (2007). Thermodynamic Parameters of Enzymes In Grassland Soils From Galicia, NW Spain. *Soil Biology & Biochemistry*, 39, pp. 311-319

Trevors, J. (1984). Effect of Substrate Concentration, Inorganic Nitrogen, O_2 Concentration, Temperature and pH On Dehydrogenase Activity In Soil. *Plant & Soil*, 77, pp.285-293

Tripathi, S.; Chakraborty, A.; Chakrabarti, K. & Bandyopadhyay, B. (2007). Enzyme Activities and Microbial Biomass In Coastal Soils of India. *Soil Biology & Biochemistry*, 39, pp. 2840-2848

Uhlirova, E.; Elhottova, D.; Triska, J. & Santruckova, H. (2005). Physiology and Microbial Community Structure In Soil At Extreme Water Content. *Folia Microbiology*, 50, pp. 161-166

Wall, A. & Heiskanen, J. (2003). Water-Retention Characteristic and Related Physical Properties of Soil On Afforested Agricultural Land In Finland. *Forest Ecology & Management*, 186, pp. 21-32

Weaver, M.; Zablotowicz, R.; Krutz, L.; Bryson, C. & Locke, M. (2012). Microbial and Vegetative Changes Associated With Development of a Constructed Wetland. *Ecological Indicators*, 13, pp. 37-45

Welp, G. (1999). Inhibitory Effects of The Total and Water-Soluble Concentrations of Nine Differents Metals On The Dehydrogenase Activity of a Loess Soil. *Biology & Fertility of Soils*, 30, pp.132-139

Włodarczyk, T. (2000). Some Aspects of Dehydrogenase Activity In Soils. *International Agrophysics*, 22, pp. 371-375

Włodarczyk, T.; Gliński, J.; Stępniewski, W.; Stępniewska, Z.; Brzezińska, M. & Kuraz, V. (2001). Aeration Properties and Enzyme Activity On The Example of Arenic Chernozem (Tisice). *International Agrophysics*, 15, pp. 131-138

Włodarczyk, T.; Stępniewski, W. & Brzezińska, M. (2002). Dehydrogenase Activity, Redox Potential, and Emissions of Carbon Dioxide and Nitrous Oxide From *Cambisols* Under Flooding Conditions. *Biology & Fertility of Soils*, 36, pp. 200-206

Wolińska, A. (2010). Dehydrogenase Activity of Soil Microorganisms and Oxygen Availability During Reoxidation Process of The Selected Mineral Soils From Poland. *Acta Agrophysica*, 180, pp. 1-88 (in Polish with English Summary)

Wolińska, A. & Bennicelli, R. (2010). Dehydrogenase Activity Response to Soil Reoxidation Process Described as Varied Condition of Water Potential, Air Porosity and Oxygen Availability. *Polish Journal of Environmental Studies*, 19, pp. 651-657

Wolińska, A. & Stępniewska, Z. (2011). Microorganisms Abundance and Dehydrogenase Activity As a Consequence of Soil Reoxidation Process, In: *Soil Tillage & Microbial Activities*, M. Miransari, (Ed.), 111-143, Research Singpost, Kerala, India

Wyszkowska, J.; Kucharski J.; Jastrzębska, E. & Hlasko, A. (2001). The Biological Properties of Soil As Influenced By Chromium Contamination. *Polish Journal of Environmental Studies*, 10, 37-42

Xiang, S.; Doyle, A.; Holden, P. & Schimel, J. (2008). Drying and Rewetting Effects On C and N mineralization and Microbial Activity In Surface and Subsurface California Grassland Soils. *Soil Biology & Biochemistry*, 40, pp. 2281-2289

Xie, W.; Zhou, J.; Wang, H.; Chen, X.; Lu, Z.; Yu J. & Chen, X. (2009). Short-Term Effects of Copper, Cadmium and Cypermethrin On Dehydrogenase Activity and Microbial Functional Diversity In Soils After Long-Term Mineral or Organic Fertilization. *Agriculture, Ecosystems & Environment*, 129, pp. 450-456

Yang, L.; Li, T. & Fu, S. (2005). Effect of Manure and Chemical Fertilizer On The Dynamics of Soil Enzymatic Activities In Vegetable Soil. *Chinese Journal of Soil Science*, 36, pp. 223-226

Yuan, B. & Yue, D. (2012). Soil Microbial and Enzymatic Activities Across a Chronosequence of Chinese Pine Plantation Development On The Loess Plateau of China. *Pedosphere*, 22, pp. 1-12

Zabaloy, M.; Garland, J. & Gomez, M. (2008). An Integrated Approach To Evaluate The Impacts of The Herbicides Glyphosate, 2,4-D and Metsulfuron-Methyl On Soil Microbial Communities In The Pampas Region, Argentina. *Applied Soil Ecology*, 40, pp. 1-12

Zhang, N.; He, X.; Gao, Y.; Li, Y.; Wang, H.; Ma, D.; Zhang, R. & Yang, S. (2010). Pedogenic Carbonate and Soil Dehydrogenase Activity In Response To Soil Organic Matter in *Artemisia ordosica* Community. *Pedosphere*, 20, pp. 229-235

Zhao, B.; Chen, J.; Zhang, J. & Qin S. (2010). Soil Microbial Biomass and Activity Response To Repeated Drying-Rewetting Cycles Along a Soil Fertility Gradient Modified By Long-Term Fertilization Management Practices. *Geoderma*, 160, pp. 218-224

Chapter 9

Succinate Dehydrogenase of *Saccharomyces cerevisiae* – The Unique Enzyme of TCA Cycle – Current Knowledge and New Perspectives

Dorota Kregiel

Additional information is available at the end of the chapter

http://dx.doi.org/10.5772/48413

1. Introduction

Glycolysis is an anabolic pathway common in both aerobic and anaerobic organisms. Sugars and polysaccharides have to be transformed into glucose or one of its phosphorylated derivatives before being processed any further. In the course of degradation, ATP is produced. Pyruvate may be regarded as the preliminary final product of the degradation - in a strictly formal sense - because it is here that the pathway ramifies: pyruvate is hydrated under anaerobic conditions resulting in either lactate (in lactic-acid bacteria) or ethanol (in yeast). If glycolysis results in these final products, it is spoken of fermentation. In the presence of oxygen pyruvate is converted to acetyl-coenzyme A and oxidized to CO_2 in the tricarboxylic acid (TCA) cycle (Figure 1). In yeasts, as well in most non-photosynthetic cells, mitochondrial oxidative phosphorylation is the main process of ATP synthesis in aerobic conditions. Aerobic pathways permit the production of 30 to 38 molecules of ATP per one molecule of glucose. Although two molecules of ATP come from glycolysis and two more directly out of the TCA cycle, most of the ATP arises from oxidative phosphorylation.

The main catalytic function of the TCA cycle is to provide reducing equivalents to the respiratory chain through the oxidative decarboxylation of acetyl–CoA (8), but every TCA cycle intermediate is commonly used by other metabolic reactions. The eight enzymes from the TCA cycle are encoded by 15 nuclear genes in *S. cerevisiae* [1]. The first reaction of TCA is catalyzed by citrate synthase (1) and it is the condensation of acetyl–CoA and oxaloacetate resulting in the formation of citrate. The second reaction of the TCA is catalyzed by aconitase (2), leading to the conversion of citrate into isocitrate. Aconitase is located both in mitochondria and in cytosol. The next step of the TCA is the oxidative decarboxylation of isocitrate to α-ketoglutarate (3). There are three known isoenzymes of isocitrate

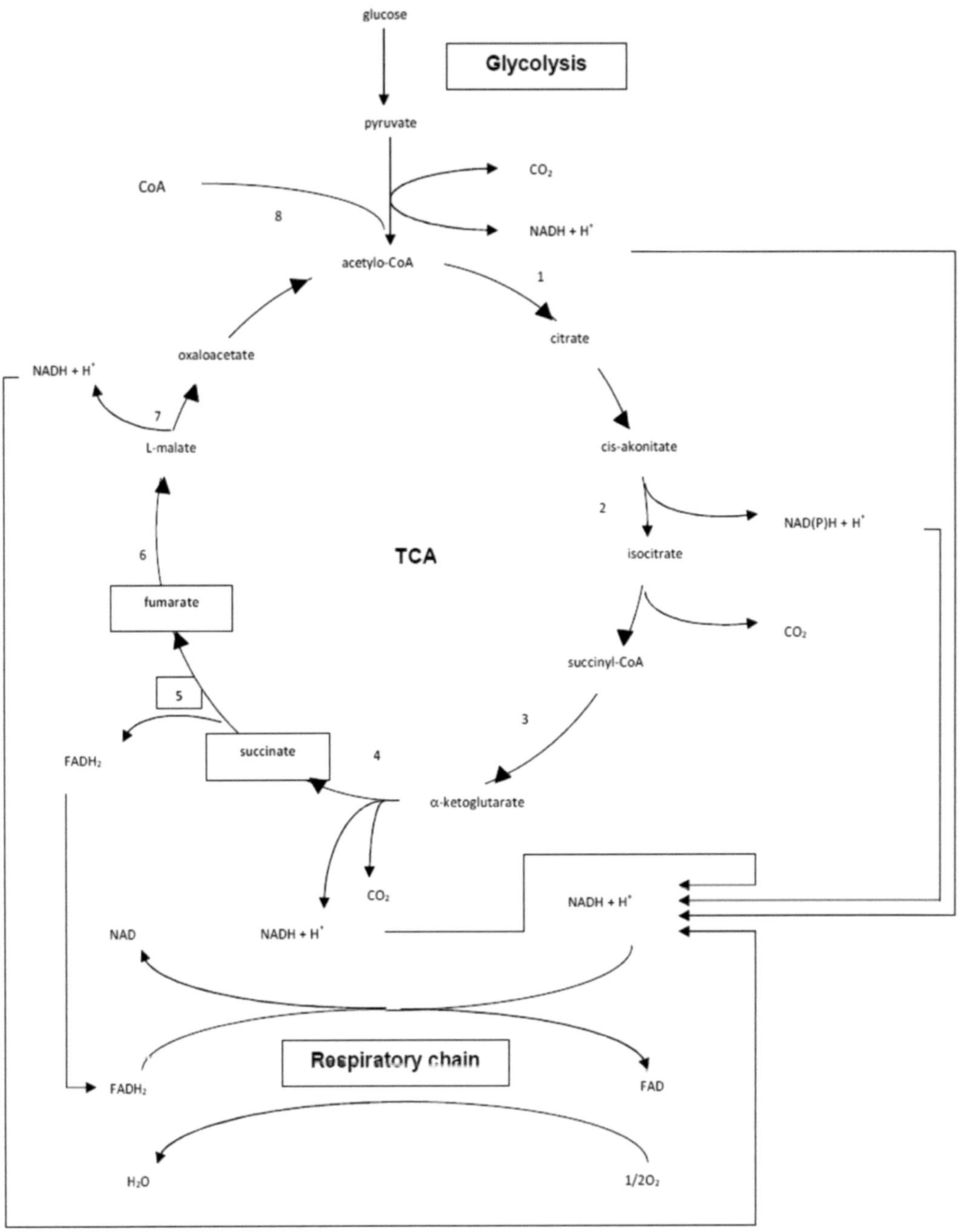

Figure 1. TCA cycle.

dehydrogenase, a mitochondrial NAD⁺-specific one and two NADP⁺-dependent ones (one mitochondrial and the other cytosolic). A number of pieces of evidence point to the role of mitochondrial NAD⁺-specific isocitrate dehydrogenase in the regulation of the rate of mitochondrial assembly besides its specific role in the TCA cycle [2]. The formation of succinate is catalyzed by α-ketoglutarate dehydrogenase (4), which promotes the oxidative decarboxylation of α-ketoglutarate via succinil–CoA to succinate, which is then converted to fumarate by succinate dehydrogenase (5). The next step of the TCA cycle is the conversion of fumarate to malate by the enzyme fumarase (6), which exists as separate cytosolic and mitochondrial forms. There is not yet a clear explanation for the existence of these two forms; however, the localization and distribution of fumarase appears to be unique because there is only one translation product, which is targeted to mitochondria [3]. Malate dehydrogenase (7) catalyzes the last step of the TCA and leads to the oxidation of malate to oxaloacetate. There are three isoenzymes of malate dehydrogenase: a cytosolic, a mitochondrial and a peroxissomal one; however, the mitochondrial one accounts for 90% of malate dehydrogenase activity when glucose is being metabolized [4].

TCA cycle flux appears to be constricted at two steps on the basis of the limited availability of the substrates oxaloacetate and α-ketoglutarate. The genes encoding TCA proteins might also be regulated by glucose levels. In *S. cerevisiae* the depletion of glucose increases 3–10 times the TCA messenger RNAs [5]. Oxygen limitation could also induce a shift in the TCA cycle, which operates as a cycle during aerobic growth [6].

All reactions of glycolysis take place in the cell cytosol. The citric acid cycle and the respiratory chain are located within mitochondrial matrix. In contrast with all of the other enzymes of the TCA cycle, which are soluble proteins found in the mitochondrial matrix, succinate dehydrogenase (SDH) is an integral membrane protein tightly associated with the inner mitochondrial membrane [7]. Membrane-bound succinate dehydrogenase (EC 1.3.99.1) or succinate-coenzyme Q reductase or Complex II, is present in all aerobic cells. It was discovered by Thunberg in 1909. Now we know that SDH has several particularly interesting properties: (1) it is a membrane-bound dehydrogenase linked to the respiratory chain and a member of the tricarboxylic acid (TCA) cycle; (2) its activity is modulated by several activators and inhibitors; and (3) it is a complex enzyme, and covalently bound flavin adenine dinucleotide (FAD) [8].

2. SDH structure

SDH is a heterotetramer composed of flavoprotein (Fp) of about 70 kDa, an iron-sulfur protein (Ip) of about 30 kDa, and two hydrophobic anchoring subunits of 7-17 kDa. The Fp contains the active site and the unusual cofactor, an 8α-N(3)-histidyl-FAD linked at a conserved histidine residue (Figure 2).

FAD attachment is stimulated by, but not dependent upon, the presence of the iron-sulfur subunit and citric acid cycle intermediates such as succinate, malate, or fumarate [9].

The Ip subunit contains three different iron-sulfur clusters: [2Fe-2S], [3Fe-4S], and [4Fe-4S] (Figure 3).

These clusters are coordinated by conserved cysteine residues: [2Fe-2S] (Cys-67, Cys-72, Cys-75, Cys-87); [4Fe-4S] (Cys-159, Cys-162, Cys-165, Cys-226) and [3Fe-4S] (Cys-169, Cys-216, Cys-222) [10].

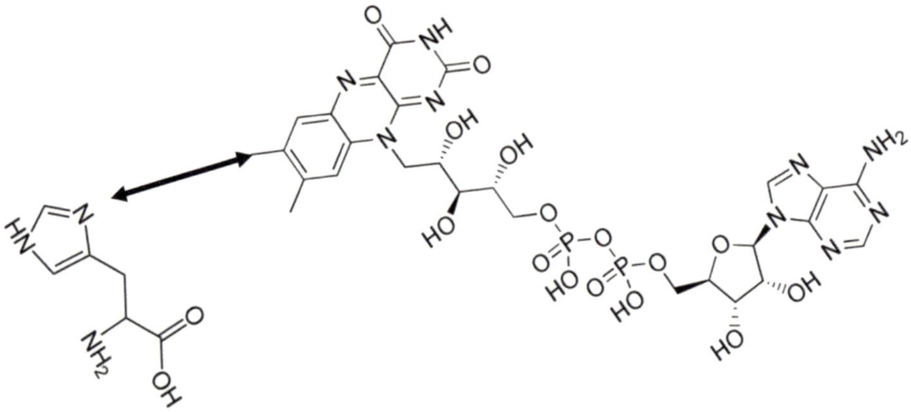

Figure 2. The covalent bond between FAD and succinate dehydrogenase.

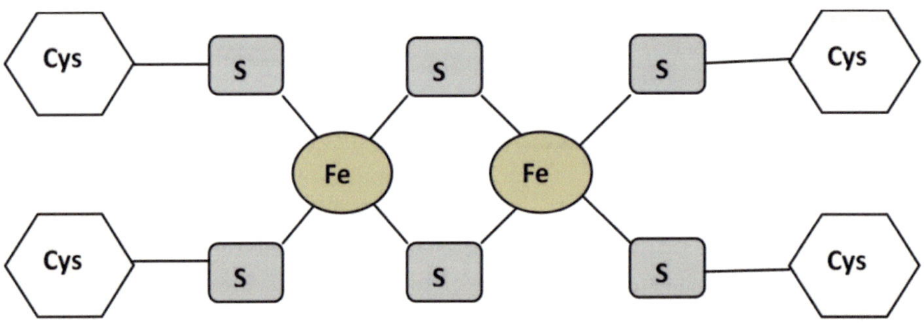

Figure 3. The 2Fe-2S cluster of succinate dehydrogenase [9].

The hydrophobic anchoring subunits are integral membrane proteins and interact with quinone substrates. The yeast and mammalian SDH also contains a *b*-type heme. Oyedotun et al. [11] demonstrated the presence of an amount of cytochrome b_{562} stoichiometric to covalent FAD. Together, the Fp and Ip form a catalytic dimer that is attached to the membrane by the anchoring subunits, thereby composing the holoenzyme. In yeast, the SDH Fp, Ip, and two anchoring subunits are encoded by the nuclear genes, *SDH1*, *SDH2*, *SDH3*, and *SDH4*, respectively, which have all been cloned and sequenced. The SDH subunits are translated in the cytoplasm, targeted to mitochondria by cleavable amino-terminal presequences, translocated across both mitochondrial membranes, and finally assembled with each other and their respective co-factors into a functional complex [10].

The quaternary structure model of the SDH for different cells, e.g. *S. cerevisiae* and mammalian, was described in study conducted by Oyedotun and Lemire [10] and Yankovskaya et al. [12]. First subunit of SDH provides the binding site for the oxidation of succinate. The side chains Thr254, His354, and Arg399 stabilize the molecule while FAD oxidizes and carries the electrons to the first of the iron-sulfur [2Fe-2S] clusters. Whereas, ubiquinone binding site is located is in a gap composed of three SDH subunits. Ubiquinone is stabilized by the side chains of His207 of second subunit, Ser27 and Arg31 of third subunit C, and Tyr83 of fourth subunit. The quinine ring is surrounded by Ile28 of third subunit and

Pro160 of second subunit B. These residues, along with Il209, Trp163, and Trp164 of second subunit B, and Ser27 (C atom) of third subunit, form the hydrophobic environment of the quinine binding pocket. The succinate binding site and ubiquinone binding site are connected by a chain of redox centres including FAD and the iron-sulfur clusters. This chain extends over 40 Å through the enzyme monomer with an edge-to-edge distance between the centres is less that 14 Å limit for physiological electron transfer [8, 11].

In the place for heme *b*, the N 2 atom of Sdh3p His106 and the S atom of Sdh4p Cys78 are correctly oriented to form coordinating bonds with the central iron atom of the heme. The distance between the iron atom and the N 2 atom is 2.21 Å, but the S atom of Cys78, the putative second heme axial ligand, is 2.96 Å away from the heme Fe. UQ can be docked into two spatially separated sites with an edge-to-edge distance of 25.4 Å. The new amino acid residues that may determine the structural or catalytic properties of each of the two quinone binding sites were identified. The model also provided insight into the unusual use of a cysteine (Sdh4p Cys78) as the second heme ligand instead of the histidine residues [10].

3. SDH function and regulation

Succinate dehydrogenase is a key enzyme in intermediary metabolism and aerobic energy production in living cells. This enzymes catalyses the oxidation of succinate into fumarate in the Krebs cycle (1), derived electrons being fed to the respiratory chain complex III to reduce oxygen and form water (2). This builds up an electrochemical gradient across the mitochondrial inner membrane allowing for the synthesis of ATP. Alternatively, electrons can be diverted to reduce the ubiquinone pool (UQ pool) and provide reducing equivalents necessary to reduce superoxide anions originating either from an exogenous source or from the respiratory chain itself (3) [13] (Figure 4).

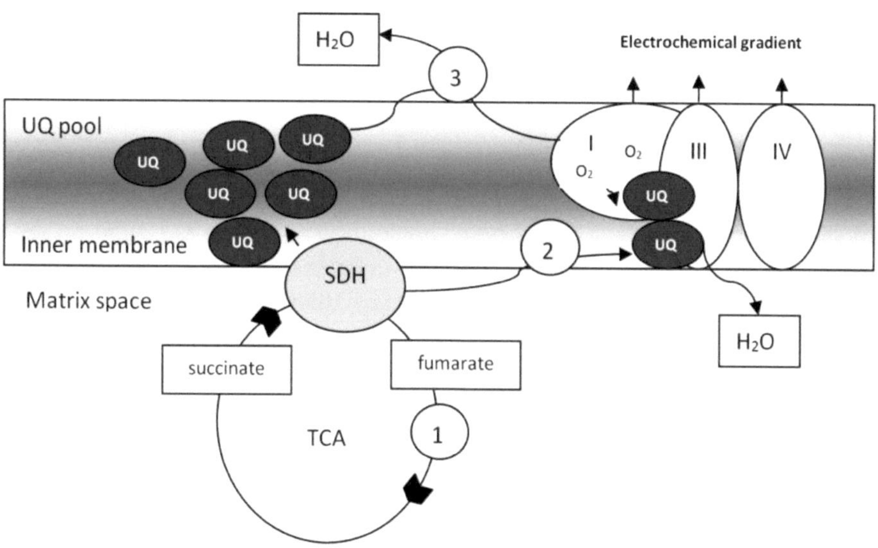

Figure 4. The functions of the succinate dehydrogenase in the mitochondria [13].

In the reaction of oxidation of succinate to fumarate, two hydrogen atoms are removed from substrate by flavin adenine dinucleotide (FAD), a prosthetic group that is tightly attached to succinate dehydrogenase (Figure 5).

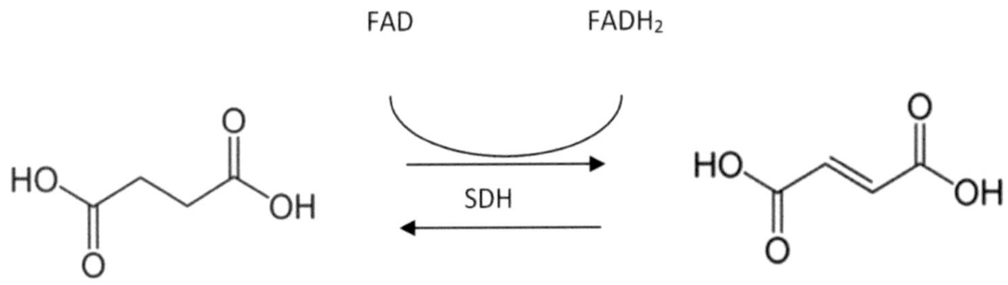

Figure 5. The succinate dehydrogenase reaction.

Two electrons from the reduced SDH-FADH2 complex are then transferred to ubiquinone (Q), a soluble component of the electron transport system complex II. Ubiquinone is then reduced to ubiquinol (QH2).

Flavin adenine dinucleotide (FAD) is an essential cofactor for SDH enzyme. The generation of adenosine triphosphate (ATP) in mitochondria is coupled to the oxidation of nicotinamide adenine dinucleotide (NADH) and FADH2 and reduction of oxygen to water within the respiratory chain and a three-dimensional structure of the mitochondrial respiratory membrane protein complex II. FAD attachment is stimulated by, but not dependent upon, the presence of the iron-sulfur subunit and citric acid cycle intermediates such as succinate, malate, or fumarate [9]. The substrate analog malonate is a competitive inhibitor of the succinate dehydrogenase complex. Malonate, like succinate, is a dicarboxylate that binds to cationic amino acid residues in the active site of the succinate dehydrogenase complex. However, malonate cannot undergo oxidation because it lacks the -CH2 - CH2- group necessary for dehydration. To study the effects of a competitive inhibitior on the activity of succinate dehydrogenase, malonate will be added to a reaction mixture; malonate is sufficiently different from succinate that it cannot de dehydrogenated, i.e. malonate is not metabolized [8, 14].

SDH is a difficult enzyme to extract from respiratory membrane whilst still retaining its *in vivo* properties. Most of the extraction procedures used in early work were rather drastic and yielded soluble preparations of rather dubious integrity. However, the recent introduction of a more gentle method, involving disruption of the membrane with chemotropic agents, has yielded an active and nearly homogeneous enzyme of relatively low molecular weight (97,000). This enzyme can be separated by freezing and thawing into two inactive subunits. One of these, an iron sulphur flavoprotein of molecular weight 70,000, contains one mole of FAD and four moles each of iron and labile sulphide per mole of protein; The other, an iron-sulphur protein of molecular weight 27,000, also contains four moles each of iron and labile sulphide. It was determined that the large subunit of SDH is essential for catalytic activity, but the function of the small subunit, be it catalytic or regulatory [8].

Electronic paramagnetic resonance measurements of SDH components indicated that at least three separate centres are present. The S-3 centre (E'o = +60 mV) is a high potential Fe-S protein and is probably identical with the 4Fe-rS centre of low molecular weight subunit. Centres S-1 and S-2 (E'o = +15 mV and -260 mV respectively), are of the 2Fe-2S ferredoxin type and are probably associated with a larger, flavin containing subunit. Thus, electron transfer from succinate to ubiquinone probably occurs in the sequence: FAD, S-1, S-3. The redox potential of S-2 is rather too low to allow this centre to be catalytically active in the forward direction [8,15].

The catalytic activity of succinate dehydrogenase is modulated both by post translational phosphorylation/acetylation and active site inhibition. For example, phosphorylation of the Sdh1 subunit leads to attenuate activity of SDH. The activity of this enzyme may be also modulated by Krebs cycle intermediates including oxaloacetate or malonate which are strong inhibitors. Mechanisms of inhibition by these compounds differ significantly because oxaloacetate, a competitive inhibitor of succinate dehydrogenase, bounds with a sulfhydryl group of the enzyme to abolish the enzyme activity [16]. It is known that SDH is sensitive to different thiol-binding reagents. Inhibition of the enzyme by these kinds of reagents resulted from the modification of a sulfhydryl group located at the active site. This thiol, although not essential for substrate binding or catalysis, could influence the binding of dicarboxylates, probably by steric hindrance when a larger group or a charged group were attached to it. The inhibition of SDH by histidine specific reagents was also reported, and the participation of an imidazole ring in the initial step of succinate oxidation was suggested. The inactivation of SDH by phenylglyoxal and 2,3 –butanedione showed the presence of an arginine-residues that interacts with dicarboxylate to form the primary enzyme-substrate complex [17].

SDH is not only known to catalyse a unique reaction, which requires the participation of its four subunits, but deleterious mutations in any of the SDH genes should invariably result in a decreased SDH activity. Therefore, the striking phenotypic differences associated with mutations in the four subunits raise puzzling questions. SDH also plays a specific role in the maintenance of the mitochondrial UQ pool reduction. Ubiquinone, beside its function as an electron carrier mediating electron transfer, is admittedly working as a powerful antioxidant in biological membranes. Then, only a portion of the UQ pool may be actually involved in electron transfer depending on dehydrogenases involved. Accordingly, the measurable redox status of the UQ pool should result from the reducing activity of the different dehydrogenases, the oxidising activity of complex III and the kinetic equilibrium in the pool. The UQ pool therefore represents an electron sink and, when reduced, an antioxidant reservoir in the mitochondrial inner membrane. However, UQ is a double-faced compound, possibly working as either an antioxidant when fully reduced to ubiquinol, or a pro-oxidant when semi-reduced to the unstable ubisemiquinone form. Possibly together with reduced cytochrome b, semi-reduced quinones constitute the prominent source of superoxides. Finally, when defective, the abnormal amount of superoxides can be produced, e.g. flavin radicals of complex I. Delivering electrons for the full reduction of UQ to UQH2 might then be of a tremendous importance for the control of oxygen toxicity in the mitochondria.

Therefore, the SDH, thanks to its unique redox properties, may be a key enzyme to control UQ pool redox poise under these conditions [13].

Also mutations in genes encoding SDH subunits lead to reduced activity of SDH enzyme. The yeast cells disrupted in *SDH2* (*sdh2*Δ) showed dramatically accumulate succinate resulting in inhibition of at least two α-ketoglutarate dependent enzymes that generate succinate as a by-product. Disruption of complex II activity should alter TCA cycle metabolite levels in the mitochondrial matrix. It was found that neither *sdh1*Δ, nor *sdh2*Δ cells have measurable SDH activity. The succinate accumulates to 8-fold higher levels in *sdh2*Δ cells relative to wild-type cells. Furthermore, complex II + III activity was completely abolished in both SDH mutants without a corresponding compensation in NADH dehydrogenase activity. As a result, complex IV activity was decreased in the SDH mutants [18] (Figure 6).

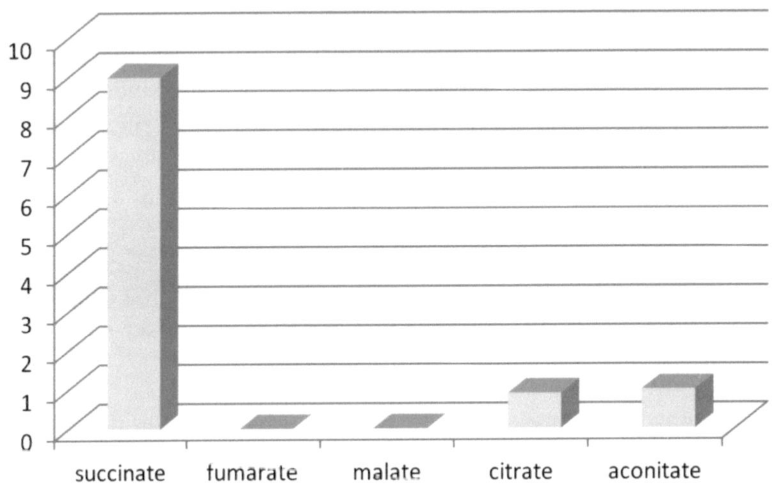

Figure 6. Relative concentration of main metabolities of TCA cycle for *sdh1*Δ, nor *sdh2*Δ yeast cells [18].

4. SDH activity assay

The succinate is the most efficient energy source, so the SDH activity assay can be an important method for measurement of the yeast vitality in scope to control, e.g. different fermentation processes [19]. SDH activities can be measured *in vitro* in cell lysates or in mitochondrial fraction as well as *in situ* in individual cells. Since SDH is bound to the inner membrane, it is easily isolated along with the mitochondria by different techniques: sucrose density gradient ultracentrifugation, free-flow electrophoresis or a commercially available kit-based method [20]. The mitochondrial fraction is the source of the enzyme. Since none of the key components can be measured directly, the reaction succinate → fumarate is measured by monitoring the reduction of an artificial electron acceptor. To use an artificial electron acceptor, the normal path of electrons through the mitochondrial electron transport system must be blocked. This is accomplished by adding either sodium azide or potassium cyanide to the reaction mixture. These poisons inhibit the transfer of electrons from cytochrome a3 to the final electron acceptor, oxygen, thus electrons cannot be passed along by the preceding cytochromes and coenzyme Q. Instead, the electrons from SDH-FADH2 can be picked up by an artificial electron acceptor, such as the dye 2,6dichlorophenolindophenol (DCIP). The reduction of DCIP can be followed spectrophotometrically since the oxidized form of the dye is blue and the reduced form is colorless. This reaction can be summarized as

$$\text{SDH-FADH2} + \text{DCIP}_{\text{oxid.}} \rightarrow \text{SDH-FAD} + \text{DCIP}_{\text{red.}} + 2\text{H}^+$$

The change in absorbance, measured at 600 nm, can be used to follow the reaction over time [21]. To use an artificial electron acceptor, the normal path of electrons in the electron transport chain must be blocked. This is accomplished by adding either potassium cyanide or sodium azide to the reaction mixture. The rate of the disappearance of the blue color is proportional to the concentration of enzyme. The change in absorbance of the mixture is measured as a function of time and the enzyme concentration is determined from these data. Enzymatic reactions in yeasts are usually studied in cell-free extracts which requires disruption of cells and as consequence, inactivation of particular enzymes often can be observed. Generally we can conclud that determination of SDH enzyme activity has proved to be a difficult enzyme to extract from respiratory membrane whilst still retaining its *in vivo* properties. Most of the described extraction procedures were rather drastic and yielded soluble preparations of rather dubious integrity [8].

In recent years quantitative histochemical procedures has been proved to be a powerful research tool, especially in microphotometric assessment *in situ* of the specific activity of dehydrogenases in individual cells. These assays are simple and valid alternative to conventional biochemical techniques. Methods *in situ* can provide the cellular resolution necessary to determine enzyme-specific activities not only in whole cell preparations but also in distinct subcellular compartments [19].

Reduction of various tetrazolium salts by dehydrogenases of metabolically active cells leads to production of highly colored end products – formazans (Figure 7). The history of the tetrazolium salts and formazans goes back 100 years, to when Friese (1875) reacted benzene diazonium nitrate with nitromethane, to produce a cherry-red "Neue Verbindung". This was the first formazan. Nineteen years later, Von Pechmann and Runge (1894) oxidised a formazan to produce the first tetrazolium salt [21].

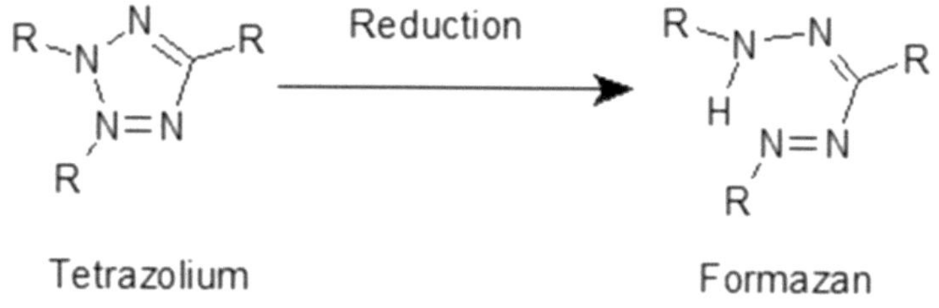

Figure 7. Tetrazolium salt and its coloured formazan.

Many hundreds of tetrazolium salts and formazans were prepared in the following years, but only a handful have found applications in biological research.There is a wide range of tetrazolium salts commonly used in the field of microbiology from the classical ones to the new generation of its derivatives. Among them are: blue tetrazolium chloride (BT), 2,3,5-triphenyl tetrazolium chloride (TTC), 3-(4,5-dimethylthiazol-2-yl)-2,5-diphenyltetrazolium bromide (MTT), 5-cyano-2,3-ditolyl tetrazolium chloride (CTC), 2,3-bis(2-methoxy-4-nitro-5-sulphophenyl)-5-[(phenylamino)carbonyl]-2H-tetrazolium hydroxide (XTT), 4-[3-(4-idophenyl)-2-(4-nitrophenyl)-2H-5-tetrazolio]-1,3-benzene disulfonate (WST1), 2-(p-iodophenyl)-3(p-nitrophenyl)-5-phenyltetrazolium chloride (INT) or 2,2'-dibenzothiazolyl-5,5'-[4-di(2-sulfoethyl)carbamoylphenyl]-3,3'-(3,3'-dimethoxy-4,4' biphenyl) ditetrazolium, disodium salt (WST-5) [19, 22, 23].

In the case of enzymatic reaction conducted *in situ* the plasma membrane forms a barrier with low degree of penetration. Therefore, cell permeabilization, e.g. by digitonin, is recommended as an alternative method for the study of intracellular enzyme activities. According to the results obtained by Berlowska et al. [23] digitonin was effective in membrane permeabilization without negative influence on cell morphology. After digitonin treatment, the visible formazan crystals were observed inside the yeast cells, but not outside them (Figures 8-10 A, B).

The formazan products are water-insoluble, but readily diffuses out of yeast cells after solubilization in DMSO. Good correlation (R^2=0,97) between BTf absorbance intensity after DMSO extraction and number of yeast cells was seen. Linear correlation was observed in the concentration range of yeast cells from 9×10^7 to5×10^8 per sample. For yeast cell concentrations below 1×10^7 per sample the formazan color intensity signals were too low to detect with good precision. The results obtained for SDH activity were in good agreement

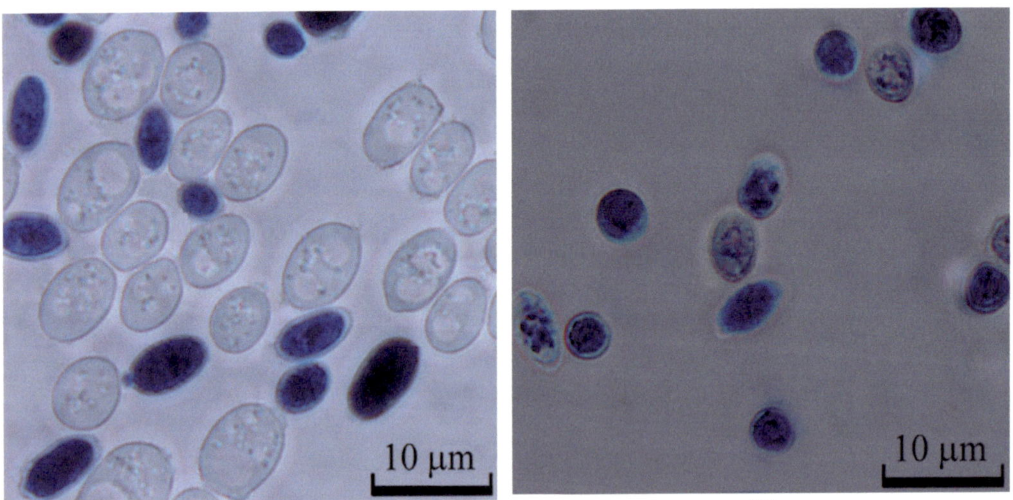

(A) (B)

Figure 8. Yeast cells after reaction with blue tetrazolium chloride (BT). A – without permeabilization; B – with permeabilization by 0.05% digitonin. Images of light microscopy.

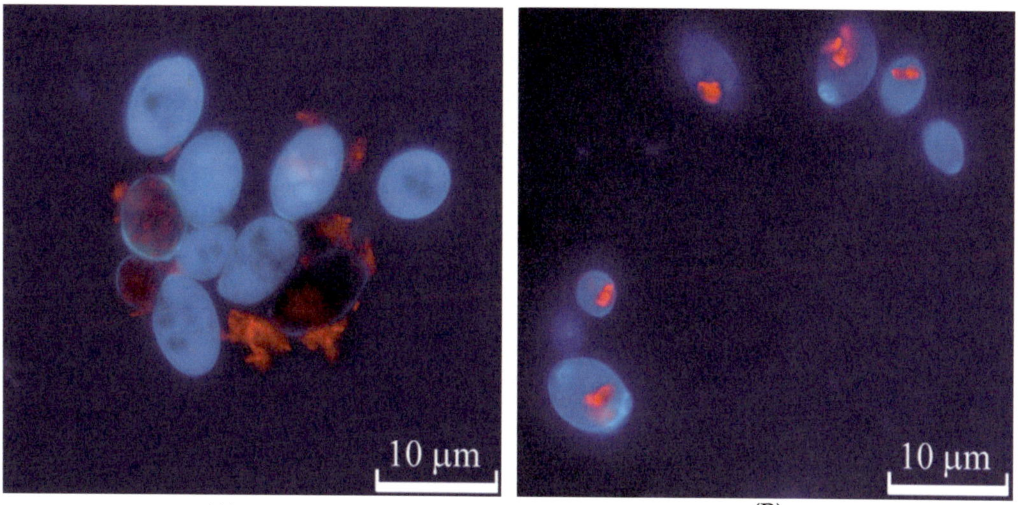

(A) (B)

Figure 9. Yeast cells after reaction with 2,3,5-triphenyl tetrazolium chloride (TTC). A – without permeabilization; B – with permeabilization by 0.05% digitonin. Images of fluorescence microscopy.

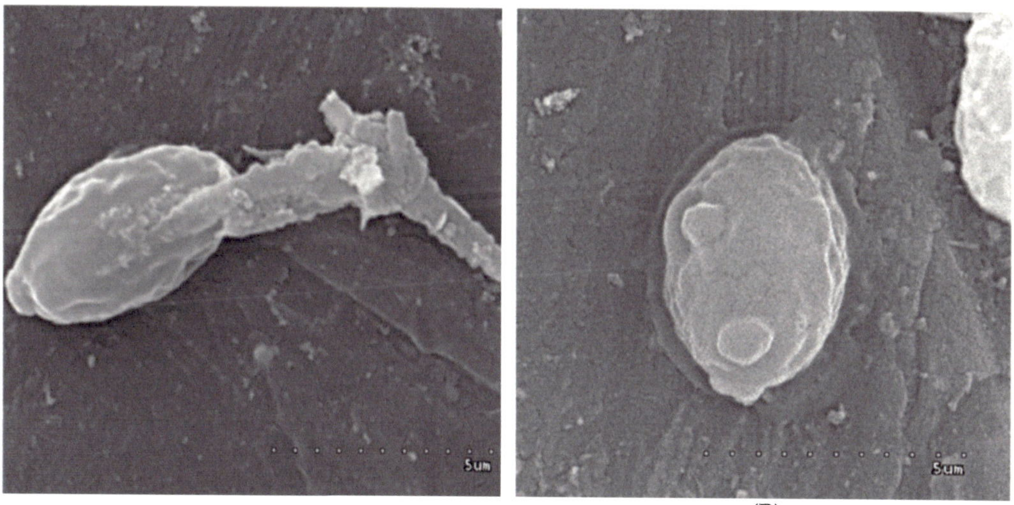

(A) (B)

Figure 10. Yeast cell after reaction with 2,3,5-triphenyl tetrazolium chloride (TTC). A – without permeabilization; B – with permeabilization by 0.05% digitonin. Images of scanning microscopy.

with that of ATP content in yeast cells. Significant decreasing of succinate dehydrogenase activity and ATP content were observed during aging of tested yeast strains [19, 23].

5. The role of SDH in human disease

Saccharomyces cerevisiae is a simple eukaryotic organism, with a complete genome sequence. Many genetic tools that have been created during these years, including the complete collection of gene deletions and a considerable number of mechanisms and pathways existing in higher eukaryotes was first studied and described in yeast. Moreover, about 40% of human genes whose mutations lead to diseases have an orthologue in yeast and genomic screens have been extended to mitochondrial diseases. The study of mitochondrial functions and dysfunction is of special interest in yeast because it is in this organism that mitochondrial genetics and recombination have been discovered and that nucleomitochondrial interactions have been studied in-depth. There are also specific reasons for choosing *S. cerevisiae* for mitochondrial studies. This organism is petite-positive, which can successfully grow in the absence of oxygen. Therefore it can lose its mitochondrial genome provided it is supplied with a substrate for fermentation. Consequently, all mutations of the mitochondrial genome can be studied without cell lethality. The frequency of homologous recombination is very high (1% recombination is considered to correspond to about 100 bp in the mitochondrial genome). It is genetically easy to transfer mitochondria from one nuclear genetic background to another via karyogamy. Additionally, mitochondria can be transformed making in vitro mutation analysis possible. The richness and ease of yeast molecular genetics opens big opportunities, and even the major difference existing between human and yeast mitochondrial genomes, i.e. the predominant heteroplasmy of human and the homoplasmy of yeast, can result in the easier definition of the pathogenic mutations. To review mitochondrial diseases may be a very difficult task because the definition might include different kinds of metabolic disorders or degenerative syndromes [24]. Moreover, some important aspects have been extensively reviewed and the reader might refer to very good recent articles by DiMauro and Garone [25] for historical aspects, by Wallace et al. [26, 27] for bioenergetics, by Spinazzola and Zeviani [28] for nucleo–mitochondrial intergenomic cross-talk. The previous review by Schwimmer et al. [29] was given an important outline of yeast models of mitochondrial diseases. SDH in yeast and human are very similar. They are composed of four subunits (SDHA-D, like SDH1-4 in yeast), all encoded by nuclear genes.

In the last ten years, deficiencies in TCA cycle enzymes have been shown to cause a wide spectrum of human diseases. For instance, mutation in the gene encoding fumarase is a rare cause of encephalomyopathy and a far more common cause of leiomyomas of the skin and uterus and of renal cancer (Table 1).

The TCA path dysfunction may also result from concurrent impairments in several steps of the cycle. The combined deficiencies in SDH and aconitase was observed in Friedreich's ataxia [22, 39]. Measures in autopsied brains from Alzheimer's Disease (AD) patients reveal a decrease in the activity of α-ketoglutarate dehydrogenase complex (KGDHC) and an increase in malate dehydrogenase (MDH) activity [33]. The ratios between TCA enzymes are consistent for each mammalian tissues presumably reflecting their metabolic demand. Consequently, in addition to the determination of residual absolute activities, estimation of ratios between enzyme activities is an effective means of detecting partial but potentially harmful deficiencies. When used to assess respiratory chain activities, this approach enabled the identification of several gene mutations, even in patients with partial respiratory chain deficiencies. At present, TCA enzyme activities are measured using a series of independent

Enzyme	Clinical presentation	References
Fumarase	Progressive encephalopathy Hereditary leiomyomatosis and renal cell cancer	[30] [31]
Malate dehydrogenase	Alzheimer's disease	[32, 33]
Citrate synthase	No disease identified so far	
Aconitase	No disease identified so far	
Isocitrate dehydrogenase	Low-grade gliomas	[34]
α-Ketoglutarate dehydrogenase	Congenital lactic acidosis	[35]
Succinyl-CoA ligase	Encephalomyopathy with mtDNA depletion	[36]
Succinate dehydrogenase	Encephalopathy (Leigh syndrome) Pheochromocytoma and paraganglioma	[37] [38]

Table 1. Primary deficiencies in TCA cycle enzymes in humans [22].

assays that are both laborious and time-consuming. The limited set of assays allowing both measurement of all TCA enzyme activities and detection of abnormalities in enzyme activity ratios were developed. These assays were used successfully to detect severe and partial isolated deficiencies in several TCA enzymes. The first assay measures succinyl-CoA ligase, SDH, glutamate dehydrogenase (GDH), fumarase, and malate dehydrogenase. This assay was performed in medium containing 50 mM KH2PO4 (pH 7.2) and 1 mg/ml BSA. The reduction of DCPIP was measured using two wavelengths (600 nm and 750 nm) with various substrates and the electron acceptors decylubiquinone and phenazine methosulfate. The second assay measured α-ketoglutarate dehydrogenase, aconitase, and isocitrate dehydrogenase activities. The pyridine nucleotide (NAD+/NADP+) reduction is measured with various substrates using wavelengths of 340 nm and 380 nm. In the third assay, citrate synthase was measured by monitoring dithionitrobenzene (DTNB; Ellman's reagent) reduction at wavelengths of 412 nm and 600 nm [22].

The deleterious mutations in any of the SDH genes should invariably result in a decreased SDH activity [13]. Hence, SDH 'inactivation' induces abnormal stimulation of the hypoxia-angiogenesis pathway. Therefore, the striking phenotypic differences associated with mutations in the four subunits raise puzzling questions. It has been reported that inherited deficiencies of SDH associated with presence of mutant protein Fp are always associated with relatively high residual activities, ranging from 25-50% of control mean values. As a comparison, less than 5% residual activity is frequently measured in patients with severe defect of complex IV or complex I. The mutations in any of the SDH cause the complex II to fully disassemble. When complex II is absent, it can be disregarded as a source of additional superoxide production. Thus, the superoxide overproduction would lead to tumour formation that should be ascribed to the decreased ability of the SDH to adequately reduce the Q pool, a necessary condition to resist oxidative stress [8].

Ubiquinone, beside its function in the respiratory chain as an electron carrier mediating electron transfer between the various dehydrogenases and the cytochrome path, is working as a powerful antioxidant in biological membranes [13]. It is possibly for this exact reason in much larger amounts compared to other electron carriers of the respiratory chain, including the sum of the dehydrogenases. When it is defective, the respiratory chain can produce an abnormal amount of superoxides involving additional respiratory chain components such as flavin radicals of complex I. Delivering electrons for the full reduction of Q to QH2 might then be of a tremendous importance for the control of oxygen toxicity in the mitochondria. Therefore, the SDH is a key enzyme to control Q pool redox poise under these conditions, due to its unique redox properties [8].

Iron-sulfur (Fe-S) proteins facilitate multiple functions, including redox activity, enzymatic function, and maintenance of structural integrity. More than 20 proteins are involved in the biosynthesis of iron-sulfur clusters in eukaryotes. Defective Fe-S cluster synthesis not only affects activities of many iron-sulfur enzymes, such as aconitase and succinate dehydrogenase, but also alters the regulation of cellular iron homeostasis, causing both mitochondrial iron overload and cytosolic iron deficiency. Fe-S cluster biogenesis takes place essentially in every tissue of humans, and products of human disease genes have important roles in the process [40].

Succinate is an oxygen sensor in the cell and can help turn on specific pathways that stimulate cells to grow in a low-oxygen environment (hypoxia). In particular, succinate stabilizes a protein called hypoxia-inducible factor (HIF) by preventing a reaction that would allow HIF to be broken down. HIF controls several important genes involved in cell division and the formation of new blood vessels in a hypoxic environment. Mutations in genes encoding SDH subunits have been linked to severe encephalopathy and, more recently, to familial paraganglioma (PGL) and pheochromocytoma (PHEO) (PHEO: adrenal gland PGL). Isolated complex II deficiency is a very rare condition, occurring in approximately 2–4% of all respiratory chain enzyme deficiencies. At least three mutations in the SDH genes have been identified in people with PGL or PHEO, which are noncancerous (benign) tumors associated with the nervous system. *SDHB-D* gene mutations are seen most commonly in people with PGL, but they were found in people with PHEO. However, a single mutation in the *SDHA* gene increases the risk that an individual will develop the condition, and additional mutation that deletes the normal copy of the gene is needed to cause tumor formation. This second mutation, called a somatic mutation, is acquired during a person's lifetime and is present only in tumor cells. The SDH genes mutations associated with nonsyndromic PGL or PHEO change single protein building blocks (amino acids) in the SDH protein sequence or result in a shortened protein. As a result, there is little or no SDH enzyme activity. Because the mutated SDH enzyme cannot convert succinate to fumarate, succinate accumulates in the cell. The excess succinate abnormally stabilizes HIF, which also builds up in cells. Excess HIF stimulates cells to divide and triggers the production of blood vessels when they are not needed. Rapid and uncontrolled cell division, along with the formation of new blood vessels, can lead to the development of tumors.

Mutations in the *SDHA* gene were identified in a small number of people with Leigh syndrome, a progressive brain disorder that typically appears in infancy or early childhood. Affected children may experience vomiting, seizures, delayed development, muscle weakness, and problems with movement. Heart disease, kidney problems, and difficulty breathing can also occur in people with this disorder. The one child died suddenly at the age of five months from a severe deterioration of neuromuscular, cardiac, and hepatic symptoms after an intermittent infection. The *SDHA* gene mutations responsible for Leigh syndrome change single amino acids in the SDHA protein or result in an abnormally short protein. These genetic changes disrupt the activity of the SDH enzyme, impairing the ability of mitochondria to produce energy. Further studies showed that several patients with complex II deficiency do not have mutations of *SDHA*. This suggested a role of additional nuclear genes involved in synthesis, assembly, or maintenance of SDH. It is not known, however, how mutations in the *SDHA* gene are related to the specific features of Leigh syndrome [41, 42].

Two plausible hypotheses have been proposed to explain the peculiar linkage between disruption of electron flow through mitochondrial complex II and tumorigenesis in neuroendocrine cells. In the reactive oxygen species (ROS) hypothesis, it is proposed that an intact, catalytically active SDHA subunit generates genotoxic ROS by uncoupled electron flow from succinate to oxygen or water in cells where one of the electron-carrying subunits (SDHB, SDHC or SDHD) is missing or inactive. The ROS model implies that genotoxic ROS mutagenize nuclear proto-oncogenes or tumor suppressors (Figure 11). This model predicts that ROS should be increased in cells lacking SDHB, SDHC or SDHD, but not when SDHA is missing. Although certain mutations in these genes result in ROS production in *Saccharomyces cerevisiae* and mammalian cell lines, it is not clear that ROS accumulate to levels that are mutagenic.

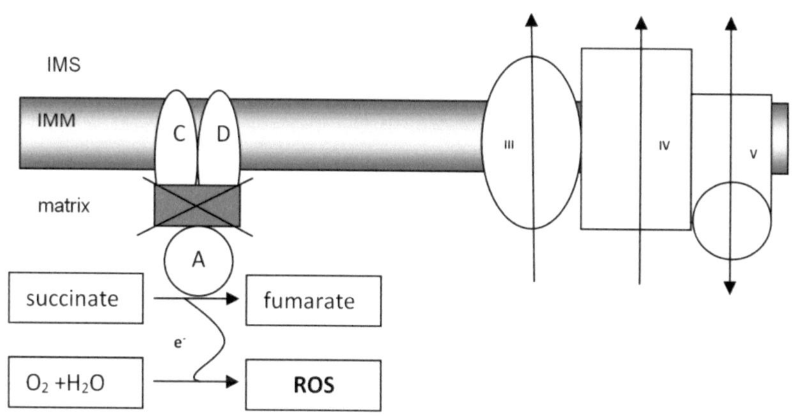

Figure 11. ROS model [18].

In the succinate accumulation hypothesis, the loss of SDHB results in loss of SDH activity and causes succinate accumulation (Figure 12).

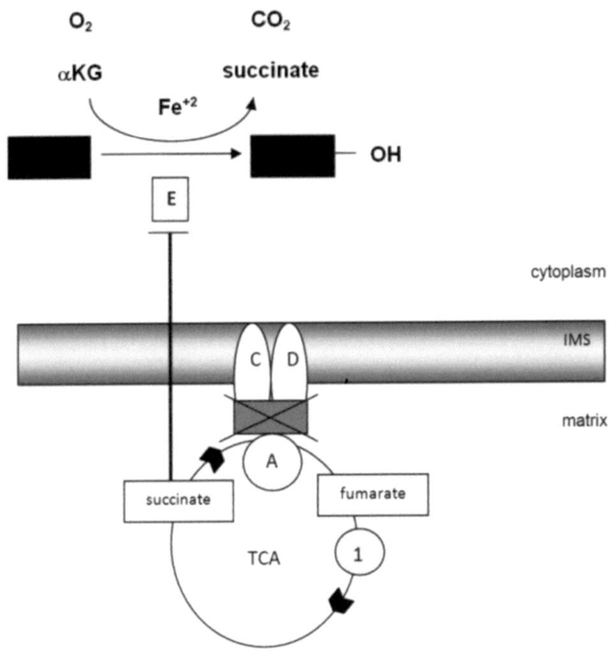

Figure 12. Succinate accumulation model [18].

Excess succinate is shuttled from the mitochondrial matrix to the cytoplasm, where it inhibits any of several aKG-dependent enzymes (E) that regulate levels or activities of important regulatory proteins (black box). The loss or inactivation of SDHB, C or D proteins yields a catalytically inactive SDHA subunit, resulting in blockade of the TCA cycle and diffusion of accumulated succinate to the cytoplasm. Succinate can then act as an inhibitor of a-ketoglutarate-dependent enzymes that use ferrous iron and molecular oxygen as cofactors to hydroxylate their substrates and generate succinate as a product. It has been demonstrated that two a-ketoglutarate -dependent enzymes, the prolyl hydroxylases, are inhibited by succinate accumulation in cells that have lost SDHD function. Smith et al [18] reported that yeast cells disrupted in *SDH2* (*sdh2*Δ) show increased in ROS production and protein oxidation without detectable increase in DNA damage. More strikingly, *sdh2*Δ cells dramatically accumulate succinate resulting in inhibition of at least two aKG-dependent enzymes that generate succinate as a by-product [18].

6. Metabolic engineering of yeast SDH for the biotechnological processes

Metabolic engineering, i.e., the intentional redirection of metabolic fluxes, plays an exceptional role in improving yeast strains for all industrial applications. In contrast to classical methods of genetic strain improvement such as selection, mutagenesis, mating, and hybridization, metabolic engineering confers two major advantages: (1) the directed modification of strains without the accumulation of unfavorable mutations and (2) the introduction of genes from foreign organisms to equip *S. cerevisiae* with novel traits. The latter is particularly crucial for industrial biotechnology to provide pathways that extend the spectrum of usable industrial media (e.g., lignocellulosic biomass) and/or to produce compounds not naturally formed by *S. cerevisiae*. Since the first introduction of metabolic engineering, there have been tremendous enhancements of its toolbox, and several related disciplines have emerged, such as inverse metabolic engineering and evolutionary engineering. These developments have strongly influenced yeast strain improvement programs in the past few years and have greatly enhanced the potential for using yeast in biotechnological production processes [43].

The main goals of metabolic engineering can be summarized in the following four categories: (1) improvement of yield, productivity and overall cellular physiology, (2) extension of the substrate range, (3) deletion or reduction of by-product formation and (4) introduction of pathways leading to new products. Commonly these goals can be achieved by a three-step procedure. Firstly, a genetic modification is proposed, based on metabolic models. After genetic modification, the recombinant strain is analysed and the results are then used to identify the next target for genetic manipulation, if necessary. Thus, the construction of an optimal strain involves a close interaction between synthesis and analysis, usually for several consecutive rounds. The rapid development and frequent success in this field is demonstrated by the large number of reviews about the theoretical and practical aspects of metabolic engineering. Knowledge of cellular and microbial physiology, as well as the underlying metabolic networks or enzymes, is an important prerequisite for successful engineering. A new term, 'inverse metabolic engineering' (IME) coins to encompass the construction of strains with a particularly desirable physiological phenotype, e.g. enhanced production of heterologous protein [44].

Recently, a computational approach for the identification of every possible biochemical reaction from a given set of enzyme reaction rules was reported. This analysis suggested that the native pathways are thermodynamically more favorable than the alternative possible pathways. The pathways generated involve compounds that exist in biological databases, as well as compounds that exist in chemical databases and novel compounds, suggesting novel biochemical routes for these compounds and the existence of biochemical compounds that remain to be discovered or synthesized through enzyme and pathway engineering [45].

Due to its importance in traditional biotechnology such as baking, brewing, and wine making, research activities historically have focused on the yeast *Saccharomyces cerevisiae.* It is relatively tolerant to low pH values and high sugar and ethanol concentrations, i.e., properties which lower the risk of contamination in industrial fermentation. These features are the major reasons for increasing *S. cerevisiae* exploration in industrial ("white") biotechnology, focusing on the fermentative production of industrially relevant biochemicals, e.g., glycerol, propanediol, sugar alcohols, organic acids, etc. Among these compounds, several organic acids may fulfill a role as platform molecules using their (multiple) functional groups as a target for enzymic or chemical catalysis [43].

In the United States were identified 10 organic acids as key chemical building blocks [44]. Similarly, the European focus group BREW identified 21 key compounds that can be produced from different, including renewable sources, a number of which were organic acids [45]. One example of such a chemical is succinic acid. Succinic acid is used as a surfactant, detergent or foaming agent, as an ion chelator, and also in the food industry as an acidulant, flavoring agent or anti-microbial agent, as well as in health-related products (such pharmaceuticals and antibiotics). Currently, it is produced from petrol and is too expensive to be used as a general building-block chemical. However, provided that its price becomes competitive, succinic acid could replace petrol-derived maleic anhydride in chemical synthesis processes in the future [46-47]. Similar chemical derivatizations can be applied to malic and fumaric acid, so that they can also be considered interesting C_4 building blocks [48-53].

The chemical behavior of the dicarboxylic acid – succinic acid is determined principally by its two carboxyl groups. This substance is either directly utilized in the pharmaceutical or chemical industry or represent building block or precursor for further chemical or enzymatic syntheses. The following reactions and derivatives are considered interesting: (1) reductions of succinic acid to 1,4-butanediol, γ-butyrolactone, tetrahydrofuran and its derivatives; (2) reductive amination of succinic acid or γ-butyrolactone to pyrrolidiones; (3) polymerization of succinic acid with diols (building block of polyesters); (4) polymerisation of succinic acid with diamines to form polyamides, etc. The examples of the substances that can be derived from succinic acid are shown in table 2.

Butanediol, tetrahydrofuran and γ-butyrolactone, are standard substances for the chemical industry. These are used as solvents, as well as for fiber and polymer production. Dimethylsuccinate is one of the so-called dibasic esters that have great potential as solvents with environmentally benign characteristics. Thus, the potential market volume for succinic acid is high, fuelling substantial efforts to establish a microbial process for succinic acid production [47].

The chemical synthesis of succinic acid is predominantly based on maleic anhydride and requires heavy metal catalysts, organic solvents, high temperatures and high pressures. It makes the conversion of maleic anhydride to succinic acid costly and ecologically questionable [49]. On the other hand, succinate is produced naturally by many microorganisms as an intermediate of the central metabolism or as a fermentation end product.

The succinate producers include bacterial strains, e.g. *Mannheimia succiniproducens* [54]. However, none of these microorganisms are currently used in industry. Some prokaryotes

No	Name	Chemical structure	Examples of uses
1	1,4 -Diaminobutane	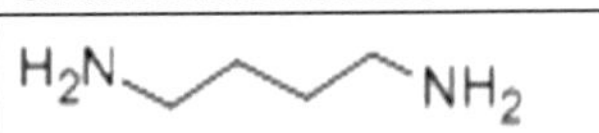	Used for polyamide production
2	Succindiamide	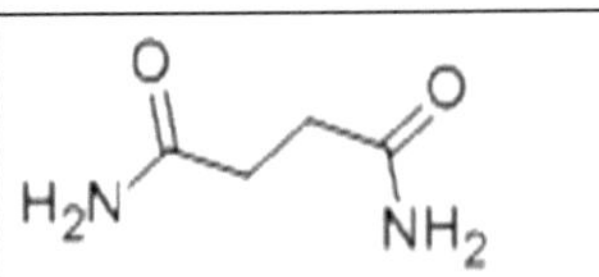	Used as anticonvulsant drugs and to form covalent bonds between proteins or peptides and plastics

No	Name	Chemical structure	Examples of uses
3	1,4 – Butanediol (BDO)	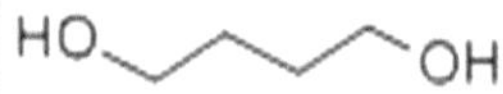	Replacement of fossil-derived BDO by bio-based BDO in polybutylene tereftalate (PBT) Replacement of fossil-derived BDO by bio-based BDO in polybutylene succinate (PBS)
4	Succinonitrile		Precurser for industrial production of polyamides
5	Dimethylsuccinate	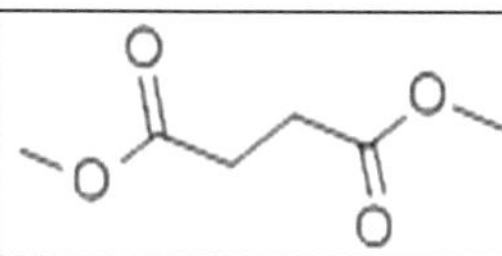	Used as fuel oxygenates and as green solvents
6	N-methyl-pyrolidone (NMP)	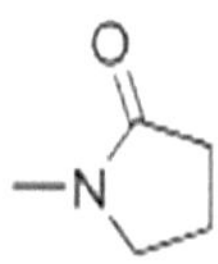	Replacement of fossil-derived NMP by bio-based NMP; NMP is an important, versatile solvent for the chemical industry
8	2-Pyrolidone	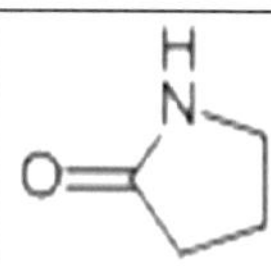	As above
7	γ-Butyrolactone (GBL)	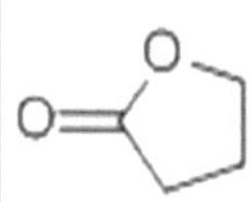	Replacement of fossil-derived BGL by bio-based BGL ; BGL is important as an intermediate in the manufacture of pyrrolidone derivatives and as a solvent for polymers and agrochemicals
9	Tetrahydrofuran (THF)	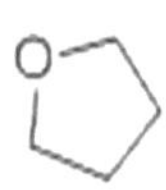	Replacement of fossil-derived THF by bio-based THF ; THF is mainly used as a solvent and as an intermediate in the production of thermoplastic polyurethanes, elastic fibers etc.

Table 2. Examples of various substances that can be derived from succinic acid by chemical conversion [47].

have complex medium requirements and they are generally unable to grow and produce organic acids at the low pH values. These restrictions provide strong incentives to integrate and optimize succinate production pathways in other microorganisms via metabolic engineering approaches.

The popularity of *S. cerevisiae* in basic and applied research is undoubtedly influenced by its classification as GRAS (generally regarded as safe) by the U.S. Food and Drug Administration (FDA). Baker's yeast *S. cerevisiae* was the first eukaryotic organism in which complete genomic sequence was determined. Several databases such as the *Saccharomyces* Genome Database (http://www.yeastgenome.org/) and the Comprehensive Yeast Genome Database (http://mips.gsf.de/genre/proj/yeast/) contain an enormous amount of information concerning *S. cerevisiae* genes, open reading frames, and gene products. The yeast *S. cerevisiae* became a well established eukaryotic model organism to study fundamental biological processes such as aging, mRNA transport, the cell cycle, and many more. *Saccharomyces cerevisiae* grows well in a simple chemically defined medium, under acidic conditions, even at pH values equal 3. At such low pH values, many weak acids, including succinate, occur predominantly in their undissociated form. This is advantageous for industrial production, as it reduces the need for titration with alkali and allows for direct recovery of undissociated acids. Consequently, there is no need for large quantities of acidifying agents, and the formation of salt byproducts (e.g. gypsum) is strongly reduced. In addition, *S. cerevisiae* robust tolerance in acidic conditions represents a major advantage in that it lowers the risk of contamination in industrial fermentation [44, 49].

The yeast-based fermentation process, which operates at a much lower pH than competing processes, allows succinic acid to be produced with a significantly higher energy efficiency compared to the traditional method. This compound is not accumulated intracellularly. It is also one of the first bio-based processes that sequesters carbon dioxide in the production process [47, 53]. This makes the yeast *Saccharomyces cerevisiae* a suitable and promising candidate for the biotechnological production of succinic acid on an industrial scale.

The metabolic engineering strategy was used for the oxidative production of succinic acid by deletion SDH1, SDH2 genes in the genome. Arikawa et al. [55] reported an increased succinic acid productivity in sake yeast strains with deletions of TCA cycle genes. In comparison to the wild-type, succinate levels were increased up to2.7 fold in a strain with simultaneous disruption of a subunit of succinate dehydrogenase (SDH1) and fumarase (FUM1) under aerobic conditions. The single deletion of gene SDH1 led to a1.6-fold increase of succinic acid production. In another study on sake yeast strains, the deletion of genes encoding for succinate dehydrogenase subunits (SDH1, SDH2, SDH3, and SDH4) also resulted inincreased succinate productivity in aerobic conditions. Raab et al. [48] reported the construction of yeast strains for the biotechnological production of succinic acid. The genes *SDH1, SDH2, IDH1* and *IDP1*, which encode mitochondrial enzymes were deleted with the aim to disrupt succinate and isocitrate dehydrogenase activity to redirect the carbon flux and to allow succinate to accumulate as an end-product. This study showed that the yeast *S. cerevisiae* is capable of synthesizing significant amounts of succinic acid, which is exported quantitatively into the culture broth and not being accumulated intracellularly.

The constructed yeast strains with disruptions in the TCA cycle produced succinic acid up to 3.62 g/L at a yield of 0.11 mol /mol glucose in shake flask cultures.

Saccharomyces cerevisiae is one of the most highly researched model organisms in different biological studies. Using this yeast we can effectively re-examine long-standing and fundamental questions regarding regulation of metabolism and prediction of dynamic models in various cells, including mammalian tissues. This is a considerable knowledge about the composition, enzymology and membrane binding of the enzyme and relatively new discoveries about its genetics and biosynthesis. Through such efforts, we are able to identify key features of cellular metabolic pathways which can be use both in medicine and in different biotechnological processes.

Author details

Dorota Kregiel
Institute of Fermentation Technology and Microbiology, Technical University of Lodz, Poland

Acknowledgement

I wish to thank dr Joanna Berlowska from Technical University of Lodz for her help in the making of figures and selection of images of yeast cells.

7. References

[1] McCammon MT, Epstein CB, Przybyla-Zawislak B, McAlister-Henn L, Butow RA (2003) Global transcription analysis of Krebs tricarboxylic acid cycle mutants reveals an alternating pattern of gene expression and effects on hypoxic and oxidative genes. Mol. Biol. Cell 14: 958–972.

[2] Kruckeberg AL, Dickinson JR (2004) Carbon metabolism. In: Dickinson JR, Schweizer M, editors. The metabolism and molecular physiology of *Saccharomyces cerevisiae*. London: CRC, pp. 42–103.

[3] Sass E, Blachinsky E, Karniely S, Pines O (2001) Mitochondrial and cytosolic isoforms of yeast fumarase are derivatives of a single translation product and have identical amino termini. J Biol Chem 276: 46111–46117.

[4] Steffan JS, McAlister-Henn L (1992) Isolation and characterization of the yeast gene encoding the MDH3 isozyme of malate dehydrogenase. J Biol Chem 267: 24708–24715.

[5] DeRisi JL, Iyer VR, Brown PO (1997) Exploring the metabolic and genetic control of gene expression on a genomic scale. Science 278: 680–668.

[6] Gombert AK, Moreira dos Santos M, Christensen B, Nielsen J (2001) Network identification and flux quantification in the central metabolism of *Saccharomyces cerevisiae* under different conditions of glucose repression. J Bacteriol 183: 1441–1451.

[7] Modica-Napolitano JS, Kulawiec M, Singh KK 2007 Mitochondria and human cancer. Curr Mol Med 7: 121-131.

[8] Hajjawi OS (2011) Succinate dehydrogenase: assembly, regulation and role in human disease. Eur J Sci Res 51: 133-142

[9] Robinson KM, Lemire BD (1996) Covalent attachment of FAD to the yeast succinate dehydrogenase flavoprotein requires import into mitochondria, presequence removal, and folding. J Biol Chem 271: 4055-4060.

[10] Oyedotun KS, Lemire BD (2004) The quaternary structure of the *Saccharomyces cerevisiae* succinate dehydrogenase. The J Biol Chem 279: 9424–9431.

[11] Oyedotun KS, Yau PF, Lemire BD (2004) Identification of the heme axial ligands in the cytochrome *b*562 of the *Saccharomyces cerevisiae* succinate dehydrogenase. The J Biol Chem 279: 9432–9439.

[12] Yankovskaya V, Horsefield R, Törnroth S, Luna-Chavez C, Miyoshi H, Léger C, Byrne B, Cecchini G., Iwata S (2003) Architecture of succinate dehydrogenase and reactive oxygen species generation. Science 299: 700-704

[13] Rustin P, Munnich A, Rötig A (2002) Succinate dehydrogenase and human diseases: new insights into a well-known enzyme Eur J Human Gen 10: 289 – 291.

[14] Hajjawi OS. and Hider RC (2009) Asymmetry of the malonate transport system in human red blood cells. Eur J Sci Res 31: 534-545.

[15] Benning MM, Meyer TF, Rayment I, Holden HM (1994) Molecular structure of the oxidized high potential iron-sulfur protein isolated from *Ectothiorhodospira vacuolata*. Biochemistry 33: 2476-2483.

[16] Huang LS, Shen JT, Wang AC, Berry EA (2006) Crystallographic studies of the binding of ligands to the dicarboxylate site of complex II, and the identity of the ligand in the 'oxaloacetate-inhibited' state. Biochim Biophys Acta 1757: 1073–1083.

[17] Jay D, Jay EG, Garcia C (1993) Inhibition of membrane-bound succinate dehydrogenase by fluorescamine. J Bioenerg Biomembr 25: 685-688.

[18] Smith EH, Janknecht R, Maher LJ (2007) Succinate inhibition of α-ketoglutarate-dependent enzymes in a yeast model of paraganglioma. Hum Mol Genet 24: 3136-3148.

[19] Kregiel D, Berlowska J, Ambroziak W (2008) Succinate dehydrogenase activity assay in situ with blue tetrazolium salt in Crabtree-positive *Saccharomyces cerevisiae* strain. Food Technol Biotechnol 46: 376–380.

[20] Hartwig S, Feckler C, Lehr S Wallbrecht K, Wolgast H, Müller-Wieland D, Kotzka J (2009) A critical comparison between two classical and a kit-based method for mitochondria isolation. Proteomics 9. 3209 3214.

[21] Altman FP (1976) Tetrazolium salts and formazans. Prog Histochem Cytochem 9: 1-56.

[22] Goncalves S, Paupe V, Dassa EP, Brière J-J, Favier J, Gimenez-Roqueplo A-P, Bénit P, Rustin P (2010) Rapid determination of tricarboxylic acid cycle enzyme activities in biological samples. BMC Biochemistry. Available: http://www.biomedcentral.com/1471-2091/11/5. Accessed: 2012 April 05.

[23] Berłowska J, Kregiel D, Klimek L, Orzeszyna B, Ambroziak W (2006) Novel yeast cell dehydrogenase activity assay in situ. Pol J Microbiol 55:127-131.

[24] Rinaldi T, Dallabona C, Ferrero I, Frontali L, Bolotin-Fukuhara M (2010) Mitochondrial diseases and the role of the yeast models. FEMS Yeast Res 10: 1006-1022.

[25] DiMauro S, Garone C (2010) Historical perspective on mitochondrial medicine. Develop Dis Res Rev Special Issue: Emerging Research in Mitochondrial Disease 16: 106-113.
[26] Wallace DC (2010) The epigenome and the mitochondrion: bioenergetics and the environment Genes Dev 24: 1571-1573.
[27] Wallace DC, Fan W, Procaccio V (2010) Mitochondrial energetics and therapeutics. Annu Rev Pathol 5: 297–348.
[28] Spinazzola A, Zeviani M (2009) Disorders from perturbations of nuclear-mitochondrial intergenomic cross-talk J Int Med 265; 174–192.
[29] Schwimmer C, Rak M, Lefebvre-Legendre L, Duvezin-Caubet S, Plane G, di Rago J-P (2006) Yeast models of human mitochondrial diseases: from molecular mechanisms to drug screening. Biotechnol J 3: 270-81.
[30] Bourgeron T, Chretien D, Poggi-Bach J, Doonan S, Rabier D, Letouze P, Munnich A, Rotig A, Landrieu P, Rustin P (1994) Mutation of the fumarase gene in two siblings with progressive encephalopathy and fumarase deficiency. J Clin Invest 93: 2514-2518.
[31] Tomlinson IP, Alam NA, Rowan AJ, Barclay E, Jaeger EE, Kelsell D, Leigh I, Gorman P, Lamlum H, Rahman S et al (2002) Germline mutations in FH predispose to dominantly inherited uterine fibroids, skin leiomyomata and papillary renal cell cancer. Nat Genet 30: 406-410.
[32] Shi Q, Gibson G (2011) Up-regulation of the mitochondrial malate dehydrogenase by oxidative stress is mediated by miR-743a. Journal of Neurochemistry118: 440–448.
[33] Shi Q, Xu H, Kleinman WA, Gibson GE (2008) Novel functions of the α-ketoglutarate dehydrogenase complex may mediate diverse oxidant-induced changes in mitochondrial enzymes associated with Alzheimer's disease. Biochim Biophys Acta 1782: 229–238.
[34] Schiff D, Purow BW (2009) Neuro-oncology: isocitrate dehydrogenase mutations in low-grade gliomas. Nat Rev Neurol 5: 303-304.
[35] Odievre MH, Chretien D, Munnich A, Robinson BH, Dumoulin R, Masmoudi S, Kadhom N, Rotig A, Rustin P, Bonnefont JP (2005) A novel mutation in the dihydrolipoamide dehydrogenase E3 subunit gene (DLD) resulting in an atypical form of alpha-ketoglutarate dehydrogenase deficiency. Hum Mutat 25: 323-324.
[36] Elpeleg O, Miller C, Hershkovitz E, Bitner-Glindzicz M, Bondi-Rubinstein G, Rahman S, Pagnamenta A, Eshhar S, Saada A (2005) Deficiency of the ADP-forming succinyl-CoA synthase activity is associated with encephalomyopathy and mitochondrial DNA depletion. Am J Hum Genet 76: 1081-1086.
[37] Bourgeron T, Rustin P, Chretien D, Birch-Machin M, Bourgeois M, Viegas-Pequignot E, Munnich A, Rotig A (1995) Mutation of a nuclear succinate dehydrogenase gene results in mitochondrial respiratory chain deficiency. Nat Genet 11: 144-149.
[38] Selak MA, Armour SM, MacKenzie ED, Boulahbel H, Watson DG, Mansfield KD, Pan Y, Simon MC, Thompson CB, Gottlieb E (2005) Succinate links TCA cycle dysfunction to oncogenesis by inhibiting HIF-alpha prolyl hydroxylase. Cancer Cell 7: 77-85.
[39] Brie`re J-J, Favier J, Gimenez-Roqueplo A-P, Rustin P (2006) Tricarboxylic acid cycle dysfunction as a cause of human diseases and tumor formation . Am J Physiol Cell Physiol 291: C1114–C1120.

[40] Ye H, Rouault TA (2010) Human iron-sulfur cluster assembly, cellular iron homeostasis, and disease. Biochemistry 49: 4945–4956.

[41] Horváth R, Abicht A, Holinski-Feder E, Laner A, Gempel K, Prokisch H, Lochmüller H, Klopstock T, Jaksch M (2006) Leigh syndrome caused by mutations in the flavoprotein (Fp) subunit of succinate dehydrogenase (SDHA). J Neurol Neurosurg Psychiatry 77: 74–76.

[42] Hensen EF, Bayley J-P (2011) Recent advances in the genetics of SDH-related paraganglioma and pheochromocytoma. Fam Cancer 10: 355–363.

[43] Nevoigt E (2008)Progress in Metabolic Engineering of *Saccharomyces cerevisiae* Microbiol. Mol Biol Rev 72: 379-412.

[44] Abbott DA, Zelle RM, Pronk JT, Van Maris AJA (2009) Metabolic engineering of Saccharomyces cerevisiae for production of carboxylic acids: current status and challenges. FEMS Yeast Res 9: 1123-1136.

[45] Assessment of current activity in the production of platform chemicals from renewable sources and horizon scan to forecast potential future developments in science and technology activity in biocatalysis (2012). Paper IB IGT T&MWG 09/06 – 57 Available: http://www.bis.gov.uk/files/file51235.pdf. Accessed: 2012 April 07

[46] Bechthold I, Bretz K, Kabasci S, Kopitzky R, Springer A (2008) Succinic acid: A new platform chemical for biobased polymers from renewable resources. Chem Eng Technol 31: 647–654.

[47] Sauer M, Porro D, Mattanovich D, Branduardi P (2007) Microbial production of organic acids: expanding the markets Trends in Biotechnology 26: 100-108.

[48] Xu G, Liu L, Chen J (2012) Reconstruction of cytosolic fumaric acid biosynthetic pathways in *Saccharomyces cerevisiae*. Microb Cell Fact. Available: http://www.microbialcellfactories.com/content/pdf/1475-2859-11-24.pdf. Accessed 2012 April 14.

[49] Raab AM, Gebhardt G, Bolotina N, Weuster-Botz D, Lang C (2010) Metabolic engineering of *Saccharomyces cerevisiae* for the biotechnological production of succinic acid. Metab Eng, 12: 518-525.

[50] Kern A, Tilley E, Hunter IS, Legiša M, Glieder A (2007) Engineering primary metabolic pathways of industrial micro-organisms. J Biotechnol 129: 6–29.

[51] Hatzimanikatis V , Li C, Ionita JA, Henry CS, Jankowski MD, Broadbelt L J (2005) Exploring the diversity of complex metabolic networks. Bioinformatics 21: 1603-1609.

[52] Zelle RM, de Hulster E, Kloezen W, Pronk JT, van Maris AJA (2010) Key process conditions for production of C_4 dicarboxylic acids in bioreactor batch cultures of an engineered *Saccharomyces cerevisiae* strain. Appl Environ Microbiol 76: 744–750.

[53] Overview of new bio-building blocks and bio-polymers. Available: http://bioplastic-innovation.com/2011/08/06/4026/ . Accessed: 2012 April 07.

[54] Lee SJ, Song H, Lee SY (2006) Genome-based metabolic engineering of *Mannheimia succiniciproducens* for succinic acid production. Appl Environ Microbiol 72: 1939-1948.

[55] Arikawa Y, Kuroyanagi T, Shimosaka M, Muratsubaki H, Enomoto K, Kodaira R, Okazaki M (1999) Effect of gene disruptions of the TCA cycle on production of succinic acid in *Saccharomyces cerevisiae*. J Biosci Bioeng 87: 28–36.

Chapter 10

Study of Succinate Dehydrogenase and α-Ketoglutarate Dehydrogenase in Mitochondria Inside Glass-Adhered Lymphocytes Under Physiological Conditions – The Two Dehydrogenases as Counterparts of Adrenaline and Acetylcholine Regulation

Marie Kondrashova, Marina Zakharchenko, Andrej Zakharchenko,
Natalya Khunderyakova and Eugen Maevsky

Additional information is available at the end of the chapter

http://dx.doi.org/10.5772/50059

1. Introduction

In the fields of biochemistry and medicine there is growing interest to the crucial role of mitochondria in changing of physiological states and disease development. This interest is fueled by rapid progress in the modern branches of mitochondriology: Mitochondrial Physiology (MiP) and Mitochondrial Medicine [1-8]. Numerous investigations demonstrated mitochondrial dysfunctions in different experimental models of stress or disease. These results were obtained under clearly pronounced functional disorders induced by strong external stimuli or pathogenic mutations. These observations increased interest in the detection of mitochondrial dysfunctions in the organism as early biomarkers of pathogenic processes. The detection of mitochondrial dysfunction is crucial at subclinical stages of disease because it can lead to more effective prevention and earlier intervention. In spite of this benefit to date the available testing methods cannot properly measure the state of mitochondria in the organism in ex vivo observations. Past successes of respiration measurements in isolated mitochondria distracted attention from obvious, methodological shortcoming of the existing method of mitochondria isolation: intentional destroy of their native structure organization into entire network. It is known nowadays that the responses of mitochondria to various physiological challenges are based on the reversible dissociation of such subtle network structure. In the network state mitochondrial respiration proceeds at a low rate while excitation leads to the dissociation of the network and an increase in respiration. We demonstrated this shortcoming in simultaneous polarographic and microscopic studies of network fragments in KCl rat liver homogenate that contains a higher density of tissue than is usual (1:1 tissue:solution) [9-11].

Since the first step of isolation of mitochondria is network destruction, this method artificially mimics the physiological signal of the activation of respiration. Thus, mitochondria isolated from non-active tissues using the standard methodology do not maintain low levels of respiration typical of the quiescent state and show rapid respiration.

A matter of great concern of biochemists is primarily not to inhibit the enzyme activity; therefore, rapid respiration of isolated mitochondria did not cause apprehension. However, the hyperactivation of functions as an initial step of their alteration preceding inhibition is well known in physiology. That is why artificially accelerated respiration of isolated mitochondria means in essence the LOSS of data for norm in biochemical investigations. The importance of preservation of the native mitochondrial network in ex vivo observations was recently highlighted in a special issue of IJBCB assembled under the editorial guidance of R. Rossignol [12].

The modern investigations clearly demonstrate strong correspondence in functional state of mitochondria and their native organization in the network. This is formulated by the impressive motto: "To be in good shape - to survive" [13]. At the beginning of mitochondria isolation coryphaei of mitochondriology considered possible loss of native properties caused by preparation. H. Krebs attracted the attention to the acceleration of respiration of mitochondria by dilution. He preferred to study mitochondria in homogenate as more native. A. Lehninger mentioned that effects observed in mitochondria in vitro can often reflect their "preparative story" but not native properties. N. Kaplan and collaborators carried out the study just of the SDH activity by the most sensitive function: the reversed electron transfer, which was at that time recently discovered by B. Chance [14]. They explained discrepancies between data of some leading laboratories by the dependence of the effects on the common experimental additions and conditions. Chance participated in discussion of results and all his life developed technics to measure mitochondria inside the organism.

However, the convenient method to study mitochondria ex vivo was not found. Meanwhile the advantages of work with isolated mitochondria by rapid recording technics stimulated biochemists to wide studies and doubts on properties of preparations were neglected.

In order to overcome the crucial methodological shortcoming of traditional procedures, we have created a novel, CytoBioChemical (CBCh) method which preserves mitochondrial network *ex vivo* using glass-adhered blood lymphocytes. Studies of various physiological and pathological states using the novel CBCh method have shown greater responses of the mitochondria to functional changes in the organism compared with mitochondria isolated in the form of single granules.

An additional, clinical benefit of the CBCh method is the ability to measure mitochondrial function from a drop of blood, substantially reducing the invasiveness of muscle biopsy that is currently required for mitochondrial disease diagnosis. Besides, CBCh method abolishes problem of liquid blood sample changing during work with a group of patients. The state of glass-adhered cells is stable for several hours and after that all samples can be activated simultaneously by insertion into media for measurement of enzyme activity.

In this review we briefly summarize our previous results on elaboration the method, focusing on recent data on its application in physiological and clinical studies. A special attention is paid to interpretation of the data because some unknown phenomena were revealed. Perspectives of further development of our method will be also considered.

2. A brief description of the cytobiochemical method. Succinate dehydrogenase and α-ketoglutarate dehydrogenase as markers of adrenaline or acetylcholine regulation in the organism

The CBCh method for the study of mitochondrial dehydrogenases was developed to avoid the loss of native mitochondrial network structure in *ex vivo* experiments. The CBCh method was based on the combination of cytochemical techniques of blood smear fixation on the glass with the use of modern biochemical media for incubation of mitochondria. The composition of the used medium was modified to resemble more closely the intracellular medium to better preserve ex vivo the quiescent state of mitochondria inside lymphocytes. We measure dehydrogenase activity by reduction of nitroblue tetrazolium (NBT) to blue formazan. The image of lymphocytes after succinate (SUC) oxidation is shown in Figure 1. Blue mitochondria are located by periphery. The bulk of the cell is occupied by large nuclei stained with neutral red after SDH measurement for identification of lymphocytes in the smear.

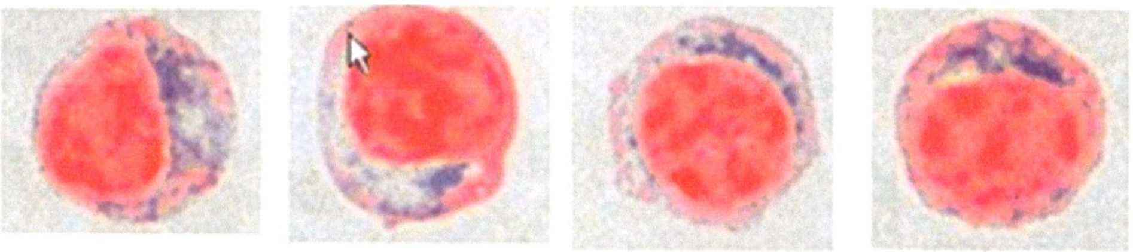

Figure 1. Images of human glass-adhered lymphocytes after SUC oxidation.
Images were collected using an oil immersion lens (100x/1.25) under a light microscope (Leica, DM 2000) equipped with a camera (Leica, DFC-425) connected to a computer.
SDH activity is registered by blue staining with formazan in mitochondria. Red staining is used for nuclei visualization.

In the search for the best conditions for network preservation, we have shown that common substances and conditions obligatory for the investigation of oxidative phosphorylation, such as inorganic phosphate, ADP, Ca^{2+}, pH 7.4, evidently sucrose, as well as dilution of tissue are not only biochemical activators but they also dissipate assemblies of mitochondria

[9-11]. Because the test of NBT reduction is more sensitive to electrons than oxygen consumption, the CBCh method first allowed us to measure low-rate respiration without artificial high concentrations of stimulators. This characteristic corresponds to a real Quiescent State of mitochondria in the cell. It is of great importance because additions distort the physiological regulation. For instance, ADP results in a shift of the ADP/GDP ratio in mitochondria and inhibits SDH. At the dawn of the polarography era, some leading mitochondriologists believed this extreme stimulation to be artificial, and the measurement of respiration without any external additions to be more physiological and desirable. However, the advantages of rapid-recording polarography were so attractive, compared with the previous monumental and slow manometric epoch that the polarographic method spread quickly over the world. Considering the aforesaid, the possibility to measure the real Quiescent State respiration is one more crucial advantage of the CBCh method. Therefore, besides the preservation of network by fixation, the quiescent state is protected by exclusion of the artificial addition of exogenous activators. Due to the combination of cytological and biochemical advantages, the novel method was called Cytobiochemical.

The measurements of the dehydrogenase (DH) activity are carried out using the cytochemical technique of glass-adhered cells by nitroblue tetrazolium reduction after 1-h incubation at 37 °C. The basal medium contains: 125 mM KCl, 10 mM HEPES, 1.22 mM NBT, pH 7.2 ± 0.05.

As distinct from the cytochemical method according to which only one sample with substrate is usually studied for DH activity measurement, the several selected samples with different additions are investigated according to CBCh procedure. The additions will be mentioned and explained further. These additions allow one to analyze the biochemical mechanisms of DH activity regulation. The regulation of the SDH activity by isocitric acid (ISC) and its combination with KGL activity were found to be most informative. A set of separate samples forms a CBCh pattern of DH activity and serves as a sensitive marker of the state of the DHs in the organism and, therefore, of the state of the organism itself. Examples of patterns and their interpretation will be described further.

The special interest to SDH is due to a great domination of SUC over other substrates in the rate of oxidation, known more than a century, and crown by the brilliant Chance's discovery of even greater domination in energy-dependent NAD^+ reduction [15, 16].

As cited below, our group demonstrated numerous effects of stimulation of mitochondrial and physiological functions by SUC. The crucial role of SDH in tumor resistance and cardioprotection and disease development is recently reported [17-20]. The special properties of KDH are less known and will be considered further.

It was shown in previous studies by our group that the oxidation of SUC and selective activation of SDH are involved in adrenergic (ADR) regulation, while the oxidation of α-ketoglutarate (KGL) and selective activation of KDH is involved in cholinergic (ACH) regulation (see also for more details [10], section 3.2.).

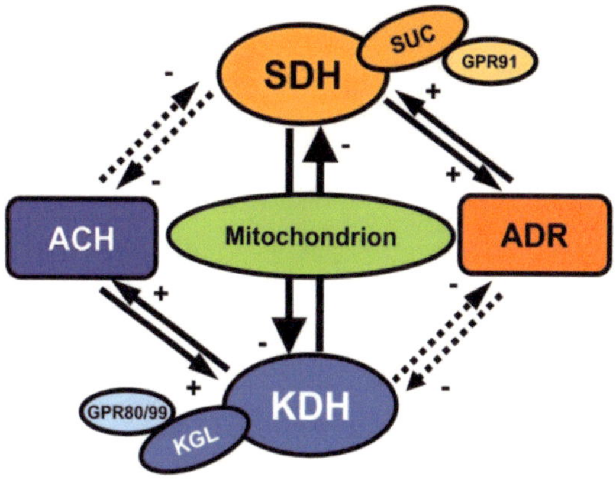

Figure 2. Substrate - hormonal system. Participation of mitochondria in physiological regulation. Reciprocal adrenergic - cholinergic regulation as a whole with oxidation of succinate - α-ketoglutarate. Link between sympathetic and parasympathetic nervous system and mitochondria, or link between non - neuronal intracellular hormones and substrates.

Therefore, the measurement of the activity of the chosen DHs, SDH and KDH, provides also information on the level of reciprocal ADR/ACH regulation in the organism. Substrates of the two DHs, SUC and KGL, were shown to influence the ADR and ACH level as well as support physiological functions under load and in recovery. It is of importance that the regulatory effects of SUC and KGL in the organism are well pronounced in concentrations far below than necessary for substrate supply. To differentiate these two activities of substrates, the low concentrations manifesting general regulatory action in the organism were called "signal" suggesting that it is realized via specific receptors [21-34]. It was an exciting event for our group when later on such receptors were found indeed, and also for only the two of many mitochondrial substrates. As the authors of the first paper stated, the cause of this selectivity was not clear [35-40]. However, it is well explained from the standpoint that just a pair of substrates is needed as a counterpart of also the paired ADR/ACH regulatory system. The novel mechanism of physiological regulation was named by our group Substrate-Hormonal System. It is presented in Figure 2. It serves as a link between the vegetative nervous system and mitochondria. This system may exist also inside cells with no innervation: in embryos, yeasts, plants, and as was shown recently, in lymphocytes [41-45].

As was mentioned above, the SDH/KDH ratio reflects in mitochondria the balance of ADR and ACH regulation in the organism.

3. Methodological meaning of statistical analysis in modern biomedical investigations

Usual desire of physician to find unique for a certain patient individual diagnosis and treatment become to be an advanced direction of the modern probative medicine. Some statistical algorithms appeared which substantiate individual conclusions made earlier intuitively. The latter usually expelled these conclusions from the area of science, although this does not correspond to the truth. Misunderstanding between practical physicians and experts in statistics is based on the mistakes in applications of statistical methods in biomedical investigations.

Very interesting and useful monograph by Thomas A. Lang and Mishelle Secic: "How to report Statistics in Medicine. Annotated Guidelines for Authors, Editors, and Reviewers" [46] has the purpose to fill the gap between mathematicians and physicians. The chapter: "The difference between clinical and statistical significance" substantiates the importance of a single observation, for example, life-saving of one person, in spite of its statistical unreliability.

In this respect the statistical validity of CBCh method for individual measurement is of importance and will be briefly considered below. We share also the idea of the authors that statistical analysis in biomedical investigations should be actively used for design of experiment but not only for evaluation of the results. We can illustrate this by our experience.

1. Statistical treatment

Besides its crucial advantage in preservation of native structure organization of mitochondria CBCh method possesses one more important property. This is high statistical validity of a single result because it is obtained by computer quantitation of a multitude of microscopic images, which measure enzyme activity. This kind of calculation finds an ever increasing application in biomedical studies. In the case of dehydrogenase activity determination by CBCh method, both in rats and in people, even in a small part of experimental samples with minimal enzyme activity, the number of objects measured was about a hundred. In most cases it was 300 - 500 images, while in a considerable part of samples with maximal activity, the number of images was more than 700. According to the standard determination of statistical significance, it is practically maximal, as judged by measurements even a hundred of images. Therefore all the data obtained in our work by computer quantitation are statistically significant at the level of 0.001 (99.99%).

We did statistical analysis for our data and illustrate this result in Fig. 3 by putting in the bars presenting ± S.D. As seen, S.D. is negligible in contrast to non-computer treatment which deals with a considerably smaller number of measurements. Therefore, we believe it possible do not overload all the figures by designation of minor bars after preliminary explanation in Methods for the presented data sampling. Such description may be convenient for presentation of other data obtained by computer quantitation.

In case of more than single individual measurement, such as data presented in Fig.8, we indicate ± S.D. in diagrams.

2. Design of experiment for the purpose to diminish "uncontrolled" statistical variations of data.

The preliminary design directed to diminishing of "uncontrolled" statistical variations of data was considered in the Guidance cited and in the valuable monograph by Glanz [47]. In these handbooks, protocols of comparison of results of the action of different substances are described, convenient even in examination of people for which the selection of conditions is limited compared with animals. We can add to them examples from our experience. A pair comparison of results in one person in contrast to group comparison belongs to these protocols. This list is continued by: performance of measurements at maximal possible identical time. Comparison with healthy control, taken on the same day and at the same hour, or at least for four seasons separately, for which the difference of reactions is known. Performance of a comparison with controls, taken in the same laboratory simultaneously with examined group instead of using averaged data for norm, which increases variations.

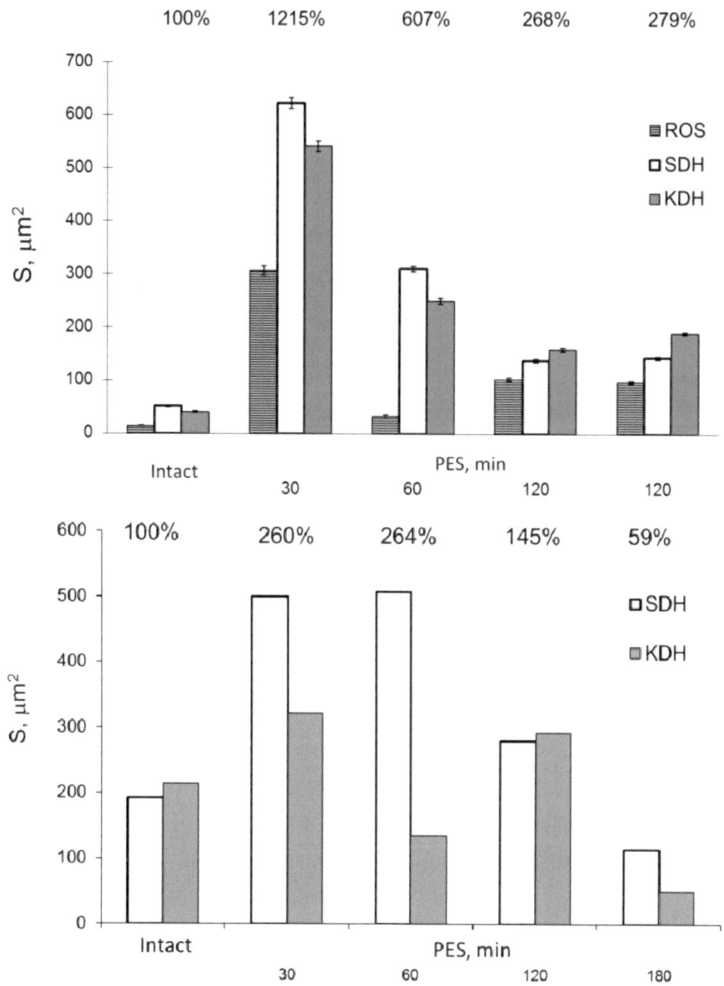

Figure 3. Changes in the activity of SDH and KDH and ROS formation in glass-adhered blood lymphocytes in rats during short-term painless psychoemotional stress (PES) in a box.
Data are presented as a total area (S) of formazan granules in 30 cells.
Owing to a great amount of objects (200-800) the presented values are statistically significant at the level of 0.001 (99.99%).

In connection with data in animal models, described in this paper, we shall consider in more detail the protocol of the selection of really statistical homogenous groups of animals, which was elaborated in our laboratory. The necessity to form statistically homogenous groups of animals for application of statistical analysis for treatment of results is well known, although often it is practically ignored. We found that, in the selection of animals assumed to be similar many factors still exist, that cause uncontrolled variations of the state of animals and preparations obtained. We select animals bred in the vivarium of our Institute within 1-3 days after birth. We do not use for one group animals from only one brood because it is known from physiological data and found by the CBCh method that at least three subgroups can be differentiated in the one brood according to physiological activity [48]. These subgroups correspond to weight distribution. For standard investigations we take animals with greater weight, close to each other on the day of selection - 4 weeks, the time of transfer of young rats to dry feed and taking off from mother milk suckling (100-110 g). It was shown in the special experiments that the prolongation of suckling to 6 weeks considerably improves CBCh data and weight increase as compared with 160-180 g, when suckling was stopped at 4 weeks. The differences are found also in animals at the age of 6 week, before maturation and at the age of 8 week, just after maturation (190-210 g). According to physiological data the differences between these ages are due not only to maturation but also to domination of adrenergic regulation before maturation and cholinergic regulation after maturation. We found manifestation of these physiological phenomena by the ratio of SDH and KDH activities. For standard investigations we have chosen the age of 8 weeks, however other ages can be used but in separate groups. Usually these details are not considered in experiments, which lead to the increase in variations of the results.

Besides careful choice of animals, maintenance of conditions of treatment of isolated tissue preparation is of importance. Usual measurement of data in so called "parallel" samples for analysis taken from liquid preparation: suspension of mitochondria, tissue homogenate or liquid blood is indeed not simultaneous but subsequent in time. It is carried out subsequently, often during prolonged time of storage, while preparation is changed, which gives variations of results. These variations reflect not only changes in the state of the preparation itself but also its response to the pronounced changes in time of cosmophysical environment. This problem is fundamentally investigated in many year studies of S. Shnoll [49]. Many investigators confess that they meet such variations when make repeated measurements. However, observed deviations seem to be obscure and are neglected. Our method of preparation of fixed smears, which are more stable in time, allows one to carry out analysis much more close to simultaneous. Elimination of the considered sources of variations affords the possibility to observe practically completely identical results for different animals in the same physiological state within the selected group of 6-8 species. Such example is given at the upper panel of Figure 3. As shown under 120 min PES CBCh data for two animals are identical. Statistical probability of data for individual organism, considered in paragraph 1, allows one to compare the influence of investigated factors on individual animals in the same group. Comparison the results obtained in one group of animals (upper panel) completely agree with results in the other group (lower panel), selected in different time. In both cases we observe maximal activation of dehydrogenases in the initial stage of stress - alarm reaction at 30 and 60 min, their further decline approaching the initial level to 120 min, however in the both cases with a small excess, which is considered in the description. Prolongation of stress to 180 min, presented at lower panel, shows increase in inhibition in time. On the reasons considered for upper panel we did not show minor bars in the lower panel as well as in the further similar figures in the text. However the number of images measured and statistical significance are given in legend.

The attention to the creative search of causes of result variations and to the real possibilities of their diminishing will improve efficiency of studies. This affords making smaller number of experiments with more definite results. Such approach presents more deep penetration of statistical analysis into biomedical investigations.

4. Demonstration of high physiological sensitivity of the CBCh method in models of increased adrenergic regulation. SDH/KDH ratio as a distinct borderline between intensive functioning and early dysfunction

Over a long initial period, the biochemical mechanisms of stress were studied under action of strong chemical or physical stressors, which are beyond the range of physiological adaptation. In spite of this, the responses of isolated mitochondria studied under pain immobilization stress for a long time, 24 h, were not pronounced, accounting for no more than 20% inhibition of succinate oxidation measured by the polarographic method [11, 50, 51]. The considerably higher sensitivity of the CBCh assay to changes in the physiological state is evident, as it reveals a great-range difference of SDH activity, 2-3 times and more, under the influence of much milder stressors compatible with life, such as the administration of ADR in submaximal dose or painless short-time restraint of rat in a narrow box for 30-120 min, psychoemotional stress (PES). A comparison of mitochondrial responses observed in isolated mitochondria and in mitochondria inside lymphocytes is convincing, considering that the state of animals after strong or physiological stress differed greatly. After strong stress, animals were exhausted and could not move, while animals after short-term stress looked like intact.

The changes in the SDH and KDH activities after the administration of ADR are presented in the upper panel of Figure 4. As shown, these activities are moderate and nearly equal in the initial quiescent state. Under the action of ADR, the initial acute phase of stress, an alarm reaction arises. It is clearly manifested by a pronounced (more than twofold!) increase in the SDH activity, and by a fall (about twentyfold!) in the KDH activity. Under 24-h restraint in a box, a stress phase of exhaustion develops. It is also clearly pronounced by a decrease in the SDH activity approximately to the initial level. It is very important that this decline does not mean a restoration of the initial state because the KDH activity nearly disappears. Therefore, the dynamics of SDH activity develops according to a bell-shaped curve with a maximum at early stages of stress. This corresponds to the classic stress dynamics followed by physiological data.

As mentioned, the SDH activity serves as a marker of ADR regulation, which supports the external work of tissues, while the KDH activity is a marker of ACH regulation which is responsible for restoration of expenditures for work and underlies the general immune resistance of the organism, including the participation of lymphocytes. Therefore, we consider the critical fall in the KDH activity as a manifestation of stress that exceeds the range of physiological adaptation. It is worth mentioning that the 24- h PES when so dramatic changes in the DH activity were observed by the CBCh method is not as severe as the widely investigated rough form of immobilization when only small changes in respiration were found in isolated mitochondria.

An extremely interesting, quite novel area of real adaptive changes in the DH activity opens during even milder PES of shorter duration: 30, 60, 120 min. As seen in the lower panel of Figure 4, during 30 and 60 min a gradual hyperactivation of the both SDH and KDH activity are observed without the critical fall of the latter. Changes in the KDH activity are only of somewhat less extent than that of SDH. By 120 min hyperactivation of the both DHs is followed by the return to the quiescent state.

The status of DHs at 120-min PES is of crucial importance.

Let us remind that in the initial completely quiescent state, the activity of both DHs is low and nearly equal. This is time to note that the KDH activity is somewhat higher than SDH, which is well reproduced in the independent observations. This small excess of KDH activity over SDH is of great importance because it corresponds to the domination of ACH regulation over ADR regulation in the quiescent state. As shown in the lower panel of Figure 4, by 120 min both DHs return nearly to the initial level. Two wonderful details of this state should be noticed. The first is that a slight prevalence of KDH activity over SDH appears which is typical of the quiescent state. The second is that the level of both is somewhat higher than in the initial state before stress.

These specific features evidence a real return to the well balanced quiescent state in spite of continuation of stress. In essence this decrease in DH activity is a real adaptation to the load during training.

Such phenomenon is well known in physiological observations and only after repeated training.

But it has been never observed in a single short-term experiment, the more so for mitochondrial processes. A search for biochemical markers of such a desirable state was only a dream of physicians and trainers. The CBCh method revealed such changes in a short-term experiment and in a wide range.

A finding these two types of early changes in responses of mitochondria to the pathogenic influence detected for the first time a distinct borderline between stimulating, adaptive action of load and the transition to its adverse action, which increases the risk of pathology.

The reliable detection of this border is particularly important for the prevention of disease or a more effective intervention.

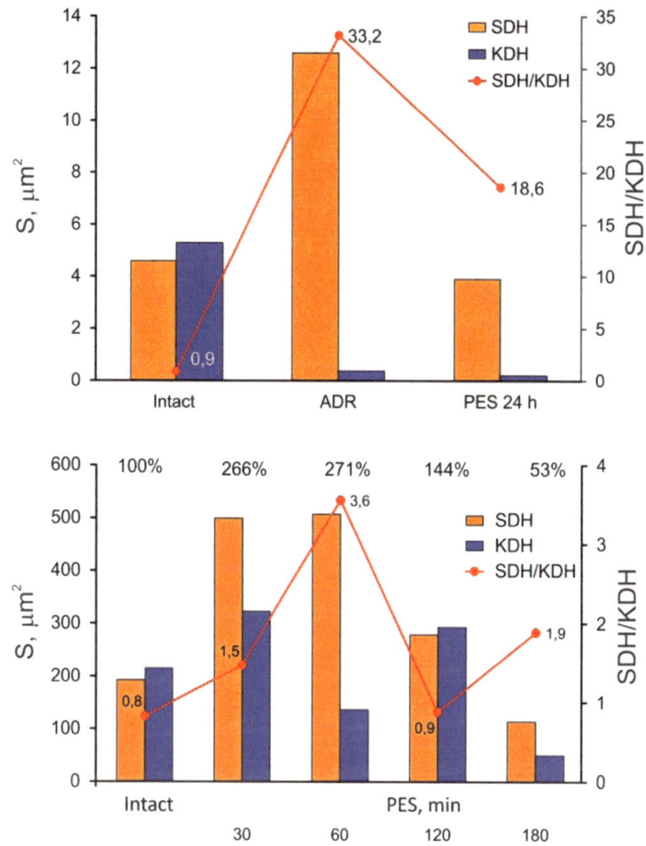

Figure 4. Changes in SDH and KDH activity and their ratio in rats.
Upper panel. Administration of moderately increased dose of adrenaline (50 μg/100 g) for 30 min and long-term (24-hour) psychoemotional stress in a box. SDH activity is expressed as a mean area (S) of formazan granules in 30 cells.
Low panel. Short-term psychoemotional stress in a box. SDH activity is expressed as a total area (S) of formazan granules in 30 cells.
Owing to a great number of objects calculated (200-800) the presented data are statistically significant at the level of 0.001 (99.99%)

As shown in the last example in the lower panel of Figure 4, the prolongation of PES to 180 min exceeds adaptive reserves as it leads to a fall in the activity of both DHs. This inhibition corresponds to the stage of exhaustion of stress. Thus the dynamics of SDH and KDH activity changes presented in both panels of Figure 4 for the first time clearly reveals mitochondrial mechanisms that underlie the physiological development of stress.

In contrast to the results obtained on isolated mitochondria, the CBCh method allows one to observe a broad range of mitochondrial states from deep quiescence up to strong excitation followed by pathological exhaustion in the model of wholly physiological stress. This area of states of mitochondria is not available under the conditions of standard biochemical investigations and is Terra Incognita for mitochondriologists.

Further we shall describe some other unknown phenomena in this unexplored area.

5. Other CBCh characteristics of mitochondria besides SDH and KDH

5.1. Endogenous substrates (ES), SDH true and SDH latent

As mentioned in Methods, for the CBCh characteristic of SDH we used additional samples besides the basal with the addition of millimolar substrate concentrations of SUC. The main of them are shown in Figure 5 and 7, which present more completely the experiments demonstrated in Figure 4. Some additional samples are also given in Figure 6.

The first column from the left shows the reduction of NBT by only ES oxidation without addition of exogenous substrates. This indicator is ignored in most cytochemical investigations. Meanwhile its consideration provides additional valuable information.

The first important characteristic which is determined with the help of ES is a more precise measurement of SDH activity with subtraction of the contribution of substrates other than SUC.

As shown in Figure 5, in intact animal the value of reduction in sample SDH is close to that in the sample ES. When subtracting the data for ES from the data for SDH, as it is accustomed in biochemistry, it turned out that the addition of 5 mM SUC absolutely does not increase the reduction in the least. This means that the reduction in the sample with SUC is not caused by SDH activity but is due to other processes. These data show that SDH in physiological quiescence of the organism is not active, keeping a latent state. In contrast, after ADR administration reduction due to ES is lower, the SDH activity is considerably higher, and the difference SDH-ES is great. We call this calculated value true SDH activity (SDHtru) because the subtraction eliminates the contribution of other substrates. To distinguish this corrected value from the common SDH activity measured without subtraction we term the latter the total SDH (SDHtot). The loss of SDH activity in the quiescent state of the organism is well explained from the standpoint of physiology. In the quiescent state, the ADR regulation is switched off reciprocally to ACH regulation. Apparently, SDH as a target of ADR in mitochondria follows hormone signal, switching activity off. As shown in the last example under prolonged -24 h stress corresponding to the exhaustion stage, the activity of both DHs declined.

The state of non-active SDH can also be termed latent or "sleeping" because it does not function in spite of SUC addition. The finding of this state of SDH is one of the most important attainments of the CBCh method. Such a state is not known in biochemical investigations of mitochondria because they lose the quiescent state during preparation and measurements. Biochemists are familiar only with initially activated SDH. The wide range of states from quiescent state to activation is lost for current observation. The loss of the initial basal level of respiration, *i.e.* the quiescent state, is in essence the absence of data for norm in current biochemical investigations.

We have found that the crucial difference in SDH activity between healthy and sick people or between calm and anxious animals is just the loss of the quiescent state. This means that, in the state of relative physiological quiescence, sick people has no quiescent state in mitochondria. This will be demonstrated further.

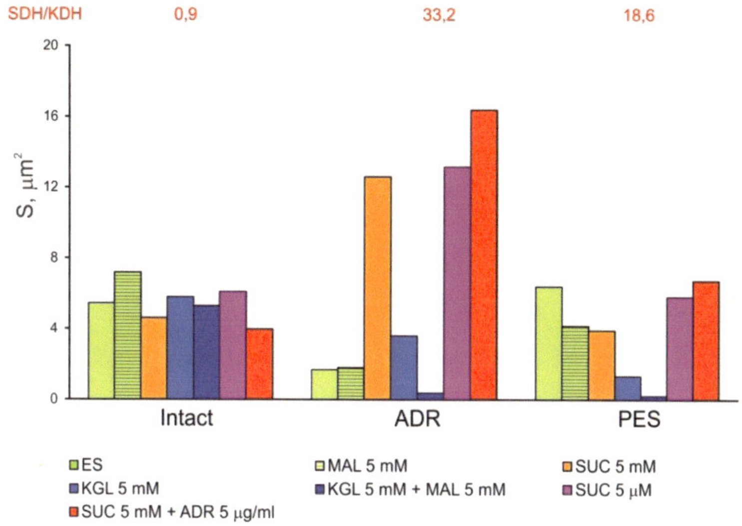

Figure 5. CBCh patterns of mitochondria measured in glass - adhered lymphocytes from intact and stressed rats. Adrenaline administration, 30 min, and PES 24 h. NBT reduction is expressed as a mean area of formazan granules in 30 lymphocytes
Owing to a great number of objects calculated (200-800) the presented data are statistically significant at the level of 0.001 (99.99%)

5.2. KDH activity measurement. Addition of malonate (MAL)

The pair of columns in Figure 5 next to SDH presents the KDH activity measurements without and with the addition of MAL. These samples contained also NAD^+ as a soluble cofactor of KDH. It is worth reminding that MAL should be obligatory added for correct determination of KDH activity. It was well known but undeservedly forgotten in many polarographic measurements that SUC generated in mitochondria during KGL oxidation may amount to a considerable part of respiration in the presence of KGL added. The SUC-dependent portion of respiration should be subtracted from total respiration with KGL. The data presented show that the SUC-dependent portion of KDH activity is absent in the intact state, when SDH is non- active. In contrast, this portion rises in excited animals. It is evident from this example that the own KGL oxidation is negligible under strong excitation. The elimination of the SUC-dependent portion of NBT reduction by MAL allows one to measure KDH activity more precisely. Reciprocal changes in KDH and SDH activities in the presence of MAL become also more evident.

5.3. Signal action of SUC and ADR

The last pairs of columns in Figure 5 presents the data on NBT reduction with the addition of 5 μM, "Signal" concentration of SUC or 5 μM of ADR with substrate concentration of

SUC - 5 mM. The effects of signal concentrations were negligible or absent when measured with subtraction of sample with MAL in the intact animal. However, both effects rise similar to the SDH activity in ADR-activated animal. Under prolonged stress, they both decline also in accordance with SDH activity. These data are the first evidence of a great SDH activity in the presence of low, signal concentration of SUC, far below the substrate ones, under the influence of ADR.

Signal concentrations of ADR act synergetically to SUC. These data serve as a further evidence for synergism between SUC and ADR.

5.4. Addition of MAL: Endogenous SUC (ESUC = ES - MAL), reactive oxygen species (ROS) formation

MAL is commonly used to exclude the contribution to oxidation of SUC, either added or endogenous (ESUC), formed in mitochondria from other substrates, particularly from KGL. Although the amount of ESUC is very low compared with that of added KGL, SUC dominates in competition for the respiratory chain due to the dominating reducing power of SDH. ESUC has additional advantage over added SUC, because it forms inside mitochondria and is better available to SDH [16]. In CBCh measurements, a decrease in formazan formation corresponds to the inhibition of respiration. The difference in reduction between sample ES and (ES+MAL) gives a measure of ESUC.

We observed an increase in ESUC content in animals excited by endogenous ADR (stress) or ADR administration. ESUC also increases in a spontaneously anxious individuals, animals or people. 12 examples of typical patterns of MAL effects on NBT reduction with ES are presented in Figure 6. They were selected from more than 20 volunteers examined. Among them we have found four types of repeating patterns and no more variations. Each type is presented by three examples. The form of pattern is well reproduced for each type. Therefore we believe the pattern to be a regular relationship but not random combination of characteristics. All volunteers were in the relatively calm state. Blood samples were taken in the laboratory between 9-10 a.m. after a very weak breakfast at about 8 at home. All of them were practically healthy, however, the level of health and the state on the day of examination varied. Probably, in accordance with these differences, we found four types of repeating patterns.

Ranging the patterns according to the increase in the SDH activity from low to high, we differentiated the following States: Deep Quiescence, Operative Quiescence, Activity, and Activation with Inhibition.

The decrease in NBT reduction, caused by MAL, is particularly pronounced in the subgroup Activation, in which the SDH activity is the highest. The members of this subgroup have personal reasons for anxiety and activity. The same type of pattern but with a lower amount of ESUC was found in the subgroup named Operative Quiescence, because the SDH activity was considerably lower than in the subgroup Activity. The decrease in NBT reduction caused by MAL evidences the contribution of ESUC to respiration when SDH is active.

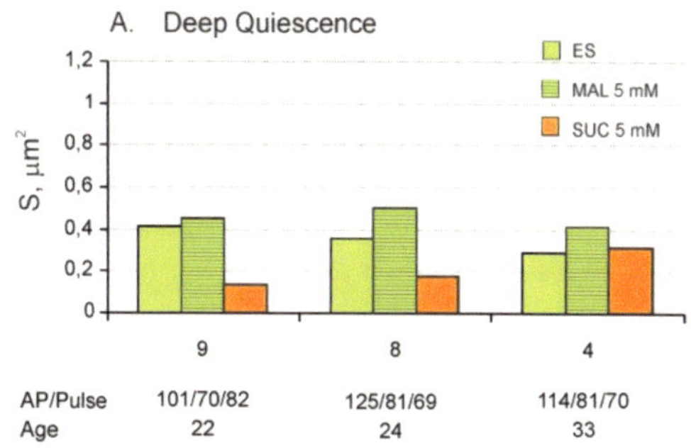

A. Deep Quiescence

Quiescent State transition to activity from which is hard.

ES NBT reduction is low, MAL increases it per 10-20%, SDH activity with 5mM SUC is low, considerably lower than ES and ES+MAL, the state of SDH - latent, “sleeping”, switched off.

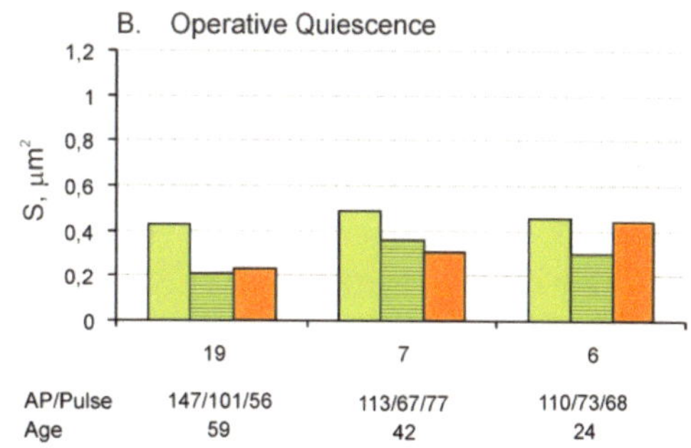

B. Operative Quiescence

Quiescent State transition to activity from which is easy.

ES reduction is low, MAL decreases it per 20 -25 %, SDH activity with 5mM SUC is low. The State of SDH is also latent as in Operative Quiescence, as measured with 5 mM SUC. However, the appearance of ESUC among ES evidences switch on fine activation of SDH and SUC oxidation.

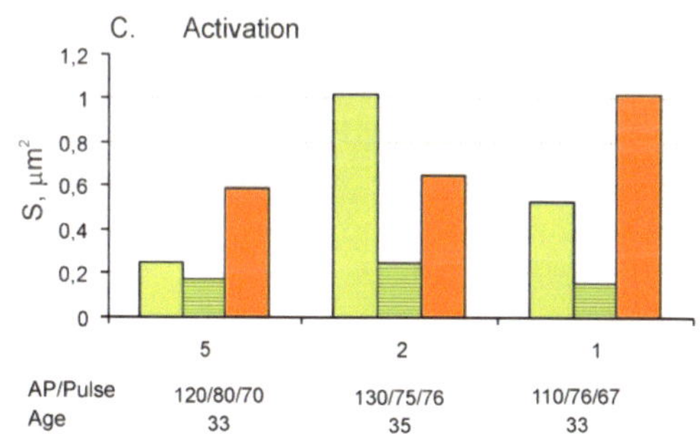

C. Activation

Activated state before work.

ES reduction is higher than in the two previous States, MAL markedly inhibits ES reduction showing great contribution of ESUC in ES, high SDH activity with 5 mM SUC.

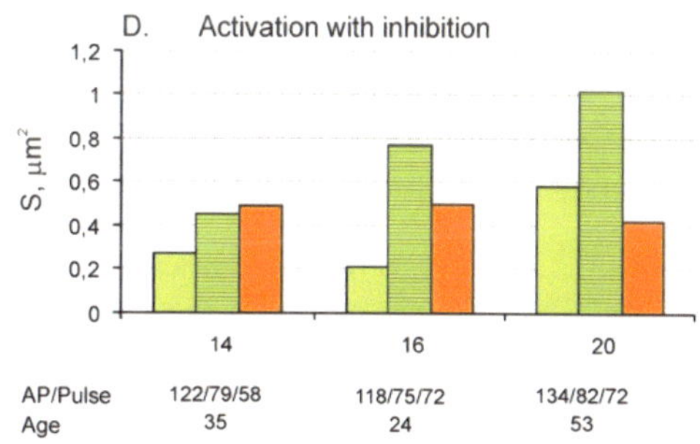

D. Activation with inhibition

Stressed, restrained Quiescence which can transform to pathology.

In such a relative, not complete Quiescence excitation is not switched off, but is suppressed by internal mechanisms and in case of their insufficiency by drugs (β - adrenoblocker for example 20).

ES reduction is low for except of example 20 where it is high, probably due to the SDH inhibition by adrenoblocker. MAL increases ES reduction 2 - 4 fold, SDH activity with SUC 5 mM is moderate and high for the example 20

Figure 6. Four (and no more) types of specific CBCh patterns of SDH found in healthy volunteers and presented by three examples. Owing to a great number of objects calculated (200-800) the presented data are statistically significant at the level of 0.001 (99.99%)

Together with this traditional effect of MAL, we revealed the opposite action of MAL, namely, an increase in NBT reduction. It was most pronounced in the subgroup Activation with Inhibition, which was distinguished according to a decreased SDH activity relative to the Activation subgroup. One cause of the inhibition (example 20) could be an elder age and initial hypertension demanding the intake small doses of adrenoblockers, which also inhibits SDH. In this case, an increase in NBT reduction by MAL is the highest. This effect is also well pronounced relatively to ES in two other examples in this subgroup. This unknown effect of MAL on NBT reduction is also observed in Deep Quiescence, in which all activities are at the lowest level.

Thus, MAL manifests the ability to increase NBT reduction in case of low SDH activity, in contrast to its traditional inhibition of respiration is great when SDH activity is high.

In our opinion, the increase in NBT reduction by MAL is a marker of ROS formation in mitochondria. As shown in Figure 7, ROS formation sharply rises at the beginning of stress. During stress, the rise of ROS formation declines as evidenced by our measurements with MAL. This agrees with other results [52]. As shown in the last example in Figure 7, ROS rise again by the exhausting stage. The biochemical mechanism of this effect of MAL is in deep inhibition of SDH, which leads to a fall of coenzyme Q reduction from full reduction to free radical form. Free radical form of Q produces superoxide in interaction with surrounding oxygen [53].

5.5. Dual regulatory effect of ISC on SDH activity in the area of physiological regulation. GTP-recoupling but not oxaloacetate (OAA)-inhibition regulates SUC oxidation under physiological conditions

ISC is a natural efficient intramitochondrial antioxidant [54]. Reducing equivalents produced by ISC NAD(P)-dependent dehydrogenases protect the most sensitive enzymes aconitase and KDH from oxidation and increase the stability of mitochondria to oxidative stress, which accompanies any stress. It was shown long ago that ISC abolishes OAA inhibition of SDH in aged mitochondria or under severe stress [50, 55]. This effect is due to OAA elimination by reduction into malate because of increase in the reduction of the respiratory chain. A similar effect is usually attained by the addition of glutamate, which eliminates OAA via transamination. As distinct from glutamate, the application of ISC for the elimination of SDH inhibition was nearly forgotten. This may be due to the fact that its fully active natural isomer is commercially hardly available, while the available chemical racemate is less active. Fortunately, due to the collaboration with microbiologists manufacturing natural ISC, we had an opportunity to investigate this substance [56].

Our investigation of the influence of natural ISC on SDH activity during PES under more physiological conditions of the CBCh method discovered an unknown effect of this substrate. In spite of acceleration of inhibited SUC oxidation, ISC strongly suppresses hyperactivated SDH. This effect is shown in Figure 7. It is clearly seen that ISC addition suppresses greatly SDH activity in the hyperactivated state, 30, 60, 120 min of PES.

However, ISC preserves the ability to stimulate inhibited SDH under 180- min PES. Besides, we have found that ISC does not influence the SDH activity in the Quiescent State. The two opposite effects of ISL: an increase and a decrease in SDH activity are easily explained by the single antioxidant mechanism. The increase in NBT reduction by MAL at the beginning of PES shows the development of oxidative stress. This agrees with other observations. The level of oxidative stress is regulated by well-balanced interactions of several processes occurring in mitochondria. Among them are initial uncoupling and its secondary recoupling by GTP [52].

Recoupling effect of GTP can explain our observation that preserving the KDH, which is a real source of GTP-formed in substrate-level phosphorylation, protects SDH from hyperactivation followed by inhibition.

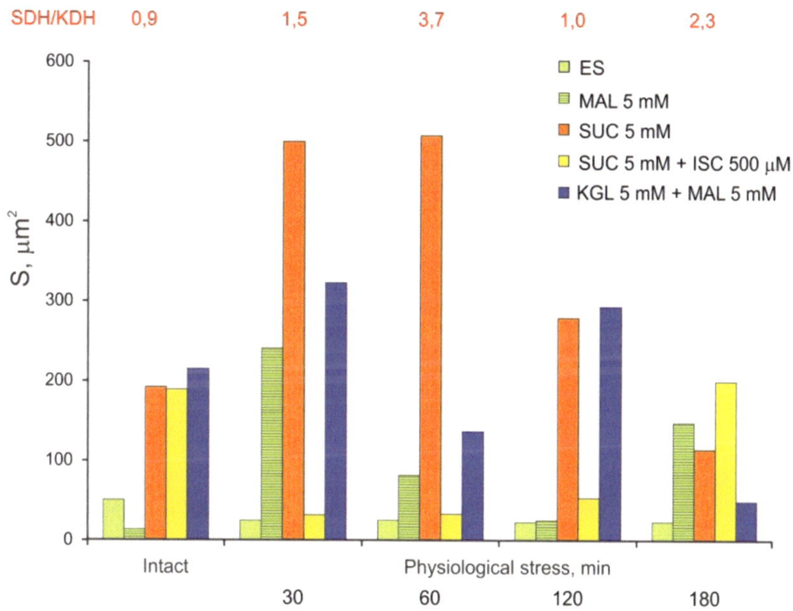

Figure 7. CBCh patterns of mitochondria measured in glass -adhered lymphocytes from intact and stressed rats under mild PES of different duration.
NBT reduction is expressed as a total area of formazan granules in 30 lymphocytes.
Owing to a great number of objects calculated (200-800) the presented data are statistically significant at the level of 0.001 (99.99%)

In this metabolic environment, protection of KDH from oxidation by ISC allows one to support GTP generation and to diminish the uncoupling that was a cause of initial SDH hyperactivation. This explanation agrees well with our repeated observations that the preservation of high KDH activity simultaneously with activation of SDH is typical of activity within physiological adaptation area. The "scissors" between SDH rise and KDH fall are observed in severe stress or pathology. Therefore, three different effects of ISL on SDH activity: suppression of hyperactivation, activation of inhibition, and absence of influence in norm, allow one to differentiate clearly functional states of SDH. The area of

SDH regulation involving KDH-dependent GTP – recoupling - is not known in biochemical investigations because SDH is initially hyperactivated under the experimental conditions. The hyperactivation of SDH elevates the risk of inhibition by OAA due to the increase in its formation. Indeed, OAA inhibition is often observed, when SUC oxidation is accelerated, for example, by uncoupling. Therefore, OAA inhibition is considered to be the only mechanism of SDH control. However, this is true for in vitro conditions in biochemical experiments in isolated mitochondria. The preservation of the area more close to physiological conditions in the CBCh method opens Terra Incognita of unknown regulation of the SDH activity by GTP-recoupling instead of usual OAA inhibition.

The last but not least, it is worth mentioning that the mutual reciprocal regulation of SDH and KDH activity is further evidence of ADR and ACH regulation in mitochondria. In essence, it corresponds to the reciprocal influence of ADR and ACH on physiological functions: stimulation, and restriction with restoration, respectively.

Before continuation, let us put in a brief insertion to memorize the better approach of CBCh determination of SDH activity to mitochondrial processes compared to pure enzymatic measurements.

Interlude: CBCh activity of SDH is not so much a characteristic of abstract DH, as a measure of the rate of SUC oxidation in mitochondria

What is the interest of physiologists and physicians in measurement of enzyme activities? Evidently they need to learn how the enzymes work in the organism in different states. Many purely enzymatic methods give exact information, although rather abstract in respect of the conditions in vivo. Preservation of the native state and native surrounding of SDH and KDH under CBCh conditions permit measurement of SUC oxidation, that approaches the real rate of respiration in mitochondria in vivo rather than an abstract activity of SDH. This is one more crucial advantage of CBCh measurements of DHs activity.

6. Revealing fine physiological differences in the state of the organism in norm by the activity of dehydrogenases as measured with the CBCh method. Differences in the activity and responses of SDH in rabbits with opposite behaviour

The CBCh method clearly detects some fine physiological differences in the state of the organism that were earlier not observed in mitochondria. Among them are some age-related differences in rats at different stages of ontogenesis, namely, in newborn animals, depending on the time of feeding with mother milk, prior to or immediately after maturation, as well as individual differences of pattern types between healthy humans that are related to the current condition or temperament [48, 57]. Both age-dependent changes and differences of temperaments reflect different ratios of ADR/ACH regulation. In the earlier age, ADR regulation dominates; after maturation, it is more balanced by ACH.

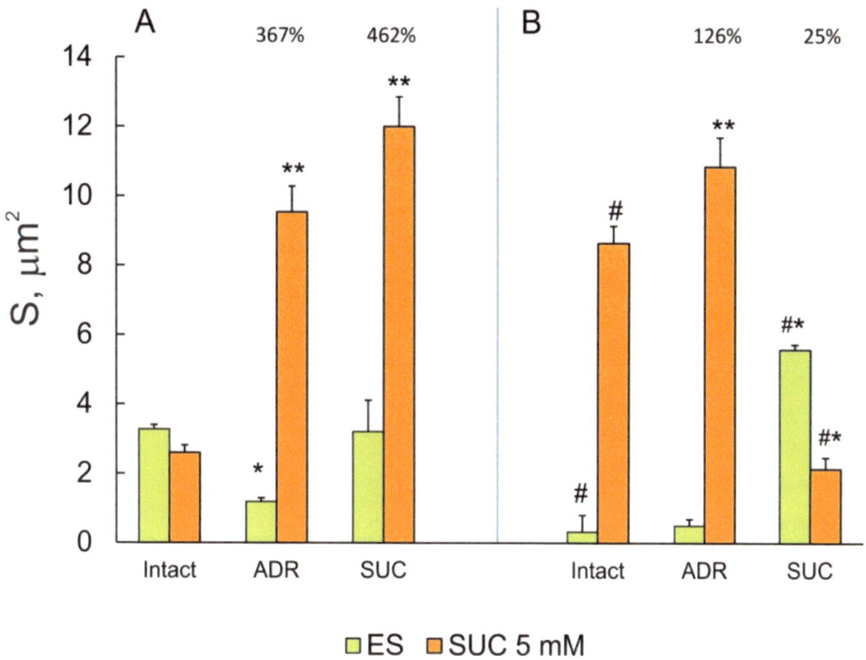

Figure 8. Responses of ES and SDH to activation by administration of adrenaline or succinate measured in lymphocytes of rabbits with different behavior.

A - calm rabbit, B - restless rabbit. SDH activity is expressed as a mean area of formazan granules in 30 cells. The mean of three measurements in different days for each rabbit are presented, M ± SEM. * - p< 0.05, ** - p<0.01, comparing to the respective initial state, # - p<0.05, comparing to the calm rabbit.

As one of the models of domination of these regulation types we studied individual differences of CBCh patterns in rabbits with different behaviour.

One rabbit was calm during the experiment; another rabbit escaped and was anxious and restless even when it was not exposed to treatment. They represented characteristic types of a cholinergic and adrenergic, respectively. Figure 8 demonstrates significant differences between the calm and restless rabbits in the level of ES and in the SDH activity.

The reserve of ES in the calm rabbit is much greater than in the restless. After the injection of ADR, the level of ES in the calm rabbit decreases; however, this level is higher than in the restless animal under the influence of ADR. In the intact restless animal, the level of ES is 15 times lower than that of the intact calm rabbit and 5 times lower than in the calm animal after the injection of ADR. This indicates an enhanced expenditure of substrates in the restless animal.

The SDH activity in the calm rabbit is low, and the injections of ADR or SUC strongly stimulates it. The stimulation of SUC is somewhat stronger, which increases the risk of the transition from hyperactivation to inhibition. By contrast to ADR, the injection of SUC also increases the level of ES.

SDH in the intact restless rabbit is hyperactivated nearly up to the level of calm under ADR influence. Thus, as judged by ES level and the SDH activity, the restless rabbit has no quiescence in the quiet state without external excitations. The same state of continuous internal metabolic excitation is inherent in all patients examined in our laboratory in contrast to healthy volunteers. The data will be considered in the next sections.

At the background of the initially high SDH activity in restless rabbit ADR administration leads to only small stimulation, which only somewhat exceeds the level under the influence of ADR in calm rabbit. The influence of SUC, which was somewhat stronger activator when injected to the calm rabbit, leads to realization of risk of transition to inhibition in restless animal.

The difference in ADR and SUC effects on ES can be explained by different mechanisms of their influence on SDH activity. ADR only activates SDH through protein kinase. The influence of SUC is more complicated. SUC is an allosteric activator of SDH [58]. Simultaneously the rise of SUC level increases OAA formation, which inhibits SDH according to the mechanism of the negative feedback.

These effects are better pronounced in a sample without SUC addition, namely ES, because SUC excess partially abolish OAA inhibition. Therefore in sample ES SDH inhibition is better observed than in sample for SDH measurement, containing 5 mM SUC, because it is not masked by the addition of substrate. Probably, the main part of ES under SUC injection is ESUC, which is accumulated due to OAA inhibition. It should be considered that measurements are made for an hour. Insight into dynamics of the processes can be achieved by a comparison of data for calm rabbit, in which oxidation is slower with those for restless one, in which oxidation is accelerated. As shown for calm rabbit, at the beginning rapid oxidation of SUC occurred, resulted in an increase in OAA, which is evidenced by the initial accumulation of ESUC. The next stage is presented by data for restless rabbit in which stronger inhibition of SDH by OAA is already developed, which is not completely abolished by SUC addition and therefore ESUC accumulation is higher than in calm animal.

The data presented show great difference of the two CBCh properties of the state of mitochondria in animals with opposite behaviour, which corresponds in magnitude to the pronounced individual peculiarities in behaviour.

We have also found the pronounced difference in the SDH and KDH activity in rats before maturation, 6 week old with domination of ADR regulation, and just after maturation, 8 week old with increase in ACH regulation. These changes are close to differences between calm and restless rabbits.

It was shown also that complicated menopause in rats and women, which is manifested under the decrease in ADR regulation in the organism, is related to the increase in the SDH activity and can be corrected by metabolic regulators, containing SUC - "Amberen" [59].

The data considered support the view of existence of the link between activity of studied DHs and the level of ADR and ACH regulation.

7. The changes in SDH and KDH activity in patients with metabolic syndrome and arterial hypertension

The stages of the SDH and KDH activity changes found under stress were observed in patients with metabolic syndrome and arterial hypertension.

Disturbance of normal metabolic flexibility between oxidation of glucose and fatty acids is the basis of many diseases. High dietary fat intake and low physical activity provoke this disorder. Complex of accompanied biochemical changes is called metabolic syndrome. Metabolic syndrome underlies various pathological symptoms, such as obesity, insulin resistance, type 2 diabetes, risk of hypertension, heart diseases and others [1, 60-65].

Metabolic syndrome is caused by increased intensity of life accompanied with persistent sympathetic hyperactivation and deficit of physical activity that leads to PES. That is why we have studied the participation of mitochondria in development of PES. The importance of mitochondria in development of metabolic syndrome is caused by their crucial role in intracellular oxidation. Aerobic fibers of skeletal muscle mitochondria make a major contribution to regulation of metabolic syndrome and support metabolic health of the whole organism.

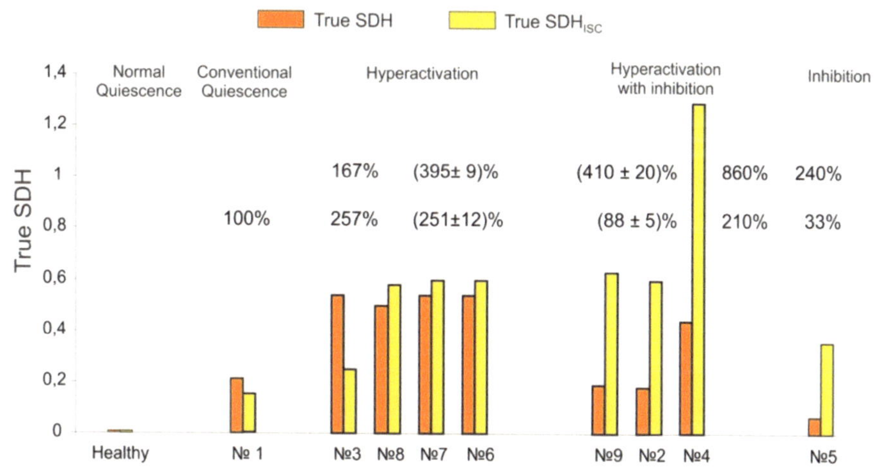

	Healthy	№ 1	№3	№8	№7	№6	№9	№2	№4	№5
Blood (arterial) pressure										
Systolic	113	138	113	118	129	152	109	128	162	156
Diastolic	67	85	67	88	92	114	79	193	128	97
Cardiac rhythm	77	58	65	89	57	73	66	80	84	73
Age	42	71	27	34	40	44	43	52	47	55

Figure 9. True SDH without and with ISC in not severe patients suffering from food allergy accompanied by arterial hypertension.
Data are expressed in mean area (S) of formazan granules in 30 cells. Owing to a great number of objects calculated (200-800) the presented data are statistically significant at the level of 0.001 (99.99%)
Inscription above - types of patterns.
SDHtru is calculated as the difference between samples SUC 5 mM - MAL 5mM.
The changes of SDH and SDHisc activity are presented in per cents respective to conditional norm of patient 1 (100%). The low row for SUC 5mM, the upper row for SUC 5 mM + ISC 5 mM.

The changes in the SDH activity in non-severe, non-hospitalized and workable patients with allergy to some food products are given in Figure 9. In this case for more precise determination of the SDH activity we subtracted ES contribution from the data for the sample with SUC addition. This value was termed as a true SDH (SDHtru). It was determined under addition of SUC and also SUC+ISL, modulator of the SDH activity. This provides determination of the functional state of SDH as considered in section 5.5. Calculation of SDHtrue help to find clearly pronounced similarity between some patients, which is masked by ES contribution. According to SDHtru the patients can be joined in subgroups with nearly identical data for members inside one subgroup and clearly different from the other.

The first example from the left is for healthy person. The next one, 1, is taken as reference control to other patients. He used diet containing no allergenic products for more than 7 years and for this period his body weight, AD and other physiological characteristics were greatly normalized. Other patients have some pathological symptoms and used recommended diet for only several months. Among them we have found increase in SDH activity.

For healthy people full absence of SDHtrue - complete quiescence state is observed. Low SDHtrue was determined for patient with normalization of the state. In all others patients different levels of activation or hyperactivation of SDH were found. In the state of hyperactivation without inhibition, ISL addition results in only the tendency to stimulation. Progressive hyperactivation is masked by internal inhibition of OAA. This is revealed by the increase in activity with ISL. For the patients with greater symptoms, presented in the right part of the Figure 9, the increase in the internal inhibition is typical, which is manifested in both, the SDH activity decrease and strong increase in activation by ISL.

Hyperactivation of the SDH activity without inhibition is more typical of patients hospitalized with hypertension. In Figure 10 data are presented on the total SDH activity (SDH tot) without subtraction of ES contribution. It is seen also, as in Figure 9, that for the healthy control very low activity of SDH is typical. The rise in the activity can be considered as hyperactivation without inhibition by both, their greater value and because ISL does not stimulate it but, in contrast, decreases. ISL addition serves as unique test of the functional state of SDH developed by our group. Additionally to the explanations of ISL influence in section 5.5 it should be mentioned that effect of ISL is dose-dependent. It can be seen from the example of patient N2. Addition of 5mM of ISL decreases hyperactivation of SDH per 2/3, while 50μM only per 1/3.

The extent of the deflection of the SDH activity from norm should be quantitatively determined using different concentrations of ISL. In Figure 10 (as in Figures 5 and 7) the KDH activity is also given, as well as the response of mitochondria to the signal concentration of SUC - 5μM. In norm the KDH activity is low and nearly equal to the SDH activity. At hyperactivation of SDH in patients N2 and N3, KDH is also greatly increased, however less than SDH. Such coupled changes in the two DH activities in patients are similar to those, which were found in rats under PES within the range of physiological adaptation. However, the changes in more severe patient, N4 are already beyond the physiological range and are analogous to that for damaging stress in Figure 4. It is also seen that the response to signal concentration of SUC corresponds to the SDH activity, measured with substrate concentration of SUC, as shown in Figure 5.

The presented examples show that the loss of full value quiescence, which is manifested in the complete switching SDH off, is a general difference of all examined patients from healthy people. Therefore mitochondria of patients in the relatively quiet state of the organism have no metabolic quiescence.

According to the influence of ISL functional state of SDH in patients with hypertension differs from that of patients with allergy. Hidden inhibition of SDH is typical of allergy, while hyperactivation of SDH is more typical of hypertension. This coincides with more pronounced sympathetic hyperactivation in hypertensive patients, demanding intake of adrenoblockers.

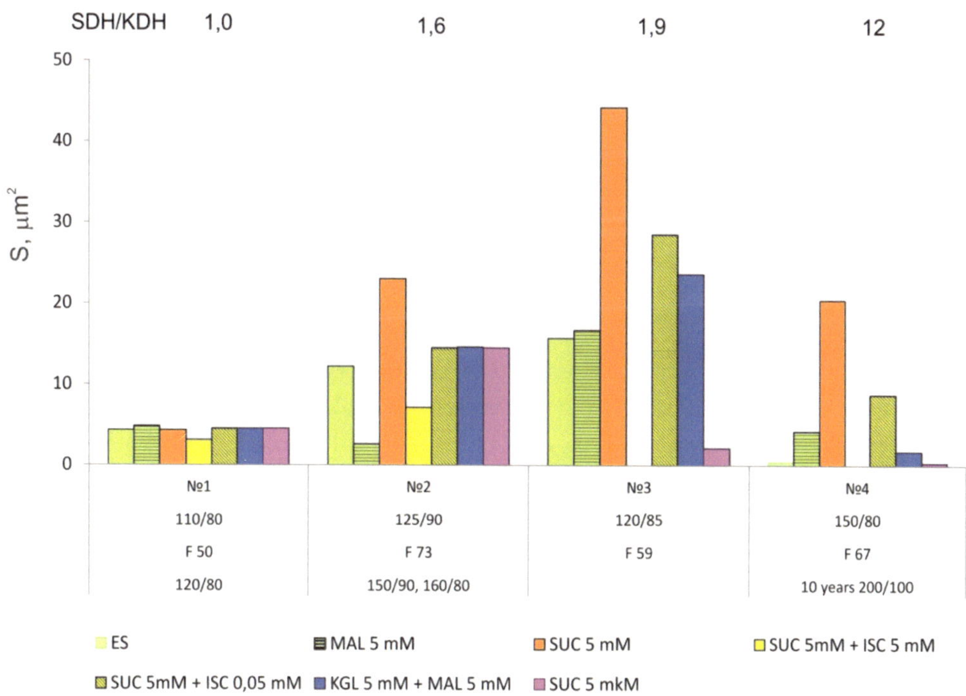

Figure 10. CBCh patterns of patients hospitalized with hypertonic disease.
The data are expressed as the total area of formazan granules in 30 lymphocytes.
AP, upper line - the data just before blood taking, low line - more stable data from History of the disease. The gender and age are indicated between.
Owing to a great number of objects calculated (200-800) the presented data are statistically significant at the level of 0.001 (99.99%)

8. Conclusion

The experience of the reviewed study evidences reliability to approach ex vivo more closely to the state of investigated tissue in the organism. In order to maintain native state of isolated tissue it is necessary to join the both, care of biochemical processes and physical structure. Only combined biochemical and biophysical approaches will provide attaining the desirable goal to study biochemical mechanisms of physiological processes ex vivo.

This approach opens unknown areas in biochemical regulation in the organism, as shown for instance, by SDH and KDH regulation. One of the mostly known phenomenons, OAA inhibition of SDH, is considered as the main regulator of the SDH activity. However, according to investigations under more physiological conditions, it seems to be in many cases rather artifact of "preparative story" of mitochondria. The wide range of SDH states under the physiological and pathological conditions in the organism, compatible with life, are DH hyperactivation, which is far from the inhibition. This state is probably controlled by GTP, produced due to KDH activity in substrate-level phosphorylation during KGL oxidation. This restriction of SDH by KDH is in essence identical to reciprocal interrelationship between parasympathetic a sympathetic nervous regulation of physiological functions.

A very convincing observation of this substrate-hormonal interaction, as judged by the CBCh method, which cannot be achieved in investigation of isolated mitochondria, can be also mentioned as the crucial result of preservation of the native structure of mitochondria under study.

9. Perspectives of the CBCh method development

The wide application in clinical practice of the proposed very sensitive method to penetrate the state of human mitochondria is the most near, reliable and useful perspective. This penetration is even deeper than in the current biochemical studies. The procedure of the measurement is not the main difficulty on this way. It is within the range of the commonly used clinical tests and may be advanced for the convenience.

In order to the CBCh analysis will bring benefit to patients, physicians should acquire deep knowledge of the investigated biochemical processes. Fundamental data from handbooks are not sufficient for this purpose. It is necessary to learn more special area, which is developed in current studies by the CBCh method. This area is advanced even in modern mitochondriology, because it is created during novel experiments. The authors are preparing the special guidance to facilitate the entrance in the new area. We are also glad to help to overcome individual difficulties in the procedure and analysis of data.

Radical approach to the state in the organism through to the preservation of the native structural state of the tissue, found in the study oh DHs, encourages widening the developed conditions on the great variety of the cyto- and histochemical determinations.

Earlier we have stated that the main advantage of the CBCh method lies in the use of lymphocytes as indicators of the overall metabolic status of the entire organism, which is especially useful as an early diagnostic tool of various mitochondrial dysfunctions. This method can be also used for individual selection of drugs in sample of blood of a certain person.

However, the CBCh method may also be used as a tool for the study of lymphocytes as an independent object of study, which carries important functions in the immune response. The use of the CBCh method for the study of dehydrogenase activity, as well as other indicators, is likely to lead to a better understanding of the dependence of the immune response on mitochondrial functions.

The novel area of CBCh investigations of lymphocytes can be related with their use as cells possessing recently discovered pronounced ACH regulation, both - vagus -dependent and, particularly interesting, self-dependent, intracellular, which is based on the internal pool of non-neuronal ACH. The simultaneous study of ACH level and dehydrogenase activity in lymphocytes can provide further penetration into substrate-hormonal regulatory system that was pioneered in our laboratory.

Author details

Marie Kondrashova*, Marina Zakharchenko, Andrej Zakharchenko,
Natalya Khunderyakova and Eugen Maevsky
Institute of Theoretical and Experimental Biophysics Russian Ac. Sci., Pushchino, Russia

Acknowledgement

M. Kondrashova wish to keep alive the memory of the teachers in physiology and biochemistry Professors Iliya Arshavsky and Sergei Severin, whose influence helped her to find the own way in the area of these interdependent branches of biology.

The more than half a century correspondence with Britton Chance encouraged persistent attempts to fill the gap between isolated mitochondria and the organism.

The authors are indebted to Rodrigue Rossignol for mutual understanding and support of our line of investigations.

The authors are grateful to Dr. Polina Schwartsburd for the systematic valuable discussions on physiological regulation.

We are thankful to Dr. Fedor Kondrashov for creative editing the manuscript and Svetlana Sidorova for the help in translation.

This work was supported by the Federal target program "Investigations and elaborations of priority directions of scientific-technical complex of Russia for 2007-2012 (project no. 16.512.11.2117).

* Corresponding Author

10. References

[1] Reaven GM (1988) Banting lecture. Role of insulin resistance in human disease. Diabetes. 37: 1595-1607.

[2] Grundy SM, Brewer HB Jr, Cleeman JI, Smith SC Jr and Lenfant C (2004) Definition of Metabolic Syndrome. Circulation. 109: 433-438.

[3] Stark R and Roden M (2007) Mitochondrial function and endocrine diseases. Eur J Clin Invest. 37: 236-248.

[4] Kim J, Wei Y and Sower JR (2008) Role of mitochondria dysfunctions in insulin resistance. Circ Res. 102: 401-414.

[5] Eberhart K, Rainer J, Bindreither D, Ritter I, Gnaiger E, Kofler R, Oefner PJ and Renner K (2011) Glucocorticoid-induced alterations in mitochondrial membrane properties and respiration in childhood acute lymphoblastic leukemia. Biochim Biophys Acta. 1807: 719-725.

[6] Mitochondrial Physiology. The Many Faces and Functions of an Organelle. Edited by. Erich Gnaiger. (2005): 1-152.

[7] Gnaiger E (2009) Capacity of oxidative phosphorylation in human skeletal muscle: new perspectives of mitochondrial physiology. Int J Biochem Cell Biol. 41: 1837-1845.

[8] Jose C, Bellance N and Rossignol R (2011) Choosing between glycolysis and oxidative phosphorylation: a tumor's dilemma? Biochim Biophys Acta. 1807: 552-561.

[9] Kondrashova MN, Fedotcheva NI, Saakyan IR, Sirota TV, Lyamzaev KG, Kulikova MV and Temnov AV (2001) Preservation of native properties of mitochondria in rat liver homogenate. Mitochondrion. 1: 249–267.

[10] Kondrashova MN, Zakharchenko MV and Khunderyakova NV (2009) Preservation of the in vivo state of mitochondrial network for ex vivo physiological study of mitochondria. Int J Biochem Cell Biol. 41: 2036-2051.

[11] Zakharchenko MV, Khunderyakova NV and Kondrashova MN (2011) Importance of Preserving the Biophysical Organization of Isolated Mitochondria for Revealing Their Physiological Regulation. Biophysics (Moscow). 56: 810–815.

[12] Rossignol R and Karbovsky M (2009) Editorial of the directed issue on mitochondrial dynamics in biology and medicine. Int J Biochem Cell Biol. 41: 1748-1749.

[13] Herzig S and Martinou J-C (2008) Mitochondrial dynamics: to be in good shape to survive. Curr Mol. 8: 131-137.

[14] Avi-Dor Y, Lamdin E and Kaplan Natan O (1963) Structural factors in the Succinate-induced Reduction of Mitochondrial Pyridine Nucleotides. Journ Biol Chem. 238: 2518-2528.

[15] Chance B and Hollunger G (1961) The interaction of energy and electron transfer reactions in mitochondria. I. General properties and nature of the products of succinate-linked reduction of pyridine nucleotide. J Biol Chem. 236: 1534 – 1543.

[16] Chance B and Hollunger G (1961) The interaction of energy and electron transfer reactions in mitochondria. III. Substrate requirements for pyridine nucleotide reduction in mitochondria. J Biol Chem. 236: 1555-1561.

[17] Wojtovich AP and Brooks PS (2008) The endogeneous mitochondrial complex II inhibitor malonate regulates mitochondrial ATP-sensitive potassium channels: implications for ischemoc preconditioning. Biochim Biophys Acta. 1777: 882-889.

[18] Selak MA, Armour SM, MacKenzie ED, Boulahbel H, Watson DG, Mansfield KD, Pan Y, Simon MC, Thompson CB and Gottlieb E (2005) Succinate links TCA cycle dysfunction to oncogenesis by inhibiting HIF-α prolyl hydroxylase. Cancer Cell. 7: 77 – 85.

[19] Pollard PJ, Wortham NC and Tomlinson IPM (2003) The TCA cycle and tumorigenesis: the examples of fumarate hydratase and succinate dehydrogenase. Ann Med. 35: 632 – 639.

[20] Rustin P, Munnich A and Rotig A (2002) Succinate dehydrogenase and human diseases: new insights into a well-known enzyme. Eur J Hum Genet. 10: 289-291.

[21] Kondrashova MN (1976) Therapeutic action of succinic (amber) acid. Pushchino: Institute of Biophysics, Academy of Sciences USSA.

[22] Dilman VM, Anisimov VN and Kondrashova MN (1976) The influence of succinic (amber) and glutamic acids on the sensitivity of hypothalamo-gonadotropic system to inhibitory action of estrogens in old rats. Pharmacol and Toxicol (In Rus) 3: 540-551.

[23] Maevsky EI, Guzar IB, Rosenfeld AS and Kondrashova MN (1982) Doesn't succinic acid mediate adrenaline stimulation in mitochondria? EBEC Reports. LBTM-CNRS, 2: 537.

[24] Kondrashova MN, Gogvadze VG and Babsky AM (1982) Succinic acid as the only energy support of intensive Ca^{2+}-uptake by mitochondria. Bioch Bioph Res Comm. 109: 376—381.

[25] Shostakovskaya IV, Doliba NM and Babsky AM (1986) Activation by acethylcholine of α-ketoglutrate oxidation in rat liver mitochondria. Ukrainian J Biochem (In Rus). 58: 54-61.

[26] Andreev AA, Vulfius EA, Budantsev AYu, Kondrashova MN AND Grishina EV (1986) Depression of neurons responses to acetylcholine by combined application of norepinephrine and substrates of the tricarboxylic acid cycle. Cell Mol Neurobiol. 6: 407-419.

[27] Kondrashova MN and Doliba NM (1989) Polarographic observation of substrate-level phosphorylation and its stimulation by acetylcholine. FEBS Lett. 243: 153—155.

[28] Kondrashova MN and Kuznetzova GD (1991) Succinic acid as a physiological signal molecule. In: Winslow W and Vinogradova OS, editors. Signal molecules and behaviour. Manchester University Press: Manchester&New York. pp. 295-301.

[29] Kondrashova MN (1991) Interaction of carbonic acid transamination and oxidation at various functional states of tissue. Biochemistry-Moscow. 56: 243—310.

[30] Kondrashova MN (2000) Mechanisms of physiological activity and cure effect of a small doses of succinic(amber) acid. Eur Journ Med Res. 5: 58-61.

[31] Kondrashova MN (2002) Hormone-similar action of succinic acid. Vopr Biol Med Pharm Chem (In Rus). 1: 7 – 12.

[32] Kondrashova MN (2005) Signal properties of succinic and α-ketoglutaric acids in oscillatory interaction of Krebs cycle and sympathetic and parasympathetic nervous

system. In: Zinchenko VP, Kolesnikova SS, Beregnova AV, editors. Reception and Cellular Signalization. Pushchino. pp. 249-253.

[33] Khunderyakova NV, Zakharchenko M, Zakharchenko AV, Simonova MA, Vasilieva AA, Romanova OI, Fedotcheva NI, Litvinova EG, Azarashvili AA, Maevsky EI and Kondrashova MN (2010) Changes in sympathetic and parasympathetic regulation connected with succinate dehydrogenase and α-ketoglutarate dehydrogenase activity in different physiological states of the organism. BBA. 1797 Suppl: 142.

[34] Zakharchenko MV, Zakharchenko AV, Khunderyakova NV, Tutukina MN, Simonova MA, Vasilieva AA, Romanova OI, Fedotcheva NI, Litvinova EG, Maevsky EI and Kondrashova MN (2011) Burst of succinate- and α-ketoglutarate dehydrogenase activity related to expression of succinate dehydrogenase subunit a, succinate receptor and other respiratory chain proteins during the short –term physiological stress in rats. MIP 2011 8th MiPconference. pp. 4-16.

[35] He W, Mlao F, Lin D, Schwandner RT, Wang Z and Gao J (2004) Citric acid cycle intermediates as ligands for orphan G-protein-coupled receptors. Nature. 429: 188 – 193.

[36] Gonzalez NS, Communi D, Hannedouche S and Boeynaems JM (2004) The fate of P2Y-related orphan receptors: GPR80/99 and GPR91 are receptors of dicarboxylic acids. Purinergic Signal. 1: 17-20.

[37] Sadagopan N, Li W, Roberds SL, Major T, Preston GM, Yu Y and Tones MA (2007) Circulating succinate is elevated in rodent models of hypertension and metabolic disease. Am J Hypertens. 20: 1209-1215.

[38] Correa PR, Kruglov EA, Thompson M, Leite MF, Dranoff JA and Nathanson MH (2007) Succinate is a paracrine signal for liver damage. J Hepatol. 47: 262-269.

[39] Regard JB, Sato IT and Coughlin SR (2008) Anatomical profiling of G protein-coupled receptor expression. Cell. 135: 561-571.

[40] Toma I, Kang JJ, Sipos A, Vargas S, Bansal E, Hanner F, Meer E and Peti-Peterdi J (2008) Succinate receptor GPR91 provides a direct link between high glucose levels and renin release in murine and rabbit kidney. J Clin Invest. 118: 2526-2534.

[41] Tracey KJ (2009) Reflex control of immunity. Nat Rev Immunol. 9: 418-428.

[42] Rosas-Ballina M, Olofsson PS, Ochani M, Valdés-Ferrer S, Levine Ya-A, Reardon C, Tusche MW, Pavlov VA, Andersson U, Chavan S, Mak TW and Tracey KJ (2011) Acetylcholine-Synthesizing T Cells Relay Neural Signals in a Vagus Nerve Circuit. Science. 334: 98-101.

[43] Kawashima K and Fujii T (2000) Extraneuronal cholinergic system in lymphocytes. Pharmacol Ther. 86: 29-48.

[44] Kawashima K and Fujii T (2003) The lymphocytic cholinergic system and its biological function. Life Sci. 72: 2101-2109.

[45] Fujii T, Takada-Takatori Y and Kawashima K (2008) Basic and clinical aspects of non-neuronal acetylcholine: expression of an independent, non-neuronal cholinergic system in lymphocytes and its clinical significance in immunotherapy. J Pharmacol Sci. 106: 186-192.

[46] Lang TA and Secic M (1997) How to report statistics in medicine. Philadelphia: American College of Physicians. 490 p.

[47] Glantz SA (2005) Primer of Biostatistics. California-San Francisco: McGraw-Hill. 520 p.
[48] Khunderyakova NV, Zakharchenko MV, Zakharchenko AV and Kondrashova MN (2008) Hyperactivation of Succinate Dehydrogenase in Lymphocytes of Newborn Rats. Biochemistry (Moscow). 73: 337-341.
[49] Shnoll S (2009) Cosmophysical factors in random processes. Stockholm: Swedish physics archive (In Rus). 388 p.
[50] Grigorenko EV and Kondrashova MN (1984) Step inhibition of succinate dehydrogenase under progressive pathology. In: Congr. Ed. Hannover. pp. 533–534.
[51] Kondrashova MN and Grigorenko EV (1984) Stress on the level of mitochondria. Zhurn Obshei Biol (In Rus). 46: 516–526.
[52] Brand MD and Nicholls DG (2011) Assessing mitochondrial dysfunction in cells. Biochem J. 435: 297 – 312.
[53] Beyer RE (1990) The participation of coenzyme Q in free radical production and antioxidation. Free Radic Biol Med. 8: 545–565.
[54] Jo SH, Son MK, Koh HJ, Lee SM, Song IH, Kim YO, Lee YS, Jeong KS, Kim WB, Park JW, Song BJ and Huh TL (2001) Control of mitochondrial redox balance and cellular defense against oxidative damage by mitochondrial NADP+-dependent isocitrate dehydrogenase. J Biol Chem. 276: 16168-16176.
[55] Wojtczak A (1969) Inhibitory action of oxaloacetate on succinate oxidation and the mechanism of its reversal. Biochim Biophys Acta. 172: 52 – 65.
[56] Finogenova TV, Shishkanova NV, Fausek EA and Eremina SS (1991) Biosynthesis of isocitric acid from ethanol by yeasts. Applied Microbiol Biotechnol. 36: 231- 235.
[57] Kondrashova MN, Khunderyakova NV and Zakharchenko M (2009) Novel cyto-bio-chemical method for revealing individual differences of physiological states of the organism by complex character (pattern) of succinate dehydrogenase activity. Ros Biomed J Medline.ru. 10:21-43.
[58] Bonomi F, Pagani S, Cerletti P and Giori C (1983) Modification of the thermodynamic properties of the electron – transferring groups in mitochondrial succinate dehydrogenase upon binding of succinate. European Journal of Biochemistry. 134: 439 – 445.
[59] Maevsky EI, Peskov AV, Uchitel ML, Pogorelov AG, Saharova NYu, Vihlyatseva EF, Bogdanova LA and Kondrashova MN (2008) A succinate-based composition reverses menopausal symptoms without sex hormone replacementv therapy. Uspekhi Gerontol (in Engl.). 21: 298-305.
[60] Meex RC, Schrauwen-Hinderling VB, Moonen-Kornips E, Schaart G, Mensink M, Phielix E, van de Weijer T, Sels JP, Schrauwen P and Hesselink MK (2010) Restoration of muscle mitochondrial function and metabolic flexibility in type 2 diabetes by exercise training is paralleled by increased myocellular fat storage and improved insulin sensitivity. Diabetes. 59: 572-579.
[61] Powelka AM, Seth A, Virbasius JV, Kiskinis E, Nicoloro SM, Guilherme A, Tang X, Straubhaar J, Cherniack AD, Parker MG and Czech MP (2006) Suppression of oxidative metabolism and mitochondrial biogenesis by the transcriptional corepressor RIP140 in mouse adipocytes. J Clin Invest. 116: 125-136.

[62] Roth RJ, Le AM, Zhang L, Kahn M, Samuel VT, Shulman GI and Bennett AM (2009) MAPK phosphatase-1 facilitates the loss of oxidative myofibers associated with obesity in mice. J Clin Invest. 119: 3817-3829.
[63] Schrauwen P (2007) High-fat diet, muscular lipotoxicity and insulin resistance. Proc Nutr Soc. 66: 33-41.
[64] Skulachev VP (1991) Fatty acid circuit as a physiological mechanism of uncoupling of oxidative phosphorylation. FEBS Lett. 294: 158-162.
[65] Mori S, Satou M, Kanazawa S, Yoshizuka N, Hase T, Tokimitsu I, Takema Y, Nishizawa Y and Yada T (2009) Body fat mass reduction and up-regulation of uncoupling protein by novel lipolysis-promoting plant extract. Int J Biol Sci. 5: 311-8.

Chapter 11

Hydroxysteroid Dehydrogenases – Localization, Function and Regulation in the Testis

Małgorzata Kotula-Balak, Anna Hejmej and Barbara Bilińska

Additional information is available at the end of the chapter

http://dx.doi.org/10.5772/47740

1. Introduction

Differentiation of the male phenotype including the outward development of secondary sex characteristics as well as the initiation and maintenance of spermatogenesis is stimulated by androgens (O'Shaughnessy et al., 2009; Verhoeven et al., 2010). There are two major androgens secreted by the testes: testosterone (T) and dihydrotestosterone (DHT). Weaker androgens: dehydroepiandrosterone (DHEA) and androstenedione are secreted in smaller amounts and converted metabolically to T and other androgens. Testosterone is the most abundant androgen. However, DHT is the most potent one.

As the result of intensive research over the last 20 years it has been confirmed that estrogens, produced by androgen aromatization, are also important in the regulation of male reproductive function (Carreau et al., 2003). In mice deficient for the estrogen receptor α gene (αERKO) infertility, increased steroid acute regulatory protein (StAR) and 17β-hydroxysteroid dehydrogenase (17β-HSD) mRNA levels together with elevated T level have been found (Akingbemi et al., 2003; Eddy et al., 1996).

Specific receptors for androgens and estrogens have been found in both somatic and germ cells of the testis (Bilinska et al., 2000; Bilinska & Schmalz-Fraczek, 1999; Sierens et al., 2005; Wang et al., 2009). It has been confirmed that these receptors act as transcription factors regulating steroidogenesis at the transcription level. Furthermore, steroidogenesis requires the coordinated expression of related proteins and steroidogenic enzymes in response to hormonal stimulation. In Leydig cells luteinising hormone (LH) induce steroidogenesis by elaborating accumulation of intracellular cyclic adenosine monophosphate (cAMP), activation of protein kinase A (PKA) and expression of StAR resulting in subsequent T biosynthesis and secretion. Intratesticular T is maintained at constantly high levels. In the rat, endogenous T concentrations are the highest at stage VIII of the spermatogenic cycle (Parvinen, 1982). In addition, this stage together with stage VII have been found to be particularly sensitive to androgen action, especially in four steps of germ cell development: spermatid adhesion and development, spermiation, progression through meiosis and spermatogonial differentiation (Verhoeven et al., 2010). Thus spermatogenesis is closely related and absolutely dependent on steroid hormone biosynthesis, action and control.

2. Steroidogenesis in the testis

In the male, Leydig cells which are present in the interstitium of the testis express all of the enzymes essential for the conversion of cholesterol to androgens and estrogens. The major pathways of steroid hormone synthesis are well established, and the sequence of the responsible steroidogenic enzymes has been elucidated (Payne & Hales, 2004). There are two major classes of steroidogenic enzymes: the cytochrome P450 heme-containing proteins and the hydroxysteroid dehydrogenases (HSDs). The first class contains: cytochrome P450 side chain cleavage (P450scc), cytochrome P450 17α-hydroxylase C17-C20 lyase (P450c17) and cytochrome P450 aromatase (P450arom), whereas the second includes both 3β-hydroxysteroid dehydrogenase/Δ5-Δ4-isomerase (3β-HSD) and 17β-hydroxysteroid dehydrogenase/ketosteroid reductase (17β-HSD/KSR), (Payne, 2007).

The initial step of T biosynthesis is the conversion of the C27 cholesterol to the C21 pregnenolone (Figure 1). This reaction is catalyzed by the cytochrome P450scc, which is located in the inner mitochondrial membrane. Pregnenolone diffuses across the mitochondrial membrane and is further metabolized by enzymes associated with the smooth

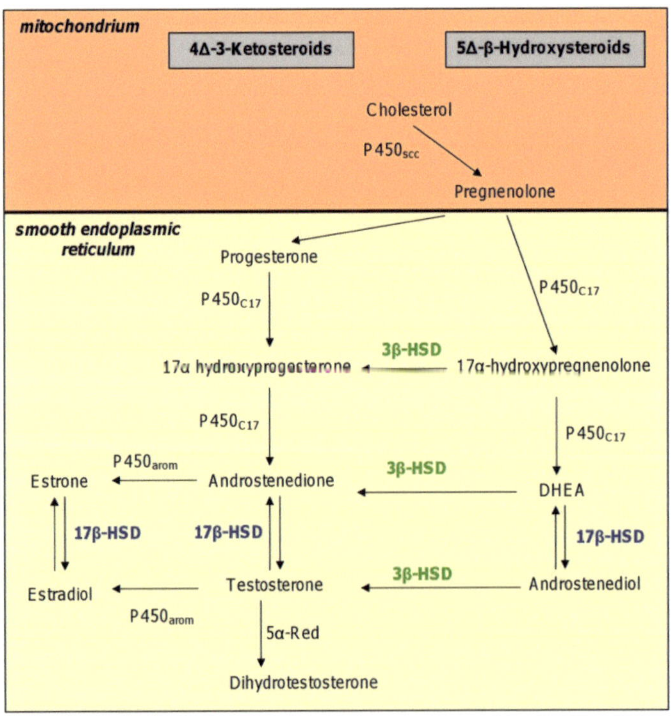

Figure 1. Steroid biosynthesis pathways in Leydig cell.

endoplasmic reticulum. P450c17 catalyzes the conversion of C21 pregnenolone or progesterone (P4) to the C19 dehydroepiandrosterone or androstenedione, respectively, while 3β-HSD catalyzes the conversion of Δ5-3β-hydroxysteroids (pregnenolone or dehydroxypregnenolone, and DHEA, respectively) to the Δ4-3-ketosteroids (P4, or 17α-hydroxyprogesterone, and androstenedione, respectively). Depending on the animal species, biosynthesis of sex hormones proceeds down either one or both of the Δ4 and Δ5 pathways. In rodents, the Δ4 pathway is primary whereas in primates, pigs and rabbits Δ5 pathway is dominant (Fluck et al., 2003; Mathieu et al., 2002). In the final step of sex hormones biosynthesis, conversion of androstenedione into T, 17β-HSD is involved. It was reported that the balance between these androgens depends on the type and activity of 17β-HSD present (Simard et al., 2005).

3β-HSDs are membrane-bound enzymes that are distributed in both mitochondrial and microsomal membranes (Payne & Hales, 2004; Pelletier et al., 2001). The relevance of dual localization of these HSDs is related to substrate accessibility (Simard et al., 2005). Coprecipitation studies have shown that, in the inner mitochondrial membrane, 3β-HSD comprises a functional steroidogenic complex with P450scc, which immediately provides substrates converted from cholesterol to the 3β-HSD (Cherradi et al., 1995).

During the past decade, multiple isoforms of 3β-HSDs have been isolated and characterized in human, mouse and rat tissues. Six, highly homologous in their amino acid sequence isoforms have been identified in the mouse, but only two of them: 3β-HSD type I (3β-HSD I), and 3β-HSD type VI (17β–HSD VI) are expressed in the testis. In human testis only 3β-HSD I has been found (Payne & Hales, 2004).

Similarly to 3β-HSDs, 17β-HSDs are membrane-bound enzymes, and their soluble forms have also been reported. To date, 14 different types of 17β–HSDs have been identified (Blanchard & Luu-The, 2007). Unlike 3β-HSDs there is very little homology among the different 17β-HSD enzymes. Only three types, 17β–HSD type 3 (17β–HSD 3), 17β–HSD type 5 and 17β–HSD type 12 (17β–HSD 12), have been detected to be exclusively expressed in the testis. 17β-HSD 3 converts androstenedione to T as well as it is an important partner of P450arom involved in conversion of C18 steroid, estrone to E2 (Andersson et al., 1995). Recently, it has been confirmed for mice, humans and primates that 17β-HSD 12 shares high homology and function with 17β-HSD 3 (Blanchard & Luu-The, 2007; Liu et al., 2007).

The hydroxysteroid dehydrogenases belong to the same phylogenetic protein family, namely the short-chain alcohol dehydrogenase reductase superfamily. These enzymes are involved in the reduction and oxidation of steroid hormones requiring NAD+/NADP+ as acceptors and their reduced forms as donors of reducing equivalents. Studies have shown that mouse 3β-HSD has different cofactor preference: 3β-HSD I requires NAD+ while 3β-HSD type IV and V requires NADP+ as cofactors. Interestingly, 17β-HSD 3 prefers NADPH as a cofactor, and its primary activity is reductive. Studies have shown that a mutation in *HSD3B3* gene leads to decreased NADPH binding to tyrosine that has been identified as a critical residue for binding. Substitution of tyrosine with different amino acids resulted in alterations in cofactor preference switching from NADPH to NADH (Andersson et al., 1995; McKeever et al., 2002). In addition, Schäfers et al. (2001) have reported different cofactor preference for 17β-HSD 3 on different days of postnatal development in rat.

Both 3β-HSD and 17β-HSD are well known Leydig cell-specific markers in different mammals, at different times of development and under different perturbation regimes (Bilinska, 1994; Hejmej et al., 2011b; Mendis–Handagama & Ariyaratne, 2001; Teerds et al., 2007). In previous studies, activity of HSDs in testis of various mammals was mostly detected using histochemical techniques (Badrinarayanan et al., 2006; Bilinska 1979, 1983, 1994; Hutson, 1989), (Figure 2). Nowadays the resolution of their localization increased with applying specific antibodies (Kotula-Balak et al., 2011; Pelletier et al., 1999; Pinto et al., 2010).

It has been reported that 3β-HSD type III (3β-HSD III) as well as 17β-HSD 3 and 17β-HSD type 10 are useful markers also for germ cells in rat, mouse, equine and black bear testis (Almeida et al., 2011; Ivell et al., 2003; O'Shaugnessy et al., 2000). Recently Scott et al. (2009) have indicated 17β-HSD 3 as a good marker for Sertoli cells in fetal mouse testis.

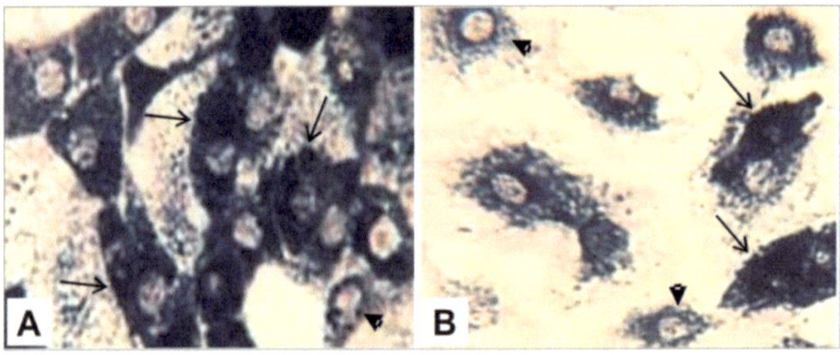

Figure 2. (A-B) Histochemical localization of 3β-HSD (A) and 17β-HSD (B) in cultured mouse Leydig cells. Note various intensity of the staining in the individual cells (arrows-strong staining, arrowheads-weak to moderate staining). Magnifications, x 320.

3. Age-dependent activity of HSDs

In testis of mammals two morphologically and functionally different Leydig cell populations have been identified. One develops prenatally (fetal Leydig cells, FLCs) and the second arises postnatally (adult Leydig cells, ALCs), (Mendis-Handagama & Aryiaratne, 2001; Pinto et al., 2010). These two generations have different gene expression profiles, which indicate that they originate from separate types of stem cell (Dong et al., 2007; O'Shaugnessy et al., 2002b). Differentiation of FLCs is initiated by human chorionic gonadotropin (hCG), whereas development of ALCs is critically dependent on LH (O'Shaughnessy et al., 1998). After birth the population of FLCs decreases in size, although some fetal-type cells persist even in the adult testis. In the rat, FLCs that are arranged in large compact clusters and contain numerous lipid droplets. In ALCs the nuclei are large and cytoplasmic content is sparse with a few lipid droplets. The cytoplasm of Leydig cells of both populations contains abundant smooth endoplasmic reticulum (SER) and tubulovascular mitochondria, which are important organelles in biosynthesis of steroid hormones.

The FLCs are fully competent steroidogenically. It has been demonstrated that in rat, FLCs start to express LH receptors (LHR) and 3β-HSD I from fetal day (fd) 15.5 (Payne & Hales, 2004). However in mouse, 3β-HSD I expression has been detected shortly before fd 11

(Baker et al., 1999). Recently with the use both histochemical and immunohistochemical methods the presence of 3α-hydroxysteroid dehydrogenase (3α-HSD) and 17β-HSD has also been confirmed in FLCs of rat (Haider, 2004).

From fd 15.5 FLCs start actively producing T and its synthesis increases gradually (Habert & Picon, 1984). Expression of hormone receptors and enzymes in FLCs arise continuously during existence of this population in the testis. Interestingly, Ivell et al. (2003) have demonstrated that 17β-HSD type 10 (17β-HSD 10) starts to be expressed at the time when FLCs begin to involute. However, the pick of oxidative activity of these enzyme has been determined on postnatal day (pd) 16 (Schäfers et al., 2001).

Fetal population of Leydig cells is the primary source of T, androstenedione and DHT in both fetal and early postnatal testis (Ariyaratne & Mendis-Handagama, 2000; Huhtaniemi & Pelliniemi, 1992). Multiple studies have shown that T-producing capacity of FLC is significantly greater than that of ALC and is calculated to be even 87 pg per cell (Aryaratne & Mendis-Handagama 2000; Huhtaniemi et al., 1982; Tapanainen et al., 1984).

During the neonatal-prepubertal period T is required for differentiation and morphogenesis of the male genital tract, activation of the hypothalamo-hypophyseal-testicular axis, completion of the testicular descent, masculinization of the brain, control of Sertoli cell number, initiation of spermatogenesis and formation of ALC precursors (Ariyaratne & Mendis-Handagama, 2000; Haider, 2004).

Steroidogenic capacity of FLCs is still high through the first postnatal week, although concentrations of circulating T are much lower due to decrease in number of these cells. Moreover, the inhibitory effects of Müllerian Inhibiting Substance (MIS) and transforming growth factor-bs (TGF-bs) on FLCs steroidogenic activity in postnatal testis have been described (Wu et al., 2007).

Testosterone production gradually increases to high levels with the development of ALCs (Benton, 1995; Chen et al., 2010; Hardy et al., 1989). The proliferation and differentiation of the adult population is regulated by an interplay of multiple regulatory factors, that can simulate, as well as inhibit, Leydig cells at each developmental stage. The development of ALCs is initiated around day 14 after birth and finishes around day 60. This process consists of multiple steps of proliferation and differentiation such as: proliferation of precursor cells; differentiation of precursor cells to Leydig cell progenitors, progenitors into newly formed adult Leydig cells, newly formed adult Leydig cells into immature adult Leydig cells; and finally, maturation of the immature adult Leydig cells to mature adult Leydig cells. In the rat, it has been reported that stem cells and mesenchymal precursor cells do not express steroidogenic enzymes however precursor cells acquire 3β–HSD III and other setroidogenic enzymes like cytochromes: P450scc and P450c17 prior to gain LHR (Hardy et al., 1989; Teerds et al., 2007; Zirkin, 2010). These cells have negligible amounts of 17β-HSD 3 while expression of steroid metabolizing enzymes 5α-reductase and 3α-HSD is high. Thus precursor cells produce androsterone as their main androgen product (O'Shaughnessy et al., 2000). Also expression of AR by early developmental stages of ALCs lineage is required for further transformations of these cells under androgen control (Ge & Hardy, 1997).

Differentiation of progenitors to newly formed adult Leydig cell is associated with the cell cytoplasm shape change from spindle shaped to polygonal. Newly formed Leydig cells move toward the central interstitium and locate near blood capillaries although they do not exclusively arrange in clusters. These cells express LHR and the levels of 3β–HSD VI, P450scc and P450c17 increase with the further steps of Leydig cell differentiation (Ariyaratne & Mendis-Handagama, 2000; Shan et al., 1993). It has been demonstrated that in mice *Hsd3b3* and *Hsd3b6,* remain fairly stable after birth but show a pubertal rise in expression around pd 20 (O'Shaughnessy et al., 2002)

5α-androstane-3α and 17β-diol is synthesized as the predominant androgen with the emergent increase in activity of 17β–HSD 3 and in a continuous presence of 5α-reductase and 3α-HSD (Hardy et al., 1990). It is worth noting, that in immature ALCs 11β-hydroxysteroid dehydrogenase type 1 (11β-HSD 1) and 11β-hydroxysteroid dehydrogenase type 2 start to be expressed. In the rat testis, the presence of 11β-HSD 1 is in coincidence with the first appearance of elongated spermatids in the seminiferous tubules (Haider, 2004).

Adult Leydig cells are the dominant cell type of the Leydig cell lineage from pd 56 (Benton et al., 1995). Transformation of immature adult Leydig cells into mature adult Leydig cells is characterized by a significant increase in the average cell size and disappearance of cytoplasmic lipid droplets. The capacity to secrete T increases significantly in mature adult Leydig cells because of their enhanced responsiveness to circulatory LH due to the acquisition of higher numbers of LHR. During this time in the mouse testis, 3β-HSD VI becomes the predominant isoform of HSDs (Payne & Hales, 2004).

Additionally, the sharp decline in 5α-reductase activity overlaps. Shan et al. (1993) have reported that the mature Leydig cells by pd 90 produce 150 times more T than progenitors, and five times more than immature Leydig cells. Such high T levels are required for initiation, maintenance and regulation of the spermatogenesis. By day 90 the secretory capacity per ALC in rat has been estimated as 1.43 pg.

During puberty ALCs are particularly sensitive to androgens and expression of AR mRNA in this time is significant. Studies have shown that in the absence of AR, there is developmental failure of ALC maturation (O'Shaughnessy et al., 2010). However, there is well known phenomenon when ALCs destroyed by ethane dimethane sulphonate (EDS) administration can proliferate to regenerate the original population of Leydig cells (Teerds & Rijnities, 2007).

In aging human testis, both serum and intratesticular T concentrations progressively decline being in correlation to decreased LH level. In rat, these changes have been reported to be strain-dependent (Harman et al., 2001). In Brown Norway rats, the decrease in T level concomitantly with an increase in FSH level and unchanged LH level have been detected (Chen et al., 2002). Several studies have demonstrated that in men decrease in T level is associated with alterations in body composition, diminished energy, muscle strength and physical function, depressed mood and decreased cognitive function (Matsumoto, 2002). These age related changes result from the loss of steroidogenic capacity of the Leydig cells and/or reduction in their number (Chen et al., 2001, 2009). It has also been found that in aging Leydig cells adenyl cyclase is maintained that results in the defect of the cAMP-LH signaling cascade. In addition, protein and mRNA levels of StAR have been significantly reduced, suggesting deficits in the transport of cholesterol to the inner mitochondrial membrane of aged Leydig cells. Moreover, the activity, protein level, and mRNA level of P450scc, P450c17, 3β–HSD and 17β–HSD have been found markedly reduced in old Leydig cells (Ivell et al., 2003; Luo et al., 1996; Midzak et al., 2009; Zirkin & Chen, 2000). Interestingly, these authors have demonstrated that long-term suppression of steroidogenesis by administration of T prevents or delays the reduced steroidogenesis that accompanies Leydig cell aging due to suppressing the production of the reactive oxygen species that are a by-product of steroidogenesis itself.

4. Regulation of HSDs function

4.1. Pituitary hormones and other peptides

Primary control of 3β-HSD expression occurs through the activation by LH its receptor and the induction of the cAMP second messenger system (Simard et al., 2005). Recent findings from our own laboratory have revealed that bank vole Leydig cells treated with LH have increased steroidogenic capacity and T secretion (Gancarczyk et al., 2003). A profound hypogonadal effect and suppression of T production has been demonstrated in boars treated with deslorelin, an agonist of gonadotropin-releasing hormone (GnRH), (Kopera et al., 2008). In Leydig cells of treated boars very weak or lack of LHR and 3β-HSD expression has been detected. In contrast, Lin et al. (2008) who treated mouse Leydig cells with GnRH agonists (I and II) have demonstrated that 3β-HSD has been stimulated directly resulting in increase of T production. On the contrary, 17β-HSD was not induced in treated cells.

Formation of cAMP activates steroidogenesis by temporally distinct manners either, acutely (minutes) due to StAR action or chronically (hours) related on P450scc, P450c17, 3β-HSD and 17β-HSD activities. The delivery of cholesterol into the inner mitochondrial membrane is the rate-determining step in steroidogenesis. Also the differences in the way that cultured Leydig cells respond to cAMP have been reported. In mouse Leydig cells, cAMP has stimulated T production which then suppressed 3β-HSD at the mRNA level, whereas addition of cAMP to cultured rat Leydig cells increases 3β-HSD activity and expression at both the mRNA and protein level after 24–72 h (Keeney & Mason 1992; Payne & Sha 1991).

Interestingly, expression of 3β-HSD has been reported to be dependent upon steroidogenesis factor 1 (SF1). SF1 response element has been detected in the proximal promoter region of the human 3β-HSD type II gene. Recently Scott et al. (2009) have confirmed that the mouse 3β-HSD I gene promoter has three potential SF1 consensus binding sites. However, it is currently unknown whether SF1 regulates the expression of 17β–HSD.

A series of studies show that pituitary hormone-prolactin (PRL) and thyroid hormones regulate activity of HSDs. In Leydig cells of hypophysectomized rats treated with PRL, a significant increase in number of 3β-HSD immunopositive cells together with an increase in T and E2 concentrations was found (Dombrowicz et al., 1992; Manna et al., 2001). Also our own studies have revealed that in bank voles treated with PRL, the levels of androgens and estrogens have increased markedly within the testis (Gancarczyk et al., 2006). These results point the role of PRL in promoting multiplication, differentiation and regulation of steroidogenic function of Leydig cells. Similar functions have been confirmed for thyroid hormones. In physiological levels these hormones have profoundly increased the number of mesenchymal precursors of ALCs and supported their further differentiation (Maran, 2003; Mendis-Handagama et al., 1998; Teerds et al., 1998).

4.2. Steroids

In Leydig cells, the action of P4, has been reported to be mediated only by non-classical receptors while the classical nuclear progesterone receptor has not been found in these cells (Oettel & Mucopadhay 2004). The direct stimulatory or inhibitory effect of P4 on steroidogenesis in Leydig cells has been demonstrated, although its mode of action remains obscure (El-Hefnawy et al., 2000; Schwarzenbach et al., 2003). In rats, exposure in utero to subnormal levels of hydroxyprogesterone suppresses testicular steroidogenesis by decreasing the HSDs levels, which in turn suppresses the reproductive activity of the male (Pushpalatha et al., 2003). Studies have shown that in elderly men P4 levels increase within the testis and the spermatic vein, having a detrimental effect on Leydig cell steroidogenic function (El-Hefnawy & Huhtaniemi, 1998). It is interesting but still not resolved if and how P4 influences HSDs.

It has been reported that both endogenous and exogenous sex hormones are able to modify steroidogenesis at the level of 3β-HSD. In rats treated with an androgen antagonist, increased induction of hCG and 3β-HSD activity has been observed, whereas treatment with an androgen agonist decreased hCG induction and 3β-HSD activity, respectively (Ruiz de Galarreta et al., 1983). Similarly, T and DHT have inhibited 3β-HSD activity in adult rat and mouse Leydig cells (Simard et al., 2005). Recent findings by Kostic at al. (2011) have demonstrated that in andorgenized rats, T upregulated P4 synthesis. In these animals prolonged treatment with high T doses caused significant increase of 3β-HSD mRNA and protein levels whereas no effect has been observed on 17β-HSD expression. Freeman (1985) has demonstrated that E2 inhibited P4 biosynthesis in a dose-dependent manner in Leydig cells via inhibition of the activity of 3β-HSD. Also studies *in vivo* on rats and bank voles treated with E2 have shown disturbances in sex hormones balance within the testis. Low T and high E2 levels in treated animals have been reported to affect spermatogenesis (Gancarczyk et al., 2004; Rao & Chinoy, 1986).

4.3. Testicular paracrine factors

It has already been accepted that the function and activity of the testis is regulated by many locally produced factors and by cell-cell interactions. The effects of cytokines and growth factors on HSDs expression has been reported to be diverse (for review see Herrmann et al., 2002). Tumor necrosis factor (TNF) and interleukin 1 (IL-1) inhibited 3β-HSD activity in mouse and rat Leydig cells. However, IL-1 only inhibited cAMP stimulated enzyme synthesis, whereas TNF also reduced basal enzyme expression. In contrast, epidermal growth factor (EGF) increased activity and expression of 3β-HSD, which has also been demonstrated for transforming growth factor (TGF). Acidic fibroblast growth factor (aFGF) and basic fibroblast growth factor (bFGF) inhibited 3β-HSD. In pig Leydig cells, bFGH has been reported to increase 17β-HSD expression (Sordoillet et al., 1992). Our own studies in bank voles have revealed that insulin like growth factor I (IGF-I) stimulates whereas IL-1 and interleukin 1α (IL-1α) inhibit testis steroidogenic and spermatogenic function in sexually active males (Gancarczyk et al., 2006; Kmicikiewicz & Bilinska, 1997; Kmicikiewicz et al., 1999). Interesting results have been recently reported by Ivell et al. (2011) who demonstrated that 17β-HSD 10 mRNA expression in mice testis is regulated by locally produced relaxin (RLN) dependently of animal age.

4.4. Photopheriod

In seasonal breeders the function of the pituitary-testicular axis undergoes annual cyclic variations. Multiple studies including our own have shown that photoperiod is an important factor regulating steroidogenesis. Changes in LH and FSH secretion depending on the light length are responsible for seasonal variations in size, structure and function of the testis (Bartke & Steger, 1992). Under laboratory conditions, bank voles kept in long light regime show higher testis weight and increased steroidogenic activity than animals exposed to short one (Bilinska et al., 2000, 2001; Tähkä et al., 1982). In several seasonal breeders, the serum and testicular concentrations of steroid hormones have exhibited seasonal fluctuations and are always lower in regressed animals (Frungieri et al., 1999; Hance et al., 2009; Kotula-Balak et al., 2003).

In addition, the Leydig cell morphology as well as localization and expression of HSDs have been found to change seasonally in the sika deer, black bear and northern fur seal (Hayakawa et al., 2010; Ibluchi et al., 2010; Tsubota et al., 1997, 2001). In black bears during their mating season, 17β-HSD 3 has been detected both in Leydig cells and in Sertoli cells. Moreover, in these animals expression of 3β-HSD was the highest in June. In Syrian hamster, specific melatonin receptors (mel1a) have been reported in Leydig cells via which melatonin down-regulated the gene expression of both 3β-HSD, and 17β-HSD (Frungieri et al., 2005).

4.5. Endocrine disrupting chemicals

A large body of information concerning the effects of endocrine disrupting compounds (EDCs) on Leydig cells steroidogenesis during fetal development and in adult mammals has been accumulated in the past decades. EDCs can disturb morphology and normal endocrine functions of the Leydig cells or oppose the actions of androgen through their estrogenic or anti-androgenic properties (Hejmej et al., 2011a). A number of compounds act directly on Leydig cells to diminish T production by interfering with the expression of steroidogenic enzymes, at the protein and/or mRNA level (Skakkebaek, et al., 2001).

Our recent results have shown that administration of the estrogenic compound, 4-*tert*-octylphenol (OP), to adult bank voles has caused the significant decrease of 3β-HSD and increase of P450arom expression concominantly with the alteration of the androgen/estrogen balance within the tesis of sexually active animals (Hejmej et al,. 2011b). Similar results have been reported by Victor-Costa et al. (2010) on rats treated with atriazine. These authors concluded that inhibition of 3β-HSD function is one of the possible mechanism through which xenoestrogens disturb spermatogenesis. *In vitro* studies on Leydig cells obtained from various mammals have revealed decrease in the activity and expression of 3β-HSD after OP, bisphenol A (BPA) and genistein administration (Hu et al., 2010; Kotula-Balak et al., 2011; Ye et al., 2011). Our study demonstrated that OP markedly disturbes morphology and steroidogenic function of the Leydig cells through direct effect on 3β-HSD expression and localization (Kotula-Balak et al., 2011). In detail, treatment with high doses of OP (10^{-4}–10^{-6} M) resulted in a reduced staining intensity and the staining was usually located near the nucleus, whereas in the low OP doses (10^{-7} and 10^{-8} M) it was manly dispersed throughout the cytoplasm (Figure 3).

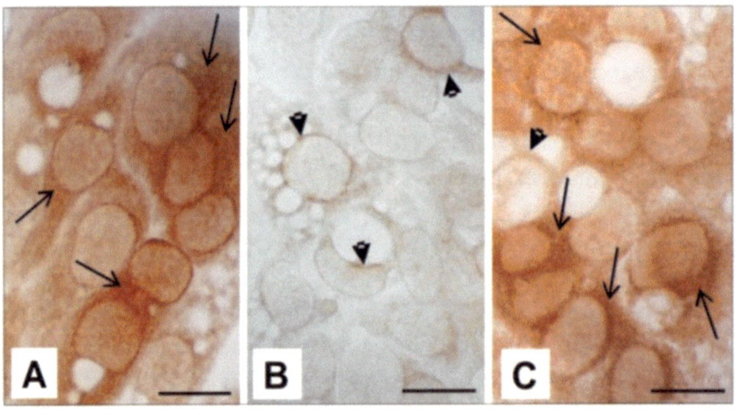

Figure 3. (A-C) Immunostaining for 3β-HSD. Positive staining of various intensity is confined to the cytoplasm of Leydig cells (arrows). Note, clearly reduced staining for 3β-HSD in Leydig cells treated with high OP dose (B). In many cells weak to moderate staining in the perinuclear region is visible (arrowheads). In Leydig cells treated with low OP dose (C) the intensity of immunostaining is similar to that of the control (A), (arrows). Only in a few cells staining in the perinuclear region is visible (arrowhead). Bars 20 μm.

It is worth noting that the effect of EDCs on HSDs function can be diverse depending on the choice of animal species, age, routes of administration and dose levels. Studies of Pogrmic-Majkic et al. (2010) have shown that in rat Leydig cells atriazine stimulated 17β-HSD, whereas other authors reported inhibition of these enzyme in rat and human microsomes treated with various xenoestrogens (Hu et al., 2010; Vaithinathan et al., 2008; Ye et al., 2011). Recently, it has also been found that antiandrogens such as tributylin, triclosan and flutamide modified HSDs expression in Leydig cells and microsomes of various mammals (Kim et al., 2008; Kumar et al., 2009; McVey & Cooke, 2003; Ohno et al., 2005; Ohsako et al., 2003).

5. HSDs in pathological conditions

5.1. Temperature and 3β-HSD activity

In recent years, disorders of human male reproductive development increased in incidence (Sharpe & Skakkebaek, 1993; Toppariet al., 1996). Cryptorchidism and hypospadias are the two most common congenital malformations that comprise a testicular dysgenesis syndrome (TDS), which arises during fetal development and manifests in adulthood (Skakkebaek et al., 2001).

Several studies have shown that increased intratesticular temperature in cryptorchid testes affects spermatogenesis, resulting in either reduced fertility or infertility. Other evidences collected over the years have indicated that increased testicular temperature negatively influences the development and differentiation of Leydig cells causing impairments in sex hormones biosynthesis (Huff et al., 2001; Kotula-Balak et al., 2001; Pinart et al., 2000).

Our recent studies have shown that in cryptorchid horses disturbances in differentiation and/or maturation of Leydig cells may be related to altered intracellular communication. In these animals decreased immunoexpression of gap junction protein, connexin 43, in testicular cells were accompanied with reduced expression of LHR, 3β-HSD and disturbed androgen/estrogen balance (Hejmej et al., 2005, 2007; Hejmej & Bilinska, 2008), (Figure 4). Altered expression of these proteins and imbalance in sex hormones level detected in cryptorchid horses suggested their additional influence on morphology and function of undescended testis. Markedly reduced expression of 3β-HSD has been also reported in rats with experimentally induced cryptorchidism (Wisner & Gomes, 1978). No significant changes in T levels have been detected in patients with cryptorchidism as well as other mammalian species (Bilinska et al., 2003; Farrer et al., 1985; Illera et al., 2003; Kawakami et al., 1999; Ren et al., 2006; Ryan et al., 1986).

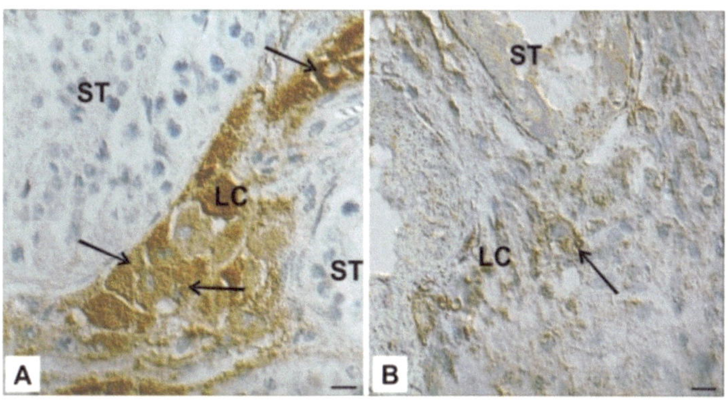

Figure 4. Immunohistochemical localization of 3β-HSD in testis of normal (A) and cryptorchid stallion (B). Counterstaining with Mayer's haematoxylin. The presence of 3β-HSD is confined to Leydig cells (arrows). Note a clearly weaker staining in the cryptorchid horse (A) than in the healthy stallion (B). LC-Leydig cells, ST-seminiferous tubules. Bars 20 μm.

5.2. 3β-HSD and 17β-HSD deficiency

The development of the male internal and external genitalia in an XY fetus requires a complex interplay of many critical genes, enzymes and cofactors. In early fetal life, in the bipotential embryo, both Wolffian ducts and Müllerian ducts are present. Testosterone produced by FLCs acts on AR to stabilize the Wolffian ducts whereas MIS causes regression of Müllerian ducts (George et al., 2010). The formation of male external genitalia is induced by T and DHT.

Disruption in androgen production and/or action leads to disorder of sex development (DSD) also known as male pseudohermaphroditism. DSD is defined as a congenital condition in which development of chromosomal, gonadal or anatomical sex is atypical in such individuals (Hughes et al., 2006). 46, XY DSD is an autosomal recessive form of DSD which was first described in 1971 (Saez et al., 1971). Disturbances of androgen production can occur in all steps of T biosynthesis and its conversion into DHT.

Around 40 various mutations have already been described in *HSD3B2* gene. Mutation in this gene results in 3β-HSD II deficiency and decreased T biosynthesis (Payne & Hales, 2004). Severe form of 3β-HSD deficiency named congenital adrenal hyperplasia (CAH) impairs also steroidogenesis in adrenals (Moisan et al., 1999). In male patients, either perineal hypospadias or perineoscrotal hypospadias and ambiguous external genitalia or microphallus have been reported (Simard et al., 2005). Also, in these individuals gynecomastia has been observed as common at puberty. Serum levels of Δ5 steroids are elevated whereas levels of Δ4 steroids are slightly increased. In adulthood, males with 3β-HSD II deficiency can reach normal levels of T due to the peripheral conversion of elevated Δ5 steroids by 3β-HSD I and/or due to testicular stimulation by high LH levels. Most of the patients with mutation in *HSD3B2* gene are raised as males and display male social sex but there are some cases described where such individuals are castrated in childhood and treated as female (Mendonca et al., 2008). Males with 3β-HSD II deficiency share common clinical features with patients deficient for 17β-HSD 3 and 5α-reductase 2.

Deficiency of the 17β-HSD 3 can be caused by either homozygous or compound heterozygous mutations in the *HSD17B3* gene (Geissler et al., 1994). This autosomal recessive disorder manifests in males as undermasculinization characterized by hypoplastic-to-normal internal genitalia (epididymis, vas deferens, seminal vesicles, and ejaculatory ducts), but female external genitalia and the absence of a prostate (Boehmer et al., 1999; Lindqvist et al., 2001; Sinnecker et al., 1996; Ulloa-Aguirre et al., 1985).

At the time of puberty, there is a marked increase in plasma LH and, consequently, in testicular secretion of androstenedione. Mendonca et al. (2010) have found that significant amounts of the circulating androstenedione are converted to T in peripheral tissues by an unidentified member of the 17β-HSD family, thereby causing virilization in many of these individuals. To date, 19 mutations in the *HSD17B3* gene have been found. Most of these patients are raised as girls during childhood but starts to display masculine behavior at puberty (Mendonca et al., 2008).

6. Conclusion

In Leydig cells, multiple factors regulate 3β-HSD and 17β-HSD function exerting very diverse effect. Nowadays further characterization of physiological and pathological conditions as well as endogenous and exogenous agents that can modify HSDs expression is becoming increasingly necessary. Especially, in light of recent reports indicating an increase in the incidences of developmental and functional disorders of the male reproductive tract. Exploration of the site and possible mechanisms of action of these agents in steroid biosynthesis is becoming important future research direction.

Author details

Małgorzata Kotula-Balak, Anna Hejmej and Barbara Bilińska
Department of Endocrinology, Institute of Zoology, Jagiellonian University, Poland

Acknowledgement

This work was financially supported by the Ministry of Science and Higher Education, Grant N N303816640.

7. References

Akingbemi, B.T.; Ge, R.; Rosenfeld, C.S.; Newton, L.G.; Hardy, D.O.; Catterall, J.F.; Lubahn, D.B.; Korach, K.S. & Hardy, M.P. (2003) Estrogen receptor-alpha gene deficiency enhances androgen biosynthesis in the mouse Leydig cell. *Endocrinology*, Vol. 144, No. 1, (January 2003), pp. 84–93, ISSN 1477-7827

Almeida, J.; Conley, A.J.; Mathewson, L. & Ball, B.A. (2011) Expression of steroidogenic enzymes during equine testicular development. *Reproduction*, Vol. 141, No. 6, (June 2011), pp. 841-878, ISSN 1470-1626

Andersrson, S.; Geissler, W.M.; Patel, S. & Ling, W. (1995) The molecular biology of androgenic 17β-hydroxysteroid dehydrogenases. *Journal of Steroid Biochemistry and Molecular Biology*, Vol. 53, No. 1-6, (June, 1995), pp. 37-39, ISSN 0960-0760

Ariyaratne, H.B.S. & Mendis-Handagama, S.M.L.C. (2000) Structural and functional changes in the testis interstitium of Sprague Dawley rats from birth to adulthood. *Biology of Reproduction*, Vol. 62, No. 3, (March 2000), pp. 680–690, ISSN 1470-1626

Badrinarayanan, S.; Rengarajan, P.; Nithya, P. & Balasubramanian, K. (2006) Corticosterone impairs the mRNA expression and activity of 3beta- and 17beta-hydroxysteroid dehydrogenases in adult rat Leydig cells. *Biochemistry and Cell Biology*, Vol. 84, No. 5, (October 2006), pp. 745–754, ISSN 0829-8211

Baker, P.J.; Sha, J.A.; McBride, M.W.; Peng, L.; Payne, A.H. & O'Shaughnessy, P.J. (1999) Expression of 3β-hydroxysteroid dehydrogenase type I and type VI isoforms in the mouse testis during development. *European Journal of Biochemistry*, Vol. 260, No. 3, (March 1999), pp. 911–917, ISSN 0014-2956

Bartke, A. & Steger, R.W. (1992) Seasonal changes in the function of the hypothalamic-pituitary-testicular axis in the Syrian hamster. *Proceedings of the Society for Experimental Biology and Medicine*, Vol. 199, No. 2, (February 1992), pp. 139-148, ISSN 0037-9727

Benton, L.; Shan, L.X. & Hardy, M.P. (1995) Differentiation of adult Leydig cells. *Journal of Steroid Biochemistry and Molecular Biology*, Vol. 53, No. 1-6, (June 1995), pp. 61-68, ISSN 0960-0760

Bilinska, B. (1979) Histochemical demonstration of Δ^5, 3β-hydroxysteroid dehydrogenase activity in cultured Leydig cells under the influence of gonadotropic hormones and testosterone. *International Journal of Andrology*, Vol. 2, Issue 1-6, (December 1979), pp. 385–394, ISNN 0105-6263

Bilinska, B. (1983) Effect of *in vitro* administration of LH, prolactin separately and LH and prolactin in mixture, on cultured Leydig cells from mouse testes: Changes of Δ^5, 3β-hydroxysteroid dehydrogenase during postnatal life. *Folia Histochemica et Cytochemica*, Vol. 21, No. 1, (1983), pp. 23-32, ISSN 0015-5586

Bilinska, B. (1994) Staining with ANS fluorescent dye reveals distribution of mitochondria and lipid droplets in cultured Leydig cells. *Folia Histochemica et Cytobiologica*, Vol. 32, No. 1, (1994), pp. 21-24, ISSN 0239-8508

Bilinska, B.; Drag, E. & Schmalz-Fraczek, B. (1999) Immunolocalization of androgen receptors intesticular cells during postnatal development of the bank vole (*Clethrionomys glareolus*, S). *Tissue & Cell*, Vol. 31, No. 6, (December 1999), pp. 621-26, ISSN 0040-8166

Bilinska, B.; Kotula-Balak, M.; Gancarczyk, M.; Sadowska, J.; Taborowski, Z. & Wojtusiak, A. (2003) Androgen aromatization in cryptorchid mouse testis. *Acta Histochemica*, Vol. 105, No. 1, (2003), pp. 157–165, ISSN 0065-1281

Bilinska, B.; Schmalz-Fraczek, B.; Kotula, M. & Carreau, S. (2001) Photoperiod-dependent capability of androgen aromatization and the role of estrogens in the bank vole testis visualized by means of immunohistochemistry. *Molecular and Cellular Endocrinology*, Vol. 10, No. 178, (June 2001), pp. 189-198, ISSN 0303-7207

Bilinska, B.; Schmalz-Fraczek, B.; Sadowska, J. & Carreau S. (2000) Localization of cytochrome P450aromatase and estrogen receptors alpha and beta in testicular cells- an immunohistochemical study of the bank vole. *Acta Histochemica*, Vol. 102, No. 2, (May 2000), pp. 167-181, ISSN 0065-1281

Blanchard, P-G. & Luu-The, V. (2007) Differential androgen and estrogen substrates specificity in the mouse and primates type 12 17β-hydroxysteroid dehydrogenase. *Journal of Endocrinology*, Vol. 194, No. 2, (August 2007), pp. 449–455, ISSN 0022-0795

Boehmer, A.L.; Brinkmann, A.O.; Sandkuijl, L.A.; Halley, D.J.; Niermeijer, M.F.; Andersson, S.; de Jong, F.H.; Kayserili, H.; de Vroede, M.A.; Otten, B.J.; Rouwe, C.W.; Mendonca, B.B.; Rodrigues, C.; Bode, H.H.; de Ruiter, P.E.; Delemarre-van de Waal, H.A. & Drop, S.L. (1999) 17beta-hydroxysteroid dehydrogenase-3 deficiency: diagnosis, phenotypic variability, population genetics, and worldwide distribution of ancient and *de novo* mutations. *Journal of Clinical Endocrinology & Metabolism*, Vol. 84, No. 12, (December 1999), pp. 4713–4721, ISSN 0021-972X

Carreau, S.; Lambard, S.; Delalande, C.; Denis-Galeraud, I.; Bilinska, B. & Bourguiba, S. (2003) Aromatase expression and role of estrogens in male gonad: a review. *Reproductive Biology and Endocrinology*, Vol. 11, No. 1, (April 2003), p. 35, ISSN 1477-7827

Chen, H.; Cangello, D.; Benson, S.; Folmer, J.; Zhu, H.; Trush, M.A. & Zirkin, B.R. (2001) Age-related increase in mitochondrial superoxide generation in the testosterone-producing cells of Brown Norway rat testes: relationship to reduced steroidogenic function? *Experimental Gerontology*, Vol. 36, No. 8, (August 2001), pp. 1361-1373, ISSN 0531-5565

Chen, H.; Ge, R.S. & Zirkin, B.R. (2009) Leydig cells: From stem cells to aging. *Molecular and Cellular Endocrinology*, Vol. 10, No. 306, Issue 1-2, (July 2009), pp. 9-16, ISSN 0303-7207

Chen, H.; Hardy, M.P. & Zirkin, B.R. (2002). Age-related decreases in Leydig cell testosterone production are not restored by exposure to LH *in vitro*. *Endocrinology*, Vol. 143, No. 5, (May 2002), pp. 1637-1642, ISSN 1477-7827

Chen, H.; Stanley, E.; Jin, S. & Zirkin, B.R. (2010) Stem Leydig cells: from fetal to aged animals. *Birth Defects Research Part C Embryo Today*, Vol. 90, No. 4, (December 2010), pp. 272-283, ISSN 1542-975X

Cherradi, N.; Chambaz, E.M. & Defaye, G. (1995) Organization of 3β-hydroxysteroid dehydrogenase/isomerase and cytochrome P450scc into a catalytically active molecular complex in bovine adrenocortical mitochondria. *Journal of Steroid Biochemistry and Molecular Biology*, Vol. 55, No. 5-6, (December 1995), pp. 507–514, ISSN 0960-0760

Dombrowicz, D.; Sente, B.; Closset, J. & Hennen, G. (1992) Dose-dependent effects of human prolactin on the immature hypophysectomized rat testis. *Endocrinology*, Vol. 130, No. 2, (February 1992), pp. 695-700, ISSN 0013-7227

Dong, L.; Jelinsky, S.A.; Finger, J.N.; Johnston, D.S.; Kope, G.S.; Sottas, C.M.; Hardy, M.P. & Ge, R-S. (2007) Gene expression during development of fetal and adult Leydig cells. *Annals of the New York Academy of Science*, Vol. 1120, (December 2007), pp. 16–35, ISSN 0077-8923

Eddy, E.M.; Washburn, T.F.; Bunch, D.O.; Goulding, E.H.; Gladen, B.C.; Lubahn, D.B. & Korach, K.S. (1996) Targeted disruption of the estrogen receptor gene in male mice causes alteration of spermatogenesis and infertility. *Endocrinology*, Vol. 137, No. 11, (November 1996), pp. 4796–4805, ISSN 1477-7827

El-Hefnawy, T.; Manna, P.R.; Luconi, M.; Baldi, E.; Slotte, J.P. & Huhtaniemi, I. (2000) Progesterone action in a murine Leydig tumor cell line (mLTC-1), possibly through a nonclassical receptor type. *Endocrinology*, Vol. 141, No. 1, (January 2000), pp. 247-255, ISSN 0013-7227

El-Hefnawy, T. & Huhtaniemi, I. (1998) Progesterone can participate in down-regulation of the luteinizing hormone receptor gene expression and function in cultured murine Leydig cells. *Molecular and Cellular Endocrinology*, Vol. 137, No. 2, (February 1998), pp. 127-138, ISSN 0303-7207

Farrer, J.H.; Sikka, S.C.; Xie, H.W.; Constantinide, D. & Rajfer, J. (1985) Impaired testosterone biosynthesis in cryptorchidism. *Fertility and Sterility*, Vol. 44, No. 1, (July 1985), pp. 125-132, ISSN 0015-0282

Fluck, C.E.; Miller, W.L. & Auchus, R.J. (2003) The 17, 20-lyase activity of cytochrome p450c17 from human fetal testis favors the delta5 steroidogenic pathway. *Journal of Clinical Endocrinology & Metabolism*, Vol. 88, No. 8, (August 2003), pp. 3762-3766, ISSN 0021-972X

Freeman, D.A. (1985) Estradiol acts as a competitive inhibitor of the 3beta-hydroxysteroid dehydrogenase/delta 5-delta 4 isomerase enzyme of cultured Leydig tumor cells. *Endocrinology*, Vol. 117, No. 5, (November 1985), pp. 2127-2133, ISSN 0013-7227

Frungieri, M.B.; Gonzales-Calvarsi, Cl.; Bartke, A. & Calendara, R.S. (1999) Influence of age and photoperiod on steroidogenic function of the testis in the golden hamster. *International Journal of Andrology*, Vol. 22, No. 4, (August 1999), pp. 243-252, ISSN 0105-6263

Frungieri, M.B.; Mayerhofer, A.; Zitta, K.; Pignataro, O.P.; Calandra, R.S. & Gonzales-Calvar, S.I. (2005) Direct effect of melatonin on Syrian hamster testes: melatonin subtype 1a receptors, inhibition of androgen production, and interaction with the local corticotropin-releasing hormone system. *Endocrinology*, Vol. 146, No. 3, (March 2005), pp. 1541-1552, ISSN 1477-7827

Gancarczyk, M.; Tworzydlo, W.; Szlachta, A.; Sadowska, J. & Bilinska, B. (2003) Hormonal regulation of estrogen formation by Leydig cells *in vitro*: immunocytochemical approach. *Folia Biologica (Krakow)*, Vol. 51, No. 3-4, (2003), pp. 181-188, ISSN 0015-5497

Gancarczyk, M.; Kuklinska, M.; Sadowska, J.; Strzezek, J. & Bilinska, B. (2006) Aromatization and antioxidant capacity in the testis of seasonally breeding bank voles: Effects of LH, PRL and IGF-I. *Theriogeneology*, Vol. 15, No. 65, Issue 7 (April 2006), pp. 1376-1391, ISSN 0093-691X

Gancarczyk, M.; Paziewska-Hejmej, A.; Carreau, S.; Tabarowski, Z. & Bilinska, B. (2004) Dose- and photoperiod-dependent effects of 17beta-estradiol and the anti-estrogen ICI 182, 780 on testicular structure, acceleration of spermatogenesis, and aromatase immunoexpression in immature bank voles. *Acta Histochemica*, Vol. 106, No. 4, (2004), pp. 269-278, ISSN 0065-1281

Ge, R.S. & Hardy, M.P. (1997) Decreased cyclin A2 and increased cyclin G1 levels coincide with loss of proliferative capacity in rat Leydig cells during pubertal development. *Endocrinology*, Vol. 138, No. 9, (September 1998), pp. 3719-3726, ISSN 1477-7827

Geissler, W.M.; Davis, D.L.; Wu, L.; Bradshaw, K.D.; Patel, S.; Mendonca, B.B.; Elliston, K.O.; Wilson, J.D.; Russell, D.W. & Andersson, S. (1994) Male pseudohermaphroditism caused by mutations of testicular 17beta-hydroxysteroid dehydrogenase 3. *Nature Genetics*, Vol. 7, No. 1, (May 1994), pp. 34–39, ISSN 1061-4036

George, M.M.; New, M.I.; Ten, S.; Sultan, C. & Bhangoo, A. (2010) The clinical and molecular heterogeneity of 17β-HSD-3 enzyme deficiency. *Hormone Research in Paediatrics*, Vol. 74, No. 4, (2010), pp. 229–240, ISSN 1663-2818

Habert, R. & Picon, R. (1984) Testosterone, dihydrotestosterone and estradiol-17beta levels in maternal and fetal plasma and in fetal testes in the rat. *Journal Steroid Biochemistry*, Vol. 21, No. 2, (August 1984), pp. 193–198, ISSN 0022-4731

Haider, S.G. (2004) Cell biology of Leydig cells in the testis. *International Review of Cytology*, Vol. 233, (2004), pp. 181-241, ISSN 0074-769

Hance, M.W.; Mason, J.I. & Mendis-Handagama, S.M. (2009) Effects of photo stimulation and nonstimulation of golden hamsters (*Mesocricetus auratus*) from birth to early puberty on testes structure and function. *Histology and Histopathology*, Vol. 24, No. 11, (November 2009), pp. 1417-1424, ISSN 0213-3911

Hardy, M.P.; Zirkin, B.R & Ewing, L.L. (1989) Kinetic studies on the development of the adult population of Leydig cells in testes of prepubertal rat. *Endocrinology*, Vol. 124, No. 2, (February 1989), pp. 762-770, ISSN 1477-7827

Hardy, M.P.; Kelce, W.R.; Klinefelter, G.R. & Ewing, L.L. (1990) Differentiation of Leydig cell precursors *in vitro*: a role for androgen. *Endocrinology*, Vol. 127, No. 1, (July 1990), pp. 488-490, ISSN 1477-7827

Harman, S.M.; Metter, E.J.; Tobin.; J.D.; Pearson, J.; Blackman, M.R. & Baltimore Longitudinal Study of Aging. Longitudinal effects of aging on serum total and free testosterone levels in healthy men. Baltimore Longitudinal Study of Aging. (2001) *Journal of Clinical Endocrinology & Metabolism*, Vol. 86, No. 2, (February 2001), pp. 724-731, ISSN 0021-972X

Hayakawa, D.; Sasaki, M.; Suzuki, M.; Tsubota, T.; Igota, H.; Kajaki, K. & Kitamura, N. (2010) Immunohistochemical localization of steroidogenic enzymes in the testis of the sika deer (*Cervus nippon*) during developmental and seasonal changes. *Journal of Reproduction and Development*, Vol. 56, No. 1, (February 2010), pp. 117-123, ISSN 0916-8818

Hejmej, A. & Bilinska, B. (2008) The effects of cryptorchidism on the regulation of steroidogenesis and gap junctional communication in equine testes. *Polish Journal of Endocrinology*, Vol. 59, No. 2, (March-April 2008), pp. 112-118, ISSN 0423-104X

Hejmej, A.; Gorazd, M.; Kosiniak-Kamysz, K.; Wiszniewska, B.; Sadowska, J. & Bilinska, B. (2005) Expression of aromatase and oestrogen receptors in reproductive tissues of the stallion and a single cryptorchid visualised by means of immunohistochemistry. *Domestic Animal Endocrinology*, Vol. 29, No. 3, (October 2005), pp. 534-547, ISSN 0739-7240

Hejmej, A.; Kotula-Balak, M. & Bilinska B. (2011a) Antiandrogenic and estrogenic compounds: Effect on development and function of male reproductive system. In: *Steroids*, Abduljabbar, H. (ed.), pp. 53-81, InTech, ISBN 978-953-307-705-5, Rijeka, Croatia.

Hejmej, A.; Kotula-Balak, M.; Galas, J. & Bilinska, B. (2011b) Effects of 4-*tert*-octylphenol on the testes and seminal vesicles in adult male bank voles. *Reproductive Toxicology*, Vol. 31, No. 1, (January 2011), pp. 95-105, ISSN 0890-6238

Hejmej, A.; Kotula-Balak, M.; Sadowska, J. & Bilinska, B. (2007) Expression of connexin 43 protein in testes, epididymides and prostates of stallions. *Equine Veterinary Journal*, Vol. 39, No. 2, (March 2007), pp. 122-127, ISSN 0425-1644

Herrmann, M.; Schölmerich, J. & Straub, R.H. (2002) Influence of cytokines and growth factors on distinct steroidogenic enzymes *in vitro*. A short tabular data collection. *Annals of New York Academy of Sciences*, Vol. 966, (June 2002), pp. 166–186, ISSN 0077-8923

Hu, G.X.; Zhao, B.H.; Chu, Y.H.; Zhou, H.Y.; Akingbemi, B.T.; Zheng, Z.Q. & Ge, R.S. Effects of genistein and equol on human and rat testicular 3beta-hydroxysteroid dehydrogenase and 17beta-hydroxysteroid dehydrogenase 3 activities. *Asian Journal of Andrology*, Vol. 12, No. 4, (July 2010), pp. 519-526, ISSN 1008-682X

Huff, D.S.; Fenig, D.M.; Canning, D.A.; Carr, M.G.; Zderic, S.A. & Snyder, H.M 3rd. (2001) Abnormal germ cell development in cryptorchidism. *Hormone Research*, Vol. 55, No. 1, (2001), pp. 11-17, ISSN 0301-0163

Hughes, I.A.; Houk, C.; Ahmed, S.F. & Lee, P.A.; & Lawson Wilkins Pediatric Endocrine Society/European Society for Paediatric Endocrinology Consensus Group (2006) Consensus statement on management of intersex disorders. *Journal of Pediatric Urology*, Vol. 2, No. 3, (June 2006), pp. 148–162, ISSN 1477-5131

Huhtaniemi, I.T.; Nozu, K.; Warren, D.W.; Dufau, M.L.; & Catt, K.J. (1982) Acquisition of regulatory mechanisms for gonadotropin receptors and steroidogenesis in the maturing rat testis. *Endocrinology*, Vol. 111, No. 5, (November 1982), pp. 1711–1720, ISSN 1477-7827

Huhtaniemi, I. & Pelliniemi, L.J. (1992) Fetal Leydig cells: Cellular origin, morphology, life span, and special functional features. *Proceedings of the Society for Experimental Biology and Medicine*, Vol. 201, No. 2, (November 1992), pp. 125–140, ISSN 0037-9727

Hutson, J.C. (1989) Leydig cells do not have Fc receptors. *Journal of Andrology*, Vol. 10, No. 2, (March-April 1989), pp. 159-165, ISSN 0196- 3635

Illera, J.C.; Silvan, G.; Munro, C.J.; Lorenzo, P.L.; Illera, M.J.; Liu, I.K. & Illera, M. (2003) Amplified androstenedione enzyme immunoassay for the diagnosis of cryptorchidism in the male horse: comparison with testosterone and estrone sulphate methods. *Journal of Steroid Biochemistry and Molecular Biology*, Vol. 84, No. 2-3, (February 2003), pp. 377–382, ISSN 0960-0760

Ivell, R.; Balvers, M.; Anand, R.J.K; Paust, H-J.; Mckinnell, C. & Sharpe, R. (2003) Differentiation-dependent expression of 17β-hydroxysteroid dehydrogenase, type 10, in the rodent testis: Effect of aging in Leydig cells. *Endocrinology*, Vol. 144, No. 7, (July 2003), pp. 3130–3137, ISSN 1477-7827

Ivell, R.; Kotula-Balak, M.; Glynn, D.; Heng, K. & Anand-Ivell, R. (2011) Relaxin family peptides in the male reproductive system a critical appraisal. *Molecular and Human Reproduction*, Vol. 17, No. 2, (February 2011), pp. 71-84, ISSN 1360-9947

Kawakami, E.; Hori, T. & Tsutsui, T. (1999) Function of contralateral testis after artificial unilateral cryptorchidism in dogs. *Journal of Veterinary Medical Science*, Vol. 61, No. 10, (1999), pp. 1107–1111, ISSN 0916-7250

Keeney, D.S. & Mason, J.I. (1992) Expression of testicular 3β-hydroxysteroid dehydrogenase/Δ5–4Δ-isomerase: regulation by luteinizing hormone and forskolin in Leydig cells of adult rats. *Endocrinology*, Vol. 130, No. 4, (April 1992), pp. 2007–2015, ISSN 0013-7227

Kim, S.K.; Kim, J.H.; Han, J.H. & Yoon, Y.D. (2008) Inhibitory effect of tributyltin on expression of steroidogenic enzymes in mouse testis. *International Journal of Toxicology*, Vol. 27, No. 2, (March-April 2008), pp. 175-182, ISSN 1091-5818

Kmicikiewicz, I. & Bilinska, B. (1997) The effect of IL-1 on basal and hCG-stimulated testosterone secretion by Leydig cells of the bank voles (*Clethrionomys glareolus*, S.). *Folia Histochemica et Cytobiologica*, Vol. 35, No. 2, (1997), pp. 95-96, ISSN 0239-8508

Kmicikiewicz, I.; Wojtusiak, A. & Bilinska, B. (1999) The effect of testicular macrophages, macrophage-conditioned medium and interleukin-1alpha on bank vole Leydig cell steroidogenesis. *Experimental and Clinical Endocrinology & Diabetes*, Vol. 107, No. 4, (1999) pp. 262-271, ISSN 0947-7349

Kopera, I.; Tuz, R.; Kotula-Balak, M.; Schwarz, T.; Koczanowski, J. & Bilinska, B. (2008) Morphofunctional alterations in testicular cells of deslorelin-treated boars: an immunohistochemical study. *Journal of Experimental Zoology*, Vol. 1, No. 309, Issue 3, (March 2008), pp. 117–126, ISSN 1932-5223

Kostic, T.S.; Stojkov, N.J.; Bjelic, M.M.; Mihajlovic, A.I.; Janjic, M.M. & Andric, S.A. (2011) Pharmacological doses of testosterone upregulated androgen receptor and 3-beta-hydroxysteroid dehydrogenase/delta-5-delta-4 isomerase and impaired Leydig cells steroidogenesis in adult rats. *Toxicological Sciences*, Vol. 121, No. 2, (June 2011), pp. 397–407, ISSN 1096-6080

Kotula-Balak, M.; Pochec, E.; Hejmej, A.; Duda, M. & Bilinska, B. (2011) Octylphenol affects morphology and steroidogenesis in mouse tumor Leydig cells. *Toxicology In Vitro,* Vol. 25, No. 5, (August 2011), pp. 1018-1026, ISSN 0887-2333

Kotula-Balak, M.; Slomczynska, M.; Fraczek, B.; Bourguiba, S.; Tabarowski, Z.; Carreau, S. & Bilinska, B. (2003) Complementary approaches demonstrate that cellular aromatization in the bank vole testis is related to photoperiod. *European Journal Histochemistry*, Vol. 47, No. 1, (2003), pp. 55-62, ISSN 1121-760X

Kotula-Balak, M.; Wojciechowska, J. & Bilinska, B. (2001) The site of aromatization in the mouse cryptorchid testis. *Folia Biologica (Krakow)*, Vol. 49, No. 3-4, (2001), pp. 279–283, ISSN 0015-5497

Kumar, V.; Chakraborty, A.; Kural, M.R. & Roy, P. (2009) Alteration of testicular steroidogenesis and histopathology of reproductive system in male rats treated with triclosan. *Reproductive Toxicology*, Vol. 27, No. 2, (April 2009), pp. 177-185, ISSN 0890-6238

Lin Y.M.; Liu, M.Y.; Poon, S.L.; Leu, S.F. & Huang, B.M. (2008) Gonadotrophin-releasing hormone-I and - II stimulate steroidogenesis in prepubertal murine Leydig cells *in vitro*. *Asian Journal of Andrology*, Vol. 10, No. 6, (November 2008), pp. 929-936, ISSN 1008-682X

Lindqvist, A.; Hughes, I.A. & Andersson, S. (2001) Substitution mutation C268Y causes 17β-hydroxysteroid dehydrogenase 3 deficiency. *Journal of Clinical Endocrinology & Metabolism*, Vol. 86, No. 2 (February 2001), pp. 921-923, ISSN 0021-972X

Liu, H.; Zheng, S.; Bellemare, V.; Pelletier, G.; Labrie, F. & Luu-The, V. (2007) Expression and localization of estrogenic type 12 17β-hydroxysteroid dehydrogenase in the cynomolgus monkey. *BMC Biochemistry*, Vol. 5, No. 8, (February 2007), p. 2, ISSN 1471-2091

Luo, L.; Chen, H. & Zirkin, B.R. (1996) Are Leydig cell steroidogenic enzymes differentially regulated with aging? *Journal of Andrology*, Vol. 17, No. 5, (September-October 1996), pp. 509-515, ISSN 0196- 3635

Manna, P.R.; El-Hefnawy, T.; Kero, J. & Huhtaniemi, I.T. (2001) Biphasic action of prolactin in the regulation of murine Leydig tumor cell functions. *Endocrinology*, Vol. 142, No. 1, (January 2001), pp. 308–318, ISSN 1477-7827

Maran, R.R. (2003) Thyroid hormones: their role in testicular steroidogenesis. *Archives of Andrology*, Vol. 49, No. 5, (September-October 2003), pp. 375-388, ISSN 0148- 5016

Mathieu, A.P.; Auchus, R.J. & LeHoux, J.G. (2002) Comparison of the hamster and human adrenal P450c17 (17alpha-hydroxylase/17,20-lyase) using site-directed mutagenesis and molecular modeling. *Journal of Steroid Biochemistry and Molecular Biology*, Vol. 80, No. 1, (January 2002), pp. 99-107, ISSN 0960-0760

Matsumoto, A.M. (2002) Andropause: clinical implications of the decline in serum testosterone levels with aging in men. *Journals of Gerontology Series A: Biological Sciences and Medical Sciences*, Vo. 57, No. 2, (February 2002), pp. M76-99, ISSN 1079-5006

McKeever, B.M.; Hawkins, B.K.; Geissler, W.M.; Wu, L.; Sheridan, R.P.; Mosley, R.T. & Andersson, S. (2002) Amino acid substitution of arginine 80 in 17β-hydroxysteroid dehydrogenase type 3 and its effect on NADPH cofactor binding and oxidation/reduction kinetics. *Biochimica et Biophysica Acta*, Vol. 1601, No. 1, (January 2002), pp. 29–37, ISSN 0006-3002

McVey, M.J. & Cooke, G.M. (2003) Inhibition of rat testis microsomal 3beta-hydroxysteroid dehydrogenase activity by tributyltin. *Journal of Steroid Biochemistry and Molecular Biology*, Vol. 86, No. 1, (July 2003), pp. 99-105, ISSN 0960-0760

Mendis-Handagama, S.M. & Ariyaratne, H.B. (2001) Differentiation of the adult Leydig cell population in the postnatal testis. *Biology of Reproduction*, Vol. 65, No. 3, (September 2001), pp. 660-671, ISSN 1470-1626

Mendis-Handagama, S.M.; Ariyaratne, H.B.; Teunissen van Manen, K.R. & Haupt, R.L. (1998) Differentiation of adult Leydig cells in the neonatal rat testis is arrested by hypothyroidism. *Biology of Reproduction*, Vol. 59, No. 2, (August 1998), pp. 351-357, ISSN 1470-1626

Mendonca, B.B.; Costa, E.M.F.; Belgorosky, A.; Rivarola, M.A. & Domenice S. (2010) 46,XY DSD due to impaired androgen production. *Best Practice & Research Clinical Endocrinology & Metabolism*, Vol. 24, No. 2, (April 2010), pp. 243–262, ISSN 1521-690X

Mendonca, B.B.; Domenice, S.; Arnhold I.J.P & Costa, E.M.F. (2008) 46,XY disorders of sexual development. In: *Pediatric Endocrinology*, New, M. (ed.), endotex.com, http://www.endotext.org/pediatrics/pediatrics11/pediatricsframe11.htm

Midzak, A.S.; Chen, H.; Papadopoulos, V. & Zirkin, B.R. (2009) Leydig cell aging and the mechanisms of reduced testosterone synthesis. *Molecular and Cellular Endocrinology*, Vol. 5, No. 299, Issue 1, (February 2009), pp. 23–31, ISSN 0303-7207

Moisan, A.M.; Ricketts, M.L.; Tardy, V.R.; Desrochers, M.; Barki, F.M.; Chaussain, J-L.; Cabrol, S.; Raux-Demay, M.C.; Forest, M.G.; Sippell, W.G.; Peter, M.; Morel, Y. & Simard, J. (1999) New insight into the molecular basis of 3β-hydroxysteroid dehydrogenase deficiency: identification of eight mutations in the *HSD3B2 gene* in eleven patients from seven new families and comparison of the functional properties of twenty-five mutant enzymes. *Journal of Clinical Endocrinology & Metabolism*, Vol. 84, No. 12 (December 1999), pp. 4410-4425, ISSN 1521-690X

Oettel, M. & Mukhopadhyay, A.K. (2004) Progesterone: the forgotten hormone in men? *Aging Male*, Vol. 7, No. 3, (September 2004), pp. 236-257, ISSN 1368-5538

Ohno, S.; Nakajima, Y. & Nakajin, S. (2005) Triphenyltin and tributyltin inhibit pig testicular 17beta-hydroxysteroid dehydrogenase activity and suppress testicular testosterone biosynthesis. *Steroids*, Vol. 70, No. 9, (August 2005), pp. 645-651, ISSN 0585- 2617

Ohsako, S.; Kubota, K.; Kurosawa, S.; Takeda, K.; Qing, W.; Ishimura, R. & Tohyama, C. (2003) Alterations of gene expression in adult male rat testis and pituitary shortly after subacute administration of the antiandrogen flutamide. *Journal of Reproduction and Development*, Vol. 49, No. 4, (August 2003), pp. 275-290, ISSN 0916-8818

O'Shaughnessy, P.J.; Baker, P.J.; Heikkilä, M.;Vainio, S. & McMahon, A.P. (2000) Localization of 17beta-hydroxysteroid dehydrogenase/17-ketosteroid reductase isoform expression in the developing mouse testis-androstenedione is the major androgen secreted by fetal/neonatal Leydig cells. *Endocrinology*, Vol. 141, No. 7, (July 2000), pp. 2631-2637, ISSN 0013-7227

O'Shaughnessy, P.J.; Baker, P.; Sohnius, U.; Haavisto, A.M.; Charlton, H.M. & Huhtaniemi, I. (1998) Fetal development of Leydig cell activity in the mouse is independent of pituitary gonadotroph function. *Endocrinology*, Vol. 139, No. 3, (March 1993), pp. 1141-1146, ISSN 477-7827

O'Shaughnessy, P.J.; Johnston, H.; Willerton, L. & Baker, P.J. (2002b) Failure of normal adult Leydig cell development in androgen-receptor-deficient mice. *Journal of Cell Science*, Vol. 1, No. 115, Issue Pt17, (September 2002), pp. 3491-3496, ISSN 0021-9533

O'Shaughnessy, P.J.; Monteiro, A.; Verhoeven, G.; De Gendt, K. & Abel, M.H. (2010) Effect of FSH on testicular morphology and spermatogenesis in gonadotrophin-deficient hypogonadal mice lacking androgen receptors. *Reproduction*, Vol. 139, No. 1, (January 2010), pp. 177-184, ISSN 1470-1626

O'Shaughnessy, P.J.; Morris, I.D.; Huhtaniemi.; I.; Baker, P.J. & Abel, M.H. (2009) Role of androgen and gonadotrophins in the development and function of the Sertoli cells and Leydig cells: data from mutant and genetically modified mice. *Molecular and Cellular Endocrinology*, Vol. 10, No. 306, Issue 1-2, (July 2009), pp. 2-8, ISSN 0303-7207

Parvinen, M. (1982) Regulation of the seminiferous epithelium. *Endocrine Reviews*, Vol. 3, No. 4, (1982), pp. 404–417, ISSN 0163-769X

Payne, A.H. (2007) Steroidogenic enzymes in Leydig cells. In: *The Leydig cell in health and disease*. Payne, A.H. & Hardy, M.P. (eds.), pp. 157-171, Humana Press Inc., ISBN 978-1-58829-754-9, Totowa, NJ

Payne, A.H. & Hales, D.B. (2004) Overview of steroidogenic enzymes in the pathway from cholesterol to active steroid hormones. *Endocrine Reviews*, Vo. 25, No. 6, (December 2004), pp. 947-970, ISSN 0163-769X

Payne, A.H. & Sha, L.L. (1991) Multiple mechanisms for regulation of 3β-hydroxysteroid dehydrogenase/Δ5–Δ4-isomerase,17β-hydroxylase/C17–20 lyase cytochrome P450, and cholesterol side-chain cleavage cytochrome P450 messenger ribonucleic acid levels in primary cultures of mouse Leydig cells. *Endocrinology*, Vol. 129, No. 3, (September 1991), pp. 1429–1435, ISSN 0013-7227

Pelletier, G. (2000) Localization of androgen and estrogen receptors in rat and primate tissues. *Histology and Histopathology*, Vol. 15, No. 4, (October 2000), pp. 1261-1270, ISSN 0213-3911.

Pelletier, G.; Li, S.; Luu-The, V.; Tremblay, Y.; Belanger, A. & Labrie, F. (2001) Immunoelectron microscopic localization of three key steroidogenic enzymes (cytochrome P450(scc), 3β-hydroxysteroid dehydrogenase and cytochrome P450((c17)) in rat adrenal cortex and gonads. *Journal of Endocrinology*, Vol. 171, No. 2, (November 2000), pp. 373–383, ISSN 0022-0795

Pinart, E.; Sancho, S.; Briz, M.D.; Bonet, S.; Garcia, N. & Badia, E. (2000) Ultrastructural study of the boar seminiferous epithelium: changes in cryptorchidism. *Journal of Morphology*, Vol. 244, No. 3, (June 2000), pp. 190–202, ISSN 62-2525

Pinto, M.E.; Egydio Fde, M.; Taboga, S.R.; Mendis-Handagama, S.M. & Góes, R.M. (2010) Differentiation of Leydig cells in the Mongolian gerbil. *Microscope Research and Techniques*, Vol. 73, No. 2, (February 2010), pp. 119-127, ISSN 1059-910X

Pogrmic-Majkic, K.; Fa, S.; Dakic, V.; Kaisarevic, S. & Kovacevic, R. (2010) Upregulation of peripubertal rat Leydig cell steroidogenesis following 24 h *in vitro* and *in vivo* exposure to atrazine. *Toxicological Sciences*, Vol. 118, No. 1, (November 2010), pp. 52–60, ISSN 1096-6080

Pushpalatha, T.; Reddy, P.R. & Reddy, P.S. (2003) Effect of prenatal exposure to hydroxyprogesteroneon steroidogenic enzymes in male rats. *Naturwissenschaften*, Vol. 90, No. 1, (January 2003), pp. 40-43, ISSN 0028-1042

Rao, M.V. & Chinoy, N.J. (1986) Effect of estradiol benzoate on rat testis and adrenal. *Experimental and Clinical Endocrinology*, Vol. 88, No. 2, (December 1986), pp. 181-184, ISSN 0232-7384

Ren, L.; Medan, M.S.; Ozu, M.; Li, C.; Watanabe, G. & Taya, K. (2006). Effects of experimental cryptorchidism on sperm motility and testicular endocrinology in adult male rats. *Journal of Reproduction and Development*, Vol. 52, No. 2, (April 2006), pp. 219–228, ISSN 0916-8818

Ruiz de Galarreta, C.M.; Fanjul, L.F.; Meidan, R. & Hsueh, A.J. (1983) Regulation of 3β-hydroxysteroid dehydrogenase activity by human chorionic gonadotropin, androgens, and anti-androgens in cultured testicular cells. *Journal of Biological Chemistry*, Vol. 25, No. 258, Issue 11, (September 1983), pp. 10988–10996, ISSN 0021-9258

Ryan, P.L.; Friendship, R.M. & Raeside, J.I. (1986) Impaired estrogen production by Leydig cells of the naturally retained testis in unilaterally cryptorchid boars and stallions. *Journal of Andrology*, Vol. 7, No. 2, (March-April 1986), pp. 100–104, ISSN 0105-6263

Saez, J.M.; De Peretti, E.; Morera, A.M.; David, M. & Bertrand, J. (1971) Familial male pseudohermaphroditism with gynecomastia due to a testicular17-ketosteroid reductase defect. I. Studies *in vivo*. *Journal of Clinical Endocrinology & Metabolism*, Vol. 32, No. 5, (May 1971), pp. 604–610, ISSN 1521-690X

Schäfers, B.A.; Schlutius, B.G. & Haider, S.G. (2001) Ontogenesis of oxidative reaction of 17β-hydroxysteroid dehydrogenase and 11β-hydroxysteroid dehydrogenase in rat Leydig cells, a histochemical study. *Histochemical Journal*, Vol. 33, No. 9-10, (September-October 2001), pp. 585–595, ISSN 0018-2214

Schwarzenbach, H.; Manna, P.R.; Stocco, D.M.; Chakrabarti, G. & Mukhopadhyay, A.K. (2003) Stimulatory effect of progesterone on the expression of steroidogenic acute regulatory protein in MA-10 Leydig cells. *Biology of Reproduction*, Vol. 68, No. 3, (March 2003), pp. 1054-1063, ISSN 1470-1626

Scott, H.M.; Mason, J.I. & Sharpe RM. (2009) Steroidogenesis in the fetal testis and its susceptibility to disruption by exogenous compounds. *Endocrine Reviews*, Vol. 30, No. 7, (December 2009), pp. 883-925, ISSN 0163-769X

Shan, L.X.; Phillips, D.M.; Bardin, C.W. & Hardy, M.P. (1993) Differential regulation of steroidogenic enzymes during differentiation optimizes testosterone production by adult rat Leydig cells. *Endocrinology,* Vol. 133, No. 5, (November *1993),* pp. 2277–2283, ISSN 1477-7827

Sharpe, R.M. & Skakkebaek, N.E. (1993) Are oestrogens involved in falling sperm counts and disorders of the male reproductive tract? *Lancet,* Vol. 29, No. 341, (May 1993), pp. 1392–1395, ISSN 0140-6736

Sierens, J.E.; Sneddon, S.F.; Collins, F.; Millar, M.R, & Saunders, P.T. (2005) Estrogens in testis biology. *Annals of the New York Academy of Sciences,* Vol. 1061, (December 2005), pp. 65-76, ISSN 0077-8923

Simard, J.; Ricketts, M-L.; Gingras, S.; Soucy, P.F.; Feltus, F.A. & Melner M.H. (2005) Molecular biology of the 3β-hydroxysteroid dehydrogenase/Δ5-Δ4 isomerase gene family. *Endocrine Reviews,* Vol. 26, No. 4, (June 2005), pp. 525–582, ISSN 0163-769X

Sinnecker, G.H.; Hiort, O.; Dibbelt, L.; Albers, N.; Dorr, H.G.; Hauss, H.; Heinrich, U.; Hemminghaus, M.; Hoepffner, W.; Holder, M.; Schnabel, D. & Kruse, K. (1996) Phenotypic classification of male pseudohermaphroditism due to steroid 5 alpha-reductase 2 deficiency. *American Journal of Medical Genetics,* Vol. 3, No. 63, Issue 1, (May 1996), pp. 223–230, ISSN 0148-7299

Skakkebaek, N.E.; Rajpert-De Meyts, E. & Main, K. (2001) Testicular dysgenesis syndrome: an increasingly common developmental disorder with environmental aspects. *Human Reproduction,* Vol. 16, No. 5, (May 2001), pp. 972-978, ISSN 0268-1161

Sordoillet, C.; Savona, C.; Chauvin, M.A.; de Peretti, E.; Feige, J.J.; Morera, A.M. & Benahmed, M. (1992) Basic fibroblast growth factor enhances testosterone secretion in cultured porcine Leydig cells: site(s) of action. *Molecular and Cellular Endocrinology,* Vol. 89, No. 1-2, (November 1992), pp. 163–171, ISSN 0303-7207

Tähkä, K.M.; Tähkä, K.M.; Teräväinen, T. & Wallgren, H. (1982) Effect of photoperiod on the testicular steroidogenesis of the bank vole (*Clethrionomys glareolus,* Schreber): an *in vitro* study. *General and Comparative Endocrinology,* Vol. 47, No. 3, (July 1982), pp. 377–84, ISSN 0016-6480.

Tapanainen, J.; Kuopio, T.; Pelliniemi, L.J. & Huhtaniemi, I. (1984) Rat testicular endogenous steroids and number of Leydig cells between the fetal period and sexual maturity. *Biology of Reproduction,* Vol. 31, No. 5, (December 1984), pp. 1027–1035, ISSN 1470-1626

Teerds, K.J. & Rijntjes, E. (2007) Dynamics of Leydig cell regeneration after EDS: A model for postnatal Leydig cell development. In: *The Leydig cell in health and disease.* Payne, A.H. & Hardy, M.P. (eds.), pp. 91-116, Humana Press Inc., ISBN 978-1-58829-754-9, Totowa, NJ

Teerds, K.J.; Rijntjes, E.; Veldhuizen-Tsoerkan, M.B.; Rommerts, F.F. & de Boer-Brouwer M. (2007) The development of rat Leydig cell progenitors *in vitro*: how essential is luteinising hormone? *Journal of Endocrinology,* Vol. 194, No. 3, (September 2007), pp. 579-593, ISSN 0022-0795

Teerds, K.J.; de Rooij, D.G.; de Jong, F.H. & van Haaster, L.H. (1998) Development of the adult-type Leydig cell population in the rat is affected by neonatal thyroid hormone levels. *Biology of Reproduction,* Vol. 59, No. 2, (August 1998), pp. 344-350, ISSN 1470-1626

Toppari, J.; Larsen, J.C.; Christiansen, P.; Giwercman, A.; Grandjean, P.; Guillette, Jr.L.J.; Jegou, B.; Jensen, T.K.; Jouannet, P.; Keiding, N.; Leffers, H.; McLachlan, J.A.; Meyer, O.;

Muller, J.; Rajpert-De Meyts, E.; Scheike, T.; Sharpe, R.; Sumpter, J. & Skakkebaek, N.E. (1996) Male reproductive health and environmental xenoestrogens. *Environmental Health Perspectives*, Vol. 104, Suppl. 4, (August 1996), pp. 741–803, ISSN 0091-6765

Tsubota, T.; Howell-Skalla, L.; Nitta, H.; Osawa, Y.; Mason, J.I.; Meiers, P.G.; Nelson, R.A. & Bahr, J.M. (1997) Seasonal changes in spermatogenesis and testicular steroidogenesis in the male black bear *Ursus americanus*. *Journal of Reproduction and Fertility*, Vol. 109, No. 1, (January 1997), pp. 21-27, ISSN 0022-4251

Tsubota, T.; Nagashima, T.; Kohyama, K.; Maejima, K.; Murase, T. & Kita, I. (2001) Seasonal changes in testicular steroidogenesis and spermatogenesis in a northern fur seal, *Callorhinus ursinus*. *Journal of Reproduction and Development*, Vol. 47, No. 6, (December 2001), pp. 415-420, ISSN 0916-8818

Ulloa-Aguirre, A.; Bassol, S.; Poo, J.; Mendez, J.P.; Mutchinick, O.; Robles, C. & Perez-Palacios, G. (1985) Endocrine and biochemical studies in a 46,XY phenotypically male infant with 17-ketosteroid reductase deficiency. *Journal of Clinical Endocrinology & Metabolism*, Vol. 60, No. 4, (April 1985), pp. 639–643, ISSN 1521-690X

Vaithinathan, S.; Saradha, B. & Mathur, P.P. (2008) Transient inhibitory effect of methoxychlor on testicular steroidogenesis in rat: an *in vivo* study. *Archives in Toxicology*, Vol. 82, No. 11, (April 2008), pp. 833–839, ISSN 0340-5761

Verhoeven, G.; Willems, A.; Denolet, E.; Swinnen, J.V, & De Gendt, K. (2010) Androgens and spermatogenesis: lessons from transgenic mouse models. *Philosophical Transactions of the Royal Society Series B, Biological Sciences*, Vol. 27, No. 1546, (May 2010), pp. 1537-1556, ISSN 1471-2970

Victor-Costa, A.B.; Carozzi Bandeira, S.M.; Oliveira, A.G.; Bohórquez Mahecha, G.A. & Oliveira, C.A. (2010) Changes in testicular morphology and steroidogenesis in adult rats exposed to atrazine. *Reproductive Toxicology*, Vol. 29, No. 3, (June 2010), pp. 323–331, ISSN 0890-6238

Wang, R-S.; Yeh, S.; Tzeng, C-R. & Chang, C. (2009) Androgen receptor roles in spermatogenesis and fertility: lessons from testicular cell-specific androgen receptor knockout mice. *Endocrine Reviews*, Vol. 30, No. 2, (April 2009), pp. 119-132, ISSN 0163-769X

Wisner, J.R Jr. & Gomes, W.R. (1978) Influence of experimental cryptorchidism on cholesterol side-chain cleavage enzyme and delta5-3beta-hydroxysteroid dehydrogenase activities in rat testes. *Steroids*, Vol. 31, No. 2, (February 1978), pp. 189–203, ISSN 0585-2617

Wu, X.; Wan, S. & Lee M.M. (2007) Key factors in the regulation of fetal and postnatal Leydig cell development. *Journal of Cellular Physiology*, Vol. 213, No. 2, (November 2007), pp. 429–433, ISSN 0021-9541

Ye, L.; Zhao, B.; Hu, G.; Chu, Y. & Ge, R-S. (2011) Inhibition of human and rat testicular steroidogenic enzyme activities by bisphenol A. *Toxicology Letters*, Vol. 30, No. 207, Issue 2 (November 2011), pp. 137– 142, ISSN 0378-4274

Zirkin, B.R. (2010) Where do adult Leydig cells come from? *Biology of Reproduction*, Vol. 82, No. 6, (June 2010), pp. 1019–1020, ISSN 1470-1626

Zirkin, B.R. & Chen, H. (2000) Regulation of Leydig cell steroidogenic function during aging. *Biology of Reproduction*, Vol. 63, No. 4, (October 2000), pp. 977–981, ISSN 0006-3363

Chapter 12

Glutamate Dehydrogenases: Enzymology, Physiological Role and Biotechnological Relevance

Eduardo Santero, Ana B. Hervás, Ines Canosa and Fernando Govantes

Additional information is available at the end of the chapter

http://dx.doi.org/10.5772/47767

1. Introduction

Glutamate dehydrogenase (GDH) is present in all domains of life and is one of the most extensively studied enzymes at the biochemical and structural levels. These enzymes are generally reversible and catalyse either the reductive amination of 2-oxoglutarate (2-OG) to yield glutamate using NAD(P) as a cofactor, or the oxidative deamination of glutamate [1] (Fig. 1). Because of the reaction it catalyses, the main role of GDH is glutamate catabolism and ammonium assimilation. However, other physiological roles for GDH have been described in some organisms, as we will see below.

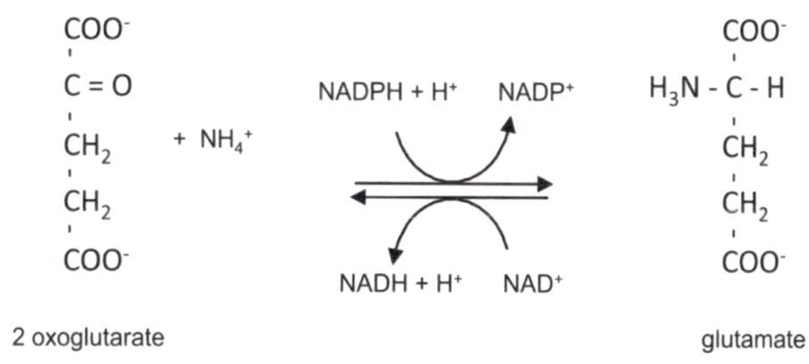

Figure 1. Reaction catalysed by glutamate dehydrogenase

The synthesis of both glutamate and glutamine are key steps in the cell metabolism in all organisms, because they represent the only means of incorporating inorganic nitrogen into carbon backbones. Inorganic nitrogen is assimilated in the form of ammonium, which is incorporated as an amino group to glutamate or an amido group to glutamine. These amino acids in turn act as amino group donors for the synthesis of most nitrogen-containing compounds in the cell. In particular, the amino group of glutamate is used in the synthesis of purines, pyrimidines, amino sugars, histidine, tryptophan, asparagine, NAD and *p*-aminobenzoate. Therefore glutamate is a key element in the nitrogen flow, as it plays a role of nitrogen donor and acceptor.

Glutamate can be synthesized by two alternative routes: one involves catalysis of GDH in the aminating direction, but ammonium assimilation is also possible by the participation of two enzymes: glutamine synthetase (GS), and glutamate synthase, also named glutamine oxoglutarate aminotransferase or GOGAT (Figure 2). The disadvantage of this pathway is its extra energy requirement. Although GDH catalyses the reductive amination of 2-OG, it is noteworthy that because of its overall high K_m for ammonium, this reaction can only be used for the synthesis of glutamate when the ammonium concentration is high (>1 mM). When the ammonium concentration is lower, ammonia is incorporated to glutamate mainly via the GS-GOGAT pathway. Generally, GDH activity is not necessary for cell growth, since most organisms can synthesize glutamate from glutamine and 2-OG using GOGAT. In fact, some bacteria naturally lack GDH and are neither glutamate auxotrophs nor affected in nitrogen assimilation. While the amination reaction provides nitrogen required for many biosynthetic pathways, the oxidative deamination reaction of GDH provides carbon to the tricarboxylic acid cycle (TCA) by conversion of L-glutamate to 2-OG and probably contributes to balancing the glutamine to glutamate ratio.

Plants and microorganisms can utilise several inorganic nitrogen sources with different oxidation states such as N_2 (by nitrogen-fixing bacteria and archaea), nitrate or nitrite, by reducing them to ammonium, which is subsequently assimilated. After formation of glutamate, the α-amino group can be transferred to a wide variety of 2-oxo acceptors to give rise to amino acids. Also, the α-amino group can be transferred again to glutamate, when 2-oxoglutarate and other amino acids are available. These reactions are carried out by the reversible activity of aminotransferases (EC 2.6.1.x) (Figure 2). Plants and microorganisms can synthesize all carbon skeletons for their amino acids and incorporate the amino group to them by transamination using glutamine and glutamate as nitrogen donors. Incorporation of ammonium in animals also occurs through the GDH and GS/GOGAT pathways. However, higher organisms are unable to reduce oxidized forms of nitrogen to ammonium, to synthesize the structures of some branched or aromatic amino acids such as tryptophan or phenylalanine, or to incorporate sulphur into covalently bonded structures. They are, therefore, totally dependent on other organisms to convert oxidized forms of nitrogen into forms available for the organism, as well as to provide some essential amino acids (Figure 2). These are supplied in the diet or are provided by bacteria from the intestinal tract.

In plants and microorganisms, the physiological roles of GDH include nitrogen assimilation [2], glutamate catabolism [1, 3], but also osmotic balance [4] and tolerance to high temperatures [5, 6]. In vertebrates, multiple biochemical pathways involve glutamate, which is also used as a neurotransmitter. The imbalance of the GDH activity may lead to disturbances of clinical relevance for humans [7]. Free ammonia is highly toxic to organisms that excrete urea as the main nitrogenous waste such as mammals, fish and adult amphibians, leading to inhibition of brain respiration and an excess ketone body formation from acetyl-CoA in the liver. To prevent these deleterious effects, GDH and some of the other enzymes that yield amino groups into the urea cycle are localized in the mitochondria. In the liver, glutamate is the source of excess ammonium release, and the concentration of glutamate modulates the rate of ammonia detoxification into urea. In pancreatic β-cells, the GDH is involved in insulin homeostasis, and oxidation of glutamate mediates amino acid-stimulated insulin secretion [8]. In the central nervous system, glutamate serves as a neurotransmitter and also as the precursor of the inhibitory neurotransmitter γ-aminobutyric acid (GABA), as well as glutamine, a potential mediator of hyperammonemic neurotoxicity [7]. Also, excessive glutamate signalling can lead to excitotoxicity, a phenomenon where over-activation of glutamate receptors initiates neuronal death [9]. The clinical importance of glutamate metabolism in β-pancreatic cells has been highlighted by the recent discovery of a dominantly expressed defect in glutamate metabolism, the hyperinsulinism/hypermmonemia syndrome (HHS). HHS was one of the first diseases that clearly linked GDH regulation to insulin and ammonia homeostasis [10]. Affected children suffer from recurrent hypoglycemia due to inappropriate secretion of insulin [10-12]. This syndrome is caused by the loss of the human glutamate dehydrogenase allosteric regulation (see below).

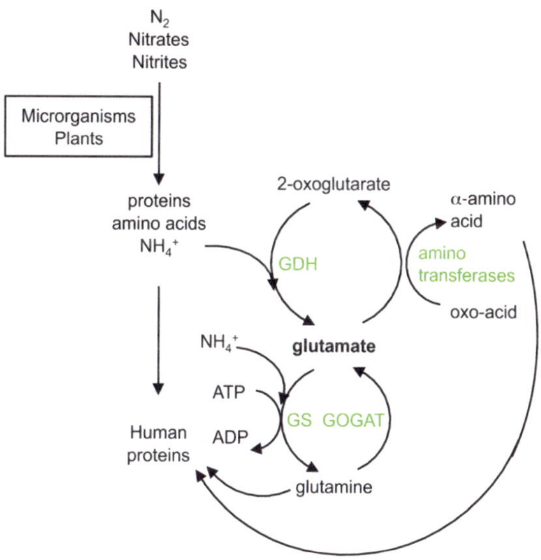

Figure 2. Flow of nitrogen in the biosphere. Molecular nitrogen, nitrites and nitrates are reduced to ammonium and assimilated by microorganisms and plants, whilst higher eukaryotes assimilate these nitrogenated compounds as protein in their diets.

2. Classification, evolution and structure of GDHs

Several GDH classifications have been done according to their size, oligomerisation state, coenzyme specificity or organism, among others. According to their cofactor specificity, there are three basic types of GDH: those that are cofactor specific for NAD (EC 1.4.1.2), those that are specific for NADP (EC 1.4.1.4) and those that can use either cofactor (EC 1.4.1.3) (dual coenzyme-specific GDHs). Lower eukaryotes and prokaryotes usually have GDHs that only function with one coenzyme whilst the enzymes that have dual coenzyme specificity are commonly found in higher eukaryotes. However, some dual-GDHs, have also been described in prokaryotes [5, 13-15]. Glutamate dehydrogenases from non-vertebrate animals differ from the GDHs of vertebrates in that they are mono-coenzyme specific and are not regulated by nucleotides [16]. In higher plants GDH is ubiquitous and also very abundant. A number of isozymes are usually present in a single species, some of them being inducible, which correlate with their abundance depending on environmental or nutritional conditions [17]. GDHs have also been characterized from a number of eukaryotic microorganisms with different coenzyme specificity such as fungi, (NAD^+ or $NADP^+$) [18], algae (NAD^+, $NADP^+$ or dual), protozoa (NAD^+ or $NADP^+$) and also different intracellular localizations (i.e. cytoplasmic, mitochondrial or in the chloroplasts) [19].

According to the molecular weight of the monomer, three groups of GDHs can be distinguished: GDH50s (MW around 50 KDa), GDH115s (MW around 115 KDa) and GDH180s (MW around 180 KDa). All NADP and dual-GDHs reported so far belong to the GDH50 group, whereas there are representatives of NAD-GDHs in all of these groups. GDH115s have been found only in lower eukaryotes [20-22], whilst the largest GDHs are present only in bacteria. GDH180s were first identified in actinomycetes [23], but recently they have been also described in other Gram positive and Gram negative bacteria (see table 1). Most GDHs reported so far are homo-oligomeric enzymes, but they differ in the number of monomers that compose them. The majority of GDHs have a hexameric structure, as is the case of vertebrate GDHs, but tetrameric and even dimeric enzymes have also been found (see table 1 and references therein). Particularly, the most recently discovered family of prokaryotic GDH180s, have representatives of either hexameric [23, 24], tetrameric [25, 26] and dimeric [27] enzymes. In addition, a couple of GDH50s composed of two different subunits in the form of a hetero-hexamer have been reported [28, 29].

Analysis of the distribution pattern of *gdh* genes from all available sequenced genomes in the three domains of life reveal that all classes of *gdh* have been found in eubacteria and archaea and all but the large GDH have been found in eukaryotes. Both NAD^+- and $NADP^+$-dependent forms of GDH have been reported in higher plants, located in mitochondria and chloroplasts, respectively. The GDH enzyme is abundant in several plant organs, and its isoenzymatic profile can be influenced by dark stress, natural senescence or fruit ripening [30]. Genes coding for GDH seem to be absent in some archaeal genomes as well as in some of the smaller eubacterial and eukaryotic genomes. Among the organisms that do encode GDH, several genes coding for GDH may be found in the same genome. However, just one or two classes are represented, no genome has yet been shown to encode all classes.

Organism	cofactor	MW enzyme (KDa)	MW subunit (KDa)	subunit number	K_m NH_4 (mM)	K_m 2-OG (mM)	K_m glu (mM)	Ref.
Archaea								
Archaeglobus fulgidus	NADP	263	47	6	4	0.5	3.9	[31]
Halobacterium halobium	NAD				450	20.2	4	[32]

Organism	cofactor	MW enzyme (KDa)	MW subunit (KDa)	subunit number	K_m NH_4 (mM)	K_m 2-OG (mM)	K_m glu (mM)	Ref.
Halobacterium halobium	NADP							[33]
Thermococcus strain AN1	NADP	204	47	4	15.5	1.7	9.12	[19]
Pyrococcus furiosus	NAD/NADP	270-290	48	6	6, 27[b]	0.33	0.6	[34]
Thermococcus profundus	NADP	263	43	6	1.6, 22[b]	0.2, 0.87[b]	6.8	[6]
Eubacteria								
Gram negative								
Capnocytophaga ochraea	NAD				3.33	1.44	2.44	[35]
Escherichia coli B/r	NADP	300	50	6	1.1	0.64		[36]
Escherichia coli PA340	NADP				2.5	0.2	2.3	[37]
Janthinobacterium lividum	NAD	1065	170	6			7.1	[24]
P. aeruginosa	NAD		180	4	15	1.6		[25, 38]
P. aeruginosa	NADP	110			7	1		[38, 39]
Psychrobacter sp TAD1	NAD	290	160	2	24.6	2.36	28.6	[27]
Psychrobacter sp TAD1	NADP	290	47	6	4	ND	67.4	[37]
Salmonella enterica	NADP				0.29	4	50	[40]
Thermus termophilus	NAD	289	46.5, 48[a]	6				[29, 41]
Thiobacillus novellus	NADP	130	50-55	2?	7.5	7.4	35.5	[42]
Thiobacillus novellus	NAD	120	50-55	2?	7.4, 0.5[d]	6.7, 0.67[d]	11.8, 13.3[d]	[43]
Gram positive								
Bacillus macerans	NADP				2.2	0.38		[44]
Bacillus polymyxa	NADP				2.9	1.4		[45]
Bacillus subtilis	NAD			6				[46, 47]
Clostridium symbiosum	NAD	282	49	6				[48, 49]
Corynebacterium glutamicum	NADP		49	6?				[50]
Mycobacterium smegmatis	NAD		180	4?				[26, 51]
Mycobacterium smegmatis	NADP	245.5	40	6	33	5	62.5	[52]
Lactobacillus fermentum	NADP	300	50	6	6.76	5.6	79	[53]

Organism	cofactor	MW enzyme (KDa)	MW subunit (KDa)	subunit number	K_m NH_4 (mM)	K_m 2-OG (mM)	K_m glu (mM)	Ref.
Peptostreptococcus asacharolyticus	NAD	266	49	6	18.4	0.82	6	[54]
Streptomyces clavuligerus	NAD	1100	179	6				[23]
Streptomyces fradiae	NADP	200	49	4	30.8	1.54	28.6	[55]

[a] NAD^+-GDH of *Thermus thermophilus* is a heterohexamer composed by two types of subunits: GdhA (46, 5 KDa) and GdhB (48 KDa)

[b] The K_m depends on the substrate concentration

[c] The kinetic constants determined for each cofactor in enzymes with dual cofactor specificity are separated by slashes

[d] The kinetic constants of NAD^+-GDH of *T. novellus* are different depending on the presence of AMP

Table 1. Some characteristics of selected prokaryotic GDHs

The distribution of *gdh* genes does not show any strong pattern that correlate with the phylogeny [56]. It was believed for some time that NAD- and NADP-GDHs were originated via single gene duplication [57], but as genomes are sequenced and more *gdh* genes are identified this hypothesis has been ruled out. The analysis of phylogenetic distribution patterns of the *gdh* gene families provides strong support for numerous horizontal gene transfer events involving prokaryotes, as well as microbial eukaryotes. Differential gene loss, on the other hand, does not seem to have played an important role in the evolution of *gdh* genes in any of the three domains of life. Sequence comparisons for GDHs from a diverse range of sources show that the hexameric enzymes are similar whatever their coenzyme specificity [58, 59]. On the other hand, the tetrameric enzymes are less well understood because of a lower number of characterized tetrameric GDHs. Organisms bearing a tetrameric GDH, which have catabolic roles, also possesses a genetically distinct hexameric NADP-linked enzyme with a biosynthetic role. Mammalian GDHs represent a clear deviation from its ancestral forms, since they have the so-called antenna, a 48 amino acid insertion near the carboxy terminus, although it is not clear when this feature evolved. Sequenced genomes from *Ciliates* show that their GDHs present a smaller *antenna* from that of mammalians, although other members of the Protista, such as trypanosomes, have GDH almost identical to the bacterial forms. *Ciliates* are an evolutionary missing link in the GDH evolution [60]

The structure of GDH of many eukaryotic and prokaryotic organisms has been considerably studied and characterized since the beginning of the 50s. As mentioned above, Most GDHs studied so far are homopolymers consisting of two to six subunits of molecular weight 40,000 to 60,000 (fungal NAD-specific GDHs and bacterial large GDH are exceptions). Most of the characterized GDHs are hexameric and the most common structure found is two trimers of subunits stacked directly on top of each other [61-63]. Some GDHs such as that from bovine liver, which is the best-characterized enzyme [1], may have higher order multimeric structures. This enzyme, which is a hexamer in solution, aggregates to form a high molecular weight species and this polymerization is promoted by a high concentration of enzyme, by high ionic strength and also by allosteric ligands or cofactors [64]. Since the local concentration of GDH in some tissues is very high, aggregation might be a regulatory mechanism of the activity *in vivo*.

Eukaryotic and prokaryotic GDHs share relatively high conservation in their primary and secondary structures [61] and the crystal structures of the bacterial [59, 65, 66] and mammalian forms [61, 63] of GDH confirm that the general architecture and the locations of the catalytically important residues have remained unchanged throughout evolution. Each subunit in this multimeric enzyme is organised into two domains separated by a deep cleft. One domain directs the self-assembly of the molecule into a hexameric oligomer with 32 symmetry. The other domain is structurally similar to the classical pyridine nucleotide-binding domain but with the direction of one of the β-strands reversed. Upon glutamate binding, the enzyme can adopt different conformations by flexing about the cleft between its two domains. NAD^+ binds in an extended conformation with the nicotinamide moiety buried deep in the cleft between the two domains [59, 61, 63, 65, 66]. The bottom domains of each trimer make wide contacts with each other, while the NAD^+-binding domains bearing the nucleotide-binding motif are poised at the top of the structure.

The largest structural difference between mammalian and bacterial GDH is the *antenna*, which has a helix-loop-helix conformation. The *antenna* ascends from the NAD^+-binding domain surface via a long, 23-residue helix and then descends back with a random coil structure. The helices of the "*antenna*" domains in each subunit of the trimer wrap around each other with a right-handed twist to form the core of the antenna protrusion. Extensive contacts between “antennae” may represent hexamer interactions in solution and, perhaps, with other enzymes within the mitochondrial matrix [61]. The fact that *antennae* are only found in the forms of GDH that are allosterically regulated by numerous ligands leads to the interpretation that it plays a major part in this regulation. In contrast to the extensive allosteric homotropic and heterotropic regulation observed in mammalian GDH (see below), bacterial forms of GDH are relatively unregulated.

3. GDH enzymology and physiological role

As a reversible enzyme, GDH has the potential for catalysing the reaction in the biosynthetic, aminating direction, or in the catabolic, deaminating direction. The actual physiological reaction of each GDH depends on several factors, as the kinetic constants of the enzyme for its different substrates or the environment where the cell is developed may widely vary. In general, $NADP^+$-GDHs usually operate in the biosynthetic direction, that is, synthesizing glutamate by the assimilation of ammonia into 2-OG [6, 31, 39, 40, 45, 67, 68], whereas NAD^+-GDHs have primarily a catabolic function, yielding ammonia and 2-oxoglutarate from the oxidative catabolism of glutamate [23, 24, 39, 68] (Table 1). Sometimes, both enzymes are present in the same organism, and play a different physiological role due to their different kinetic properties or their different time or place of expression [27, 51, 69, 70]. *Pseudomonas aeruginosa* and presumably other members of the genus *Pseudomonas* have a $NADP^+$-specific and a NAD^+-specific GDH, and it has been hypothesized than the latter acts specifically in arginine catabolism by converting glutamate, a product of the ammonia-producing arginine succinyl transferase (AST) pathway, into 2-OG, since it is allosterically modulated by arginine (positively) and citrate (negatively) [25]. Similarly, the only active GDH from *Bacillus subtilis* (RocG, NAD^+-dependent) appears to be involved in arginine and proline catabolism [46]. On the other hand, despite the catabolic function assigned to NAD^+-GDHs, the existence of an NAD^+-specific GDH with an unusual biosynthetic role has been reported in the oral bacterium *Capnocytophaga ochraea* [35]. In this case, it was found that only the NAD^+-GDH ammonium assimilating activity could be detected in cell free extracts, probably due to the high concentration of ammonium and ammonium precursors that can be found in the gingival crevicular fluid. Interestingly, GDHs have been shown to play a substantial and even predominant role in nitrogen assimilation in conditions of N_2 fixation in the Gram positive bacteria *Bacillus macerans* and *Bacillus polymyxa* [44, 45]. Nitrogen fixation only occurs under extreme nitrogen-limiting conditions, when nitrogen from other sources is very scarce. In these conditions nitrogen is always assimilated using the GS/GOGAT pathway, since the K_m of GS for ammonium is much lower than that of GDH. This is not the case in *B. macerans* and *B. polymyxa,* as in these organisms the GOGAT activity is much lower than GDH activity in nitrogen-fixing cells. A NAD^+-GDH involved in glutamate fermentation has also been described in the anaerobic Gram-positive bacteria *Peptostreptococcus asaccharolyticus* [54]. In this organism, GDH is the first enzyme of the glutamate fermentation via the hydroxyglutarate pathway, and can represent as much as 10% of total protein when grown on glutamate. Very high levels of GDH production has also been reported in some hyperthermophilic archaea like *Pyrococcus furiosus* or some *Thermococcus* strains [5, 6, 71]. These preferentially biosynthetic enzymes represent an exceptionally high percentage of total soluble protein of the cell, in some cases up to 20%, which suggests an important role of these enzymes in these organisms at an extremely high temperature for life.

Determination of the GDH enzymatic structure has allowed the elucidation of the mechanisms for allosteric regulation and negative cooperativity. The activity of glutamate dehydrogenase in animals is allosterically regulated by purine nucleoside phosphates and other metabolic intermediates. In brief, GTP and ATP are allosteric inhibitors whereas GDP and ADP are allosteric activators. Hence, a lowering of the energy charge accelerates the oxidation of amino acids. Their *in vivo* regulation may be dependent on the metabolic status of the cell, or in the tissue they are located. The intracellular compartmentalization of the cofactors, and the GDH itself, may also drive the reaction in one or the other direction. In vertebrate cells, GDH appears to be localized primarily in the mitochondrial matrix [1]. The effects of nucleosides on mammalian GDH are complex. For the bovine liver GDH, four binding sites per subunit have been described, being the active site (site I), the adenine nucleotide regulatory site (site II), the guanine nucleotide regulatory site (site III) and the reduced coenzyme regulatory site (site IV). Some substrate and effectors bind just the active site (glutamate, oxoglutarate, ammonia, $NADP^+$) while others are able to bind two different sites (NAD^+, ADP, NAD(P)H) [19]

One characteristic of microbial GDHs is the absence of the *antenna* structure that functions as a heterotropic allosteric site. In agreement with this, the vast majority of microbial GDHs do not appear to have this level of complexity in GDH modulation by purine nucleoside phosphates. However, some microbial GDHs also show homotropic and even heterotropic allosteric control, especially those from the GDH180 family. GDHs from *Psychrobacter* sp. TAD1, *Streptomyces clavuligerus* or *Pseudomonas aeruginosa* show positive cooperativity of substrate binding, a common feature associated with the complex regulation in vertebrate GDHs, but unusual in bacteria [23, 25, 37]. Furthermore, the Gram-positive bacterium *Clostridium symbiosum* displays an apparent negative cooperativity and inhibitory effect of the enzyme cofactor in certain conditions of pH and concentration [49]. On the other hand, heterotropic control, either positive or negative, has been found in an increasingly number of microorganisms. Accordingly, some aminoacids such as L-aspartate or L-arginine are positive allosteric effectors of NAD+-GDH from the psychrophilic bacterium *Janthinobacterium lividum* and *P.aeruginosa* [24, 25], while nucleotides such as ATP or AMP modulate NADP+-GDH from *Salmonella enterica* sv. Typhimurium and NAD+-GDH from *S. clavuligerus* and *Thiobacillus novellus* [23, 40, 43]. In the latter case, AMP has been found to be actually an essential activator for *S. clavuligerus* GDH activity. Conversely, some microbial GDHs have allosteric inhibitors, such as TCA cycle intermediates in the archaeon *Halobacterium halobium*, in *Salmonella enterica* sv Typhimurium, and in *P. aeruginosa* [25, 40, 72], or nucleotides such as ADP in the NAD^+-GDH of *Capnocytophaga ochraea* [35]. Finally, the NAD^+-GDH from the actinomycete *Mycobacterium smegmatis* is modulated by the small protein kinase GarA [26].

4. Regulation of bacterial GDH gene expression

The diverse roles of bacterial GDH in different organisms provide for a variety of regulatory mechanisms. Here we show a few examples for selected bacteria in which transcriptional regulation of GDH genes has been characterized.

4.1. Regulation of GDH synthesis in the enterobacteria

Transcriptional regulation of the *gdhA* gene, encoding NADP-GDH, was first described in the diazotrophic enterobacterium *Klebsiella pneumoniae,* and then in *Escherichia coli* [73]. Transcription of *gdhA* is repressed in both enteric bacteria under nitrogen limitation by the general nitrogen control system (Figure 3,A). This is consistent with the fact that low affinity for ammonium limits the use of GDH for glutamate synthesis at ammonium concentrations below 1 mM. The enterobacterial general nitrogen control (Ntr) system is a very well characterized signal transduction and regulatory network encompassing seven elements: the alternative σ factor σ^{54}, encoded by *rpoN*, the uridylyl transferase-uridylyl removing enzyme GlnD, two PII signal transduction proteins, GlnB and GlnK, the two-component system NtrB-NtrC and the LysR-type transcriptional regulator Nac. Nitrogen status is signalled by the intracellular pools of glutamine (indicative of nitrogen sufficiency), and 2-OG (indicative of nitrogen limitation). These signals are perceived by the PII proteins by means of their reversible GlnD-dependent uridylylation in response to decreased glutamine levels and by allosteric modulation via 2-oxoglutarate binding. Interaction with PII proteins in either their uridylylated or deuridylylated states is responsible for posttranslational regulation of the activities of a variety of proteins involved in nitrogen metabolism, including glutamine synthetase, nitrogenase and the sensor kinase/phosphatase NtrB. Deuridylylated GlnB promotes NtrB-dependent dephosphorylation of NtrC under nitrogen excess [2, 74]. In nitrogen-limiting conditions, phosphorylated NtrC (NtrC-P) activates multiple σ^{54}-dependent promoters, controlling the expression of over one hundred genes related in *E. coli* [75]. NtrC-regulated genes encode multiple functions related to nitrogen metabolism, including transport and utilization pathways for diverse nitrogen sources, the nitrogen fixation sensor-regulator pair NifL-NifA (in the diazotrophic *Klebsiella pneumoniae*), and the nitrogen regulator Nac. Transcription of Nac is initiated from an NtrC-activated σ^{54}-dependent promoter and autorepressed [76, 77]. Nac in turn activates and represses a set of genes whose products are mostly related to nitrogen metabolism. Nac-activated genes include those involved in the catabolism of histidine, proline, urea and alanine, among others [78-80]. Notably, Nac represses the genes coding for the two enzymes that synthesize glutamate, *gdhA,* encoding GDH, and *gltAB,* encoding GOGAT [81]. Unlike most LysR-type transcriptional regulators, the activity of Nac is not modulated by the presence of a small ligand, and the protein is synthesized in its active form [79]. Thus, the nitrogen limitation dependency of Nac regulation exclusively reflects the increase in concentration under nitrogen limitation due to its transcriptional regulation. Nac is not present in many bacteria containing the Ntr system, and is conspicuously absent in the closely related and well-characterized *S. enterica* [82]. As the need for two different regulators (NtrC and Nac) has been questioned, Nac has been proposed to act as an "adaptor" that integrates genes transcribed from σ^{70}-dependent promoters (which cannot be directly regulated by the activator of σ^{54}-dependent promoters NtrC) into the general nitrogen control network [83].

Transcription of the *gdhA* gene is repressed by Nac under nitrogen limitation in both *E. coli* and *K. pneumoniae* (in contrast, *gdhA* expression is not nitrogen-regulated in *S. enterica*). Nac exhibits two modes of transcriptional repression of the *gdhA*p promoter. "Weak" repression involves Nac binding as a dimer to the promoter region in a single site located at -100 to -75 relative to the transcriptional start site [84]. "Strong" repression is observed when a Nac tetramer is simultaneously bound to the aforementioned site, centered at -89, and a downstream site centered at +47. The proposed regulatory mechanism for "strong" repression involves Nac bending DNA, looping out the intervening region, and forming a repressor loop reminiscent of those formed by LacI and GalR at the *lac* and *gal* promoters, respectively [79].

In addition to Nac-mediated repression, *gdhA* is subjected to positive regulation by a second LysR-type transcriptional regulator, ArgP. ArgP is an activator whose activity is antagonized by lysine. ArgP interacts with the *gdhA* promoter region at a single site between -100 and -50. Lysine inhibits interaction of ArgP with this site, thus preventing activation [85]. Activation is also prevented by Nac interaction with the overlapping binding site centered at -89, as both regulatory proteins bind in a mutually exclusive fashion. Thus "weak" repression is in fact the result of the antagonist action of Nac on ArgP-mediated activation [82, 85]. The role of lysine as a signal can be rationalized if one considers that glutamate must serve two roles in the cell, as both an amino group donor for over 80% of the nitrogenous compounds in the cell, and as a counterion to K^+ in osmotic pressure homeostasis. In order to sense the amount of glutamate being used for biosynthesis, the pools of one or more amino group acceptor may be used instead of glutamate itself. Lysine is likely one of those nitrogenous compounds that serve as surrogates for glutamate in signaling the extent of glutamate overflow from osmotic pressure homeostasis into the biosynthetic pathways for other nitrogenous compounds [85]. Consistently, *gdhA*p activity is decreased in rich media containing amino acids [81], and lysine levels may be one of several signals that mediate the feedback regulation of *gdhA*p. However, the identity of other compounds that may fulfill this role is as of yet unknown.

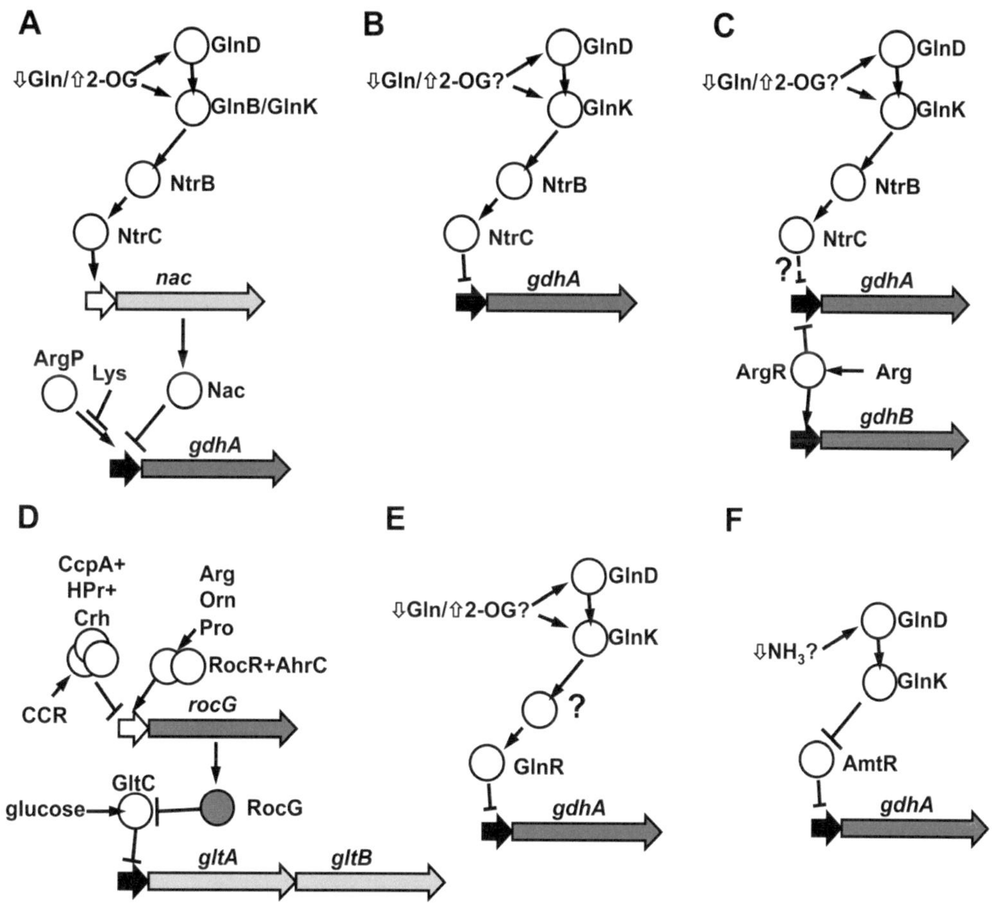

Figure 3. Regulatory circuits for bacterial glutamate dehydrogenase genes. The cartoons represent the known regulatory circuits for *K. pneumoniae* (A), *P. putida* (B), *P. aeruginosa* (C), *B. subtilis* (D), *S. coelicolor* (E) and *C. glutamicum* (E). Sigma 54- or SigL-dependent promoters are displayed as white arrows. Other promoter types are displayed as black arrows. Question marks represent aspects that are not well characterized. CCR: carbon catabolite repression.

4.2. Regulation of GDH synthesis in pseudomonads

Little is known about the transcriptional regulation of GDH synthesis in most Gram-negative bacteria. However, the regulatory mechanisms behind *gdhA* regulation were recently revealed in the soil bacterium *Pseudomonas putida* [86]. Similarly to its enteric counterparts, *P. putida* represses *gdhA* expression during nitrogen-limited growth in a Ntr system-dependent fashion (Figure 3,B). *P. putida* harbors a simplified general nitrogen control system that lacks the major PII protein, GlnB, and the transcriptional regulator Nac. In this scheme, the only PII protein, GlnK, controls the phosphorylation/dephosphorylation balance of NtrC in response to nitrogen availability using mechanisms likely similar to those found in the enterics [87]. In the absence of Nac, NtrC has been shown to directly regulate some of the functions that are controlled by Nac in the enterics, including activation of *codB* and *ureD*, encoding cytosine deaminase and urease respectively [88], and repression of the GDH gene *gdhA* [86]. NtrC binds the σ^{70}-dependent *gdhA*p promoter region cooperatively at four different sites, centered at positions -118, -95, -21 and +12 relative to the transcriptional start. While simultaneous occupancy of all four sites yields the maximal levels of repression, only the promoter-proximal I and II sites were found to be absolutely required for repression. A mechanism based on a repressor loop involving four NtrC dimers has been proposed for negative regulation of the *gdhA*p promoter in *P. putida*. Although repression is partially sensitive to the phosphorylation state of NtrC *in vitro, in vivo* evidence indicated that phosphorylation of NtrC is not required for efficient repression of the *gdhAp* promoter. Since NtrC synthesis is itself nitrogen regulated *via* the *glnA*p promoter, the increase in repressor concentration under nitrogen limitation appears to be the major determinant of nitrogen-regulated expression of *gdhA* in *P. putida* [86].

Pseudomonas aeruginosa encodes two GDH enzymes, a NAD-dependent GDH encoded by *gdhB*, primarily involved in glutamate deamination to 2-oxoglutarate, and a NADP-dependent GDH, encoded by *gdhA*, primarily involved in ammonium assimilation. The catabolic enzyme is linked to the AST pathway for aerobic utilization of arginine as a carbon source. Glutamate is the end product of the AST pathway, and NAD-GDH serves the purpose of connecting arginine catabolism to the central metabolism *via* 2-oxoglutarate. Transcription of *gdhB* is activated by the arginine regulatory protein ArgR in response to the presence of arginine, which is also an allosteric modulator of the enzyme activity (Figure 3,C). An ArgR binding site is found immediately upstream from the -35 box of the *gdhB* promoter. Induction of the expression of *gdhB* and activation of the encoded dehydrogenase by arginine serve to direct the flow of glutamate into the TCA cycle [25]. Conversely, transcription of *gdhA* is repressed by ArgR in the presence of arginine. Such repression could serve to minimize the operation of an energy-consuming futile cycle involving the simultaneous function of *gdhA* and *gdhB* when *P. aeruginosa* uses arginine as a carbon source. Repression of *gdhA* expression is exerted from a single ArgR binding site centered at position -41, and a simple steric hindrance mechanism has been proposed at this promoter [89]. Downregulation of *gdhA* expression by the Ntr system under nitrogen limitation has also been reported [90], but the factors and mechanisms involved are uncharacterized. Since *P. aeruginosa* also lacks Nac, direct repression by NtrC similar to that observed in *P. putida* [86], may also occur in this organism.

4.3. Regulation of GDH synthesis in *Bacillus subtilis*

The *Bacillus subtilis* genome contains two genes encoding GDHs, *rocG* and *gudB*. While RocG is an enzymatically active GDH, GudB is inactive, due to a duplication of three amino acid residues at its active center. Decryptification of *gudB* in a *rocG* background is achieved by high-frequency acquisition of a suppressor mutation consisting of the precise deletion of part of the 9-bp direct repeat that prevents activity [46, 47]. Both RocG and decryptified GudB are primarily catabolic dehydrogenases, and *de novo* glutamate synthesis in *B. subtilis* is performed exclusively by GOGAT.

Similarly to *P. aeruginosa* GdhB, RocG is not nitrogen-regulated and is linked to arginine and proline catabolism, as its expression is induced by arginine, ornithine and, to a lesser extent, proline. Transcription from the *rocG* promoter depends on the product of the *Bacillus subtilis* *sigL* gene, an ortholog of the alternative σ factor σ^{54} [91], and is activated by the UAS binding protein RocR in response to the presence of arginine, ornithine, or proline with the assistance of the arginine-dependent activator AhrC (Figure 3,D). SigL, RocR and AhrC also control transcription of the two operons, *rocABC* and *rocDEF*, involved in arginine conversion into glutamate. In that regard, RocG-dependent deamination of glutamate to 2-oxoglutarate can be viewed as the final step in the use of arginine, ornithine, and proline as carbon or nitrogen, providing rapidly metabolizable carbon- or nitrogen-containing compounds for biosynthesis [92]. Interestingly, inducibility of RocG synthesis by arginine precludes growth on glutamate as the sole carbon source.

The most salient feature of the *rocG*p SigL-dependent promoter is the absence of an upstream activator sequence (UAS) for RocR. Instead, the UAS present at the *rocA*p promoter, located immediately downstream from the *rocG* coding sequence, is the *cis*-acting element essential for RocR-dependent activation of *rocG*p. This sequence has been shown to be active when placed upstream or downstream and as far as 15 kb away from the target promoter [93]. According to the general model of σ^{54}-dependent promoter activation, RocR bound to its target sequence activates transcription by interacting with promoter-bound σ^{54}-RNA polymerase by a mechanism that involves looping out of the intervening DNA sequences. The AhrC protein is also required for activation of *rocG* (as well as *rocABC* and *rocDEF*), and it apparently modulates the activity of RocR by means of protein-protein contacts [94].

Unlike the arginine utilization operons *rocABC* and *rocDEF*, *rocG* transcription is subjected to carbon catabolite repression in the presence of glucose. Such repression is mediated by the regulatory protein CcpA. A *cre* (catabolite responsive element) site is located at positions +39 and +51, and is required for CcpA binding and *rocG*p repression. CcpA binding at this location prevents transcription from the *rocG*p promoter, but also acts as a roadblock for low-level readthrough transcription from an upstream promoter, which is relevant in the absence of both glucose and the cognate inducers [92]. As in other CcpA-regulated genes, CcpA-mediated repression requires the assistance of the accessory proteins HPr and Crh.

A final feature of the *B. subtilis* GDH RocG worth mentioning is its role in the regulation of the assimilatory glutamate synthase operon, *gltAB*. RocG belongs to a group of proteins designated "trigger enzymes", which have catalytic as well as regulatory functions. RocG interacts with the LysR-type transcriptional regulator GltC, which is the cognate activator of the *gltAB* operon. RocG-GltC interaction results in inactivation of the latter, which in turn prevents activation of the assimilatory glutamate synthase in conditions in which glutamate is already being synthesized from arginine-related amino acids or proline. This mechanism allows tight control of glutamate metabolism by the availability of carbon and nitrogen sources [95-97].

4.4. Regulation of GDH synthesis in other Gram-positives

The *Streptomyces coelicolor* gene *gdhA*, encoding a NADPH-dependent GDH, is negatively regulated under ammonium limitation by GlnR (Figure 3,E). GlnR is an OmpR-like transcriptional factor, which is the master regulator of nitrogen metabolism in *S. coelicolor*. This response includes activation of *glnA*, *glnII* genes, encoding two glutamine synthetases, and repression of *gdhA*, among others [98]. The GlnR regulon appears to be conserved in *Mycobacterium* and other actinomycetes. The nitrogen signal is transduced by a variation of the enterobacterial Ntr signal transduction system, including the PII protein GlnK and GlnD, which catalyzes adenylylation and deadenylylation of GlnK in response to the nitrogen status. GlnR activity is presumably regulated by phosphorylation, as it displays the conserved phosphorylatable aspartate residue present in many response regulators, but the identity of its sensor kinase is currently unknown [99]. GlnR exerts repression of the *gdhA* promoter region by binding at a conserved site centered at position -73, but the underlying mechanisms of repression are not yet understood [98]. Interestingly, synthesis of the regulator GlnR is repressed by the response regulator PhoP under phosphate limitation, thus providing a link between phosphorus and nitrogen metabolism in *S. coelicolor* [100].

AmtR is a repressor of the TetR family, which acts as the global nitrogen regulator in the industrially relevant *Corynebacterium glutamicum* [Burkowsky, 2003]. Because of the high basal intracellular concentrations of glutamate and glutamine (up to 200 mM and up to 50 mM, respectively), *C. glutamicum* does not use these amino acid pools to sense nitrogen availability. Instead, ammonium is probably used to modulate the activity of the adenylyl transferase GlnD [101]. Adenylylated GlnK interacts with AmtR to release the repressor from its target promoters under nitrogen limitation. AmtR represses the expression of at least 35 genes, including *glnA*, encoding glutamine synthetase, *gltBD*, encoding glutamate synthase and the *amtB-glnK-glnD* operon, encoding a high-affinity ammonium transporter and the signal transduction proteins, GlnK and GlnD. GDH activity is high and relatively constant in different growth conditions [50]. Transcription from the *gdhA*p promoter is repressed 2-fold by AmtR (Figure 3,F) under nitrogen excess and interaction of AmtR with the *gdhA*p promoter region has been detected [102]. Intriguing as it may be, upregulation of GDH under nitrogen limitation does not appear to be physiologically relevant, as sufficient activity is present under the high ammonium concentration conditions in which GDH contributes to ammonium assimilation. Other transcription factors (FarR, WhiH and OxyR) have been documented to bind the *gdhA*p promoter region, but the relevance of these interactions is so far unknown [102].

5. Involvement of GDH in biotechnological processes

In addition to the different physiological roles of GDHs from different bacteria in nitrogen assimilation, amino acid catabolism, osmotic balance or tolerance to high temperatures, GDH catalysis is crucial for a number of biotechnological processes. These include industrial production of glutamate by *C. glutamicum* and related species, which involves catalysis in the aminating direction, and production of aromas by lactic acid bacteria during cheeses ripening, which involves catalysis in the opposite direction.

5.1. Production of L-Glutamate by *Corynebacterium glutamicum*

C. glutamicum is a facultatively anaerobic, nonpathogenic, non-motile, biotin-auxotrophic, Gram-positive soil bacterium that was isolated more than 50 years ago in a screen for bacteria that excrete glutamate. Since then, derivative strains of this bacterium and related species have been isolated as glutamate producers, and industrial processes for biological glutamate production have been developed. Annual L-glutamate production is estimated to be over 2 million tons, with a continuously increasing demand (3% per year), above all in developing countries. L-glutamate is mainly used in food as a flavour enhancer.

In addition to its importance in industrial biotechnology, *C. glutamicum* has gained interest as a model organism for the *Corynebacterineae*, an industrially relevant suborder of the actinomycetes. Because of the academic and industrial interest, its physiology and metabolism have been deeply characterised, three research groups have independently determined the genomic sequence of *C. glutamicum* strains [103-105], and global analysis techniques such as proteomics, transcriptomics, metabolomics and metabolic flux analysis have been used to obtain a holistic view of the glutamate production process.

C. glutamicum exponentially growing cells do not accumulate glutamate. However, glutamate production and excretion can be easily induced by an astonishing variety of treatments, which allow accumulation of glutamate in the culture medium to a concentration as high as 80 g L^{-1}. These include biotin limitation [106], which was the first identified condition for glutamate overproduction, addition of fatty acid ester surfactants such as Tween 40 (polyoxyethylene sorbitan monopalmitate) or Tween 60 (polyoxyethylene sorbitan monoestearate) [107], addition of certain β-lactam antibiotics such as penicillin [108], use of glycerol-auxotrophic or fatty acid-auxotrophic strains [109], use of temperature-sensitive strains cultured at higher temperature [110], or addition of ethambutol, an inhibitor of cell wall arabinogalactan synthesis [111]. In spite of the apparent diversity of treatments to induce glutamate production and secretion, all of them have features in common and there is a unified view that glutamate production is triggered by environmental conditions that produce damage of the cell surface structures [112-114].

Since *C. glutamicum* is a biotin auxotroph, supplementation of a defined medium with a limiting concentration of biotin in batch cultures results in reduced total biomass and concomitant glutamate production. Biotin is a co-factor of acetyl-CoA carboxylase, the first enzyme for fatty acid biosynthesis, and a reduction of this enzyme activity leads to changes in the fatty acid composition of the membrane. In support of this view, a mutation in *dtsR1*, which codes for the beta subunit of acetyl/propionyl-CoA carboxylase, requires fatty acids for growth and, when limited, overproduces glutamate even when grown with an excess of biotin [109]. On the contrary, amplification of its gene dosage resulted in reduction of glutamate production induced by biotin limitation, detergents or penicillin [115]. Glycerol limitation in mutants unable to produce it provokes similar membrane alterations due to the limitation of membrane lipid precursors. Besides the ubiquitous cell membrane, members of the genus *Corynebacterium*, together with *Mycobacteria* and *Nocardia*, have a special cell envelope structure, which consists of a second lipid layer containing mycolic acids, which has a highly ordered structure and plays an important role in determining solute fluxes [116]. Interestingly, mycolic acids of the outer lipid layer are covalently linked to an arabinogalactan layer, which in turn is covalently linked to the underlying peptidoglycan of the cell wall. This structure may thus be envisioned as one large macromolecule, the mycolyl-arabinogalactan-peptidoglycan complex. Treatments with ethambutol or penicillin clearly affect the structure of the cell envelope. The triggering effect of Tween 40 or Tween 60 (but not Tween 20 or Tween 80) may be due to alterations of the mycolic layer structure [112] although it may also affect the activity of acetyl-CoA carboxylase [113], thus leading to cell envelope alterations.

It was thought for some time that all these treatments affecting cell envelope structure increased the membrane permeability, which would in turn allow leakage of glutamate. However, it is obvious that a specific carrier-mediated export is required, as the increase in the membrane permeability is specific for glutamate, not for other solutes, and glutamate is still exported against a concentration gradient. The glutamate transporter was identified by characterising mutants that produced and accumulated glutamate. All these mutants had substitution mutations or even an insertion mutation in the NCgl1221 gene, which codes for a product showing homology to mechanosensitive channels such as the *E. coli yggB* gene product [113]. The mutant alleles appear to code for a constitutively opened glutamate channel. Exchange of the wild-type allele for these mutant alleles led to glutamate overproduction and excretion without any inducing treatment, and rendered cells resistant to the L-glutamate analog 4-fluoroglutamate. Overexpressing wild type NCgl1221 did not result in constitutive glutamate excretion but led to increased glutamate production and excretion after the induction treatments. On the contrary, a deletion mutant lacking NCgl1221 could not excrete glutamate [113]. Therefore, opening the glutamate channel appears to be essential for glutamate production and export. In addition, this is the first response of *C. glutamicum* to membrane tension alterations, which triggers all other metabolic adaptations leading to efficient glutamate production. The NCgl1221 (*yggB*) product has 4 transmembrane segments, is located in the cytoplasmic membrane [117] and has been recently shown to work as a mechanosensitive channel able to increase the cell survival rate of *Bacillus subtilis* after osmotic down-shock [118].

Although *C. glutamicum* has other potential ammonium assimilating enzymes such as alanine dehydrogenase or diaminopimelate dehydrogenase, their contribution to ammonium assimilation is very limited, according to the *in vivo* flux analyses [119]. *C. glutamicum* and related glutamate-producing species have GDH, GS and GOGAT. Thus, *C. glutamicum* assimilates ammonium, the preferred carbon source of most bacteria, either via GS/GOGAT or via GDH. As in many other bacteria, GS/GOGAT is the main ammonium assimilation pathway when its concentration is limiting, whilst GDH assimilate most of the ammonium when it is present in concentrations above 5 mM [120]. Under glutamate production conditions, which implies high ammonium concentration, the *gltB* and *gltD* genes encoding the GOGAT subunits, are fully repressed by ammonium and no GOGAT activity is detected [119, 121]. Therefore, the main glutamate producing enzyme in these conditions is GDH.

Production and excretion of large quantities of glutamate require glutamate export to the culture medium but also modification of metabolic fluxes to produce high amounts of glutamate. Metabolic flux analysis under exponential growth vs. glutamate producing conditions [119, 122-125], indicated significant changes in catabolic pathways (Figure 4). Under non-growing, glutamate-producing conditions, glucose catabolism either via glycolysis or the pentose phosphate pathway was clearly reduced. Additionally, there was a significant redistribution of the fluxes at the 2-OG branch point between the TCA cycle and the glutamate biosynthetic pathway, that reduced the flux towards succinyl CoA formation about 2/3, with a concomitant increase in the reductive aminating reaction catalysed by GDH. The increase of the metabolic flux towards glutamate is not due to an increase in the 2-OG-producing isocitrate dehydrogenase (ICDH) or GDH activities because they remain constant [125]. In fact, increasing either ICDH or GDH activity by overexpressing their coding genes, did not affect glutamate production [123].

The factor with greatest impact on glutamate production is a reduction of the 2-oxoglutarate dehydrogenase complex (ODHC) activity [123, 125]. During exponential growth, although sufficient GDH specific activity was observed, the flux catalyzed by GDH was very small because the K_m value of GDH for 2-OG is much higher (approximately 50-fold higher) than that of ODHC. Once the ODHC specific activity was decreased after the triggering signals, 2-OG accumulated, and consequently, glutamate was overproduced by GDH. In spite of the initial report claiming that an *odhA* deletion mutant, which has no ODCH activity, overproduced and excreted glutamate [126], there are a number of reports on *odhA* mutants with contradictory results. It appears that those OdhA mutants that overproduced glutamate had an additional mutation in Ncgl1221 (*yggB*) [113], which clearly indicates that reduction or elimination of ODHC activity is necessary but not sufficient for glutamate overproduction.

Obviously, GDH activity requires the presence of sufficient concentration of its substrates. Paradoxically, most of the NADPH required for anabolic reactions is generated at the pentose phosphate pathway, which is reduced under these conditions. However, it appears that the reaction catalysed by ICDH provides sufficient NADPH to fulfill the demands of the GDH reaction. Similarly, glutamate production and excretion requires a continuous supply of carbon. This is achieved by an increase in the anaplerotic reactions that produce oxaloacetate. In *C. glutamicum* two reactions can yield oxaloacetate. One is the standard anaplerotic reaction catalysed by the phosphoenol pyruvate carboxylase (PEPc). The other is pyruvate carboxylase (Pc), which catalyses oxaloacetate production from pyruvate. Metabolic flux of PEPc remains constant in both conditions whilst the Pc flux, undetectable during exponential growth, is clearly increased in the glutamate-producing conditions upon addition of Tween-40 [124]. This suggests a relevant role of Pc in providing sufficient oxaloacetate for glutamate production. Despite this evidence, pyruvate carboxylase cannot be relevant for glutamate production when induced by biotin limitation, as biotin is also a prosthetic group of Pc. As expected, *pyc* disruptants lacking Pc can produce and excrete glutamate as the wild type strain [127], indicating that PEPc may be sufficient to provide the required carbon under this particular inducing regime.

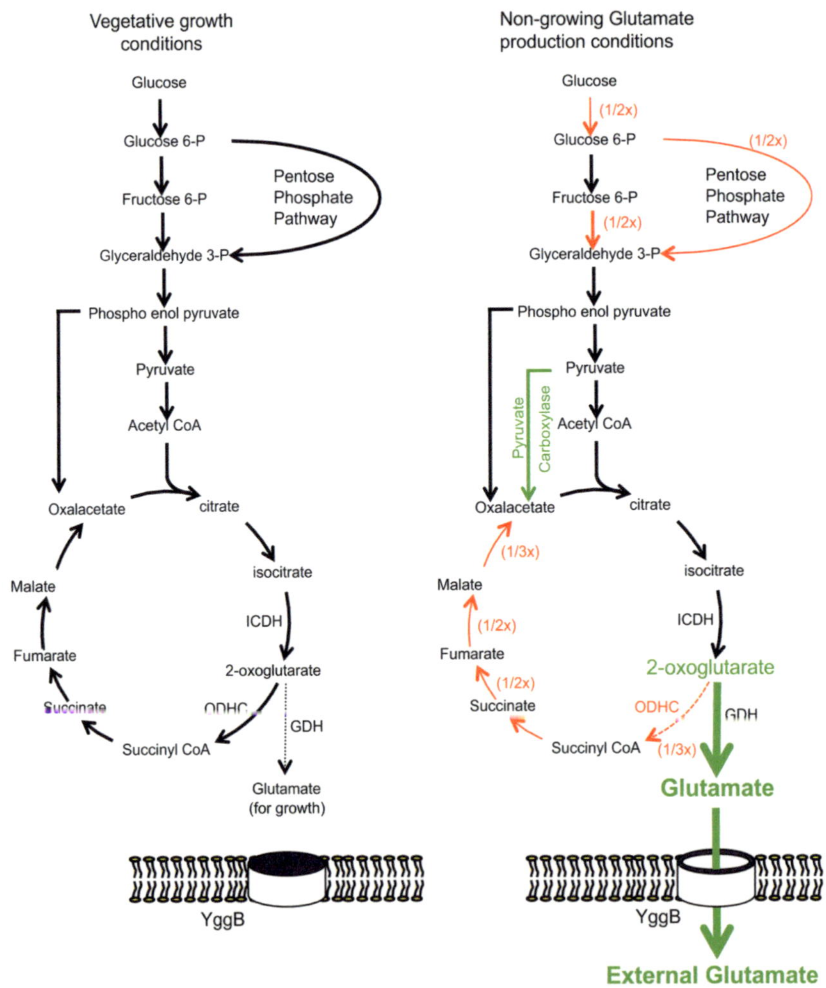

Figure 4. Metabolic fluxes from glucose to glutamate under vegetative growth or non-growing glutamate-producing conditions. Red arrows show metabolic fluxes reduced during glutamate production. Green arrows show those reactions that appear or are increased under the same condition.

How is ODHC activity decreased during glutamate production? ODHC is a complex composed of three different subunits: OdhA (E1), SucB (E2) and LpdA (E3). Many TCA cycle enzymes are regulated at transcriptional or posttranscriptional levels [128]. Activity of this complex is controlled by OdhI, a small protein (15,4 KDa) that binds OdhA (the E1 subunit). Induced glutamate excretion is virtually abolished in a OdhI deletion mutant, thus indicating that inhibition of ODHC activity through OdhI is critical for glutamate overproduction [129]. The OdhI function is regulated by phosphorylation by PknG, which phosphorylates and inactivates OdhI [130], and dephosphorylation by Ppp. PknG deletion mutants showed higher glutamate production when induced by some treatments [129], which is consistent with its role in controlling ODHC activity.

The main question is why do *C. glutamicum* cells excrete large amounts of L-glutamate when exposed to the mentioned induction treatments? Certainly, it does not appear that it is a response to metabolic changes since the inducing treatments do not imply changes in nutritional status. Since these treatments all affect cell surface structures and may therefore alter membrane tension, L-glutamate production may be the response to this membrane stress, as glutamate might function as a compatible solute to prevent cells from bursting. Thus, L-glutamate production by *C. glutamicum* may be a mechanism of adaptation to environmental changes affecting cell surface structures that starts by opening a glutamate export mechanosensitive channel, which in turn somehow triggers the metabolic adaptations (reduction of pentose phosphate pathway, reduction of ODHC activity, increase of the anaplerotic reactions) required to produce and excrete high amounts of glutamate. A similar role of L-glutamate in osmoprotection was also described for *E. coli* [131], and confirmed by the osmosensitive phenotype provoked by mutations unable to activate *gdhA* in a GOGAT deficient background during osmotic upshift [4]. However, there are no reports on metabolic adaptations leading to glutamate production and excretion upon osmotic shock in this bacterium.

5.2. Aroma and flavour production by Lactic Acid Bacteria (LAB)

Accelerating or diversifying flavour development in cheese is of major economical interest since final flavour of cheeses partly determines the consumer's choice. Flavour formation occurs during cheese ripening, and is associated to non-starter lactic acid bacteria (NSLAB), adventitious microflora that occur in the milk or appear later during cheese manufacturing. After characterisation, some strains have been selected as flavour-producing adjunts and shown that cheeses with these adjunts are richer in free amino acids and have enhanced flavour intensity [132]. However, flavour development is a time consuming and expensive process that is still not well mastered, and selection of strains for cheese ripening is still an empirical process based on cheese trials with different strains and sensory analyses, that has had varying success [133].

Amino acid catabolism, particularly that of aromatic amino acids, branched chain amino acids, and methionine, is a major player in flavour formation in cheese, especially in cheeses containing only LAB. Amino acid conversion to aroma compounds proceeds by two different pathways: (i) elimination reactions catalysed by amino acid lyases that produce different alcohols, and (ii) transamination reactions leading to different 2-oxoacids. The latter is the major pathway in LAB. The resulting 2-oxoacids are not responsible for flavour production but are then transformed to aldehydes, alcohols, carboxylic acids, hydroxy acids or methanethiol by additional steps that may be catalysed by the 2-oxoacid-producing LAB or by the inoculated starter LABs [134, 135]. Figure 5 shows the transaminations and further reactions leading to aroma compounds. It appears that the different flavours produced by different LAB strains depend on the proportion of the different 2-oxoacid produced and, therefore, on the relative aminotransferases activities [136].

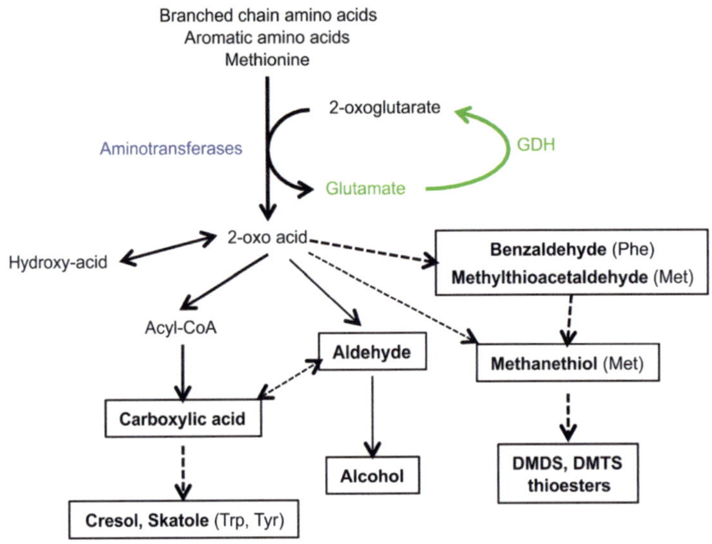

Figure 5. Main amino acid catabolic pathways in LAB leading to aroma compounds (boxed). Doted arrows represent chemical reactions.

Although lactococci have high aminotransferase activity, only very low and slow amino acid degradation occurs. Cheeses are rich in amino acids and peptides but the concentration of 2-OG is low. Supplementation of several types of cheeses with 2-OG led to a decrease in the levels of amino acids important for aroma development, which indicated that catabolism of these amino acids had been enhanced [137, 138], and sensory analyses indicated that addition of 2-OG resulted in changes in creamy character, aroma intensity and fruity notes. These results clearly established that 2-OG availability is the limiting factor for the transamination reactions that convert aminoacids to aroma compounds and prompted scientist to look for sources of 2-OG.

Ocurrence of GDH in starter and non-starter LAB is heterogeneous, as just a few LAB strains show GDH activity. Some strains of the starter *Lactococcus lactis* have GDH activity but it is coded in a plasmid (pGDH442) rather than in its chromosome [139]. However, these GDH^{+} strains are glutamate auxotrophs and cannot synthesize glutamate because of 2-OG limitation. The oxidative TCA cycle in these strains is interrupted at the isocitrate dehydrogenase step [140], thus blocking the major source of 2-OG for most bacteria. Because of this, GDH can only be used for glutamate catabolism, and this reaction constitutes the major source of 2-OG in these strains. As shown in Fig. 5, the GDH role in aroma development by LAB in cheeses is to catalyse the oxidative deamination of glutamate in order to replenish the 2-OG consumed by the transamination reactions. The importance of GDH in aroma production was initially shown by cloning and heterologous expression of the GDH gene from *Peptostreptococcus asaccharolyticus* into a GDH^- strain of *L. lactis*, which resulted in an increase of amino acids degradation and, more importantly, an increase in the proportion of carboxylic acids, which are important aroma compounds [141]. This result is the proof of concept that GDH can substitute the exogenous 2-OG. Other reports describing the effect of GDH on aroma production have followed, including the natural transfer by conjugation of the *gdh* gene coded in the pGDH442 plasmid among LAB strains [142]. The relevance of this is that the resulting transconjugants are not considered genetically modified organisms and can be used in the food industry as starters or adjuncts.

An evident correlation between GDH activity and aroma production has been established for both mesophilic and thermophilic LAB [143, 144]. Because of this, the presence of GDH activity has been proposed as a criterion to select flavour-producing LAB strains for cheese ripening. The use of different LAB strains with different aminotransferases especificities together with the use of GDH to enhance the transamination reactions may represent an effective way of intensifying and diversifying cheese aromas.

Author details

Eduardo Santero*, Ines Canosa and Fernando Govantes

Centro Andaluz de Biología del Desarrollo, CSIC-Universidad Pablo de Olavide-Junta de Andalucía, Seville, Spain

Ana B. Hervás

Department of Microbiology, Faculty for Biology. University of Freiburg, Freiburg, Germany

Acknowledgement

We are very grateful to Nuria Pérez and Guadalupe Martín-Cabello for their technical help and all members of the laboratory for their insights and suggestions. Work in the authors' laboratories is supported by the Spanish Ministry of Economy and Competitivity, grants BIO2008-01805, BIO2010-17853, BIO2011-24003 and CSD2007-00005, and by the Andalusian government, grants P05-CVI-131 and P07-CVI-2518.

6. References

[1] Smith EL, Austen, BM, Blumenthal, KM, Nyc, JF (1975) Glutamate dehydrogenases. In: Boyer PD, editor. The Enzymes: Academic Press. p. 293–367.

---

* Corresponding Author

[2] Reitzer L (2003) Nitrogen assimilation and global regulation in *Escherichia coli*. Annu Rev Microbiol. 57:155-176.

[3] Tanous C, Chambellon E, Sepulchre AM, Yvon M (2005) The gene encoding the glutamate dehydrogenase in *Lactococcus lactis* is part of a remnant Tn3 transposon carried by a large plasmid. J Bacteriol. 187:5019-5022.

[4] Nandineni MR, Laishram RS, Gowrishankar J (2004) Osmosensitivity associated with insertions in argP (*iciA*) or *glnE* in glutamate synthase-deficient mutants of *Escherichia coli*. J Bacteriol. 186:6391-6399.

[5] Consalvi V, Chiaraluce R, Politi L, Vaccaro R, De Rosa M, Scandurra R (1991) Extremely thermostable glutamate dehydrogenase from the hyperthermophilic archaebacterium *Pyrococcus furiosus*. Eur J Biochem. 202:1189-1196.

[6] Kobayashi T, Higuchi S, Kimura K, Kudo T, Horikoshi K (1995) Properties of glutamate dehydrogenase and its involvement in alanine production in a hyperthermophilic archaeon, *Thermococcus profundus*. J Biochem. 118:587-592.

[7] Kelly A, Stanley CA (2001) Disorders of glutamate metabolism. Ment Retard Dev Disabil Res Rev. 7:287-295.

[8] Smith TJ, Stanley CA. Glutamate dehydrogenase. In: D'Mello JPF, editor. Amino Acids in Human Nutrition and Health Wallingford, Oxfordshire, UK: CAB International; 2012. p. 1-19.

[9] Aarts MM, Tymianski M (2004) Molecular mechanisms underlying specificity of excitotoxic signaling in neurons. Curr Mol Med. 4:137-147.

[10] Stanley CA, Lieu YK, Hsu BY, Burlina AB, Greenberg CR, Hopwood NJ, *et al.* (1998) Hyperinsulinism and hyperammonemia in infants with regulatory mutations of the glutamate dehydrogenase gene. N Engl J Med. 338:1352-1357.

[11] MacMullen C, Fang J, Hsu BY, Kelly A, de Lonlay-Debeney P, Saudubray JM, *et al.* (2001) Hyperinsulinism/hyperammonemia syndrome in children with regulatory mutations in the inhibitory guanosine triphosphate-binding domain of glutamate dehydrogenase. J Clin Endocrinol Metab. 86:1782-1787.

[12] Stanley CA, Fang J, Kutyna K, Hsu BY, Ming JE, Glaser B, *et al.* (2000) Molecular basis and characterization of the hyperinsulinism/hyperammonemia syndrome: predominance of mutations in exons 11 and 12 of the glutamate dehydrogenase gene. HI/HA Contributing Investigators. Diabetes. 49:667-673.

[13] Yarrison G, Young DW, Choules GL (1972) Glutamate dehydrogenase from *Mycoplasma laidlawii*. J Bacteriol. 110:494-503.

[14] Maulik P, Ghosh S (1986) NADPH/NADH-dependent cold-labile glutamate dehydrogenase in *Azospirillum brasilense*. Purification and properties. Eur J Biochem. 155:595-602.

[15] Consalvi V, Chiaraluce R, Politi L, Pasquo A, De Rosa M, Scandurra R (1993) Glutamate dehydrogenase from the thermoacidophilic archaebacterium *Sulfolobus solfataricus*: studies on thermal and guanidine-dependent inactivation. Biochim Biophys Acta. 1202:207-215.

[16] Hoffmann RJ, Bishop SH, Sassaman C (1978) Glutamate dehydrogenase from coelenterates is NADP specific. J Exp Zool. 203:165-170.

[17] Srivastava HS, Singh, R. P. (1987) Role and regulation of L-glutamate dehydrogenase activity in higher plants. Phytochemistry. 26:597–610.

[18] Le John HB (1971) Enzyme regulation, lysine pathways and cell wall structures as indicators of major lines of evolution in fungi. Nature , . 231:164-168.
[19] Hudson RC, Daniel RM (1993) L-glutamate dehydrogenases: distribution, properties and mechanism. Comp Biochem Physiol B. 106:767-792.
[20] Hemmings BA (1980) Purification and properties of the phospho and dephospho forms of yeast NAD-dependent glutamate dehydrogenase. J Biol Chem. 255:7925-7932.
[21] Uno I, Matsumoto K, Adachi K, Ishikawa T (1984) Regulation of NAD-dependent glutamate dehydrogenase by protein kinases in *Saccharomyces cerevisiae*. J Biol Chem. 259:1288-1293.
[22] Veronese FM, Nyc JF, Degani Y, Brown DM, Smith EL (1974) Nicotinamide adenine dinucleotide-specific glutamate dehydrogenase of *Neurospora*. I. Purification and molecular properties. J Biol Chem. 249:7922-7928.
[23] Minambres B, Olivera ER, Jensen RA, Luengo JM (2000) A new class of glutamate dehydrogenases (GDH). Biochemical and genetic characterization of the first member, the AMP-requiring NAD-specific GDH of *Streptomyces clavuligerus*. J Biol Chem. 275:39529-39542.
[24] Kawakami R, Sakuraba H, Ohshima T (2007) Gene cloning and characterization of the very large NAD-dependent l-glutamate dehydrogenase from the psychrophile *Janthinobacterium lividum*, isolated from cold soil. J Bacteriol. 189:5626-5633.
[25] Lu CD, Abdelal AT (2001) The gdhB gene of *Pseudomonas aeruginosa* encodes an arginine-inducible NAD(+)-dependent glutamate dehydrogenase which is subject to allosteric regulation. J Bacteriol. 183:490-499.
[26] O'Hare HM, Duran R, Cervenansky C, Bellinzoni M, Wehenkel AM, Pritsch O, *et al.* (2008) Regulation of glutamate metabolism by protein kinases in mycobacteria. Mol Microbiol. 70:1408-1423.
[27] Camardella L, Di Fraia R, Antignani A, Ciardiello MA, di Prisco G, Coleman JK, *et al.* (2002) The Antarctic Psychrobacter sp. TAD1 has two cold-active glutamate dehydrogenases with different cofactor specificities. Characterisation of the NAD+-dependent enzyme. Comp Biochem Physiol A Mol Integr Physiol. 131:559-567.
[28] Prunkard DE, Bascomb NF, Molin WT, Schmidt RR (1986) Effect of Different Carbon Sources on the Ammonium Induction of Different Forms of NADP-Specific Glutamate Dehydrogenase in Chlorella sorokiniana Cells Cultured in the Light and Dark. Plant Physiol. 81:413-422.
[29] Tomita T, Miyazaki T, Miyazaki J, Kuzuyama T, Nishiyama M (2010) Hetero-oligomeric glutamate dehydrogenase from *Thermus thermophilus*. Microbiology. 156:3801-3813.
[30] Lam HM, Coschigano KT, Oliveira IC, Melo-Oliveira R, Coruzzi GM (1996) The Molecular-Genetics of Nitrogen Assimilation into Amino Acids in Higher Plants. Annu Rev Plant Physiol Plant Mol Biol. 47:569-593.
[31] Aalen N, Steen IH, Birkeland NK, Lien T (1997) Purification and properties of an extremely thermostable NADP+-specific glutamate dehydrogenase from *Archaeoglobus fulgidus*. Arch Microbiol. 168:536-539.
[32] Bonete MJ, Camacho, M. L., Cadenas, E. (1986) Purification and some properties of NAD^+-dependent glutamate dehydrogenase from *Halobacterium halobium*. Int. J. Biochem. 18:785-789.

[33] Bonete MJ, Camacho, M. L., Cadenas, E. (1987) A new glutamate dehydrogenase from *Halobacterium halobium* with different coenzyme specificity. Int. J. Biochem. 19:1149-1155.

[34] Hudson RC, Ruttersmith LD, Daniel RM (1993) Glutamate dehydrogenase from the extremely thermophilic archaebacterial isolate AN1. Biochim Biophys Acta. 1202:244-250.

[35] Grantham WC, Brown AT (1983) Ammonia utilization by a proposed bacterial pathogen in human periodontal disease, *Capnocytophaga ochracea*. Arch Oral Biol. 28:327-338.

[36] Sakamoto N, Kotre AM, Savageau MA (1975) Glutamate dehydrogenase from *Escherichia coli*: purification and properties. J Bacteriol. 124:775-783.

[37] Di Fraia R, Wilquet V, Ciardiello MA, Carratore V, Antignani A, Camardella L, *et al.* (2000) NADP+-dependent glutamate dehydrogenase in the Antarctic psychrotolerant bacterium *Psychrobacter* sp. TAD1. Characterization, protein and DNA sequence, and relationship to other glutamate dehydrogenases. Eur J Biochem. 267:121-131.

[38] Janssen DB, op den Camp HJ, Leenen PJ, van der Drift C (1980) The enzymes of the ammonia assimilation in *Pseudomonas aeruginosa*. Arch Microbiol. 124:197-203.

[39] Brown CM, Macdonald-Brown DS, Stanley SO (1973) The mechanisms of nitrogen assimilation in pseudomonads. Antonie Van Leeuwenhoek. 39:89-98.

[40] Coulton JW, Kapoor M (1973) Studies on the kinetics and regulation of glutamate dehydrogenase of *Salmonella typhimurium*. Can J Microbiol. 19:439-450.

[41] Ruiz JL, Ferrer, J., Camacho, M., Bonete, M. J. (1998) NAD-specific glutamate dehydrogenase from *Thermus thermophilus* HB8: purification and enzymatic properties. FEMS Microbiology Letters. 159:15-20.

[42] LeJohn HB, Suzuki, I. and J. A. Wrights (1968) Glutamate Dehydrogenases of *Thiobacillus novellus,* kinetic properties and a possible control mechanism. J Biol Chem. 243:118-128.

[43] Kanamori K, Weiss RL, Roberts JD (1987) Role of glutamate dehydrogenase in ammonia assimilation in nitrogen-fixing Bacillus macerans. J Bacteriol. 169:4692-4695.

[44] Kanamori K, Weiss RL, Roberts JD (1987) Ammonia assimilation in *Bacillus polymyxa*. 15N NMR and enzymatic studies. J Biol Chem. 262:11038-11045.

[45] Belitsky BR, Sonenshein AL (1998) Role and regulation of *Bacillus subtilis* glutamate dehydrogenase genes. J Bacteriol. 180:6298-6305.

[46] Gunka K, Tholen S, Gerwig J, Herzberg C, Stulke J, Commichau FM (2012) A high-frequency mutation in *Bacillus subtilis*: requirements for the decryptification of the gudB glutamate dehydrogenase gene. J Bacteriol. 194:1036-1044.

[47] Rice DW, Baker PJ, Farrants GW, Hornby DP (1987) The crystal structure of glutamate dehydrogenase from Clostridium symbiosum at 0.6 nm resolution. Biochem J. 242:789-795.

[48] Hamza MA, Engel PC (2008) Homotropic allosteric control in clostridial glutamate dehydrogenase: different mechanisms for glutamate and NAD+? FEBS Lett. 582:1816-1820.

[49] Bormann ER, Eikmanns BJ, Sahm H (1992) Molecular analysis of the *Corynebacterium glutamicum* gdh gene encoding glutamate dehydrogenase. Mol Microbiol. 6:317-326.

[50] Harper CJ, Hayward D, Kidd M, Wiid I, van Helden P (2010) Glutamate dehydrogenase and glutamine synthetase are regulated in response to nitrogen availability in *Myocbacterium smegmatis*. BMC Microbiol. 10:138.

[51] Sarada KV, Rao NA, Venkitasubramanian TA (1980) Isolation and characterisation of glutamate dehydrogenase from *Mycobacterium smegmatis* CDC 46. Biochim Biophys Acta. 615:299-308.
[52] Misono H, Goto, N., Nagasaki, S. (1985) Purification, Crystallization and Properties of NADP+ -Specific Glutamate Dehydrogenase from *Lactobacillus fermentum*. Agric. Biol Chem. 49:117-123.
[53] Hornby DP, Engel PC (1984) Characterization of *Peptostreptococcus asaccharolyticus* glutamate dehydrogenase purified by dye-ligand chromatography. J Gen Microbiol. 130:2385-2394.
[54] Miñambres B, Olivera ER, Jensen RA, Luengo JM (2000) A new class of glutamate dehydrogenases (GDH). Biochemical and genetic characterization of the first member, the AMP-requiring NAD-specific GDH of *Streptomyces clavuligerus*. J Biol Chem. 275:39529-39542.
[55] Vancurova I, Vancura A, Volc J, Kopecky J, Neuzil J, Basarova G, *et al.* (1989) Purification and properties of NADP-dependent glutamate dehydrogenase from *Streptomyces fradiae*. J Gen Microbiol. 135:3311-3318.
[56] Andersson JO, Roger AJ (2003) Evolution of glutamate dehydrogenase genes: evidence for lateral gene transfer within and between prokaryotes and eukaryotes. BMC Evol Biol. 3:14.
[57] Benachenhou-Lahfa N, Forterre P, Labedan B (1993) Evolution of glutamate dehydrogenase genes: evidence for two paralogous protein families and unusual branching patterns of the archaebacteria in the universal tree of life. J Mol Evol. 36:335-346.
[58] Lilley KS, Baker PJ, Britton KL, Stillman TJ, Brown PE, Moir AJ, *et al.* (1991) The partial amino acid sequence of the NAD(+)-dependent glutamate dehydrogenase of *Clostridium symbiosum*: implications for the evolution and structural basis of coenzyme specificity. Biochim Biophys Acta. 1080:191-197.
[59] Baker PJ, Britton KL, Engel PC, Farrants GW, Lilley KS, Rice DW, *et al.* (1992) Subunit assembly and active site location in the structure of glutamate dehydrogenase. Proteins. 12:75-86.
[60] Smith TJ, Stanley CA (2008) Untangling the glutamate dehydrogenase allosteric nightmare. Trends Biochem Sci. 33:557-564.
[61] Peterson PE, Smith TJ (1999) The structure of bovine glutamate dehydrogenase provides insights into the mechanism of allostery. Structure. 7:769-782.
[62] Smith TJ, Peterson PE, Schmidt T, Fang J, Stanley CA (2001) Structures of bovine glutamate dehydrogenase complexes elucidate the mechanism of purine regulation. J Mol Biol. 307:707-720.
[63] Smith TJ, Schmidt T, Fang J, Wu J, Siuzdak G, Stanley CA (2002) The structure of apo human glutamate dehydrogenase details subunit communication and allostery. J Mol Biol. 318:765-777.
[64] Bitensky MW, Yielding KL, Tomkins GM (1965) The Effect of Allosteric Modifiers on the Rate of Denaturation of Glutamate Dehydrogenase. J Biol Chem. 240:1077-1082.
[65] Yip KS, Stillman TJ, Britton KL, Artymiuk PJ, Baker PJ, Sedelnikova SE, *et al.* (1995) The structure of *Pyrococcus furiosus* glutamate dehydrogenase reveals a key role for ion-pair networks in maintaining enzyme stability at extreme temperatures. Structure. 3:1147-1158.

[66] Stillman TJ, Baker PJ, Britton KL, Rice DW (1993) Conformational flexibility in glutamate dehydrogenase. Role of water in substrate recognition and catalysis. J Mol Biol. 234:1131-1139.

[67] Noor S, Punekar NS (2005) Allosteric NADP-glutamate dehydrogenase from aspergilli: purification, characterization and implications for metabolic regulation at the carbon-nitrogen interface. Microbiology. 151:1409-1419.

[68] Joseph AA, Wixon RL (1970) Ammonia incorporation in *Hydrogenomonas eutropha*. Biochim Biophys Acta. 201:295-299.

[69] Sanwal BD, Lata M (1961) The occurrence of two different glutamic dehydrogenases in *Neurospora*. Can J Microbiol. 7:319-328.

[70] Lawit SJ, Miller PW, Dunn WI, Mirabile JS, Schmidt RR (2003) Heterologous expression of cDNAs encoding Chlorella sorokiniana NADP-specific glutamate dehydrogenase wild-type and mutant subunits in *Escherichia coli* cells and comparison of kinetic and thermal stability properties of their homohexamers. Plant Mol Biol. 52:605-616.

[71] LeJohn HB, McCrea BE (1968) Evidence for two species of glutamate dehydrogenases in *Thiobacillus novellus*. J Bacteriol. 95:87-94.

[72] Bonete MJ, Perez-Pomares F, Ferrer J, Camacho ML (1996) NAD-glutamate dehydrogenase from *Halobacterium halobium*: inhibition and activation by TCA intermediates and amino acids. Biochim Biophys Acta. 1289:14-24.

[73] Bender RA (2010) A NAC for regulating metabolism: the nitrogen assimilation control protein (NAC) from *Klebsiella pneumoniae*. J Bacteriol. 192:4801-4811.

[74] Leigh JA, Dodsworth JA (2007) Nitrogen Regulation in Bacteria and Archaea. Annu Rev Microbiol.

[75] Zimmer DP, Soupene E, Lee HL, Wendisch VF, Khodursky AB, Peter BJ, *et al.* (2000) Nitrogen regulatory protein C-controlled genes of *Escherichia coli*: scavenging as a defense against nitrogen limitation. Proc Natl Acad Sci U S A. 97:14674-14679.

[76] Feng J, Goss TJ, Bender RA, Ninfa AJ (1995) Activation of transcription initiation from the nac promoter of *Klebsiella aerogenes*. J Bacteriol. 177:5523-5534.

[77] Feng J, Goss TJ, Bender RA, Ninfa AJ (1995) Repression of the *Klebsiella aerogenes nac* promoter. J Bacteriol. 177:5535-5538.

[78] Frisch RL, Bender RA (2010) An Expanded Role for the Nitrogen Assimilation Control protein (NAC) in the Response of *Klebsiella pneumoniae* to Nitrogen Stress. J Bacteriol. Published online ahead of print.

[79] Goss TJ, Bender RA (1995) The nitrogen assimilation control protein, NAC, is a DNA binding transcription activator in *Klebsiella aerogenes*. J Bacteriol. 177:3546-3555.

[80] Macaluso A, Best EA, Bender RA (1990) Role of the nac gene product in the nitrogen regulation of some NTR-regulated operons of *Klebsiella aerogenes*. J Bacteriol. 172:7249-7255.

[81] Bender RA, Snyder PM, Bueno R, Quinto M, Magasanik B (1983) Nitrogen regulation system of Klebsiella aerogenes: the nac gene. J Bacteriol. 156:444-446.

[82] Goss TJ, Janes BK, Bender RA (2002) Repression of glutamate dehydrogenase formation in *Klebsiella aerogenes* requires two binding sites for the nitrogen assimilation control protein, NAC. J Bacteriol. 184:6966-6975.

[83] Bender RA (1991) The role of the NAC protein in the nitrogen regulation of *Klebsiella aerogenes*. Mol Microbiol. 5:2575-2580.

[84] Rosario CJ, Bender RA (2005) Importance of tetramer formation by the nitrogen assimilation control protein for strong repression of glutamate dehydrogenase formation in *Klebsiella pneumoniae*. J Bacteriol. 187:8291-8299.
[85] Goss TJ (2008) The ArgP protein stimulates the *Klebsiella pneumoniae gdhA* promoter in a lysine-sensitive manner. J Bacteriol. 190:4351-4359.
[86] Hervas AB, Canosa I, Santero E (2010) Regulation of glutamate dehydrogenase expression in *Pseudomonas putida* results from its direct repression by NtrC under nitrogen-limiting conditions. Mol Microbiol. 78:305-319.
[87] Garcia-Gonzalez V, Jimenez-Fernandez A, Hervas AB, Canosa I, Santero E, Govantes F (2009) Distinct roles for NtrC and GlnK in nitrogen regulation of the *Pseudomonas* sp. strain ADP cyanuric acid utilization operon. FEMS Microbiol Lett. 300:222-229.
[88] Hervas AB, Canosa I, Little R, Dixon R, Santero E (2009) NtrC-dependent regulatory network for nitrogen assimilation in *Pseudomonas putida*. J Bacteriol. 191:6123-6135.
[89] Hashim S, Kwon DH, Abdelal A, Lu CD (2004) The arginine regulatory protein mediates repression by arginine of the operons encoding glutamate synthase and anabolic glutamate dehydrogenase in *Pseudomonas aeruginosa*. J Bacteriol. 186:3848-3854.
[90] Li W, Lu CD (2007) Regulation of carbon and nitrogen utilization by CbrAB and NtrBC two-component systems in *Pseudomonas aeruginosa*. J Bacteriol. 189:5413-5420.
[91] Debarbouille M, Martin-Verstraete I, Kunst F, Rapoport G (1991) The *Bacillus subtilis* sigL gene encodes an equivalent of sigma 54 from gram-negative bacteria. Proc Natl Acad Sci U S A. 88:9092-9096.
[92] Belitsky BR, Kim HJ, Sonenshein AL (2004) CcpA-dependent regulation of *Bacillus subtilis* glutamate dehydrogenase gene expression. J Bacteriol. 186:3392-3398.
[93] Belitsky BR, Sonenshein AL (1999) An enhancer element located downstream of the major glutamate dehydrogenase gene of *Bacillus subtilis*. Proc Natl Acad Sci U S A. 96:10290-10295.
[94] Gardan R, Rapoport G, Debarbouille M (1997) Role of the transcriptional activator RocR in the arginine-degradation pathway of *Bacillus subtilis*. Mol Microbiol. 24:825-837.
[95] Commichau FM, Herzberg C, Tripal P, Valerius O, Stulke J (2007) A regulatory protein-protein interaction governs glutamate biosynthesis in *Bacillus subtilis*: the glutamate dehydrogenase RocG moonlights in controlling the transcription factor GltC. Mol Microbiol. 65:642-654.
[96] Gunka K, Newman JA, Commichau FM, Herzberg C, Rodrigues C, Hewitt L, *et al.* (2010) Functional dissection of a trigger enzyme: mutations of the *Bacillus subtilis* glutamate dehydrogenase RocG that affect differentially its catalytic activity and regulatory properties. J Mol Biol. 400:815-827.
[97] Picossi S, Belitsky BR, Sonenshein AL (2007) Molecular mechanism of the regulation of *Bacillus subtilis gltAB* expression by GltC. J Mol Biol. 365:1298-1313.
[98] Tiffert Y, Supra P, Wurm R, Wohlleben W, Wagner R, Reuther J (2008) The *Streptomyces coelicolor* GlnR regulon: identification of new GlnR targets and evidence for a central role of GlnR in nitrogen metabolism in actinomycetes. Mol Microbiol. 67:861-880.
[99] Wray LV, Jr., Fisher SH (1993) The *Streptomyces coelicolor glnR* gene encodes a protein similar to other bacterial response regulators. Gene. 130:145-150.
[100] Rodriguez-Garcia A, Sola-Landa A, Apel K, Santos-Beneit F, Martin JF (2009) Phosphate control over nitrogen metabolism in *Streptomyces coelicolor*: direct and

indirect negative control of *glnR, glnA, glnII* and *amtB* expression by the response regulator PhoP. Nucleic Acids Res. 37:3230-3242.

[101] Nolden L, Ngouoto-Nkili CE, Bendt AK, Kramer R, Burkovski A (2001) Sensing nitrogen limitation in *Corynebacterium glutamicum*: the role of *glnK* and *glnD*. Mol Microbiol. 42:1281-1295.

[102] Hanssler E, Muller T, Palumbo K, Patek M, Brocker M, Kramer R, *et al.* (2009) A game with many players: control of gdh transcription in *Corynebacterium glutamicum*. J Biotechnol. 142:114-122.

[103] Ikeda M, Nakagawa S (2003) The *Corynebacterium glutamicum* genome: features and impacts on biotechnological processes. Appl Microbiol Biotechnol. 62:99-109.

[104] Kalinowski J, Bathe B, Bartels D, Bischoff N, Bott M, Burkovski A, *et al.* (2003) The complete *Corynebacterium glutamicum* ATCC 13032 genome sequence and its impact on the production of L-aspartate-derived amino acids and vitamins. J Biotechnol. 104:5-25.

[105] Yukawa H, Omumasaba CA, Nonaka H, Kos P, Okai N, Suzuki N, *et al.* (2007) Comparative analysis of the *Corynebacterium glutamicum* group and complete genome sequence of strain R. Microbiology. 153:1042-1058.

[106] Shiio I, Otsuka SI, Takahashi M (1962) Effect of biotin on the bacterial formation of glutamic acid. I. Glutamate formation and cellular premeability of amino acids. J Biochem. 51:56-62.

[107] Duperray F, Jezequel D, Ghazi A, Letellier L, Shechter E (1992) Excretion of glutamate from *Corynebacterium glutamicum* triggered by amine surfactants. Biochim Biophys Acta. 1103:250-258.

[108] Nunheimer TD, Birnbaum J, Ihnen ED, Demain AL (1970) Product inhibition of the fermentative formation of glutamic acid. Appl Microbiol. 20:215-217.

[109] Kimura E, Abe C, Kawahara Y, Nakamatsu T, Tokuda H (1997) A dtsR gene-disrupted mutant of *Brevibacterium lactofermentum* requires fatty acids for growth and efficiently produces L-glutamate in the presence of an excess of biotin. Biochem Biophys Res Commun. 234:157-161.

[110] Delaunay S, Gourdon P, Lapujade P, Mailly E, Oriol E, Engasser JM, *et al.* (1999) An improved temperature triggered process for glutamate production with *Corynebacterium glutamicum*. . Enzyme Microb Technol. 25:762–768.

[111] Radmacher E, Stansen KC, Besra GS, Alderwick LJ, Maughan WN, Hollweg G, *et al.* (2005) Ethambutol, a cell wall inhibitor of *Mycobacterium tuberculosis*, elicits L-glutamate efflux of *Corynebacterium glutamicum*. Microbiology. 151:1359-1368.

[112] Eggeling L, Krumbach K, Sahm H (2001) L-glutamate efflux with *Corynebacterium glutamicum*: why is penicillin treatment or Tween addition doing the same? J Mol Microbiol Biotechnol. 3:67-68.

[113] Nakamura J, Hirano S, Ito H, Wachi M (2007) Mutations of the *Corynebacterium glutamicum* NCgl1221 gene, encoding a mechanosensitive channel homolog, induce L-glutamic acid production. Appl Environ Microbiol. 73:4491-4498.

[114] Sano C (2009) History of glutamate production. Am J Clin Nutr. 90:728S-732S.

[115] Kimura E (2002) Triggering mechanism of L-glutamate overproduction by DtsR1 in coryneform bacteria. J Biosci Bioeng. 94:545-551.

[116] Belisle JT, Vissa VD, Sievert T, Takayama K, Brennan PJ, Besra GS (1997) Role of the major antigen of *Mycobacterium tuberculosis* in cell wall biogenesis. Science. 276:1420-1422.

[117] Hashimoto K, Nakamura K, Kuroda T, Yabe I, Nakamatsu T, Kawasaki H (2010) The protein encoded by NCgl1221 in *Corynebacterium glutamicum* functions as a mechanosensitive channel. Biosci Biotechnol Biochem. 74:2546-2549.

[118] Tesch M, de Graaf AA, Sahm H (1999) In vivo fluxes in the ammonium-assimilatory pathways in *Corynebacterium glutamicum* studied by 15N nuclear magnetic resonance. Appl Environ Microbiol. 65:1099-1109.

[119] Tesch M, Eikmanns BJ, de Graaf AA, Sahm H (1998) Ammonia assimilation in *Corynebacterium glutamicum* and a glutamate dehydrogenase-deficient mutant. Biotechnol. Lett. 20:953–957.

[120] Beckers G, Nolden L, Burkovski A (2001) Glutamate synthase of *Corynebacterium glutamicum* is not essential for glutamate synthesis and is regulated by the nitrogen status. Microbiology. 147:2961-2970.

[121] Sahm H, Eggeling L, de Graaf AA (2000) Pathway analysis and metabolic engineering in *Corynebacterium glutamicum*. Biol Chem. 381:899-910.

[122] Shimizu H, Tanaka H, Nakato A, Nagahisa K, Kimura E, Shioya S (2003) Effects of the changes in enzyme activities on metabolic flux redistribution around the 2-oxoglutarate branch in glutamate production by *Corynebacterium glutamicum*. Bioprocess Biosyst Eng. 25:291-298.

[123] Shirai T, Fujimura K, Furusawa C, Nagahisa K, Shioya S, Shimizu H (2007) Study on roles of anaplerotic pathways in glutamate overproduction of *Corynebacterium glutamicum* by metabolic flux analysis. Microb Cell Fact. 6:19.

[124] Shirai T, Nakato A, Izutani N, Nagahisa K, Shioya S, Kimura E, *et al.* (2005) Comparative study of flux redistribution of metabolic pathway in glutamate production by two coryneform bacteria. Metab Eng. 7:59-69.

[125] Asakura Y, Kimura E, Usuda Y, Kawahara Y, Matsui K, Osumi T, *et al.* (2007) Altered metabolic flux due to deletion of *odhA* causes L-glutamate overproduction in *Corynebacterium glutamicum*. Appl Environ Microbiol. 73:1308-1319.

[126] Sato H, Orishimo K, Shirai T, Hirasawa T, Nagahisa K, Shimizu H, *et al.* (2008) Distinct roles of two anaplerotic pathways in glutamate production induced by biotin limitation in *Corynebacterium glutamicum*. J Biosci Bioeng. 106:51-58.

[127] Bott M (2007) Offering surprises: TCA cycle regulation in *Corynebacterium glutamicum*. Trends Microbiol. 15:417-425.

[128] Schultz C, Niebisch A, Gebel L, Bott M (2007) Glutamate production by *Corynebacterium glutamicum*: dependence on the oxoglutarate dehydrogenase inhibitor protein OdhI and protein kinase PknG. Appl Microbiol Biotechnol. 76:691-700.

[129] Niebisch A, Kabus A, Schultz C, Weil B, Bott M (2006) Corynebacterial protein kinase G controls 2-oxoglutarate dehydrogenase activity via the phosphorylation status of the OdhI protein. J Biol Chem. 281:12300-12307.

[130] Csonka LN, Epstein W. Osmoregulation. In: Neidhardt C, Curtiss R, Ingraham JL, Lin ECC, Low KB, Magasanik B, *et al.*, editors. *Escherichia coli* and *Salmonella*: cellular and molecular biology,. 2nd ed. Washington, DC.: American Society for Microbiology Press,; 1996. p. 1210–1217.

[131] Swearingen PA, O'Sullivan DJ, Warthesen JJ (2001) Isolation, characterization, and influence of native, nonstarter lactic acid bacteria on Cheddar cheese quality. J Dairy Sci. 84:50-59.

[132] Crow V, Curry B, Hayes M (2001) The ecology of non-starter lactic acid bacteria (NSLAB) and their use as adjuncts in New Zealand Cheddar. International Dairy Journal. 11:275-283.

[133] Yvon M Rijnen L (2001) Cheese flavour formation by amino acid catabolism. Int. Dairy J. 11:185-201.

[134] Curtin A, McSweeney P. Catabolism of Amino Acids in Cheese during Ripening. In: Patrick F. Fox PLHM, Timothy M. Cogan and Timothy P. Guinee, editor. Cheese: Chemistry, Physics and Microbiology. 3rd ed: Elsevier; 2004. p. 435-454.

[135] Kieronczyk A, Skeiea S, Langsruda T, Le Barsb D, Yvon M (2004) The nature of aroma compounds produced in a cheese model by glutamate dehydrogenase positive *Lactobacillus* INF15D depends on its relative aminotransferase activities towards the different amino acids. International Dairy J. 14:227-235.

[136] Banks JM, Yvon M, Gripon JC, de la Fuente MA, Brechany EY, Williams AG, *et al.* (2001) Enhancement of amino acid catabolism in Cheddar cheese using α-ketoglutarate: amino acid degradation in relation to volatile compounds and aroma character. Int. Dairy J. 11:215-243.

[137] Shakeel-Ur-Rehman, Fox RE (2002) Effect of added α-ketoglutaratic acid, pyruvic acid or pyridoxal phosphate on proteolysis and quality of Cheddar cheese. Food Chem. 76:21-26.

[138] Tanous C, Chambellon E, Yvon M (2007) Sequence analysis of the mobilizable lactococcal plasmid pGdh442 encoding glutamate dehydrogenase activity. Microbiology. 153:1664-1675.

[139] Morishita T, Yajima M (1995) Incomplete operation of biosynthetic and bioenergetic functions of the citric acid cycle in multiple auxotrophic lactobacilli. Bios. Biotech. Biochem. 59:251-255.

[140] Rijnen L, Courtin P, Gripon JC, Yvon M (2000) Expression of a heterologous glutamate dehydrogenase gene in *Lactococcus lactis* highly improves the conversion of amino acids to aroma compounds. Appl Environ Microbiol. 66:1354-1359.

[141] Tanous C, Chambellon E, Le Bars D, Delespaul G, Yvon M (2006) Glutamate dehydrogenase activity can be transmitted naturally to *Lactococcus lactis* strains to stimulate amino acid conversion to aroma compounds. Appl Environ Microbiol. 72:1402-1409.

[142] Tanous C, Kieronczyk A, Helinck S, Chambellon E, Yvon M (2002) Glutamate dehydrogenase activity: a major criterion for the selection of flavour-producing lactic acid bacteria strains. Antonie Van Leeuwenhoek. 82:271-278.

[143] Helinck S, Le Bars D, Moreau D, Yvon M (2004) Ability of thermophilic lactic acid bacteria to produce aroma compounds from amino acids. Appl Environ Microbiol. 70:3855-3861.

Chapter 13

Amperometric Glucose Sensors for Whole Blood Measurement Based on Dehydrogenase Enzymes

Marco Cardosi and Zuifang Liu

Additional information is available at the end of the chapter

http://dx.doi.org/10.5772/48491

1. Introduction

Self-monitoring blood glucose (SMBG) is an important component of modern therapy for diabetes mellitus. SMBG has been recommended for people with diabetes and their health care professionals in order to achieve a specific level of glycaemic control and to prevent hypoglycaemia. The goal of SMBG is to collect detailed information about blood glucose levels at many time points to enable maintenance of a more constant glucose level by more precise regimens. It can be used to aid in the adjustment of a therapeutic regimen in response to blood glucose values and to help individuals adjust their dietary intake, physical activity, and insulin doses to improve glycaemic control on a day-to-day basis.

SMBG can aid in diabetes control by: (http://www.diabetes.co.uk/blood-glucose/blood-glucose-self-monitoring.html)

- facilitating the development of an individualized blood glucose profile, which can then guide health care professionals in treatment planning for an individualized diabetic regimen;
- giving people with diabetes and their families the ability to make appropriate day-to-day treatment choices in diet and physical activity as well as in insulin or other agents;
- improving patients' recognition of hypoglycaemia or severe hyperglycaemia;
- enhancing patient education and patient empowerment regarding the effects of lifestyle and pharmaceutical intervention on glycaemic control.
- helps to determine which foods or diet are best for one's control
- helps inform the patient and doctor about how well the medication regime is working
- reduces anxiety about, and increases understanding of, hypoglycaemia. This is because if untreated hypoglycaemia can result in coma for diabetic patients and is therefore a

condition that diabetics try to avoid through proper action. To this end, regular testing can predict dangerous drop in blood glucose concentration which can lead to hypoglycaemia.

- is important for undertaking dangerous tasks which could be influenced by high or low blood sugar, such as driving and handling dangerous machinery

Disadvantages are mainly seen when either the patient lacks motivation to test or does not have sufficient education on how to interpret the results to make sufficient use of home testing equipment. Where this is the case, the following disadvantages may outweigh the potential benefits:

- anxiety about one's blood sugar control and state of health
- the physical pain of finger pricking
- expense to the NHS or other medical body

Numerous trials have been carried out to determine the true impact of SMBG on glycaemic control. Some, including randomized, controlled trials, have demonstrated the efficacy of SMBG. Among patients with type 1 diabetes, SMBG has been associated with improved health outcomes. [1] Specifically, increasing frequency of SMBG was linearly correlated with reductions in HbA1c among type 1 patients in Scotland. [2] Among patients with type 2 diabetes, a higher frequency of SMBG was associated with better glycaemic control among insulin-treated patients who were able to adjust their regimen. [3]

SMBG works by having patients perform a number of glucose tests each day or each week. The test most commonly involves pricking a finger with a lancet device to obtain a small blood sample, applying a drop of blood onto a reagent strip (typically an enzyme electrode), and determining the glucose concentration by inserting the strip into an electronic meter for an automated reading. Test results (a measure of the glucose concentration in the blood sample) are then recorded either in a logbook or stored in the glucose meter's electronic memory. People with diabetes can be taught to use their SMBG results to correct any deviations out of a desired target range by changing their carbohydrate intake, exercising, or using more or less insulin.

The frequency with which patients with diabetes should monitor their blood glucose level varies from person to person. Most experts agree that insulin-treated patients should monitor blood glucose at least four times a day, most commonly fasting, before meals, and before bed. In addition, patients using insulin can benefit by obtaining postprandial blood glucose readings to help them more accurately adjust their insulin regimen. A positive correlation between frequency of SMBG and glycaemic control among patients with insulin-treated type 1 or type 2 diabetes has been demonstrated. [1-3] Patients treated with intermediate, short-acting, or rapid-acting insulin may benefit from SMBG data to make adjustments in their regimen.

For patients with type 2 diabetes, optimal SMBG frequency varies depending on the pharmaceutical regimen and whether patients are in an adjustment phase or at their target for glycaemic control. If a patient is on a stable oral regimen with HbA1c concentration within the target range, infrequent SMBG monitoring is appropriate. In such cases, patients can use SMBG data as biofeedback at times of increased stress or changes in diet or physical activity. [4, 5]

It is important to point out however that this debate is not straightforward, however. One factor that is often overlooked is that the numbers obtained by testing are only one part of the picture, which requires additional data to be complete. For example, it is important to relate the number to what and when the patient last ate. The patient's exercise regime must also be considered, as well as when and how much medication has been taken. If this were not complicated enough, the patient (and physician) need to consider factors such as whether the patient has recently been ill, or even subjected to high levels of stress, which can distort the picture.

Blood glucose meters which utilise an enzyme electrode (a term first coined by Updike and Hicks in the late sixties [6]) as the glucose sensing element are particularly suitable medical devices for SMBG.

All the commercial electrochemical meter systems typically comprise two components. The meter - applies potential differences in a programmed sequence to the sensor, collects current data and analyses the current time response of the sensor, records and displays results. The enzyme electrode (or biosensor) test strip - collects the blood sample, the sample undergoes an enzymatic chemical reaction followed by an electrochemical detection step. The patient simply inserts the enzyme electrode into the meter and applies a small drop of blood to the sensor. After a short delay (typically 5 seconds) the blood glucose value is displayed by the meter in mg/dL or mmol/L.

The advantages offered by biosensors in SMBG arise for the following reasons. Blood is a complex fluid and glucose levels vary widely over time in a single patient, many factors besides glucose vary in blood from healthy, patients (haematocrit, oxygen levels, and metabolic by-products) therefore great specificity is a prime requirement. In addition, patients with diabetes may have a wide range of other medical problems creating even greater variation in their blood. Finally, biosensors can be used directly in the blood without requiring major modifications to the biological sample (increased temperature or pressure, dramatic pH changes, addition of highly reactive chemicals, etc).

The enzyme electrodes commonly used in SMBG can be defined as a combination of any electrochemical probe with a thin layer of enzyme based reagent that is selective for glucose (*or other important analytes such as β-hydroxybutyrate*). In these devices, the function of the enzyme is to provide selectivity by virtue of its biological affinity for a particular substrate molecule. For example, an enzyme is capable of catalysing a particular reaction of a given substrate even though other isomers of that substrate or similar substrates may be present.

Typically, the progress of the enzyme reaction (which is related to the concentration of analyte) is monitored by the rate of formation of product or the disappearance of a reactant. If either the product or reactant is electroactive, then the progress of the reaction can be monitored directly using amperometry. In this technique, current flow is measured in response to an applied voltage. The resultant current is monitored by the meter and then interpolated into an accurate measurement of glucose using on-board software algorithms giving the user a concentration value in typically less than 7 seconds.

The final method of analysis used will ultimately depend on several properties of the enzyme. The main considerations are;

1. does the enzyme contain redox active groups
2. are the products of the biochemical reaction electroactive
3. is one of the substrates or cofactors electroactive
4. what is the required speed of response
5. what will be the final application of the sensor

The answer to the first three criteria will depend largely on the system under investigation. The answer to the latter three depends on the requirements and application of the sensor under consideration. If the enzyme does not contain any redox groups, then the method of analysis will be restricted to monitoring either the release of products or the consumption of substrate by their reaction at the transducing electrode. The current produced can then be related to the concentration of analyte. Of particular relevance to this article, are the devices that incorporate nicotinamide adenine dinucleotide (NAD), flavin adenine dinucleotide (FAD) or pyrroloquinoline quinone (PQQ) dependent dehydrogenase enzymes.

2. Electrochemical oxidation of NAD(P)H

Given that the nicotinamide coenzymes are electron carriers, and therefore by definition electroactive, it would appear at first sight (points 2 and 3 above) that these systems would be ideal candidates for commercial enzyme electrode devices. The electrochemical oxidation of NADH to NAD^+ is however both chemically and kinetically complicated at common electrode surfaces such as gold, platinum or glassy carbon.

The electrochemical detection of NADH has generated great interest because the pyridine nucleotides NAD^+ and $NADP^+$ are ubiquitous in all living systems and are required for the reactions of more than 450 oxidoreductases. [7] Although the formal potential of NADH/ NAD^+ couple in neutral pH at 25 °C is estimated to be 0.56 vs. SCE [8, 9] significant over-potential is often required for the direct oxidation of NADH at bare electrodes. [10] Unlike in nature where the oxidation of NADH occurs as a 1-step hydride transfer, on bare electrodes the reaction has been shown to occur via a different and higher energy pathway which produces biologically inactive $NAD^{\cdot}$ radicals as intermediates. The large amount of energy required to produce these intermediates is the origin of the large overpotential (typically 1 Volt) required at bare electrodes. As a result, the direct electrochemical oxidation of NADH has been shown to produce a mixture of products including biologically active NAD^+, $(NAD)_2$ dimers and products from the side reactions of the electrogenerated $NAD^{\cdot}$ radicals.

In addition, the direct oxidation of NADH is often accompanied by electrode fouling due to the polymerisation oxidation products on the electrode surface. [11] The fouling of the electrode surface can occur by two mechanisms. First, the $NAD^{\cdot}$ radicals interact directly with oxide functionalities on the surface of the electrode and second the reaction product, NAD^+, adsorbs onto the electrode surface. Both of these processes are irreversible and result in the gradual blocking of the electrode during continued oxidation. It is this electrode fouling that results in the irreproducibility of the analytical signal from bare electrodes. In addition, if this method were to be used in commercial glucose sensors for SMBG, the high over potentials required would result in the response being a combination of signals from the oxidation of both NADH and common interferents, e.g. ascorbic acid, uric acid, paracetamol etc. found in blood.

The problems mentioned above can be overcome by using small catalytic molecules called mediators. These molecules can transfer charge from the reduced coenzyme directly to the electrode thereby bypassing the direct oxidation of NADH. The use of this approach has three potential advantages when compared to the direct oxidation of the reduced nicotinamide. First, by judicious choice of the mediator problems associated with electrode fouling or competing reactions can be avoided (especially true if he chosen mediator has a site for hydride transfer). Virtually 100% of the oxidation product via this mechanism is biologically active NAD. Second, the rate of electron transfer between the NADH molecule and the mediator can be enhanced resulting in a more stable increased signal. The rate at which NADH is chemically oxidised will governed to a large extent by the potential difference between the formal redox potential of the two species. Finally, a mediator can be selected with a redox potential that will limit the effects of interference. This is particularly relevant in cases where the bio-analyte is present in blood at low concentrations as in the case of beta-hydroxybutyrate (0.0 – 0.3mEq/L). The electrocatalytic oxidation of NADH at a modified electrode is shown schematically in Figure 1. Of particular

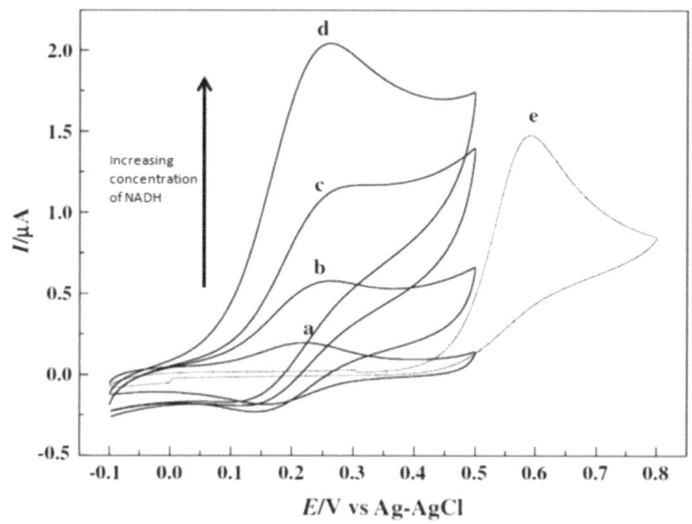

Figure 1. Cyclic voltammograms of the o-AP modified GC electrode in phosphate buffer (a) in the absence and in the presence of increasing concentrations of NADH/mM: (b) 0.1, (c) 0.3 and (d) 0.5; and (e) 0.6 at bare GC in phosphate buffer (0.1 M, pH 7.0). Scan rate = 10 mV/s. Note, the peak oxidation for NADH at the bare electrode occurs at 0.6 Volts. When the electrode is modified, in this case with o-amino phenol the oxidation of NADH is concomitant with the oxidation of the immobilised redox couple (traces b – d). *Figure and data adapted from reference 11.*

interest are voltammograms a and e. These show the direct oxidation of NADH (e) at the glassy carbon electrode and the cyclic voltammogram (a) of the immobilised species. Addition of NADH to the solution causes an increase in 1. Note that the signal due to the oxidation of NADH now occurs at the potential of the mediator resulting in a decrease in the operating voltage of approximately 0.3 V.

Among the mediators used so far are quinones, diimines, ferrocene, thionine oxometalates, polymetallophthalocyanines, ruthenium complexes, pyrroloquinoline quinone, fluorenones, and quinonoid redox dyes such as indamines, phenazines, phenoxazines and phenothiazines.[11] A generalised reaction schematic for the mediated oxidation of NADH is shown below in Figure 2.

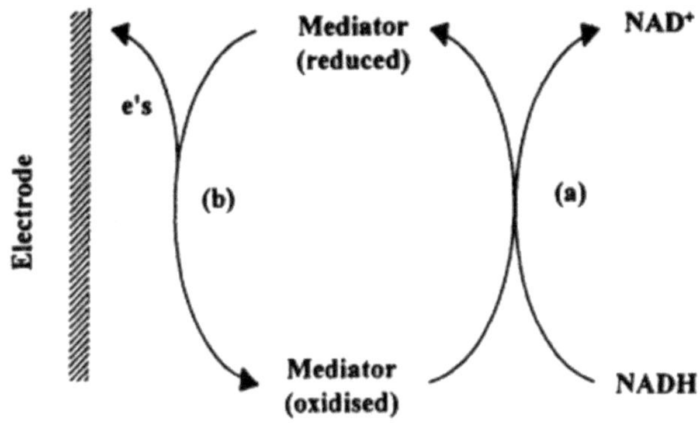

Figure 2. Scheme showing the mediated oxidation of NADH. In this context, the mediator reacts with NADH in a chemical step to oxidise it (a) and is itself reduced. The mediator is then itself reoxidised at the electrode surface (step b). The chemical oxidation of NADH by this mechanism bypasses the problems associated with the direct oxidation at a bare electrode surface.

To design an NADH sensor, the mediator is normally immobilized on the electrode surface or within the electrode material. Mediators can be immobilized by chemisorption by covalent attachment directly electrode surface or by electrochemical polymerisation of the mediators at the electrode surface or, alternatively via covalently attached/physically entrapment in polymers, incorporation in carbon paste grown at electrode surface or deposited on the electrodes by drop coating. The method that is ultimately chosen to produce the modified electrode depends upon the method of mass production (for commercial sensors) and the materials used in the device. For example, in sensors that utilise screen printed carbon based electrodes (*screen-printing technology is a kind of low-cost thick film technology which allows to deposit thick films, a few to hundreds micrometers and is well suited for mass production and portable devices. Such a micro fabrication route offers high-volume production of extremely inexpensive and yet highly reproducible disposable enzyme electrodes – this will be discussed further in the text*) it is convenient to incorporate the mediator directly into the carbon ink particularly if the mediator contains delocalised aromatic rings, as found in quinine and phenoxazine dye based mediators, which form strong chemisorbed bonds with the carbon and graphite plates. The mediator loading, activity stability etc can all be investigated using conventional electrochemical techniques such as DC cyclic voltammetry.

It is important that modified electrodes designed for the reoxidation of NADH are stable over their stated shelf-life, respond only to NADH and not to any other species present in the blood sample satisfy the following kinetic requirements; i) the reaction between the NADH molecule and the mediator is fast, ii) the transfer of charge within the mediating layer is fast and iii) the electron transfer between the reduced mediator and the electrode is also fast.

3. Kinetic modelling of NADH oxidation at chemically modified electrodes.

In the scientific literature, NADH oxidation at chemically modified electrodes is most commonly suggested to occur via a two-step reaction mechanism. In the first step, NADH forms a charge transfer complex with the oxidised form of the mediator bound to the electrode surface. In the second step, electron exchange takes lace and the complex breaks down producing NAD^+ and a reduced mediator site. Because the electrode is polarised, the reduced mediator site is reoxidised in a non-rate limiting electron exchange to the bulk electrode material. This scheme is shown in Figure 3. The important kinetic constants are also represented.

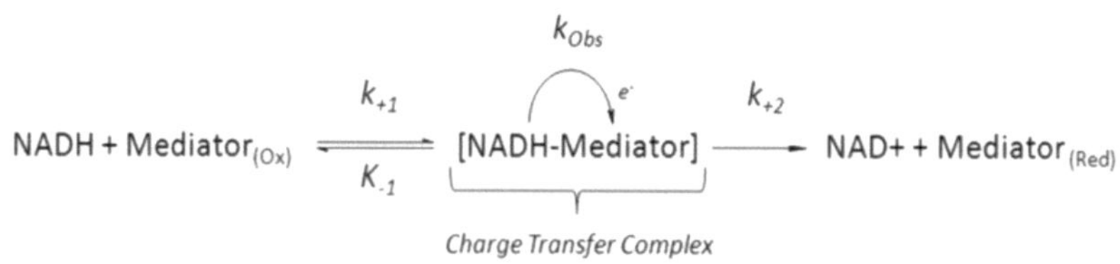

Figure 3. Two-step mechanism commonly proposed for the oxidation of NADH at chemically modified electrodes. The mediator and NADH form a charge transfer complex that dissociates to give rise to the reduced mediator and biologically active NAD^+.

For this type of mechanism, the catalytically limiting current (i_k) observed under controlled hydrodynamic conditions can be expressed as: $i_k = nFAK_{cat}\Gamma C_{NADH}/K_M$ where C_{NADH} is the bulk concentration of NADH and Γ represents the concentration of binding sites in the immobilised film. This type of model assumes that the rate of electron transfer between the mediating species in the film and the NADH is sufficiently fast so as not to be rate limiting, the NADH freely diffuses into the film whereupon it adsorbs to the catalytic site and it undergoes oxidation to NAD^+. Also, the expression for i_k is valid only for thin films where the concentration of NADH is insufficient to saturate the binding sites.[12]

Due to the formation of the charge-transfer complex, this reaction scheme is commonly analysed using Michaelis-Menten kinetics. From figure 3 it is possible to construct the following kinetic argument.

This is now a straight line plot. From the slope of such a plot, values of k_{+2} can be calculated. By extrapolation of zero NADH concentration, i.e. the intercept, values of K_M can be estimated. Values for k_{Obs} can be obtained from Koutecky-Levitch plots under steady state

k_{+1} *and* k_{-1} *represent the formation and dissociation rate constants for the complex*

k_{+2} *represents the catalytic rate constant for the break down of the complex*

K_M *, which is analogous to the Michaelis – Menten constant, is defined as* $K_M = \frac{k_{-1} + k_{+2}}{k_{+1}}$

The second order reaction rate, k_{Obs}, for any given concentration of NADH can thus be expressed as

$$k_{obs} = \frac{k_{+2}}{K_M + [NADH]} \qquad \text{(eq 1)}$$

Inverting eq 1 gives the following expression;

$$\frac{1}{k_{obs}} = \frac{K_M}{k_{+2}} + \frac{[NADH]}{k_{+2}} \qquad \text{(eq 2)}$$

Scheme 1.

oxidation conditions as described by Compton and Hancock.[12] Typically, values of k_{Obs} tend to be in the range 10^{-3} to 10^{-1} cm s^{-1}. Thus it is possible using such laboratory techniques to ensure that the surface coverage of mediator (moles/cm^2) is optimised to achieve favourable measurement linearity and speed of response.

4. Commercial sensors for SMBG incorporating NAD(P)-linked dehydrogenase enzymes

Commercial examples of this type of device are the "Abbott Optium Xceed glucose and β-ketone test strips" (http://www.abbottdiabetescare.com/precision-xtra-blood-glucose-and-ketone-monitoring-system.html). These strips make use of the NAD^+ dependent glucose dehydrogenase (EC 1.1.1.47) and NAD^+ dependent β-hydroxybutyrate dehydrogenase (EC 1.1.1.30). Exogenous NAD^+ is incorporated into the reagent ink which is deposited onto the individual electrode by screen printing. The mediator molecule that is used to recycle the reduced form of the coenzyme is phenanthroline quinone (*Manufacturers own vial insert data sheet*). Advantages often cited with this type of chemistry include good selectivity and no reaction with oxygen.

5. Screen printing as a means of mass-manufacturing enzyme sensors

Screen printing is arguably the most versatile of all printing processes. It can be used to print on a wide variety of substrates, including paper, paperboard, plastics, glass, metals, fabrics, and many other materials including paper, plastics, glass, metals, nylon and cotton. Some common products from the screen printing industry include posters, labels, decals, signs, and all types of textiles and electronic circuit boards. The advantage of screen printing over other print processes is that the press can print on substrates of any shape, thickness and size.

A significant characteristic of screen printing is that a greater thickness of the ink can be applied to the substrate than is possible with other printing techniques. This allows for some very interesting effects that are not possible using other printing methods. Because of the simplicity of the application process, a wider range of inks and dyes are available for use in screen printing than for use in any other printing process. The major chemicals used include screen emulsions, inks, and solvents, surfactants, caustics and oxidizers used in screen reclamation.

Screen printing consists of three elements: the screen which is the image carrier; the squeegee; and ink. The screen printing process uses a porous mesh stretched tightly over a frame made of wood or metal. Proper tension is essential to accurate colour registration. The mesh is made of porous fabric or stainless steel mesh. A stencil is produced on the screen either manually or photochemically. The stencil defines the image to be printed in other printing technologies this would be referred to as the image plate.

Screen printing ink is applied to the substrate by placing the screen over the material. Ink with a paint-like consistency is placed onto the top of the screen. Ink is then forced through the fine mesh openings using a squeegee that is drawn across the screen, applying pressure thereby forcing the ink through the open areas of the screen. Ink will pass through only in areas where no stencil is applied, thus forming an image on the printing substrate. The diameter of the threads and the thread count of the mesh will determine how much ink is deposited onto the substrates. Figure 4 shows an example of an image for an electrochemical cell consisting of a working, a counter and a reference element. This type of structure can be easily produced by screen printing.

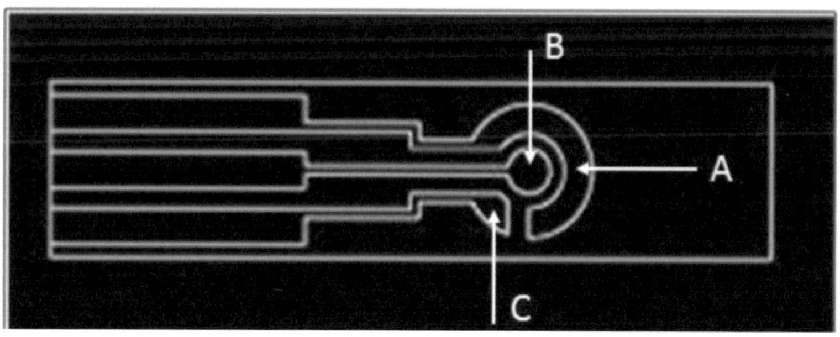

Figure 4. Example of a screen printed electrochemical cell (screen image) showing the Reference element (A), the working electrode (B) and the counter electrode (C). Comercially available screen printed electrodes for research purposes can be obtained from DropSens, Edificio CEEI - Parque Tecnológico de Asturias - 33428 Llanera (Asturias) Spain. http://www.dropsens.com/en/screen_printed_electrodes_pag.html

Many factors such as composition, size and form, angle, pressure, and speed of the blade (squeegee) determine the quality of the impression made by the squeegee. At one time most blades were made from rubber which, however, is prone to wear and edge nicks and has a tendency to warp and distort. While blades continue to be made from rubbers such as neoprene, most are now made from polyurethane which can produce as many as 25,000 impressions without significant degradation of the image.

If the item was printed on a manual or automatic screen press the printed product will be placed on a conveyor belt which carries the item into the drying oven or through the UV curing system. Rotary screen presses feed the material through the drying or curing system automatically. Air drying of certain inks, though rare in the industry, is still sometimes utilized.

The rate of screen printing production was once dictated by the drying rate of the screen print inks. Due to improvements and innovations the production rate has greatly increased. Some specific innovations which affected the production rate and have also increased screen press popularity include:

- Development of automatic presses versus hand operated presses which have comparatively slow production times.
- Improved drying systems which significantly improves production rate.
- Development and improvement of U.V. curable ink technologies
- Development of the rotary screen press which allows continuous operation of the press. This is one of the more recent technology developments.

There are three types of screen printing presses. The flat-bed (probably the most widely used), cylinder, and rotary. Flat-bed and cylinder presses are similar in that both use a flat screen and a three step reciprocating process to perform the printing operation. The screen is first moved into position over the substrate, the squeegee is then pressed against the mesh and drawn over the image area, and then the screen is lifted away from the substrate to complete the process. With a flat-bed press the substrate to be printed is positioned on a horizontal print bed that is parallel to the screen. Rotary screen presses are designed for continuous, high speed web printing. The screens used on rotary screen presses are seamless thin metal cylinders. The open-ended cylinders are capped at both ends and fitted into blocks at the side of the press. During printing, ink is pumped into one end of the cylinder so that a fresh supply is constantly maintained. The squeegee is a free floating steel bar inside the cylinder and squeegee pressure is maintained and adjusted by magnets mounted under the press bed.

Screen printing inks are moderately viscous inks which exhibit different properties when compared to other printing inks such as offset, gravure and flexographic inks though they have similar basic compositions (pigments, solvent carrier, toners, and emulsifiers). There are five different types of screen ink to include solvent, water, and solvent plastisol, water plastisol, and UV curable.

5.1. UV curable inks

UV curable inks consist of liquid pre-polymers, monomers, and initiators which upon being exposed to large doses of U.V. Radiation instantly polymerize the vehicle to a dry, tough thermosetting resin. They also require less energy, overall, to dry or "cure" compared to gas or electric driers. The down side of UV inks is they can cost as much as three times that of regular inks and must be handled differently than conventional inks due to safety issues. Additionally, solvents are required for clean-up which results in some VOC emissions.

5.2. Plastisol Inks

Plastisol inks (both solvent and water based) are used in textile screen printing. Plastisol ink is a PVC (polyvinyl chloride) based system that essentially contains no solvent at all. Along with UV ink used in graphic screen printing, it is referred to as a 100% solid ink system. Plastisol is a thermoplastic ink in that it is necessary to heat the printed ink film to a temperature high enough to cause the molecules of PVC resin and plasticizer to cross-link and thereby solidify, or cure. The temperature at which most plastisol for textile printing cures at is in the range of 149 °C to 166 °C (300 °F to 330 °F). Plastisol inks are commonly used for printing graphics on articles such as tee shirts.

5.3. Solvent inks & water inks

Solvent and water based screen printing inks are formulated with primarily solvent or water. The solvent evaporates and results in VOC emissions. Water based inks, though they contain significantly less, may still emit VOC's from small amounts of solvent and other additives blended into the ink. The liquid waste material may also be considered hazardous waste. Water-based inks are a good choice when a "soft hand" is desirable. Water-based inks also have the advantage of being an excellent ink system for high speed roll-to-roll yardage printing. Such printing is done on large sophisticated equipment that has very large drying (curing) capacity. Finally, because of the fragility of the components used in the manufacture of enzyme electrodes, e.g. enzymes, co-enzymes and mediators water based inks tend to be the ink of choice when formulating the reagent ink component of the device.

Screen (or image transfer) preparation includes a number of steps. First the customer provides the screen printer with objects, photographs, text, ideas, or concepts of what they wish to have printed. The printer must then transfer a "picture" of the artwork (also called "copy") to be printed into an "image" (a picture on film) which can then be processed and eventually used to prepare the screen stencil. Once the artwork is transferred to a positive image that will be chemically processed onto the screen fabric (applying the emulsion or stencil) and eventually mounted onto a screen frame that is then attached to the printing press and production begins. Screen mesh refers to the number of threads per inch of fabric. The more numerous the threads per inch the finer is the screen. Finer mesh will deposit a thinner ink deposit. This is a desirable affect when printing a very fine detail and halftones. Typically a fabric should be 200-260 threads per inch. Water based inks work best on finer mesh. These are generally used in graphic and industrial printing. Course mesh will deposit a heavier ink deposit. This type of screen is used on flatter, open shapes. Typically a course screen mesh will be 160-180 threads per inch. These are generally used in textile printing. An example artwork for a multi-layered screen printed device. Each colour represents a different layer requiring different screens and artworks.

Enzyme electrodes are normally built up in layers using different art works and different inks. The inks range from; conducting carbon and silver inks (to generate the conducting pathways and/or the reference half cell of the device), inks containing the various chemical and biochemical components of the device (enzyme, buffer salts, mediator, stabilisers etc) and inks that are used to define structural components of the device (insulating inks to define the electrode areas and geometries, adhesive inks to provide three dimensional elements such as capillary spacers and coloured inks to provide branding and product identification.) Each of these individual layers normally requires a separate screen and is carried out at a separate printer/dryer station.

The blood monitors (the systems that the user receives) are made up of three main parts: the Optium meter, the Optium Plus blood glucose electrodes, and the Optium blood ß-Ketone electrodes. When the blood sample is applied to the electrode, the glucose or ketone (ß-hydroxybutyrate) in the blood reacts with the chemicals on the electrode. This reaction produces a small electrical current that is measured and the result is displayed by the sensor. Optium Xceed monitors are designed for testing blood obtained from a finger prick, but you can also use it to test blood from other, less painful, sites such as the base of the thumb, forearm or upper arm. A recent study published in *Clinical Chemistry and Laboratory Medicine* [13] concluded that the Optium Blood glucose test strip had a within-run imprecision coefficient of variation (CV) of 4.2%. Good response linearity was found in glucose concentrations in the range 31–444 mg/dL (1.7–24.7 mmol/L). In the concentration range studied, the glucose meter error was 5.14% and the linear regression equation was y = 0.91x +6.2 (r=0.984) against a Modular P clinical analyzer. The Passing-Bablok agreement test indicated good concordance of results. However, for glucose concentrations <100 mg/dL (5.6 mmol/L) (n=69) the error was 6.82% with regression equation y=0.86x+5.9 (r=0.757). Between-lot differences amounted to 0.7%–18.2%. The authors concluded that meter had good precision and accuracy when compared to the laboratory method and met the quality recommendations of the National Committee of Clinical Laboratory Standards (NCCLS, currently the Clinical Laboratory Standards Institute), the National Academy of Clinical Biochemistry (NACB) and the International Organization for Standardization (ISO).

6. Ketone sensors which incorporate a NAD-linked dehydrogenase

In addition to glucose, another important analyte which is of particular relevance to diabetic patients is the ketone body, • -hydroxybutyrate. Diabetic ketoacidosis (DKA), a condition were the level of ketone bodies in the blood are elevated, is a problem that occurs in people with diabetes. It occurs when the body cannot use sugar (glucose) as a fuel source because there is no insulin or not enough insulin. Fat is used for fuel instead. People with type 1 diabetes do not have enough insulin, a hormone the body uses to break down sugar (glucose) in the blood for energy. When glucose is not available, fat is broken down instead.

As fats are broken down, acids called ketones build up in the blood and urine. In high levels, ketones are poisonous. This condition is known as ketoacidosis. Blood glucose levels rise (usually higher than 300 mg/dL) because the liver makes glucose to try to combat the problem. However the cells cannot pull in that glucose without insulin. Diabetic ketoacidosis is often the first sign of type 1 diabetes in people who do not yet have other symptoms. It can also occur in someone who has already been diagnosed with type 1 diabetes. Infection, injury, a serious illness, or surgery can lead to diabetic ketoacidosis in people with type 1 diabetes. Missing doses of insulin can also lead to ketoacidosis in people with diabetes. People with type 2 diabetes can develop ketoacidosis, but it is rare. It is usually triggered by a severe illness. In ketoacidosis, the body fails to adequately regulate ketone production causing such a severe accumulation of keto acids that the pH of the blood is substantially decreased. In extreme cases ketoacidosis can be fatal. [14] Despite considerable advances in diabetes therapy, key epidemiological figures related to DKA remained nearly unchanged during the last decades at a global level. Prevention of DKA – especially in sick day management – relies on intensive self-monitoring of blood glucose and subsequent, appropriate therapy adjustments. Self-monitoring of ketone bodies during hyperglycemia can provide important, complementary information on the metabolic state. Both methods for self-monitoring of ketone bodies at home are clinically reliable and there is no published evidence favouring one method with respect to DKA prevention.

Ketone sensitive test strips can be manufactured using NAD-linked enzyme, β-hydroxybutyrate dehydrogenase and chemically modified electrodes like the ones described above. It is interesting to note however that in the case of this particular enzyme representatives of the common classes of quinoid NADH redox mediator, including Meldola Blue, 4-methyl-1,2-benzoquinone, 1-methoxy phenazine methosulphate and 2,6-dichloroindophenol, were shown to inhibit the NAD-dependent enzyme β-hydroxybutyrate dehydrogenase, severely limiting their utility in the construction of a stable biosensor electrode for the ketone body β-hydroxybutyrate. [14] The authors speculated that this was due to 1,4-nucleophilic addition with enzyme amino acid residues such as cystine present on the enzyme. Consequently, this mode of inhibition is overcome through the use of mediators such as 1, 10-phenanthroline quinone (Figure 5.)

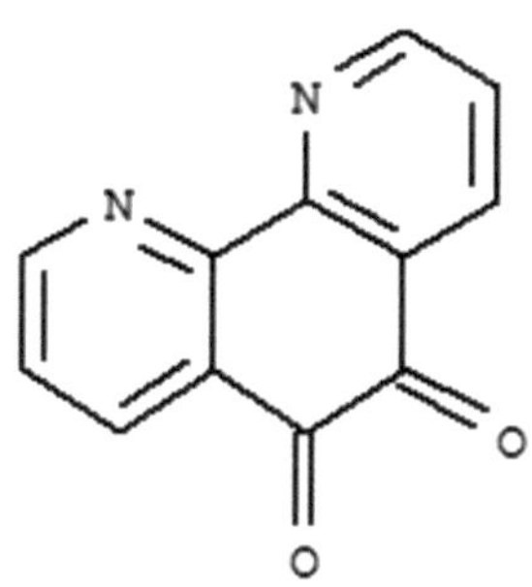

Figure 5. Chemical structure of 1,10 Phenanthroline quinone (oxidised form). The oxidation potential of this molecule is 0.1 Volt vs Ag/AgCl. The electrochemical reduction of 1,10-phenanthroline-5,6-quinone, like other quinones, is reversible and occurs by 2e- transfer in a single step in aqueous solution and by two 1 e-transfer steps in aprotic media.

This technology resulted in the launch of the MediSense® Optium™ β-Ketone electrode. The strip was stable, (≤10% loss in response at 30 °C versus 4 °C) with a long shelf life of 18 months. Diabetics were able to determine their • -hydroxybutyrate level with good precision (0.43 mM 3-OHB, 10.5% CV; 1.08 mM, 5.9%; 3.55 mM, 3.2%; n = 20 per level) and accuracy (versus reference assay: slope = 0.98; intercept = 0.02 mM, r = 0.97, n = 120) over the range 0.0–6.0 mM in 30 s using a small volume of blood (5 μl). The electrode had a low operating potential (+200 mV versus Ag/AgCl) such that the effect of electroactive agents in blood was minimised.[15-18]

7. Alternative approaches to ketone sensing using Diaphorase

The diaphorases (EC1.6.99.3) are a ubiquitous class of flavin-bound enzymes that catalyze the reduction of various dyes which act as hydrogen acceptors from the reduced form of di- and tri- phosphopyridine nucleotides, i.e. NADH, NADPH. They catalyse the following reaction:

$$NADPH + H^{+} + acceptor \longrightarrow NADP^{+} + reduced\ acceptor$$

Either NADH or NADPH may be used as reductants. However, no exchange of hydrogen between the coenzymes is catalysed. Typical acceptor molecules include dyes such as 2, 6- dichlorophenolindophenol and tetrazolium dyes and redox couples such as ferricyanide anions and ferricinium cations. The Expasy entry for diaphorase is http://enzyme.expasy.org/EC/1.6.99.3

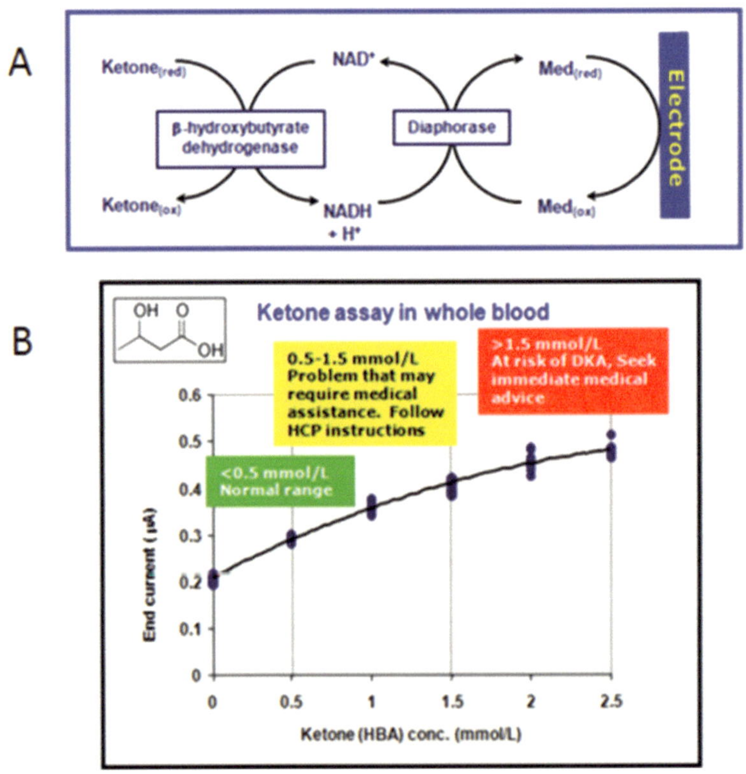

Figure 6. Calibration curve (B) for the diaphorase based ketone sensor developed and manufactured in house (unpublished results). The calibration characteristics are shown over the physiologically relevant range 0 – 2.5 mmol/L β-hydroxybutyrate spiked into a whole blood sample. Each concentration determination is 16 repetitions. The reaction scheme for the sensor is shown in (A). Here, the mediator is the ferricyanide anion. HBA in the blood was measured using the Randox RX Monza Chemistry Analyser, http://www.randox.com/rx%20monza.php. The test time was 7 seconds.

In this configuration, NAD^+, Diaphorase, acceptor molecule and β-hydroxybutyrate dehydrogenase are all formulated together in the enzyme ink and laid down on the test strip using an appropriate manufacturing method. Data generated in-house (unpublished results) using this kind of prototype test strip (with ferricyanide as the acceptor) is shown below for the clinically relevant concentration range of β-hydroxybutyrate. The strip was manufactured by screen printing (*cf* above). Ferricyanide, buffer salts, NAD^+, binders, diaphorase and β-hydroxybutyrate dehydrogenase were mixed into a suitable enzyme ink and printed onto carbon electrodes. The operating voltage was 0.4 Volts.

8. Glucose test strips using PQQ linked and FAD-linked glucose dehydrogenase

PQQ-GDH (pyrrolo quinoline quinone glucose dehydrogenase) belongs to a class of enzymes called quinoproteins which require ortho-quinone cofactors to oxidize a wide variety of alcohols and amines to their corresponding aldehydes and ketones. The soluble quinoprotein Glucose Dehydrogenase, EC 1.1.99.17, (sGDH), systematic name D-glucose: (pyrrolo-quinoline-quinone)1-oxidoreductase, uses pyrrolo-quinoline quinone (PQQ) as a cofactor. sGDH is a strongly basic (pI = 9.5) dimeric enzyme of identical subunits. One monomer (50 kDa, 454 residues) has been reported to bind one PQQ molecule and two Ca(II) ions. One of the Ca(II) ions is required for activation of the cofactor; the other is needed for functional dimerisation of the protein. sGDH oxidizes a wide range of mono-and disaccharides and is able to donate electrons to several neutral or cationic artificial electron acceptors, including short artificial ubiquinone homologues. The natural electron acceptor of PQQ-GDH is ubiquinone although the enzyme will react with a variety of artificial acceptors such as the ferricyanide anion ion and the ferricinum cation. The oxidised form of PQQ can be converted into the reduced form $PQQH_2$ by the transfer of 2 electrons and two protons from the substrate molecule. [19-21]

There are two types of PQQ-GDH enzymes that can be considered for biosensor design. One is intracellular and soluble (sPQQ-GDH) whereas the other molecule is insoluble and firmly bound to the outer surface of the cytoplasmic membrane (mPQQ-GDH). mPQQ-GDH is very selective for glucose but has the disadvantage that it requires extensive solubilisation and stabilisation with detergent molecules[22] . For these reasons, mPQQ-GDH has not been successfully commercialised for biosensor application. It can oxidise a number of monosaccharides, in addition to glucose, such as maltose, mannose and lactose. Consequently, patients that have high levels of, for example maltose in the blood (which could result as a side effect of peritoneal dialysis) or have an inbred genetic disorder resulting in impaired carbohydrate metabolism, would obtain an inaccurate high reading when testing with glucose electrodes incorporating this enzyme. The increasing demand for dialysis and slower growth in capacity for haemodialysis has reinforced the need for an integrated approach to providing dialysis. Peritoneal dialysis is the preferred option for a proportion of patients with end stage renal failure. Peritoneal dialysis fluid usually contains glucose as an osmotic agent to enable water to pass across the peritoneum. Some patients lose the osmotic effect of glucose quickly, but large icodextrin molecules, which are not easily transported across the peritoneal membrane, maintain an osmotic gradient. Icodextrin is not metabolised in the peritoneal cavity, but the polymer can move into the blood stream via the lymphatic system. During systemic circulation, icodextrin is mainly metabolised into maltose which accumulates due to a lack of circulating maltase. It is the accumulation in the systemic circulation of these metabolites of icodextrin that may lead to the disparity between finger stick and formal blood glucose measurement. Maltose interferes with glucose assays that use glucose dehydrogenase with cofactor pyrroloquinolinequinone (PQQ-GDH) leading to falsely increased readings.

Because of the oxygen insensitivity of PQQ-GDH however, there is much commercial interest in producing a mutant form of the enzyme that retains its non-reactivity to oxygen but improving its specificity with respect to D-glucose. Recently, Roche have reported the successful production of a mutant strain of PQQ-GDH which shows no cross-reactivity with maltose. [23]

Finally, it is worth mentioning that PQQ containing proteins lend themselves to an interesting electrode configuration. Whilst flavoproteins such as glucose oxidase exchange electrons with an electrode surface via small molecular weight mediators such as ferrocene, ferricyanide etc, PQQ containing enzymes can exchange electrons via cytochrome b562. This cytochrome will exchange electrons directly with the electrode surface without the need for any mediator molecule. [24] Information on PQQ-GDH can be found on the EXPASY Proteomics server at: http://expasy.org/enzyme/1.1.5.2.

A new enzyme on the market which has just recently become commercially available to the biosensor industry is the flavo-protein FAD-GDH [25] (Toyobo Develops FAD-GDH Enzyme, produced by Aspergillus for SMBG http://www.toyobo.co.jp/e/press/press31072009.htm). As its name suggests, the enzyme catalyses the oxidation of glucose but does not utilise dioxygen as a co-reactant. It can react with a number of artificial electron acceptors such as the ferricyanide anion and the ferricinium cation. The enzyme, which is isolated from *Aspergillus terreus* shows good thermal stability and high selectivity for glucose. Its absorption spectrum is typical of flavoproteins showing two distinctive peaks corresponding to the oxidised flavin cofactor at 465nm and 385 nm. In the presence of glucose the enzyme is bleached and these characteristic absorption bands disappear. [26] According to the data sheet from the manufacturer (Toyobo, Japan) the enzyme has a pH optimum of 7.0 and a KM for glucose of 67.6mM. The information on the EXPASY proteomics server relating to this enzyme is at: http://expasy.org/enzyme/1.1.99.10.

This enzyme is currently used in a number of glucose test strips including the OneTouch Verio family of test strips produced and supplied by LifeScan. The design of the OneTouch Verio BG test strip is shown in Figure 7. The test strip incorporates gold and palladium electrodes which are orientated in a co-facial manner. The dimensions of the two electrodes are defined and controlled during the manufacturing process by a die-punch process. The electrodes are separated from each other by a thin plastic spacer that has a nominal thickness of 95 μm. The glucose-sensitive reagents, citraconate buffer salts, potassium ferricyanide mediator and flavo-protein glucose dehydrogenase (FAD-GDH), are laid down on the 'bottom' palladium electrode. FAD-GDH enzyme was selected for its high substrate-specificity and non-reactivity towards oxygen. The strip may be defined as a 'side-fill strip' because blood may be applied to the 400 nL sample chamber from either the left or the right side of the test strip (Figure 7). The levels of glucose in the sample are determined within 5 seconds, the BG value being shown on the meter display in geographically appropriate units (mg/dl or mM).

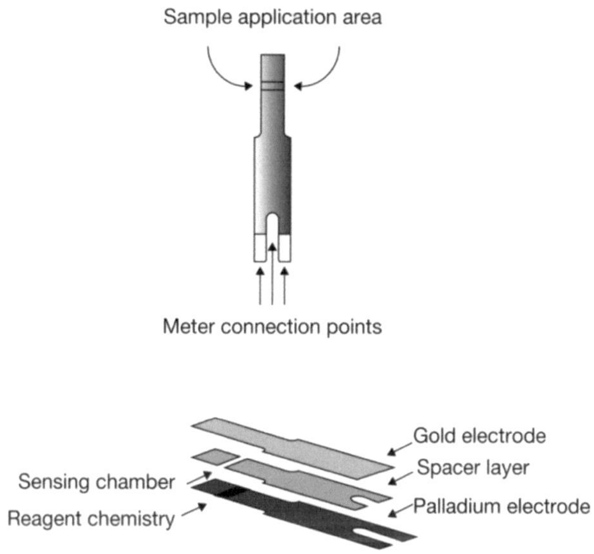

Figure 7. Architecture of the OneTouch Verio glucose test strip. The test strip incorporates gold and palladium electrodes that are orientated in a cofacial manner. The dimensions of the two electrodes are defined and controlled during the manufacturing process by a die-punch process. The electrodes are separated from each other by a thin plastic spacer that has a nominal thickness of 95 μm. The glucose-sensitive reagents, citraconate buffer salts, potassium ferricyanide mediator, and flavoprotein glucose dehydrogenase, are laid down on the "bottom" palladium electrode. The user can apply blood to the test strip either from the right hand side or the left hand side. The glucose level is reported within 5 seconds of the start of the test procedure.

Manufacturing process controls and built-in signal processing compensation mechanisms eliminate the need for user calibration coding, thus reducing the potential for user error. The meter uses a novel multi-pulse signal and has an improved glucose-Hct-temperature-antioxidant compensation algorithm for higher accuracy and precision over a wide range of blood samples. The OneTouch Verio BG test strip is designed to work with meters that are technically equivalent but have different user interfaces. Currently available meters include OneTouch Verio, OneTouch VerioPro, and OneTouch VerioIQ. The performance characteristics of the OneTouch VerioPro BGMS are summarized as follows: plasma equivalent calibration; 0.4 μl sample size; FAD-GDH enzyme; 20–600 mg/dl glucose range; 20–60% Hct range; 10–90% humidity range (non-condensing); 6–44°C temperature range; and up to 3048 m altitude use. A summary of the dehydrogenase enzymes currently used in commercial enzyme electrodes is shown in Table 1.

Enzyme	Reacts with O_2?	Reacts with maltose?
PQQ-Glucose Dehydrogenase	No	Yes (Not Genetically Engineered variant [23])
NAD-Glucose Dehydrogenase	No	No
NAD-β-hydroxybutyrate dehydrogenase	No	No
FAD-Glucose Dehydrogenase	No	No

Table 1. Commonly used dehydrogenase enzymes for Commercial Self testing Biosensors

Shown below are examples of commercially available test strips for home use which incorporate dehydrogenase based enzymes.

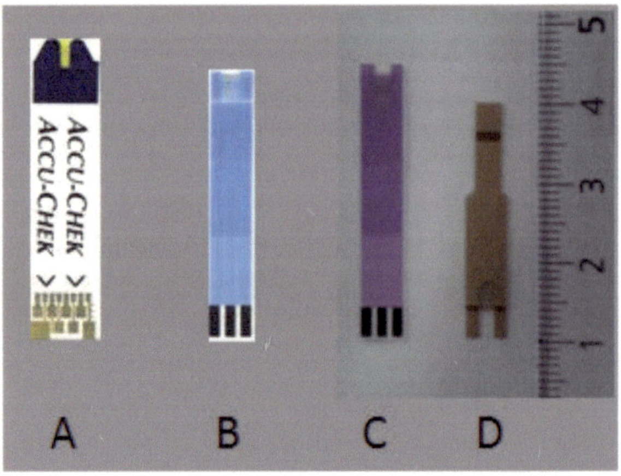

a. The Roche Accu-Check Aviva test strip. This test strip uses PQQ-GDH as the enzyme of choice. http://www.roche.com/products/product-details.htm?type=product&id=2
b. The Abbott Optium Xceed glucose test strip. This strip uses NAD-Glucose dehydrogenase in the sensing chemistry. http://www.abbott-diabetescare.com/AU/ProductDetail.aspx?product=57
c. The Abbott Optium Xceed ketone test strip. This strip uses NAD^+ linked β-hydroxybutyrate dehydrogenase. Because it uses the same mediator chemistry as B, and hence operates at the same potential, the two strips can be used in the same meter. http://www.abbott-diabetescare.com/AU/ProductDetail.aspx?product=57
d. The LifeScan OneTouch Verio test strip. This strip incorporates FAD-GDH in the strip chemistry to measure glucose in whole blood. http://www.lifescan.co.uk/ourproducts/teststrips/onetouchveriopro

(Photograph courtesy of Mr Christopher Leach, LifeScan Scotland Ltd., Inverness, UK).

Figure 8. Examples of commercially available test strips for self-testing utilising dehydrogenases.

9. Biosensors using dehydrogenase enzymes for continuous monitoring

9.1. FAD-GDH biosensor for continuous glucose monitoring

It has been well established that many of the diabetic complications leading to both chronic and acute health problems, such as adult blindness, end-stage renal disease, lower-limb amputations, and heart disease or stroke, can be reduced or even prevented through intensive blood glucose control. [27-29] Effective glycaemic control requires frequent measurements of blood glucose in order to take necessary therapeutic interventions. Such an approach is exemplified by the use of so-called 'continuous glucose monitoring' (CGM) apparatus and methodologies that are used by a growing number of patients to monitor their diabetes condition. Such systems are composed of a probe that is inserted into the body such that it contacts glucose containing liquids in the body, such as interstitial fluid. These CGM systems are designed to operate over extended periods of time, typically over a number of days or longer. In reality CGM is a misnomer, inasmuch as the device samples in an episodic manner, but on a sufficiently high frequency to distinguish such devices from single measurement episodic systems. Nevertheless, compared with episodic self-monitoring blood glucose (SMBG), CGM follows blood glucose dynamics and hence, provides patients and healthcare professionals with not only current blood glucose levels, but also real-time rate and direction of changes. Blood glucose thresholds can be set to alert for possible dangerous trends, for instance rapid blood glucose descents that may lead to hypoglycemia. Advances in CGM research and development are also critical to realize "artificial pancreas" of a closed-loop system in conjunction with an insulin pump.

Enzyme catalysed electrochemical biosensors have been the most successful technology for the commercialized SMBG products (as discussed in the preceding part of this chapter). For CGM, enzymes are also employed as the means of target analyte recognition, coupled to electrochemical transduction methods for determination of the analyte of interest. Correspondingly, such systems have been so far limited to the use of a redox enzyme, such as glucose oxidase (GOx) in which the prosthetic group is intimately associated with the enzyme, such that it cannot diffuse or leach away over the duration of sensor operation.[30] Recently commercialized flavin adenine dinucleotide-dependent glucose dehydrogenase (FAD-GDH, EC 1.1.99.10), where the FAD moiety forms an integral part of the enzyme, has attracted great attention for blood glucose monitoring because of its advantages over GOx in terms of insensitivity to molecular oxygen. However, like many redox enzymes, direct electron transfer between FAD-GDH and electrode surface cannot occur because the active centre FAD is insulated by the large proteins. Therefore, mediators are usually employed to shuttle electron between the FAD moiety and electrode surface. An example of such mediators is potassium ferricyanide used for LifeScan OneTouch Verio family of SMBG products. However, use of small molecular mediators in biosensors faces tremendous technical challenges for CGM which requires essentially all the reactive reagents immobilized without leaching out from the electrodes to achieve long-term measurement stability and to meet biocompatibility requirements, in particular for in-vivo applications. An attractive approach to tackle this challenge is to use a polymeric mediator which has mediator moieties chemically attached to polymer chains. Because of its large molecular size, the polymeric mediator can be co-immobilized with enzyme at electrode by various means, including surface grafting, [31-34] layer-by-layer surface adsorption, [35-37] retention behind semi-permeable dialysis membranes,[38-40] physical entrapment [41-43] or cross-link in hydrogels, [44-47] entrapment in electropolymerized [48,49] or chemically formed layers [50] or in inorganic layers,[51,52] and blend in carbon pastes. [53]

At LifeScan Scotland Limited, we have synthesized a ferrocene polymeric mediator which is a copolymer of vinylferrocene, acrylamide and 2-(diethylamino)ethyl methacrylate.[54] Like small molecular mediators, an effective polymeric mediator for biosensors should be able to exchange electrons with enzyme prosthetic group and then be re-oxidized/re-reduced at electrode in a reversible manor at a sufficient low potential to avoid or minimize interferences resulting from oxidation/reduction of other components in the sample fluid, such as uric acid, ascorbic acid etc. in a bodily fluid. Figure 9 shows cyclic voltammograms of the ferrocene polymeric mediator and FAD-GDH (from Toyobo Co. Ltd., Japan) in phosphate buffered saline (PBS). In the absence of glucose, cyclic voltammetry shows almost symmetric anodic and cathodic peaks which are attributed to oxidation and reduction of the ferrocene moieties of the polymeric mediator. After adding 2.5 mM glucose, the cyclic voltammogram changed dramatically, with a large increase in the anodic peak and a significant decrease in the cathodic peak. This was a typical phenomenon of enzyme-dependent catalytic reduction of ferrocenium generated during the oxidation half-cycle in the presence of glucose. The cathodic peak increased further as glucose concentration increased to 5.0 mM. These results clearly indicated that the ferrocene polymeric mediator exhibited preferential redox properties with respect to transfer of electrons from the reduced enzyme prosthetic group $FADH_2$ to the glassy carbon electrode surface. The ferrocene moieties are reduced to ferrocenium moieties upon accepting the electrons from $FADH_2$ and then are re-oxidized on transfer of electrons to the electrode.

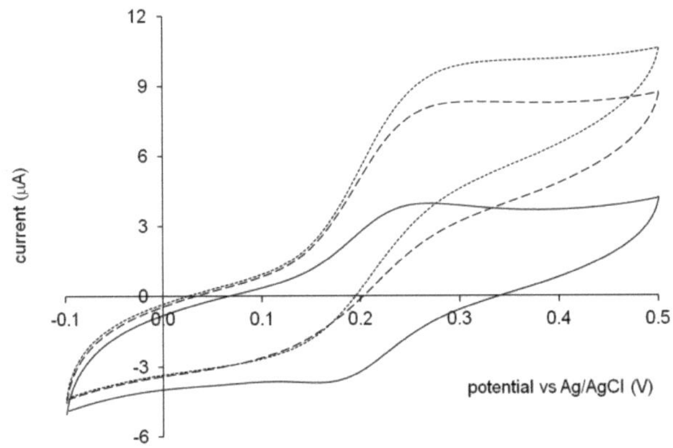

Figure 9. Cyclic voltammograms: 1.5mg/mL FAD-GDH, 9 mg/mL ferrocene polymeric mediator in 0.01M pH7.4 PBS, scan rate 5 mV/s, without (solid line), with 2.5 mM glucose (broken line) & 5.0 mM glucose (dotted line)

We have also developed a technique for co-immobilization of FAD-GDH and the ferrocene polymeric mediator in modified electrodes which were fabricated by screen-printing a water-based carbon ink containing both the enzyme and the ferrocene polymeric mediator. The ink contained graphite particles as conductive pigments and a pH sensitive copolymer as a binder. [55] The copolymer binder was water-soluble in the presence of ammonium hydroxide and hence, the ink was miscible with dissolved FAD-GDH and the ferrocene polymeric mediator during the ink formulation. As ammonia evaporated upon drying a screen-printed ink pad, the binder copolymer became water-insoluble and the ink layer evolved to form a 3-dimensional nano-porous structure which effectively entraps the large molecules of the enzyme and the ferrocene polymeric mediator whilst allows free diffusion of glucose and water molecules. [54] Figure 10 shows scanning electron microscopy image of the electrode surface.

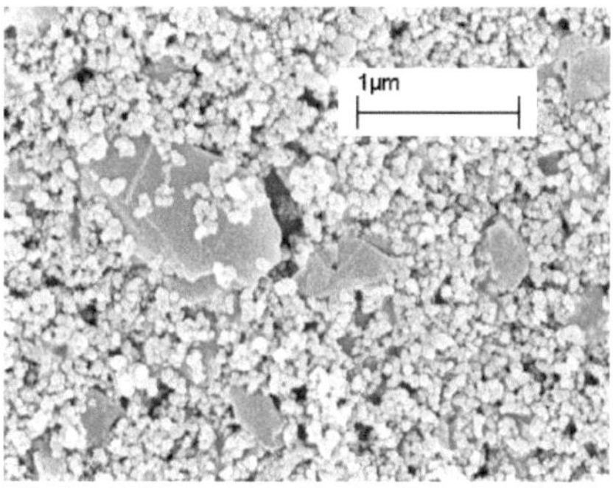

Figure 10. Scanning electrode microscopy image of an electrode with immobilized FAD-GDH and the ferrocene polymeric mediator

The main advantages of these electrodes are

- The enzyme and ferrocene polymeric mediator are co-immobilized in vicinity of the electrode graphite particles, which is beneficial for fast electron transfer.
- The 3-dimensional nano-porous structure enables higher loading of the redox species and provides larger electrode surface area than a planar electrode with an enzyme layer.
- No chemical reactions are involved throughout the immobilization and hence, potential enzyme denaturing is avoided.
- The screen-printing technique suits mass production and has low manufacturing cost.

Because of the presence of ammonium hydroxide, the water-based carbon ink was basic in nature. FAD-GDH stability in the wet ink was studied by comparing glucose sensitivity of the modified electrodes fabricated from the same batch of formulated ink with varied stand-out time. The glucose sensitivity of the electrodes was assessed by testing current responses of a 3-electrode setup to glucose in a concentration range of 0 to 20 mM. The test conditions and results are presented in Figure 11. It is clearly seen that 2-day stand-out of the wet ink significantly reduced glucose calibration slope of the test. During the stand-out period, the wet ink containing FAD-GDH and the polymeric mediator was kept in a fridge at 4 °C and negligible changes in ink rheology and solid content were detected. For comparison, the stand-out was also investigated for an ink using GOx as an enzyme and insignificant effect on glucose calibration slope was found. Therefore, the reduction in glucose sensitivity of the FAD-GDH working electrodes is attributed to decrease in the enzyme activity probably resulting from the enzyme denaturing in the basic wet ink.

Since Clark and Lyons reported the first enzymatic electrode for glucose measurement in 1962, [56] molecular oxygen has been involved in the enzymatic redox reactions as an electron acceptor for the first generation of biosensors. However, this type of biosensors is based on measuring generation of hydrogen peroxide or depletion of oxygen and hence, exhibits

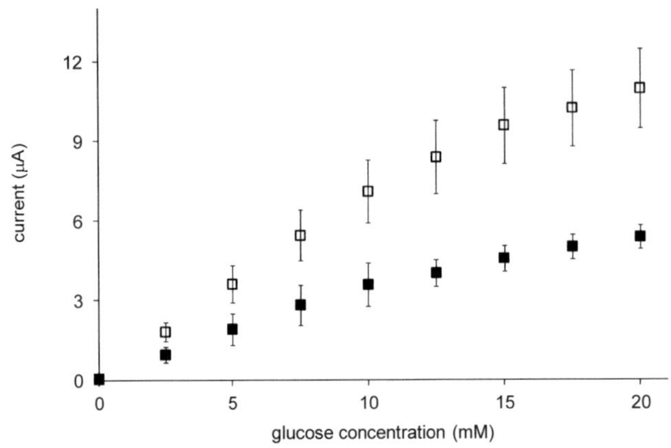

Figure 11. Calibration plots for screen-printed electrodes prepared by using the ink printed after ink formulation without delay (□) and the ink printed 2 days (■) after ink formulation, tested in 0.01M pH7.4 PBS by using an Ag/AgCl reference electrode and a platinum counter electrode, 0.3 V potential, room temperature, the error bars are first standard deviation of 7 electrodes

fundamental limitations. Measurement of hydrogen peroxide requires relatively high operational potential (normally >0.45V) and suffers from significant interference resulting from oxidation of other substances in the bodily fluid. [57, 58] For oxygen measurement, the test result is sensitive to the variations in oxygen supply and test conditions. [59] The second generation biosensors use mediators to shuttle electrons between the enzyme prosthetic groups and electrodes. Typically, mediators have the attractive property of being selected for a particular desirable redox potential at which the mediators readily undergo redox reactions at the electrode whilst the redox reactions of the interferents are insignificant. However, for some enzymes, such as GOx, oxygen can compete with the mediators to accept electrons from the reduced prosthetic group $FADH_2$ to form hydrogen peroxide which cannot be oxidized at the electrode under the applied potential for the mediator re-oxidation. As a result, the biosensor response to the analyte is dampened in the presence of oxygen. As shown in Figure 12, tested in PBS with a 3-electrode setup, the current response to 5 mM glucose for the screen-printed electrode modified with GOx and the ferrocene polymeric mediator decreased almost 60% as blood oxygen content increased from 8 Kpa to 23.14 Kpa. In contrast, the screen-printed electrode modified with FAD-GDH and the ferrocene polymeric mediator had little change in the current response in a range of the blood oxygen content from 9.8 Kpa to 20.1 Kpa.

The oxygen insensitivity of FAD-GDH makes the enzyme very attractive for the development of CGM biosensors. This is because the biosensors using an oxygen-sensitive enzyme in general has more profound response dampening effect at low glucose concentrations due to the generation of hydrogen peroxide, which can contribute to significant accuracy error for the CGM and impose critical challenges for CGM to provide patients and healthcare professionals with a true picture of ambient glycemia profile which is critical for reliable detection and/or prediction of hypoglycemia, an important step to good diabetes management.

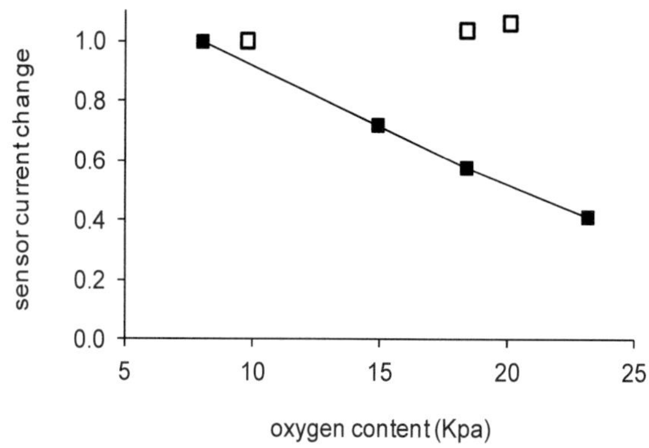

Figure 12. Effect of oxygen content on current response to 5 mM glucose in 0.01M pH7.4 PBS for screen-printed electrodes containing FAD-GHD (□) and GOx (■), tested by using an Ag/AgCl reference electrode and a platinum counter electrode

Like many medical devices, CGM biosensors normally need sterilization, in particular for *in vivo* or *ex-vivo* applications to eliminate any harmful contaminants such as fungi, bacteria, viruses, and spore forms etc. There are different established methods for sterilizing medical devices. Some of them involve the use of high energy means which can cause damages to materials in certain circumstances. It is essential that the key redox species, including FAD-GDH and the ferrocene polymeric mediator in the biosensor are not subject to any significant damages during a chosen sterilization process. Figure 13 shows comparison of glucose calibration plots for the same batch electrodes fabricated by screen-printing the water-based carbon ink containing FAD-GDH and the ferrocene polymeric mediator before and after 25 KGy e-beam sterilization. The results indicate 10-16% reduction in steady state current in the tested range of glucose concentrations. A sample of the ferrocene polymeric mediator also went through the sterilization process alongside the electrodes. ^{1}H nuclear magnetic resonance and size exclusion chromatography indicated no change to the composition and molecular weight of the ferrocene polymeric mediator after the e-beam sterilization. The reduction in the current responses to glucose is probably attributed to e-beam induced decrease in the enzyme activity.

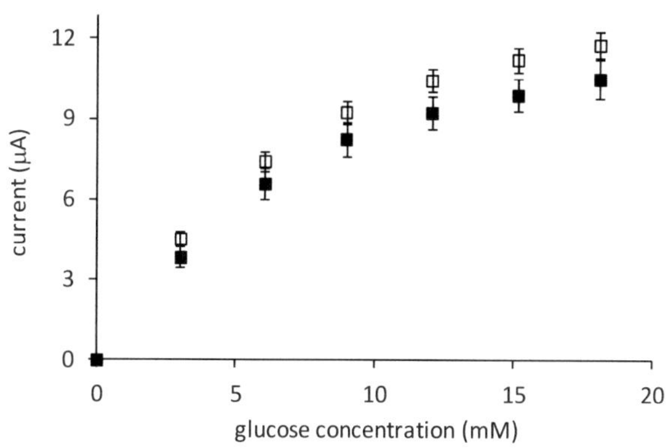

Figure 13. Calibration plots for screen-printed electrodes before (□) and after (■) 25 KGy e-beam sterilization, tested in 0.01M pH7.4 PBS by using an Ag/AgCl reference electrode and a platinum counter electrode, the error bars are first standard deviation of 7 electrodes

CGM biosensors were fabricated by sequentially screen-printing and drying a carbon ink, an Ag/AgCl ink and the water-based carbon ink containing FAD-GDH and the ferrocene polymeric mediator on the same plane of a plastic foil to form two carbon tracks, a pseudo-reference electrode and a working electrode, respectively. Then the plastic foil was heat-laminated onto a plastic disc which had pre-formed micro-channels (typical channel dimension: width x depth = 0.3 x 0.3 mm) in a way that the micro-channel runs over the two electrodes with the working electrode upstream of the pseudo-reference electrode. Surface areas exposed to the micro-channel are 0.3 x 2.0 mm and 0.3 x 5.5 mm for the working electrode and the pseudo-reference electrode, respectively. [54] For comparison, biosensors with GOx were also fabricated in the same way except for replacing FAD-GDH with GOx in the water-based carbon ink. Loadings of the two types of enzymes in the water-based carbon inks were identical in weight.

The two types of biosensor discs were tested side-by-side at room temperature. Human blood samples were collected on the same day for the test. Upon testing, a continuous flow of the human blood samples were pumped from blood sample reservoirs through the channels of each biosensor at 15 μL per minute by using a peristaltic pump (Ismatec). Sensor current responses to blood glucose were recorded by using a potentiostat (Uniscan Instruments) operated at 0.3 V potential. Step changes of blood glucose concentration were realized by switching the blood sample reservoirs which were open to the atmosphere and under gentle magnetic agitation. Low glucose blood samples were prepared by standing the blood sample reservoirs in a 37 °C water-bath to accelerate glucose consumption by blood cells, whilst high glucose blood samples were prepared by spiking with a 20% wt/wt glucose stock (from Sigma-Aldrich). YSI measurements (YSI 2300 STAT Glucose Analyzer) of glucose concentrations were performed for the blood samples as references. In order to minimize error caused by blood mixing in the disc channel upon changing glucose concentration, the YSI measurements were performed by taking a sample from the reservoir when the previous blood sample in the biosensor micro-channel was completely depleted indicated by steady sensor current. Given the slow decrease in blood glucose concentration with time at room temperature, two YSI measurements were normally performed for each blood sample during the period it flowed through the biosensor disc channel.

In a typical test, the two types of biosensors were tested continuously over 54 hours with daily changes of blood samples from different donors. The disc channels were flushed with PBS (0.01 M, pH7.4) prior to the blood changes. Figure 14 shows typical recorded current responses of one FAD-GDH and one GOx biosensors to the same blood samples with varied glucose concentrations in a range of 2.29 to 25.64 mM between 23 and 30 hour test time. YSI measured blood glucose concentrations are also shown. For both biosensors, the currents clearly followed the step-changes in blood glucose concentration and reached "steady-state" in 3 to 5 minutes. It is expected that the real response time of the biosensors to the blood glucose concentration changes is shorter than 3 to 5 minutes because the recorded steady-state currents can only be obtained after complete depletion of the "old" blood samples from the disc channels upon the reservoir switching. By a close look, the current response of the GOx biosensor clearly drifted whilst the current response of FAD-GDH sensor did not between 28.5

and 30 hour test time. This is probably due to the oxygen effect on the GOx biosensor as oxygen content of the blood sample gradually increased under continuous agitation.

The steady-state current responses of the two biosensors over the 54-hour test are plotted against the YSI measured blood glucose concentration in Figure 15. For both of the biosensors, the test results demonstrated good linear correlations between the current responses and blood glucose concentration over the tested glucose concentration range. However, the FAD-GDH biosensor had significantly higher current responses than the GOx biosensor, leading to higher sensitivity to glucose concentration variations and hence, potentially better measurement accuracy. This suggests that FAD-GDH has higher activity than GOx given that the enzyme loadings were the same for the two types of biosensors.

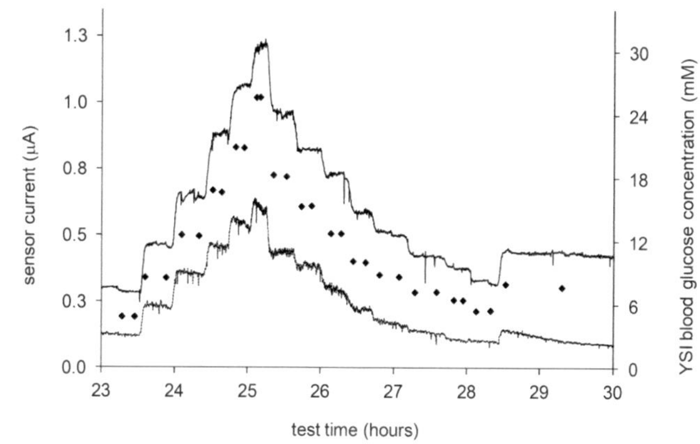

Figure 14. Current response variations of sensors with screen-printed working electrodes containing FAD-GDH (solid line) and GOx (dotted line) and Ag/AgCl reference electrode, with varied blood glucose concentrations (◆), tested with continuous flow of human blood at a flow rate of 15 μL/min

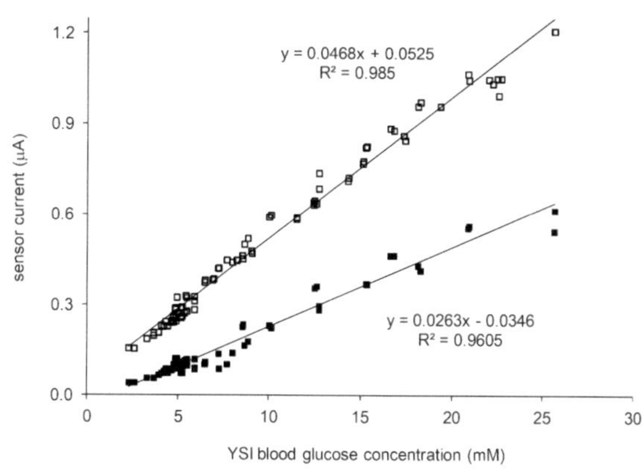

Figure 15. Calibration plots of sensors with screen-printed working electrodes containing FAD-GDH (□) and GOx (■) respectively and Ag/AgCl reference electrode, tested with continuous flow of human blood at a flow rate of 15 μL/min

Because the blood cells continuously consumed glucose and result in gradual decrease in blood glucose concentration at room temperature, direct assessment on measurement stability of the biosensors by recording sensor response change with time at a fixed glucose concentration proved to be unreliable. Nevertheless, closely aligned data points to the linear regression line in Figure 15 illustrate stable biosensor response during the 54-hour test, suggesting good FAD-GDH stability during the continuous measurement.

9.2. NAD and ferrocene polymeric mediator redox couple

As a coenzyme for hundreds of oxidoreductases, NAD has attracted great attention for biosensor development. Because its direct oxidation at electrode surface requires high applied potential and involves side reactions leading to the formation of electrochemically inactive by-products (see the preceding section), NAD is normally coupled with mediators for its application in biosensors. In the literature, ferrocenium and its derivatives have been reported as mediator for oxidation of NADH to NAD^+. [60-63] As mentioned in the preceding section, small molecular mediators are not normally suitable for continuous monitoring biosensors. The effective immobilization of the ferrocene polymeric mediator in the screen-printed carbon electrode developed at LifeScan Scotland Limited holds potentials for development of continuous monitoring biosensors using NAD-dependant oxidoreductases. To investigate electrochemical communication between NAD and the immobilized ferrocene polymeric mediator, 0.3 V potential (ECOChemie PGSTat Autolab, Type III) was applied to an electrochemical cell containing 3.0 mL 0.01M pH7.4 PBS and equipped with a screen-printed carbon electrode (approximately dimension length x width = 10 x 5 mm) with immobilized ferrocene polymeric mediator (about 12.5% wt/wt), an Ag/AgCl reference electrode and a platinum counter electrode to oxidize the ferrocene moieties of the polymeric mediator to ferrocenium moieties. After the current of ferrocene oxidation reached a steady level, PBS in the cell was replaced with 0.3 mM NADH (from Sigma-Aldrich) in 3.0 mL 0.01M pH7.4 and a step increase in current was detected, which indicated oxidation of NADH by the ferrocenium moieties of the polymeric mediator. The NADH oxidation was followed by measuring absorbance of the solution at 340 nm wavelength by UV/vis spectroscopy (Cecil Instruments, CE9500) at different time intervals. The result shown in Figure 16 illustrates gradual decrease in NADH concentration (E_{340} = 6,330) with the oxidation time, which can be attributed to continuous oxidation of NADH by the ferrocene polymeric mediator immobilized in the carbon electrode.

0.65 mL above oxidized NADH solution was collected and mixed with 0.25 mL 99.5% ethanol (EtOH, Sigma-Aldrich) in a UV cuvette (l = 1cm). UV spectra of the solution in the cuvette were recorded before and after addition of 0.4 mg alcohol dehydrogenase (ADH, Sigma-Aldrich). As shown in Figure 17, the addition of EtOH and ADH increased the absorbance peak at around 340 nm to almost the same level as the control solution which contained 0.3 mM NADH in 0.01M pH7.4 PBS and did not subject to the oxidation process. This suggests that NADH oxidation by the immobilized ferrocene polymeric mediator predominantly, if not completely produced electrochemically active NAD^+ that was reduced to NADH by EtOH in the presence of ADH.

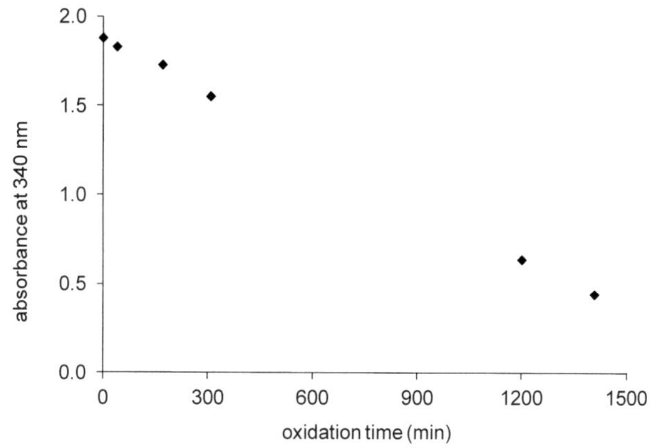

Figure 16. Variation of UV absorbance at 340 nm of 0.3 mM NADH in 0.01 M pH7.4 PBS by carbon electrode with immobilized ferrocene polymeric mediator at 0.3 V potential, by using an Ag/AgCl reference electrode and a platinum counter electrode, at room temperature

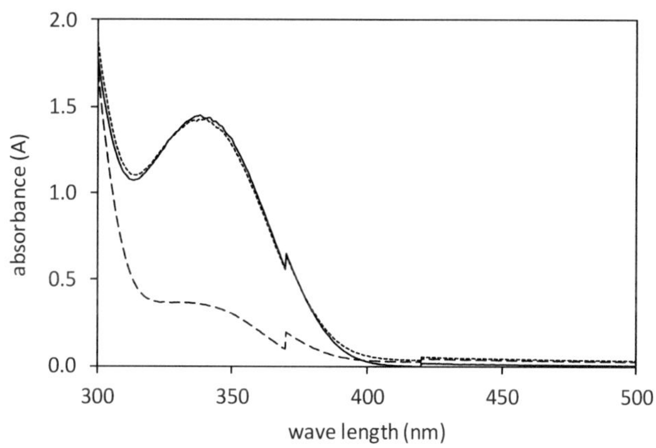

Figure 17. UV spectroscopy monitoring reduction of oxidized NADH in the presence of ADH and EtOH, 0.65ml NADH (0.2 mM) + 0.25ml EtOH (99.5%) + 20ul ADH (control) (solid line), 0.65ml oxidized NADH + 0.15mL EtOH (before reduction) (broken line), 0.65ml oxidized NADH + 0.25ml EtOH + 20 ul ADH (4mg/ml) (dotted line)

For continuous monitoring, use of native NAD as a coenzyme apparently is not ideal due to its low molecular weight and high water-solubility. Direct immobilization of free NAD at electrode has been a long-standing challenge. [64, 65] There are a large number of publications in the literature concerned with NAD retention for various purposes. In the field of biosensors, one promising approach is covalently attaching NAD moieties to polymer chains which can be immobilized at electrode by different means. This can be achieved by directly coupling NAD to an electrode modified with a polymer bearing functional groups,[66] entrapping polymeric NAD in semi-permeable membranes [67, 68] or synthesizing a charged polymeric NAD which was then physically adsorbed at electrode surface in conjunction with counter-charged polymer(s) in a manner of layer-by-layer. [69] At LifeScan Scotland Limited, we intend to develop continuous monitoring biosensors using NAD-dependent enzymes by immobilizing all the redox species, i.e.

an enzyme(s), the ferrocene polymeric mediator and a polymeric NAD in an electrode with the 3-dimentional nano-porous structure (see Figure 10). A polymeric NAD is normally synthesized by two routes. One is coupling NAD or NAD analogue with a polymer bearing functional groups. [66, 70-73] The other route involves synthesis of a NAD monomer and then its copolymerisation with another co-monomer(s). One of the challenges for NAD immobilization is to keep NAD coenzymic activity while achieving effective retention at electrode. Yamazaki *et al* synthesized three NAD monomers (N^6-[N-(6-methacrylamidohexyl)carbamoylmethyl]-, N^6[N-[2-[N-(2-methacrylamidoethyl)carbamoyl]ethyl] carbamoylmethyl]-, and N^6-[N-[N-(2-hydroxy-3-methacrylamidopropyl) carbamoylmethyl] carbamoylmethyl]-NAD) and then copolymerized them with various co-monomers (acrylamide, N-(2-hydroxyethyl)-, N-ethyl-, N,N-diethyl-, and N,N-dimethylacrylamide, acrylic acid, and 6-methacrylamidohexylammonium) by free radical polymerisation to form a series of polymeric NADs. [74] Their studies revealed that hydrophilicity and length of the spacers linking NAD moieties and the polymer backbone had the most important effects on coenzymic activity of the polymeric NADs. This suggests that keeping mobility of the NAD moieties covalently attached to a polymer chain is critical to NAD coenzymic activity. Chemically modification to native NAD is required to tailor chemical properties of the spacers. N^6-amino group on the adenine ring of NAD (see Figure 18) is normally selected as the site for this purpose. [75, 76] Lindberg *et al* reported alkylation of NAD^+ with iodoacetic acid followed by alkaline rearrangement to give N^6-carboxymethyl-NAD^+.[75] However, the reaction between NAD^+ and iodoacetic acid took 10 days in the dark at room temperature. We successfully synthesized N^6-carboxymethyl-NAD^+ by a modified method with a dramatic reduction in the reaction time.

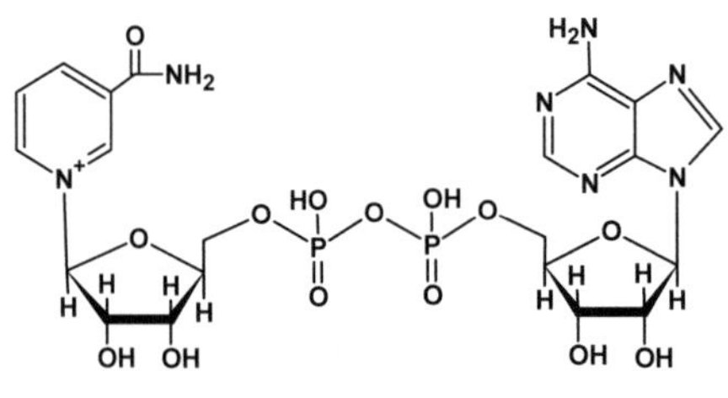

Figure 18. Chemical structure of NAD^+

9.4. Synthesis of N^1-carboxymethyl-NAD^+ (compound 1 in Figure 19)

1.0g NAD^+ (1.51mmol) was dissolved in 3.5mL 0.1M pH 7.0 sodium phosphate buffer in a 5ml Biotage microwave reaction tube. Then, 1.5 g (8.06mmol, 5.34eq) iodoacetic acid was added and pH was adjusted to 7.0 by using 5.0M NaOH aqueous solution. The reaction vessel was sealed and the mixture was heated to 50°C for 10 minutes by using microwave irradiation. After that, the pink solution (c.a. 5.0 mL) was acidified to pH3.0 using 5M HCl aqueous solution before being poured into 25 mL pre-cooled (-5°C) mixture of acetone/IMS (1:1). The resulting precipitate was filtered, washed first with 5.0 mL IMS, then 15 mL dry diethyl ether before air drying under dry nitrogen for 10 minutes. Further drying overnight in a desiccator over fused $CaCl_2$ gave 1.62 g crude N^1-carboxymethyl-NAD^+ as a pink amorphous solid.

9.5. Synthesis of N^6-carboxymethyl-NADH (compound 2 in Figure 19)

9.1g (c.a. 10.57mmol) above prepared crude N^1-carboxymethyl-NAD^+ was dissolved in 1.3% w/v $NaHCO_3$ in 450 mL aqueous solution and the solution was deoxygenated by sparging with nitrogen for 10 minutes. 3.5 g (20.1mmol) sodium dithionite was added in one portion and the mixture was stirred at ambient temperature to affect reduction of the nicotinamide moiety. After 1.0 hour, the solution colour changed from pink to yellow. The solution was then sparged with air for 10 minutes to destroy any excess dithionite and the pH was brought to 11.0 by using 5M NaOH aqueous solution. The mixture was heated at 70°C for 90 minutes, to promote Dimroth rearrangement to N^6-carboxymethyl-NADH, before cooling to 25°C. Thin-layer chromatography (silica gel, isobutryic acid/water/32% NH_4OH (aq), 66/33/1.5 by volume) showed no evidence for the presence of N^1-carboxymethyl-NADH.

9.6. Oxidation of N^6-carboxymethyl-NADH to N^6-carboxymethyl-NAD^+ (compound 3 in Figure 19)

The reaction mixture containing N^6-carboxymethyl-NADH was treated with 17.5 mL 3M Tris buffer (pH7.0) and the pH was adjusted to 7.5 using 5M HCl aqueous solution. 3.5 mL acetaldehyde (62.6mmol) was added, immediately followed by 10.5 mg yeast alcohol dehydrogenase (*saccharomyces cerevisiae*) (~300U/mg) before allowing agitating at ambient temperature to deoxidize the nicotinamide moiety. After 18 hours, the reaction mixture (c.a. 485 mL) was concentrated *in vacuo* (30°C/10-15bar) to approximately 1/3 volume and poured into 1800 mL pre-cooled (-5°C) mixture of acetone/IMS (1:1). The fine slurry was left to age for 18 hours at 3°C. The resulting precipitate was collected by centrifugation and washed on a glass sinter with 40 mL IMS then 120 mL dry diethyl ether before air-drying under dry nitrogen for 10 minutes. Further drying overnight in a desiccator over fused $CaCl_2$ afforded 3.99g crude N^6-carboxymethyl-NAD^+ as a tan coloured hygroscopic solid.

1.0 g of the above-prepared crude N^6-carboxymethyl-NAD^+ was taken up in 20 mL water and passed through a Sephadex G10 gel filtration column (2x10cm, 20 mL). All eluted fractions containing UV active material were combined (60 mL total volume) and added to a column of Dowex 1-X2 ion exchange resin (Cl^-, 4x50cm, 200 mL) which had been pre-equilibrated with water. A linear gradient of 0-50 mM LiCl (buffered to pH 3.0), at 10 mL per minute over 65 minutes, was applied using *"Presearch Combiflash Companion"* chromatography equipment. The fractions eluted between 25-35 mM were combined (c.a. 100 mL), neutralized to pH 7.0 with 5M LiOH and evaporated to approximately 1/3 volume and poured into 300 mL pre-cooled (-5°C) mixture of acetone/IMS (1:1). The fine slurry was left to age for 18 hours at 3°C. The resulting precipitate was collected by centrifugation and washed on a glass sinter with 30 mL IMS then 50 mL dry diethyl ether before air-drying under dry nitrogen for 10 minutes. Further drying overnight in a desiccator over fused $CaCl_2$ afforded 0.307g purified N^6-carboxymethyl-NAD^+ as a cream coloured hygroscopic solid.

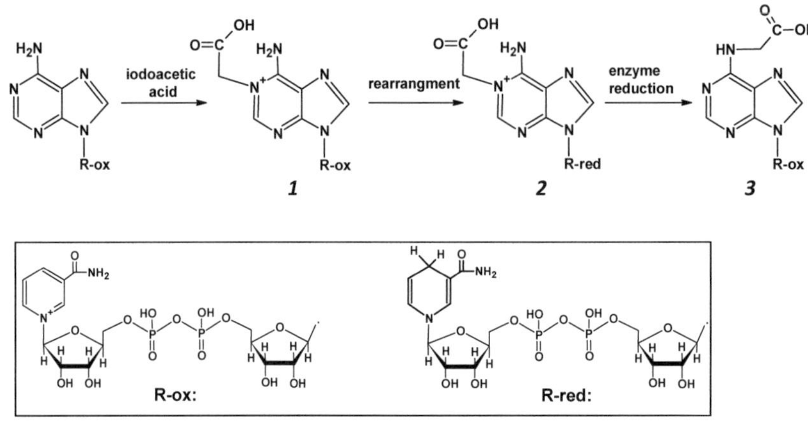

Figure 19. Synthesis of N[6]-carboxymethyl-NAD+

The synthesized N^6-carboxymethyl-NAD^+ is an important intermediate for NAD immobilization at electrodes for continuous monitoring biosensors. An extension of this work could involve synthesis of various polymeric NADs with tailor-made chemical properties to meet biosensor requirements for continuous monitoring of different analytes.

10. Conclusions

In this article we have outlined various strategies for the electrochemical exploitation of dehydrogenase enzymes in sensor devices. The techniques used ultimately depend upon the class of dehydrogenase enzyme used. For enzymes that are $NAD(P)^+$ linked it is essential to develop a base transducer (a modified electrode) that efficiently reoxidises the reduced coenzyme. Alternatively, sensing schemes can be designed which utilize Diaphorase thereby facilitating the biochemical oxidation of reduced coenzyme. With FAD and PQQ dependent enzymes the most successful strategy has been to utilize mediator molecules such as the ferricyanide anion to couple the enzymatic activity to the electrode. Although not yet exploited commercially, dehydrogenase enzymes could also have a role in continuous monitors. With the NAD^+ dependent enzymes there is the additional complication of immobilizing the chemical components of the sensor to prevent drift in the device over time. This also includes the coenzyme molecule and we have illustrated how the synthesis of the N^6-carboxymethyl derivative of NAD can be an important intermediate in achieving this. Attachment of suitable ligands at this position, with sufficient flexibility, should allow the development of stable reagents which will facilitate the development of continuous devices.

The electrochemistry of NADH oxidation has been well researched over the last 30 years. Also, over 250 enzymes use this ubiquitous coenzyme so schemes which utilize NAD dependent enzymes should allow for the measurement of a range of analytes in blood. In spite of this, it is interesting to note that with the exception of Abbott, none of the other major biosensor manufacturers have embraced this technology. The reason for this could be due to the fact that glucose SMBG is still the largest biosensor market worldwide and it is adequately served by enzymes such as glucose oxidase, FAD-GDH and PQQ-GDH, none of which require any exogenous coenzyme to function. This makes the manufacture of the sensor relatively straight forward compared to those that require a modified electrode, an enzyme and in addition exogenous coenzyme added to the enzyme ink. In addition, switching to a NAD-dehydrogenase system may be incompatible with the manufacturing equipment currently used by manufacturers and thus prevent the adoption of this technology.

Author details

Marco Cardosi and Zuifang Liu
LifeScan Scotland Limited, a Johnson & Johnson Company, UK

11. References

[1] Evans JMM, Newton RW, Ruta DA, MacDonald TM, Stevenson RJ, Morris AD (1999) Frequency of blood glucose monitoring in relation to glycemic control: observational study with diabetes database. BMJ 319:83–86,
[2] Benjamin EM (2002) Self-Monitoring of Blood Glucose: The Basics. Clinical Diabetes January. 20:45-47
[3] Franciosi M, Pellegrini F, De Bernardis G, Belfiglio M, Nicolucci A (2001) The impact of blood glucose self-monitoring on metabolic control and quality of life in type 2 diabetic patients. Diabetes Care. 24:1870–1877
[4] Diabetes Education Study Group of the European Association for the Study of Diabetes (2003) Type 2 diabetes patient education basics. Blood glucose monitoring: a must in diabetes management.
[5] Peel E (2004) Blood glucose self-monitoring in non-insulin treated Type 2 diabetes: a qualitative study of patients' perspectives. British Journal of General Practice. 54:183-8
[6] Updike SJ, Hicks GP (1967) The Enzyme Electrode. Nature. 214:986 - 988
[7] White HB (1982) Evolution of coenzymes and the origin of pyridine nucleotides, in: Everse J, Anderson B, You K-S, editors. The Pyridine Nucleotide Cofactors, Academic Press, New York, 1982, pp. 1 -17
[8] Rodkey FL (1955) Oxidation-reduction potentials of the diphosphopyridine nucleotide system, J. Biol. Chem., 213:777–786
[9] Rodkey FL (1959) The effect of temperature on the oxidation-reduction potential of the diphosphopyridine nucleotide system. J. Biol. Chem. 234:188 - 90
[10] Blaedal WJ, Jenkins RA (1975) Study of the electrochemical oxidation of reduced nicotinamide adenine dinucleotide. Anal. Chem. 47:1337–1343
[11] Nassef HM, Radi A-E, O'Sullivan CK (2006) Electrocatalytic sensing of NADH on a glassy carbon electrode modified with electrografted o-aminophenol film Electrochemistry Communications. 8:1719–1725
[12] Compton RG, Hancock G (1999) Comprehensive Chemical Kinetics, Vol 37, Application of Kinetic modelling (eds), Elsevier, pp.49 -85

[13] Bogdan S, Jerzy N, Wojciec G (2008) Analytical evaluation of the Optium Xido blood glucose meter. Clinical Chemistry and Laboratory Medicine. Vol 46, Issue 1, pp.143–147
[14] Eisenbarth GS, Polonsky KS, Buse JB (2008) Type 1 Diabetes Mellitus. In: Kronenberg HM, Melmed S, Polonsky KS, Larsen PR. Kronenberg: Williams Textbook of Endocrinology. 11th ed. Philadelphia, Pa: Saunders Elsevier; chapter 31, S13 – S61
[15] Forrow NJ, Sanghera GS, Walters SJ, Watkin JL (2005) Development of a commercial amperometric biosensor electrode for the ketone β-3-hydroxybutyrate. Biosenors and Bioelectronics. 20:1617-1625
[16] Vanelli M (2003) The direct measurement of 3-beta-hydroxybutyrate enhances the management of diabetic ketoacidosis in children and reduces time and costs of treatment. Diabetes Nutrition and Metabolism. 16:312-316
[17] Guerci B (2003) Accuracy of an electrochemical sensor for measuring capillary blood ketones by fingerstick samples during metabolic deterioration after continuous subcutaneous insulin infusion interruption in type 1 diabetic patients. Diabetes Care. 26:1137-1141
[18] Bektas F (2004) Point of care blood ketone testing of diabetic patients in the emergency department. Endocrine Research. 30:395-402
[19] Duine JA, Frank J, van Zeeland JK (1979) Glucose dehydrogenase from Acinetobacter calcoaceticus: a 'quinoprotein'". FEBS. Lett. 108:443 – 446 Yamada M, Sumi K, Matsushita K, Adachi O, Yamada Y (1993) Topological analysis of quinoprotein glucose dehydrogenase in Escherichia coli and its ubiquinone-binding site. J. Biol. Chem. 268:12812 – 12817
[20] Yamada M, Sumi K, Matsushita K, Adachi O, Yamada Y (1993) Topological analysis of quinoprotein glucose dehydrogenase in Escherichia coli and its ubiquinone-binding site. J. Biol. Chem. 268:12812 –12817
[21] Oubrie A, Rozeboom HJ, Dijkstra BW (1999) Active-site structure of the soluble quinoprotein glucose dehydrogenase complexed with methylhydrazine: a covalent cofactor-inhibitor complex". Proc. Natl. Acad. Sci. U. S. A. 96:11787–11791
[22] Dewanti AR, Duine JA (1998). Reconstitution of membrane-integrated quinoprotein glucose dehydrogenase apoenzyme with PQQ and the holoenzyme's mechanism of action. Biochemistry. 37: 6810–6818
[23] Yamaoka H, Sode K (2007) SPCE Based Glucose Sensor Employing Novel Thermostable Glucose Dehydrogenase, FADGDH: Blood Glucose Measurement with 150nL Sample in One Second J Diabetes Sci Technol. 1:28–35
[24] Ramanavicius K, Habermüller E, Laurinavicius V, Schuhmann W (1999) Polypyrrole entrapped quinohemoprotein alcohol dehydrogenase. Evidence for direct electron transfer via conducting polymer chains. Anal. Chem. 71: 3581–3586
[25] Fraeyman A, Claeys G, Zaman Z (2010) Effect of non-glucose sugars and haematocrit on glucose measurements with Roche Accu-Chek Performa glucose strips Ann Clin Biochem September. 47:494—496
[26] Tsujimura S, Kojima S, Ikeda T, Kanon K (2009) Potential-step coulometry of D-glucose using a novel FAD-dependent glucose dehydrogenase. Analytical and Bioanalytical Chemistry. 386:645-651

[27] DCCT (1993) The effect of intensive treatment of diabetes on the development and progression of long-term complications in insulin-dependent diabetes mellitus. New England Journal of Medicine. 329:977-986

[28] DDCI/EDCI (2005) Intensive diabetes treatment and cardiovascular disease in patients with type 1 diabetes. New England Journal of Medicine, 353:2643-2653

[29] American diabetes association (2002) Implications of the United Kingdom prospective diabetes study. Diabetes Care. 25:528-532

[30] Liu, Z, Cardosi M (2011) Electrochemical measurement of blood glucose for diabetic patients. In: Berhardt, Leon V editor, Advances in Medicine and Biology, vol 25, pp.1-39

[31] Raitman OA, Katz E, Buckmann AF, Willner I (2002) Integration of Polyaniline/Poly(acrylic acid) Films and Redox Enzymes on Electrode Supports: An in Situ Electrochemical/Surface Plasmon Resonance Study of the Bioelectrocatalyzed Oxidation of Glucose or Lactate in the Integrated Bioelectrocatalytic Systems, J. American Chemical Society. 124:6487-6496

[32] Yoon HC, Hong M-Y, Kim H-S (2000) Functionalization of a Poly(amidoamine) Dendrimer with Ferrocenyls and Its Application to the Construction of a Reagentless Enzyme Electrode, Anal. Chem. 72:4420-4427

[33] Zhang S, Yang W, Niu Y, Sun C (2004) Multilayered construction of glucose oxidase and poly(allylamine)ferrocene on gold electrodes by means of layer-by-layer covalent attachment, Sensors and Actuators B. 101:387–393

[34] Boguslavsky L, Kalash H, Xu Z, Beckles D, Geng L, Skotheim T, Laurinavicius V, Lee HS (1995) Thin film bienzyme amperometric biosensors based on polymeric redox mediators with electrostatic bipolar protecting layer, Analytica Chimica Acta. 311:15-21

[35] Hoshi T, Sagae N, Daikuhara K, Takahara K, Anzai J (2007) Multilayer membranes via layer-by-layer deposition of glucose oxidase and Au nanoparticles on a Pt electrode for glucose sensing, Materials Science and Engineering: C. 27:890-894

[36] Zheng H, Okada H, Nojima S, Suye S, Hori T (2004) Layer-by-layer assembly of enzymes and polymerized mediator on electrode surface by electrostatic adsorption, Science and Technology of Advanced Materials. 5:371–376

[37] Gao Q, Guo Y, Zhang W, Qi H, Zhang C (2011) An amperometric glucose biosensor based on layer-by-layer GOx-SWCNT conjugate/redox polymer multilayer on a screen-printed carbon electrode, Sensors and Actuators B. 153:219–225

[38] Bohm S, Pijanowska D, Olthuis W, Bergveld P (2001) A flow-through amperometric sensor based on dialysis tubing and free enzyme reactors, Biosensors & Bioelectronics. 16:391–397

[39] Higson SPJ, Vadgama PM (1993) Diamond-like carbon coated microporous polycarbonate as a composite barrier for a glucose enzyme electrode, Anal. Chim. Acta, 271:125-133

[40] Eisele S, Ammon HPT, Kindevater R, Grijbe A, Gopel W (1994) Optimized biosensor for whole blood measurements using a new cellulose based membrane, Biosens & Bioelectron. 9:119-124

[41] Jiménez C, Bartrolí J, Rooij NF de, Koudelka-Hep M (1995) Glucose sensor based on an amperometric microelectrode with a photopolymerizable enzyme membrane, Sensors and Actuators B: Chemical. 27:421-424
[42] Chen X, Hu Y, Wilson GS (2002) Glucose microbiosensor based on alumina sol–gel matrix/electropolymerized composite membrane, Biosensors & Bioelectronics. 17:1005-1013
[43] Vilkanauskyte A, Erichsen T, Marcinkeviciene L, Laurinavicius V, Schuhmann W (2002) Reagentless biosensors based on co-entrapment of a soluble redox polymer and an enzyme within an electrochemically deposited polymer film, Biosensors & Bioelectronics. 17:1025-1031
[44] Heller A (2006) Electron-conducting redox hydrogels: design, characteristics and synthesis, Current Opinion in Chemical Biology. 10:664–672
[45] Himuro Y, Takai M, Ishihara K (2009) Poly(vinylferrocene-co-2-hydroxyethyl methacrylate) mediator as immobilized enzyme membrane for the fabrication of amperometric glucose sensor, Sensors and Actuators B: Chemical, 136:122-127
[46] Feldman B, Brazg R, Schwartz S, Weinstein R (2003) A Continuous Glucose Sensor Based on Wired Enzyme™ Technology-Results from a 3-Day Trial in Patients with Type 1 Diabetes, Diabetes Technology & Therapeutics. 5:769-779
[47] Kan S, Bae YH (2003) A sulfonamide based glucose-responsive hydrogel with covalently immobilized glucose oxidase and catalase. J of Controlled Release. 86:115-121
[48] Lange U, Roznyatovskaya NV, Mirsky VM (2008) Conducting polymers in chemical sensors and arrays , analytica chimica acta. 614 :1–26
[49] Pan D, Chen J, Yao S, Nie L, Xia J, Tao W (2005) Amperometric glucose biosensor based on immobilization of glucose oxidase in electropolymerized o-aminophenol film at copper-modified gold electrode, Sensors and Actuators B. 104:68–74
[50] Njagi J, Andreescu S (2007) Stable enzyme biosensors based on chemically synthesized Au–polypyrrole nanocomposites, Biosensors and Bioelectronics. 23:168–175
[51] Cosnier S, Le Lous K (1996) A new strategy for the construction of amperometric dehydrogenase electrodes based on laponite gel-methylene blue polymer as the host matrix. J. Electroanal. Chem. 406:243– 246
[52] Mousty C (2004) Sensors and biosensors based on clay-modified electrodes—new trends. Applied Clay Science. 27.159– 177
[53] Saito T, Watanabe M (1998) Poly(vinylferrocene-co-hydroxyethyl methacrylate) for use as electron mediator in enzymatic glucose sensor, Reactive & Functional Polymers. 37:263-269
[54] Liu Z, Cardosi M, Rodgers J, Lillie G, Simpson L (2010) Synthesis and study of copolymer of vinylferrocene, acrylamide and 2-(diethylamino)ethyl methacrylate as a polymeric mediator for electrochemical biosensors, Reactive & Functional Polymers. 70:715–725
[55] Rodgers J, Liu Z, McNeilage A, MacLennan M, Moffat J, Lillie G, MacDonald M (2008) Water-miscible conductive ink for use in enzymatic electrochemical-based sensors. US Patent 7465380

[56] Clark LC Jr., Lyons C (1962) Electrode systems for continuous monitoring in cardiovascular surgery. Ann. N. Y. Acad. Sci. 102:29–45

[57] Wang J (2008) Electrochemical glucose biosensors. Chem. Rev. 108:814-825

[58] Dzyadevych SV, Arkhypova VN, Soldatkin AP, El'skaya AV, Martelet C, Jaffrezic-Renault N (2008) Amperometric enzyme biosensors: Past, present and future. ITBM-RBM. 29:171

[59] Cardosi MF, Turner APF (1987) The realization of electron transfer from biological molecules to electrodes. In: Turner, A.P.F., Karube, I., Wilson, G.S. (Eds.), Biosensors: Fundamentals and Applications. Oxford University Press, London and New York, pp. 257–275

[60] Serban S, Murr NE (2006) Redox-flexible NADH oxidase biosensor: A platform for various dehydrogenase bioassays and biosensors. Electrochimica Acta. 51:5143–5149

[61] Serban S, Murr NE (2004) Synergetic effect for NADH oxidation of ferrocene and zeolite in modified carbon paste electrodes New approach for dehydrogenase based biosensors. Biosensors and Bioelectronics 20:161–166

[62] Bu H, Mikkelsen SR, English AM (1998) NAD(P)H Sensors Based on Enzyme Entrapment in Ferrocene-Containing Polyacrylamide-Based Redox Gels. Anal. Chem. 70:4320-4325

[63] Kato D, Iijima S, Kurita R, Sato Y, Jia J, Yabuki S, Mizutani F, Niwa O (2007) Electrochemically amplified detection for lipopolysaccharide using ferrocenylboronic acid. Biosens. Bioelectron. 22:1527–1531

[64] Mano N, Kuhn A (2005) Molecular lego for the assembly of biosensing layers. Talanta 66:21–7

[65] Maines A, Prodromidis MI, Tzouwara-Karayanni SM, Karayannis MI, Ashworth D, Vadgama P (2000) Reagentless enzyme electrode for malate based on modified polymeric membranes. Anal Chim Acta. 408:217–24

[66] Suye S, Aramoto Y, Nakamura M, Tabata I, Sakakibara M (2002) Electrochemical reduction of immobilized $NADP^+$ on a polymer modified electrode with a co-polymerized mediator. Enzyme Microb. Technol. 30:139–144

[67] Montagné M, Marty JL (1995) Bi-enzyme amperometric d-lactate sensor using macromolecular NAD+. Anal Chim Acta. 315:297–302

[68] Leca B, Marty JL (1997) Reusable ethanol sensor based on a NAD+ dependent dehydrogenase without cofactor addition. Anal Chim Acta. 340:143–8

[69] Suye S, Zheng H, Okada H, Hori T (2005) Assembly of alternating polymerized mediator, polymerized coenzyme, and enzyme modified electrode by layer-by-layer adsorption technique. Sensors and Actuators B. 108:671–675

[70] Wykes JR, Dunnill P, Lilly MD (1972) The preparation of soluble high molecular weight NAD derivative active as a cofactor. Biochimica Biophysica Acta. 2686:260-268

[71] Nakamura Y, Suye S, Kira J, Tera H, Tabata I, Senda M (1996) Electron-transfer function of NAD+-immobilized alginic acid. Biochimica Biophysica Acta. 1289: 221 -225

[72] Zapelli P, Pappa R, Rossodivita A, Re L (1977) Preparation and Coenzymic Activity of Soluble Polyethyleneimine-Bound NADP+ Derivatives. Eur. J. Biochem. 72:309-315

[73] Fuller CW, Rubin JR, Bright HJ (1980) A simple procedure for covalent immobilization of NADH in a soluble and enzymically active form. Eur J Biochem. 103:421–30

[74] Yamazaki Y, Maeda H (1981) The Synthesis of New Polymer Derivatives of NAD by Radical Copolymerization and Their Coenzymic Activity. Agric. Biol Chem. 45:2277-2288

[75] Lindberg M, Larsson P, Mosbach K (1973) A New Immobilized NAD+ Analogue, Its Application in Affinity Chromatography and as a Functioning Coenzyme. Eur. J. Biochem. 40:187-193

[76] Muramatsu M, Urabe I, Yamada Y, Okada H (1977) Synthesis and Kinetic Properties of a New NAD^+ Derivative Carrying a Vinyl Group. Eur. J. Biochem. 80:111-117